THE DUXBURY SERIES IN STATISTICS AND DECISION SCIENCES

Applications, Basics, and Computing of Exploratory Data Analysis, Velleman and Hoaglin
A Course in Business Statistics, Second Edition, Mendenhall
Elementary Statistics, Fifth Edition, Johnson
Elementary Statistics for Business, Second Edition, Johnson and Siskin
Essential Business Statistics: A Minitab Framework, Bond and Scott
Fundamentals of Biostatistics, Third Edition, Rosner
Fundamentals of Statistics in the Biological, Medical, and Health Sciences, Runyon
Fundamental Statistics for the Behavioral Sciences, Second Edition, Howell
Introduction to Probability and Statistics, Seventh Edition, Mendenhall
An Introduction to Statistical Methods and Data Analysis, Third Edition, Ott
Introductory Business Statistics with Microcomputer Applications, Shiffler and Adams
Introductory Statistics for Management and Economics, Third Edition, Kenkel
Mathematical Statistics with Applications, Fourth Edition, Mendenhall, Wackerly,
    and Scheaffer
Minitab Handbook, Second Edition, Ryan, Joiner, and Ryan
Minitab Handbook for Business and Economics, Miller
Operations Research: Applications and Algorithms, Winston
Probability and Statistics for Engineers, Third Edition, Scheaffer and McClave
Probability and Statistics for Modern Engineering, Second Edition, Lapin
Statistical Experiments Using BASIC, Dowdy
Statistical Methods for Psychology, Second Edition, Howell
Statistical Thinking for Behavioral Scientists, Hildebrand
Statistical Thinking for Managers, Second Edition, Hildebrand and Ott
Statistics: A Tool for the Social Sciences, Fourth Edition, Ott, Larson, and Mendenhall
Statistics for Business and Economics, Bechtold and Johnson
Statistics for Management and Economics, Sixth Edition, Mendenhall, Reinmuth, and Beaver
Understanding Statistics, Fourth Edition, Ott and Mendenhall

THE DUXBURY ADVANCED SERIES IN STATISTICS AND DECISION SCIENCES

Applied Nonparametric Statistics, Second Edition, Daniel
Applied Regression Analysis and Other Multivariable Methods, Second Edition, Kleinbaum,
    Kupper, and Muller
Classical and Modern Regression with Applications, Second Edition, Myers
Elementary Survey Sampling, Fourth Edition, Scheaffer, Mendenhall, Ott
Introduction to Contemporary Statistical Methods, Second Edition, Koopmans
Introduction to Probability and Its Applications, Scheaffer
Introduction to Probability and Mathematical Statistics, Bain and Englehardt
Linear Statistical Models, Bowerman and O'Connell
Probability Modeling and Computer Simulation, Matloff
Quantitative Forecasting Methods, Farnum and Stanton
Time Series Analysis, Cryer
Time Series Forecasting: Unified Concepts and Computer Implementation, Second Edition,
    Bowerman and O'Connell

THIRD EDITION

Probability and Statistics for Engineers

Richard L. Scheaffer
University of Florida

James T. McClave
Infotech, Inc.

PWS-KENT PUBLISHING COMPANY, BOSTON

PWS–KENT
Publishing Company

20 Park Plaza
Boston, Massachusetts 02116

PWS-KENT Publishing Company is a division of Wadsworth, Inc.

Library of Congress Cataloging-in-Publication Data

Scheaffer, Richard L.
    Probability and statistics for engineers / Richard L. Scheaffer,
James T. McClave.—3rd ed.
      p. cm.
    Includes bibliographical references.
    ISBN 0-534-92184-1 : $35.00 (est.)
    1. Statistics. 2. Probabilities. I. McClave, James T.
II. Title.
TA340.S33   1990
519.2′02462—dc20   89-22861
                                        CIP

ISBN 0-534-92184-1

Printed in the United States of America

1 2 3 4 5 6 7 8—94 93 92 91 90

Sponsoring Editor: *Michael Payne*
Production Editor: *Eve B. Mendelsohn*
Manufacturing Coordinator: *Margaret Sullivan Higgins*
Production: *Greg Hubit Bookworks*
Interior Designer: *John Edeen*
Cover Designer: *Jean Hammond*
Compositor: *Doyle Graphics Limited*
Cover Printer: *John P. Pow Company, Inc.*
Text Printer and Binder: *R. R. Donnelley & Sons Company*
Cover Photo: *Steven Hunt / The ImageBank*

Preface

This textbook is a calculus-based introduction to probability and statistics with emphasis on techniques and applications that are useful in engineering and the physical and biological sciences.

Although many of the examples and exercises deal with problems related to engineering, the book is suitable for students in many other areas of science. Problems dealing with energy, pollution, weather, chemical processes, size and abundance of animals, and earthquakes show the breadth of statistical applications in the sciences and provide interesting examples for students from a wide variety of sciences. These examples tend to be based upon real problems.

The notion of *probabilistic models* for real phenomena is emphasized throughout the text. This idea contains two key components, as the two words suggest. First, *probability* is used in model building since many, if not most, experiments do not produce outcomes that can be precisely predetermined. When that is the case, one should look at the probability, or relative frequency, distribution of the possible outcomes. Second, the word *model* is used to suggest that what we finally achieve is only an approximate, and usually oversimplified, description of the real phenomenon.

After a brief introductory chapter, Chapters 2 through 5 develop the basic ideas of probability and probability distributions for random variables. These probability distributions are, in fact, used as models for outcomes of experiments. Chapter 6 presents a transition between probability and statistics, as it deals with the probability distributions of certain common statistics. Chapters 7 through 12 develop the basic ideas of statistics, which is, in one sense, the process of fitting models to data and then using the models in making inferences. Chapter 13 is concerned with special applications to quality control.

The book is fairly compact in style, but contains many motivational examples and worked examples. Random variables are introduced early in Chapter 3 since physical scientists commonly deal with numerical outcomes of experiments. Multiple regression is presented with both quantitative and qualitative independent variables so that analysis of variance models can be easily embedded into the discussion.

The entire textbook can be covered in a two-quarter (6 quarter hour) sequence. For a one-semester (3 hour) or a one-quarter (4 hour) course, a good outline would be to cover all of Chapters 1 and 2, followed by a more cursory coverage of Chapters 3 and 4. Emphasis should be placed upon the binomial and Poisson distributions in Chapter 3 and the uniform, exponential, and normal distributions in Chapter 4. Chapter 5 can be skipped unless one wants to include a detailed look at the properties of estimators in later chapters. Certain of the important ideas presented in Chapter 5 are reviewed in later chapters as they are needed for developing estimators and test statistics. Chapters 6, 7, and 8 form the heart of basic statistical inference, and should be covered completely (except for those sections marked as optional). These chapters form the basis for the more interesting methods of regression analysis and the analysis of designed experiments found in Chapters 9, 10, and 11. Covering the first five sections of both Chapter 9 and Chapter 10 will provide the student with a solid introduction to regression analysis. The emphasis in Chapter 11 can be placed either on completely randomized versus randomized block designs or on factorial experiments, depending upon the interest of the instructor. Chapters 12 and 13 may or may not be covered, as time and interest dictate.

Many new exercises and some new examples, most coming from real data or real experimental situations, are included in the third edition. The review exercises in Chapter 14, all of which are based upon real data, are not tied to any particular section or chapter. More emphasis is placed upon graphical techniques for data analysis, including the use of normal scores plots to check for normality of a data set, residual plots to check on the underlying assumptions in regression, and box plots to compare centrality and variability among data sets. To allow for easy handling of more complex data analysis techniques, use of statistical packages for the computer is introduced early, beginning with the simple one- and two-sample analyses of Chapter 7. Minitab is used throughout the text, but other packages (SAS and SPSS) are also explained in the later sections on regression analysis. The introduction of a statistical package eases the computational burden for the student and allows more detailed (and more useful) analysis to be performed. In particular the use of a computer will aid in learning how to fit appropriate regression models, with careful study of the assumptions used and the pitfalls to be avoided.

Some sections are marked optional and can be skipped with no loss of continuity in the remaining sections. The material on moment-generating functions is concentrated in Sections 3.9, 4.9, and 5.6. A method for approximating probability distributions is discussed only in Section 6.6. Maximum likelihood (Section 7.6) and Bayesian methods (Section 7.7) are not essential for the development of statistical methods presented in later chapters. As mentioned above, all of Chapter 5 can be skipped in a course that is to emphasize statistical data analysis.

The writing of this book has been an enjoyable task, and the authors hope that students in engineering and the sciences find the book both enjoyable and informative as they seek to learn how statistics can be applied to their field of interest.

We'd like to thank the following reviewers of this edition for their helpful comments and suggestions: Bill Fulkerson (Deere & Co.), John Kinney (Rose-Hulman Institute), Huseyin Sarper (University of Southern Colorado), and Larry Stephens (University of Nebraska).

Richard L. Scheaffer
James T. McClave

Contents

*Sections marked with an asterisk are optional and may be omitted without loss of continuity.

9 Simple Regression 364

10 Multiple Regression Analysis 421

11 The Analysis of Variance 483

12 Nonparametric Statistics 552

1

Statistics in Engineering

ABOUT THIS CHAPTER

What impression does the term *statistics* bring to mind? Is it merely a list of facts and figures on sporting events, the national economy, or a group of diseases? Certainly facts and figures (or data) will be used in subsequent chapters. However, our basic concern is not the sets of data themselves, but rather the use of the information contained in these sets of data to make intelligent decisions. In this chapter we introduce the fundamental idea of using sample data to infer some property of the larger set of data from which the sample was drawn. We also present four methods for displaying sets of data.

CONTENTS

1.1   WHAT IS STATISTICS?

We live in an age in which we are deluged by facts and figures, or "statistics," on almost every subject imaginable. We hear or read government reports on the number of people on welfare and the number of armaments that the U.S. has available, business reports on quarterly earnings, and sports reports on batting averages and yards gained. In the midst of all of these facts and figures, most of them of little interest to us, why would anyone want to study a discipline called statistics? What is statistics, anyway?

Quantitative facts such as those indicated above will form the basis of our study, but we are about to study more than just sets of data. In this text we present an introduction to inferential statistics, as distinct from the layman's view of statistics as merely observable facts. By the term *inferential statistics* we mean that we want to use the data to make intelligent, rigorous statements (inferences) about a much larger phenomenon from which the data were selected.

Let's take a look at a few examples of how inferential statistics might aid an engineer. Suppose a civil engineer wants to study the traffic intensity at a certain intersection. He might observe the traffic flow at some representative times, including mornings, evenings, weekdays, and weekends, and then come up with an estimate of the average number of vehicles passing through the intersection per hour. The actual observations are important to him, but only as they help him form his estimate of the intensity over a wide time span.

We might wish to compare various combinations of drying times and amount of aggregate as to their effect on the strength of concrete. Certain test samples will have to be formed and their strengths measured. The measurements are listed, but they are important only as they relate to the inferential process of determining which combination of drying time and amount of aggregate produces the best product.

Computer components might be tested until they fail, with their lifelengths being recorded. This set of measurements might then be applied to the inferential problem of estimating the reliability of the entire computer system.

A quality control engineer might periodically sample a few manufactured items coming off an assembly line and count the number of defectives. This number is important to him only as it helps him to decide whether or not the line is operating within nominal standards.

The figures or measurements are important, but they are being collected for a well-defined inferential purpose. It is this inferential procedure that is the subject of modern statistics, and the main theme of this text.

1.2   ELEMENTS OF INFERENTIAL STATISTICS

The ideas in Section 1.1 form a general notion of inferential statistics, but we now make these ideas more specific. Any problem in statistics has as its starting point a *population of interest.*

DEFINITION 1.1	A **population** is the total set of measurements of interest in a particular problem.

In a reliability study the population may consist of the lifelengths of all the components of a certain type produced by factor A. In a quality control study the population may consist of a status report (which could be numerical) on the quality of each item contained in a large shipment.

Of course, the population measurements are unknown at the outset, and in many cases they cannot be completely determined. We cannot, for example, test all of factory A's components to get a complete list of lifelengths, for this would be too time-consuming and leave the factory with nothing to sell. We can, however, obtain a representative set of measurements from the population by performing an *experiment*.

For example, we could obtain a measurement from a population of component lifelengths by performing an experiment that consists of testing a component from factory A until it fails and recording its lifelength. Or we could examine a set of four manufactured items and record whether or not each is defective. This experiment would yield four measurements from the population of defective/nondefective classifications of all the manufactured items. The set of measurements yielded by an experiment is called the *sample*.

DEFINITION 1.2	A **sample** is a subset of the population that contains measurements obtained by an experiment.

The sample might consist of a number of lifelength measurements or recorded observations on frequency of defectives. Our objective is to use these sample data for purposes of making inferences about the population from which the sample was obtained. We might estimate the average lifelength of all components in the population, or estimate the reliability of a system made up of such components. We might estimate the proportion of defective items in a population, or decide whether to accept the lot from which the sample was drawn as meeting the standards of our factory.

Questions as to the appropriate number and type of sample observations are important in any statistical investigation. A civil engineer wants to estimate the traffic intensity on a particular road. He might measure traffic flow by a mechanical device, such as the cables commonly seen stretched across a roadway. For how many days should he record the data to obtain a reliable estimate?

A mechanical engineer wants to test the strength of a certain material. The test is destructive and the material is expensive. Thus he wants to test as few specimens as possible. When can he stop testing and still make a reliable decision?

An environmental engineer wants to estimate the bacterial count in a lake. She will take samples from the lake and use a special culture to count the bacteria in the sample. How many samples does she need, and what should be the volume of water in each one?

An industrial engineer wants to measure quality of output from a continuous production line. How large should each sample be, and how often should he take samples?

A chemical engineer wants to improve a production yield by selecting the best combination of temperature, pressure, and amount of catalyst. How can he set up an experiment to aid his decision making?

These examples illustrate the importance of the sampling process in making decisions. Methods of answering the questions raised here will be addressed throughout the remainder of this text.

1.3 DETERMINISTIC AND PROBABILISTIC MODELS

To develop formally the methodology for inference making, we must start with a model for the phenomenon under study. A *model*, as we use the term, may be thought of as a theoretical, and usually oversimplified, explanation of a complex system. In the sciences these models usually take the form of a mathematical equation, such as the equations for heat transfer in physics or the equations for various chemical reactions. It is important to discuss the concept of models in some detail, as we will be referring to specific models throughout the remainder of this text.

The models most familiar in the physical sciences and engineering are called *deterministic models*, which have the distinguishing feature that specific outcomes of experiments can be accurately predicted. One such model, for example, is *Ohm's Law, $I = E/R$*. The law states that electric current I is directly proportional to the voltage E and inversely proportional to the resistance R in a circuit. Once the voltage and resistance are known, the current is determined. If many circuits with identical voltages and resistances are studied, the current measurements may differ by small amounts from circuit to circuit, owing to inaccuracies in the measuring equipment or other uncontrollable influences. These discrepancies will be negligible, and for most practical purposes Ohm's Law provides a useful deterministic model of reality.

The amounts of certain chemicals to be produced in a controlled experiment can likewise be fairly accurately predicted, given the initial conditions, although again there is almost always some slight variation among observed results.

On the other hand the models that arise from statistical investigations, called *probabilistic models*, are characterized by the fact that although specific outcomes of an experiment cannot be predicted with certainty, *relative frequencies* for various possible outcomes are predictable. Suppose that ten items are drawn from a large lot and examined for defects. We cannot say specifically how many defective items we will see, but given an appropriate model and some assumptions on the nature of the lot, we can give figures for the *chance* (or probability) that we will see, say, more than 8 or fewer than 3 defectives. The formal definition of probability will be given in Chapter 2.

Deterministic models leave no room for statistical inference; probabilistic models are the building blocks upon which a theory for inference making is based.

It is not always clear which type of model is best for certain situations, and deterministic models are sometimes used for situations in which probabilistic models might have been more appropriate. Consider a traffic flow problem again. Suppose that, over a long period of time, the number of automobiles entering a certain intersection averages ten per minute. A deterministic model might interpret this figure as a constant, and predict that exactly ten automobiles enter the intersection during every minute. A probabilistic model would treat the arrivals as random events with some minutes having perhaps no arrivals and some perhaps 25 arrivals, but with 10 arrivals per minute as the average. Which model is appropriate? The answer, no doubt, depends on the use to which the model is to be put. If one wants only to assess a rough figure for total arrivals over a month, then the deterministic model might be adequate. If, however, one wants to study peak loads and answer questions related to how often the arrivals exceed 20 per minute, then the deterministic model is of absolutely no value. *Our philosophy is to employ a probabilistic model whenever there is more than negligible variation among outcomes of an experiment. Probabilistic models will be used throughout this text.*

To review, our study of statistical inference will depend on a clear definition of the population under investigation, on the availability of an experimental method to produce sample measurements, and on the postulating of a probabilistic model to allow us to use the sample measurements for inference making.

This book not only emphasizes probabilistic models but also discusses how appropriate models can be chosen from a combination of theoretical factors and empirical evidence. The empirical evidence usually comes in the form of data, and the following section presents some methods of displaying data sets informatively.

1.4 DISPLAYING SETS OF DATA

The use of sample data to draw conclusions or make decisions should begin by obtaining a clear understanding of the important features of the observed data. We try to display the data in some helpful form, other than in a table, since most people do not readily grasp patterns in large lists of numbers. The visual display of data helps us gain insights into important questions to be asked and helps us formulate appropriate probabilistic models to aid in answering these questions. This section contains a number of graphical and numerical methods for displaying data and summarizing key information.

1.4.1 Stem-and-Leaf Displays

Table 1.1 shows percentages of investment for pollution control for fifteen manufacturing industries.

A much more informative picture of these data can be obtained by a different type of listing, called a *stem-and-leaf* display. The leftmost digit in each number

TABLE 1.1
*Percentage of Total Plant and Equipment
Investment for Pollution Control
(15 manufacturing industries for 1977)*

17	02	07	04	08
17	04	04	14	03
03	02	04	10	01

Source: *World Almanac and Book of Facts*, 1979.

provides a convenient starting point for breaking the data into two groups, one group starting with 0 and another starting with 1. Thus we use 0 and 1 to form the *stem*, writing them in a vertical array as follows.

$$
\begin{array}{c|}
0 \\
1 \\
\end{array}
$$

For each observation we then record the second digit as a *leaf* on the appropriate row of the stem, to the right of the vertical bar. The first observation, 17, gives us

$$
\begin{array}{c|c}
0 & \\
1 & 7 \\
\end{array}
$$

and adding the second and third yields

$$
\begin{array}{c|c}
0 & 3 \\
1 & 7\ 7 \\
\end{array}
$$

The complete stem-and-leaf display is shown in Figure 1.1.

FIGURE 1.1
Stem-and-Leaf Display for the Data of Table 1.1

$$
\begin{array}{c|l}
0 & 3\ 2\ 4\ 2\ 7\ 4\ 4\ 4\ 8\ 3\ 1 \\
1 & 7\ 7\ 4\ 0 \\
\end{array}
$$

Since most of the observations have a first digit of zero, we could stretch the stem to two 0 categories and two 1 categories with second (leaf) digits of 0 through 4 going on the upper row and digits of 5 through 9 on the lower row. The stretched display is shown in Figure 1.2. Since it is easy to order the observations in any row, we also show the ordered stem-and-leaf display.

In most cases order will be numerical, from low to high. Now we can see easily that the observations range from 01 to 17, with nine of the fifteen

FIGURE 1.2
Stretched Stem-
and-Leaf Display for
the Data of
Table 1.1

```
0 | 3 2 4 2 4 4 4 3 1        0 | 1 2 2 3 3 4 4 4 4
0 | 7 8                      0 | 7 8
1 | 4 0                      1 | 0 4
1 | 7 7                      1 | 7 7
    Unordered                    Ordered
```

observations being 4 or less. Figure 1.2 is more informative than Table 1.1, and can be produced with little extra effort.

Stem-and-leaf displays of data with more than two digits can be constructed in a variety of ways. The data in Table 1.2 show the number of days of unhealthful air in 1979, 1980, and 1981 for selected U.S. cities. (The pollutant standard index (PSI) is based on five pollutants, and air is judged to have adverse effects on human health if the index exceeds 100.)

TABLE 1.2
Number of Days with Pollutant Standard Index (PSI) 100 or Above

	1979	1980	1981
Los Angeles	248	221	248
Riverside-Ontario	208	171	184
New York	113	131	104
Denver	128	89	79
San Diego	106	not available	69
Houston	118	101	67
Seattle	60	33	35
Philadelphia	79	63	32
Portland	55	56	30
Salt Lake City	47	54	30
St. Louis	88	55	26
Washington, D.C.	47	69	21
Louisville	58	59	20
Chicago	82	48	18
Milwaukee	33	16	11

Source: *Statistical Abstract of the U.S.*, 1984.

We will construct a stem-and-leaf display of the 1979 data by allowing the hundreds and tens digits to form the stem, and the units digit the leaves. The result is as follows.

```
03 | 3
04 | 77
05 | 58
06 | 0
07 | 9
08 | 28
09 |
10 | 6          10|6 is read as 106
11 | 38
12 | 8
13 |
14 |
15 |
16 |
17 |
18 |
19 |
20 | 8
21 |
22 |
23 |
24 | 8
```

This display is very spread out, so we tighten it up by allowing the hundreds digit to form the stem and the other digits the leaves. The two-digit leaves must be separated to avoid confusion.

```
0 | 33, 47, 47, 55, 58, 60, 79, 82, 88
1 | 06, 13, 18, 28
2 | 08, 48
```

The result is now very compact, and can be spread slightly by dividing each group of two-digit leaves into those under 50 and those at 50 or above, as follows.

```
0 | 33, 47, 47
0 | 55, 58, 60, 79, 82, 88
1 | 06, 13, 18, 28
1 |
2 | 08, 48
2 |
```

We now have a compact, yet informative, display. It is still easy to spot, for example, the large gap between 128 and 208, which puts the Los Angeles-Riverside-Ontario region of California in a class by itself. Yet we also see the variation among the regions with lower indices.

Sometimes another column is added to a stem-and-leaf display to provide cumulative frequency counts as we move toward the middle from either end. This is sometimes called a *depth* column, and is illustrated as follows for the stem-and-leaf plot developed above.

```
  3    0 | 33, 47, 47
 (6)    0 | 55, 58, 60, 79, 82, 88        (6)        depth ??
  6    1 | 06, 13, 18, 28
  2    1 |
  2    2 | 08, 48                                     3/9
  0    2 |
```

Depth

TABLE 1.3
Fuel Economy of 1983 Autos (as determined by the EPA)

Make	Mpg	Make	Mpg	Make	Mpg
*Alpha Romeo Spider	23	*Datsun Nissan Sentra	40	Mercury Zephyr	19
AMC Concord	21	*Datsun 200SX	25	Oldsmobile Cutlass Sup.	18
AMC Spirit	21	*Datsun Pulsar	20	Oldsmobile Delta 88	17
*Audi 4000	39	*Datsun 810	30	Oldsmobile 98	17
*Audi 5000	19	Dodge Aries	28	Oldsmobile Omega	27
*BMW 320	25	Dodge Challenger	23	Oldsmobile Toronado	16
*BMW 633 CSI	19	Dodge Colt	39	Plymouth Colt	39
*BMW 7331	18	Dodge Charger	32	Plymouth Gran Fury	19
Buick Century	26	Dodge Diplomat	19	Plymouth Horizon	28
Buick Electra	18	Dodge Miranda	17	Plymouth Reliant	29
Buick LeSabre	18	Dodge Omni	29	Plymouth Sapporo	24
Buick Regal	26	Dodge 400 Convertible	26	Plymouth Turismo	32
Buick Riviera	17	*Fiat Spider 2000	22	Pontiac Bonneville	18
Buick Skyhawk	25	*Fiat X1/9	26	Pontiac Grand Prix	21
Buick Skylark	21	Ford Escort	33	Pontiac 2000	28
Cadillac Deville/Brougham	22	Ford Fairmount Futura	21	Pontiac Phoenix	21
Cadillac Eldorado, Seville	17	Ford EXP	29	*Porsche 911	16
Chevrolet Celebrity	26	Ford LTD	21	*Porsche 944	21
Chevrolet Cavalier	25	Ford Mustang	26	*Subaru	30
Chevrolet Chevette	31	Ford Thunderbird	21	*Toyota Celica	25
Chev. Chevette Diesel	37	*Honda Accord	29	*Toyota Celica Supra	21
Chevrolet Citation	22	*Honda Civic	46	*Toyota Corolla	31
Chevrolet Impala/Caprice	17	*Jaguar XJ	17	*Toyota Cressida	22
Chevrolet Malibu	18	*Mazda GLC	34	*Toyota Starlet	42
Chevrolet Monte Carlo	20	*Mazda Rx7	20	*Toyota Tercel	32
Chrysler Cordoba/300	18	Mercury Capri	22	*Volkswagen Diesel Rabbit	48
Chrysler LeBaron	17	Mercury Cougar	21	*Volkswagen Jetta	30
Chrysler New Yorker	24	Mercury Lynx	27	*Volkswagen Rabbit	32
*Datsun 2-Seater 280zx	19	Mercury Marquis	21	*Volkswagen Scirocco	29

Source: *World Almanac and Book of Facts*, 1983.
*Non-U.S. manufacturers.

The depth column counts and accumulates the number of observations in each row until it gets to the row overlapping the middle observation. For that row it shows only the frequency count in the row (usually in parentheses).

The value of a stem-and-leaf display is much greater for larger data sets. We show this, and illustrate the use of back-to-back plots on the same stem, with the data in Table 1.3.

Since it is of some interest to compare U.S. and foreign manufacturers, we place the leaves for U.S. manufacturers to the left and those for foreign manufacturers to the right of a common stem.

```
            U.S.                             Foreign

                                      1
9 9 9 8 8 8 8 8 7 7 7 7 7 7 7 6  1  6 7 8 9 9 9
  4 4 3 2 2 2 1 1 1 1 1 1 1 1 0  2  0 0 1 1 2 2 3
    9 9 9 8 8 8 7 7 6 6 6 6 6 5 5  2  5 5 5 6 9 9
                      3 2 2 1  3  0 0 0 1 2 2 4
                        9 9 7  3  9
                               4  0 2
                               4  6 8
```

min 16   mid

From this display we can see that both groups have minimum values of 16, but the foreign autos have a more even spread up to the mid thirties, and more values above the mid thirties. The back-to-back display certainly gives more insight into the comparison of gas mileages than does Table 1.3.

1.4.2 Histograms and Bar Graphs

A stem-and-leaf display can be quickly and easily turned into a bar graph by, essentially, turning it onto its side. The final stem-and-leaf display for the 1979 air pollution data, given in Section 1.4.1, produces the bar graph, or histogram, shown in Figure 1.3.

FIGURE 1.3
Number of Days with PSI 100 or above, 15 Selected U.S. Cities (frequency is number of cities out of the 15)

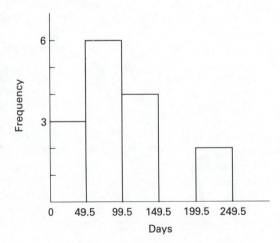

The interval in which the data fall, approximately 0 to 250, is broken into a convenient number of subintervals, or classes, and the class boundaries are chosen so that no observation falls on a boundary. The height of the bar over each class represents the frequency of observations found in that class. Thus three of the observations fall into the first class, six fall into the second class, and so on. Figure 1.3 is called a *frequency histogram*. If the vertical axis showed the relative frequencies of 3/15, 6/15, and so on, the result would be called a *relative frequency histogram*.

Histograms are widely used as a simple, but informative, method of data display. A particularly informative graph is given as Figure 1.4, which shows the frequency histogram of results from quality control measurements on 500 steel rods. These rods had a lower specification limit (LSL) of 1 centimeter, and rods with diameters less than the LSL were to be declared defective.

FIGURE 1.4
Diameters of 500 Steel Rods

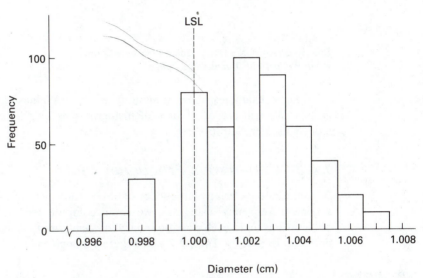

Source: Deming, W. E. (1978). "Making Things Right," Statistics: A Guide to the Unknown, *2nd ed., Holden-Day, San Francisco.*

The high bar at 1.0 cm, coupled with the gap at 0.999 cm, should cause some serious questioning of those generating the data. It appears that rods with diameters slightly under 1.0 cm were being passed as acceptable, thus inflating the frequency at 1.0 cm. This turned out to be the case.

The bars of a histogram must appear in a specific order because of the numerical scale on the horizontal axis, but this is not true of other types of bar graphs. A very useful quality control device is the *Pareto diagram*, which graphs the frequencies of defects with the order of bars determined by the size of the frequency. Figure 1.5 shows a Pareto diagram for defects in inspected machined parts. Pareto diagrams show quality control engineers which types of defects occur most frequently, and are helpful in studying ways to improve quality.

FIGURE 1.5
Defects in Inspected
Machined Parts

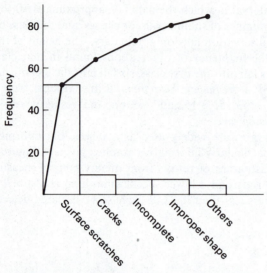

Source: Ishikawa, K. (1980). Guide to Quality Control,
Asian Productivity Organization, Tokyo.

These examples show something of the usefulness and versatility of bar graphs. We will see other uses of histograms as aids to selecting probabilistic models in later chapters.

1.4.3 Cumulative Frequency Plots

In addition to the relative frequency of observations within each class, as shown on a relative frequency histogram, we could compute the cumulative frequency of observations up to each class boundary. A *cumulative frequency plot* of the 1979 air pollution data from Table 1.2 and Figure 1.3 is given in Figure 1.6.

FIGURE 1.6
Cumulative Relative
Frequencies for
Number of Days
with PSI 100 or
above, 15
Selected Cities

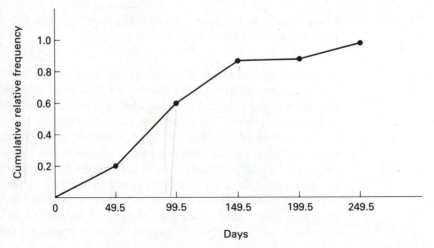

The graph in Figure 1.6 shows that $3/15 = 0.20$ of the observations are less than 49.5, $9/15 = 0.60$ of the observations are less than 99.5, and $13/15 = 0.87$ of the observations are less than 149.5.

A cumulative frequency plot attached to a Pareto diagram, as in Figure 1.7, is a convenient way to keep track of the cumulative effect of a number of causes of defects. Here more than 60% of the defects are due to surface scratches or cracks. Correcting these two problems would improve quality tremendously.

FIGURE 1.7
Defects in Inspected Machined Parts

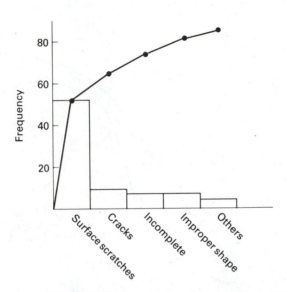

Cumulative frequency plots form a convenient way of detecting shifts in distributions of data. This phenomenon is illustrated in Figure 1.8 by plotting the 1979 and 1981 air pollution data, from Table 1.2, on the same graph. Notice that the dashed line, representing 1981, is entirely above the solid line, representing 1979; also the dashed line rises much more steeply from zero. This indicates that the

FIGURE 1.8
Cumulative Relative Frequencies for Number of Days with PSI 100 or above, 15 Selected U.S. Cities

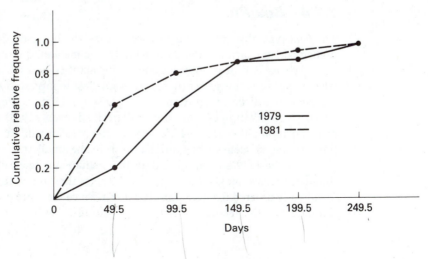

1981 data set has more small observations. One could suggest, from these data, that air quality improved between 1979 and 1981. However, a more rigorous statistical analysis is necessary to substantiate the amount of improvement.

As sample sizes get larger and larger, the cumulative frequency plots can become closer to smooth curves, as in Figure 1.9. Notice how the work start times bunch around 8:00 A.M., whereas arrival times are more spread out. What are the workers saying about desired work start times?

FIGURE 1.9
*Work Starts,
Arrivals, and
Desired Starts,
Pittsburgh Central
Business District*

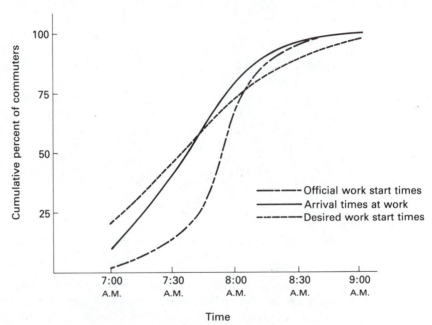

Source: Hendrickson, C., and E. Plank (1984). Transportation Research, 18A, No. 1, p. 29.

1.4.4 Box Plots

The next type of graphical display we discuss, the box plot, depends upon certain numerical summaries of the data, namely, the median, quartiles, and range.

The *median* of n observations is the number located at position $(n + 1)/2$ when the observations are ordered from smallest to largest. If n is odd, say 31, then the median will be at position $(31 + 1)/2 = 16$. That is, the median will be the sixteenth observation up from the smallest. If n is even, say 30, then the median will be at position $(30 + 1)/2 = 15.5$. That is, the median will be halfway between the fifteenth and sixteenth observations up from the smallest.

For the 1979 air pollution data the median can be easily found from any of the stem-and-leaf plots following Table 1.2 since the data on the stem-and-leaf plots are already ordered from smallest to largest. Since there are $n = 15$ observations in the data set, the median is found at position $(15 + 1)/2 = 8$. The value of the median for this data set is 82. We will use the notation Q_2 to denote a median. Thus $Q_2 = 82$ for the data under discussion.

The median splits a data set in the middle, and it is sometimes informative to split each half of the data again to find the *quartiles*. The *lower quartile, Q_1,* can be thought of as the median of the lower half of the values in a data set. The *upper quartile, Q_3,* is the median of the upper half of the values in the set. The quartiles and median—Q_1, Q_2, and Q_3—split the data set into four sets of roughly equal frequency. When n is odd, there is some question as to whether or not to include the median itself in the set of observations from which Q_1 is determined. We will include it. Thus if $n = 31$, then the median, Q_2, is at location 16 and Q_1 is at location $(16 + 1)/2 = 8.5$, or midway between the eighth and ninth observations. Q_3 is then found by counting down to the eighth and ninth observations *from the largest.* Under this rule there will be 8 observations smaller than Q_1, which is as close as we can come to one-fourth of the $n = 31$ original data points.

The 1979 air pollution data have a median at position 8, and thus Q_1 is at position $(8 + 1)/2 = 4.5$. Since the fourth observation from the smallest is 55 and the fifth is 58, $Q_1 = (55 + 58)/2 = 56.5$, which can be rounded to 56. Similarly, $Q_3 = (118 + 113)/2 = 115.5$ or 116.

The *range* of a set of observations is the difference between the largest and smallest observations. For the 1979 air pollution data the range is $248 - 33 = 215$.

A *box plot* is a device for showing the median, quartiles, and range all on one simple display. The median, quartiles, and extreme values are ranked on a real-number line. Then a narrow box connects Q_1 and Q_3, and a line is drawn from each end of the box to the extreme values. Q_2 is shown by a line across the box. Figure 1.10 shows a box plot for the 1979 air pollution data.

FIGURE 1.10
Number of Days with PSI 100 or above, 15 Selected U.S. Cities

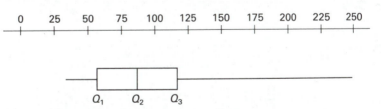

The box plot clearly shows that most of the observations are below 116, and that the upper extreme of 248 is far away from the bulk of the data. Roughly 50% of the data lie inside the box (between 56 and 116). The range of the data is clearly seen to be the length of the entire plot.

The range is one measure of the variability within a set of data, but this measure depends only on the extreme observations, which may be quite far from the others in the data set. A more stable measure of variability is the *interquantile range* (IQR) defined by

$$IQR = Q_3 - Q_1$$

The IQR is simply the length of the box in a box plot. Sometimes the extreme observations are so far from the others that we call them *outliers*, observations that seem to behave differently from the remainder of the data set. A handy rule for determining outliers is to mark off a distance of 1.5 IQR below Q_1 and above Q_3. Any observation less than $Q_1 - 1.5$ IQR or greater than $Q_3 + 1.5$ IQR will be called an outlier, and may command special attention. It behooves the good data analyst to try to find a reason for the presence of outliers.

For the 1979 air pollution data,

$$\text{IQR} = Q_3 - Q_1 = 115.5 - 56.5 = 59.0$$

and

$$Q_3 + 1.5 \text{ IQR} = 204$$

($Q_1 - 1.5$ IQR is negative.) Thus Los Angeles and Riverside-Ontario are outliers and merit special attention in air pollution studies. By this same rule Riverside-Ontario is not an outlier in 1980, but becomes an outlier again in 1981 because there was so much improvement (decreased variation and lower indices) among most of the remaining cities. Los Angeles is an outlier in each of the three years. Outliers are sometimes indicated by asterisks (*) on the box plots.

Box plots are very effective when used for comparing data sets. Figure 1.11 shows the three box plots generated from the data in Table 1.2 for 1979, 1980, and 1981. Notice how the data tend to concentrate more and more toward the lower end of the scale except for the high extreme, which remains high for all three years. The series of box plots nicely depicts the major changes in the air quality data over the years.

FIGURE 1.11
Number of Days with PSI 100 or above, 15 Selected U.S. Cities

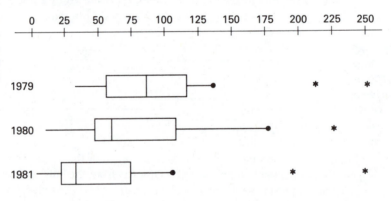

1.4.5 Transformations of Data

As we have seen, two key features of a set of data are its *centrality* (possibly measured by the median) and its *variability* (possibly measured by the interquantile range). These features will remain important components of our study of statistics throughout this textbook, and other measures will be developed for them. However centrality and variability are measured, it is important to look at both features when studying any set of data.

Often these two features, centrality and variability, are related to each other. Typically as the centrality increases for comparable data sets, so does the variability. For example, heights of adult females in the U.S. will vary much more around their median than will heights of first grade girls around theirs. This fact causes problems in data analysis because we often want to compare measures of centrality for two data sets, but if the centralities differ, the variabilities may differ as well. Then one may have to seriously question whether or not it makes sense to compare measures of centrality when the data sets differ substantially in variability. For an extreme example suppose a certain chemical fertilizer is ideally 50% water. Company A is doing a fine job of quality control and all of its test samples show exactly 50% water. Company B, on the other hand, has a very poor mixing procedure, resulting in half of its test samples showing 0% water and half showing 100% water, for a median of 50%. The measures of centrality are the same, but the comparison of the two makes no practical sense because of the huge difference in variability. Similar but perhaps less extreme cases occur regularly.

A basic principle to keep in mind, then, is to *compare centralities only if variabilities are similar*. If variabilities are not similar, it often happens (fortunately) that the data can be transformed to a new measurement scale on which the variabilities will be similar, as in the following example. Time intervals between successive failures of the air conditioning system on a Boeing 720 jet airplane were recorded in order. (Source: F. Proschan, *Technometrics*, 5, No. 3, 1963, p. 376.) The thirty measures are divided into 2 groups of 15 to see if the later time intervals seem to be shorter than the earlier ones (possibly indicating wear out of the system). The data (in hours) are as follows:

Group I   23, 261, 87, 7, 120, 14, 62, 47, 225, 71, 246, 21, 42, 20, 5

Group II   12, 120, 11, 3, 14, 71, 11, 14, 11, 16, 90, 1, 16, 52, 95

Box plots for the two groups are shown in Figure 1.12. For Group I the median is 47 and IQR = 83. For Group II the median is 14 and IQR = 50.5. Obviously the later times appear to be shorter, but is it fair to measure this difference by looking at the median of 47 versus the median of 14? Probably not, since the two variabilities are also quite different.

FIGURE 1.12
Times Between Successive Failures

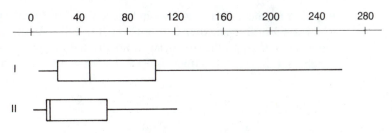

Instead of analyzing the original data, suppose we transform it by taking square roots. Box plots of the transformed data are shown in Figure 1.13. For the transformed data, Group I has a median of 6.8 and an IQR of 5.6, while Group II has a median of 3.7 and an IQR of 4.5. Notice that the IQRs (and, indeed, other

FIGURE 1.13
*Square Roots of
Times Between
Successive Failures*

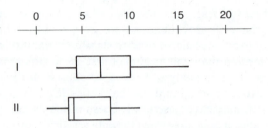

features of variability displayed in the box plots) are now quite similar. A comparison of the medians (6.8 versus 3.7) is now a fair way to describe a key difference in centrality between these data sets.

How did we come up with the square root transformation here? We knew a general rule for transformations to stabilize variabilities that runs as follows. If variability is proportional to $(\text{centrality})^{1-p}$, then transform data point x to x^p to stabilize the variability. If $p = 0$, use the transformation $\ln x$. For the data on the air conditioners the IQRs have ratio $83/50.5 = 1.64$ and the medians have ratio $47/14 = 3.36$. However, the square roots of the medians are in ratio $\sqrt{47/14} = 1.83$, very close to the ratio on the IQRs. Therefore the variability (as measured by IQR) is roughly proportional to centrality (as measured by the median) to the $1/2$ power. Since $1 - p = 1/2$, $p = 1/2$ and the square transformation works.

To see another example of the use of a transformation, consider the following two data sets on LC50 measurements for certain species of fish as determined for two chemical toxicants that can be found in water. Environmental impact studies often involve such comparisons. (An *LC50* is the concentration lethal to 50% of the organisms of a particular species under a given set of conditions and specified time period.)

Methyl Parathion (MP)   2.8, 3.1, 4.7, 5.2, 5.2, 5.3, 5.7,
                         5.7, 6.6, 7.1, 8.9, 9.0

Baytex (B)   0.9, 1.2, 1.3, 1.3, 1.4, 1.5, 1.6,
                         1.6, 1.7, 1.9, 2.4, 3.4

Box plots for these data are shown in Figure 1.14. For MP the median is 5.5 and the IQR is 1.8. For B the median is 1.6 and the IQR is 0.5. Again, the variability for MP is much greater than that for B, so an unadjusted comparison of medians may not be fair. In this case the ratio of IQRs is $1.8/0.5 = 3.6$, while the ratio of medians is

FIGURE 1.14
*LC50s for Two
Toxicants*

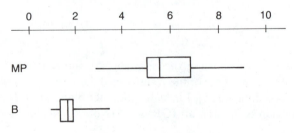

5.5/1.6 = 3.44. Since these ratios are approximately equal, the variability appears to be approximately proportional to the centrality. (That is, $1 - p = 1$ or $p = 0$.) The ln transformation should be used to stabilize variability, and the box plots of the natural logarithms of the original data are shown in Figure 1.15. For the transformed data the median for MP is 1.70 and the median for B is 0.44, while both IQRs equal 0.32. On the transformed scale, the variabilities are stabilized and it is fair to compare the medians as measures of centrality.

Other important uses of transformations will be shown in later chapters.

FIGURE 1.15
LC50s for Two Toxicants after Logarithmic Transformation

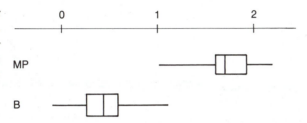

1.4.6 Plotting Data over Time

When data are being collected over time—such as production figures per month, total defects observed per hour, or net earnings per quarter—it is often helpful to keep track of both the frequency information (as in a stem-and-leaf plot) and the time order of these observations. This can easily be accomplished by simply putting a time line off to one side of a stem-and-leaf plot, and placing a dot along the line whenever an observation is added to the array. An illustration is given in Figure 1.16 for data on U.S. imports of petroleum.

FIGURE 1.16
U.S. Net Imports of Petroleum Energy

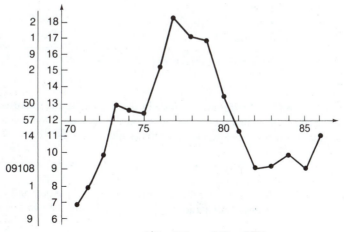

9|6 = 6.9 quadrillion BTUs

Source: World Almanac and Book of Facts, *1988*

Interpreting the stem-and-leaf plot of Figure 1.16, take note of the clumping of data around 9 and in the 11 through 13 range. Also note the extreme values of 6.9 and 18.2. From the time line we can see that these clumps are actually spread over two periods, as imports rose rapidly in the 70s, declined in the early 80s, and then began to rise again.

Figure 1.16 is called a *digidot plot* and is due to J. Stuart Hunter (*The American Statistician*, vol. 42, no. 1, 1988). The digidot plot is an effective way to plot data from quality control tests over time.

EXERCISES

1.1   The data given below show the consumer price index (CPI) for each of 16 U.S. cities as of July 1983.
   (a)   Construct a stem-and-leaf display of these data.
   (b)   Construct a relative frequency histogram of these data.
   (c)   Construct a cumulative frequency plot of these data.
   (d)   Do any cities appear to have unusually high or low CPIs?

Consumer Price Indexes, Selected U.S. Cities

Anchorage	258	Miami	325
Baltimore	297	New York	286
Boston	288	Scranton	291
Buffalo	296	Pittsburgh	286
Chicago	308	Portland	297
Dallas	332	St. Louis	320
Denver	304	San Francisco	294
Los Angeles	163	Seattle	300

Source: *World Almanac and Book of Facts*, 1984.

1.2   The amount of hydrogen chloride in the stratosphere is important because it may catalyze the removal of ozone. Data on the amount of hydrogen chloride in sections of the stratosphere before and after the eruptions of El Chichon Volcano in Mexico are given below.
   (a)   Construct back-to-back stem-and-leaf plots for the preeruption and posteruption data.
   (b)   Construct parallel box plots of the preeruption and posteruption data.
   (c)   Construct cumulative frequency plots of the two sets of data on the same axes.
   (d)   What are the most striking features of the two data sets?

*Amount of HCl in Columns
above 12 km at Various Locations
(in 10^{15} molecules per square
centimeter)*

Preeruption	Posteruption
0.78	1.46
1.26	1.40
0.75	1.29
0.69	1.22
0.88	1.63
1.56	1.90
1.18	1.24
	1.45
	1.79

Source: Mankin, W. G., and M. T. Coffey
(1984). *Science*, 226, October, p. 171.

1.3 Using the fuel economy data of Table 1.3, construct parallel box plots for autos from U.S. and foreign manufacturers. What questions are prompted by the plot? What conclusions are apparent?

1.4 Data on the average life of AA batteries under continuous use are given below.
  (a) Construct back-to-back stem-and-leaf plots of the alkaline versus the nonalkaline batteries. (Group the heavy-duty and regular batteries together.)
  (b) Construct parallel box plots for the alkaline and nonalkaline batteries. What pattern do you detect?

Brand and Model	Average Life
Alkaline AA	
Duracell MM 1500	4.1 hr
Eveready Energizer E91	4.5
K-Mart Super Cell	4.3
Panasonic AM3	3.8
Radio Shack 23552	4.8
Ray-O-Vac 815	4.4
Sears 3090	4.7
Heavy-Duty AA	
Eveready 1215	1.8
Radio Shack 23582	2.0
Ray-O-Vac 5AA	0.6
Sears 344633	0.7
Regular AA	
Everready 1015	0.8
K-Mart K15S	0.7
Radio Shack 23468	1.0
Ray-O-Vac 7AA	0.5

Source: *Consumer Reports*, November 1983, p. 592.

1.5 Who was the greatest New York Yankee home run hitter? The home run outputs of four contenders are shown below. Construct parallel box plots for the four players. Who appears to be the best home run hitter? Who appears to be second best? What considerations affect your choice?

Babe Ruth		Lou Gehrig		Mickey Mantle		Roger Maris	
Year	Home Runs	Year	Home Runs	Year	Home Runs	Year	Home Runs
1920	54	1923	1	1951	13	1960	39
1921	59	1924	0	1952	23	1961	61
1922	35	1925	20	1953	21	1962	33
1923	41	1926	16	1954	27	1963	23
1924	46	1927	47	1955	37	1964	26
1925	25	1928	27	1956	52	1965	8
1926	47	1929	35	1957	34	1966	13
1927	60	1930	41	1958	42		
1928	54	1931	46	1959	31		
1929	46	1932	34	1960	40		
1930	49	1933	32	1961	54		
1931	46	1934	49	1962	30		
1932	41	1935	30	1963	15		
1933	34	1936	49	1964	35		
1934	22	1937	37	1965	19		
		1938	29	1966	23		
		1939	0	1967	22		
				1968	18		

Source: *The Baseball Encyclopedia*, 4th Ed. Joseph L. Reicher, Ed., 1979.

2

Probability

ABOUT THIS CHAPTER

What is it that allows us to make inferences from a sample to a population, and then measure the degree of accuracy in that inference? The answer is the theory of *probability*. Outcomes of most experiments cannot be predetermined, but certain outcomes may be more likely than others. This leads us to the notion of probability, which is fundamental to our development of statistical inference.

CONTENTS

2.1 AN ELEMENTARY DEFINITION OF PROBABILITY

If one flips a balanced coin into the air, what is the chance that it will land heads up? Most people would say that this chance, or probability, should be 0.5, or something very close to 0.5. Upon further questioning, one can determine that to most people the meaning of the 0.5 is that approximately one-half of the time, in repeated flipping, the coin should land heads up. From there on, the reasoning gets more fuzzy. Will 10 tosses result in exactly 5 heads? Will 50 tosses result in exactly 25 heads? Probably not. So the 0.5 is assumed to be a long-run relative frequency, or *limiting relative frequency*, as the number of flips gets large.

To see what might happen in actual practice, we will look at coin-flipping data actually generated by John Kerrich, a mathematician interned in Denmark during World War II. Mr. Kerrich actually tossed a coin 10,000 times, keeping a tally of the number of heads. After 10 tosses he had 4 heads, a relative frequency of 0.4; after 100 tosses he had 44 heads (0.44); after 1000 tosses he had 502 heads (0.502); and after 10,000 tosses he had 5067 heads (0.5067). The relative frequency of heads remained very close to 0.5 after 1000 tosses (although the figure at 10,000 is further from 0.5 than the figure at 1000).

If n is the number of trials of an experiment (such as the number of flips of a coin), then it seems that one might define the probability of an event E (such as observing a head) by

$$P(E) = \lim_{n \to \infty} \frac{\text{number of times } E \text{ occurs}}{n}$$

But will this limit always converge? If so, can we ever determine what it will converge to without actually conducting the experiment many times? For these and other reasons, this is not an acceptable mathematical definition of probability, but it is a property that should hold, in some sense. Another definition of probability must be found that allows such a limiting result to hold as a consequence. This definition will be given in Section 2.3.

2.2 A BRIEF REVIEW OF SET NOTATION

Before going into a formal discussion of probability, it is necessary to outline the *set notation* we will use. Suppose we have a *set S* consisting of points labeled 1, 2, 3, and 4. We denote this by $S = \{1, 2, 3, 4\}$. If $A = \{1, 2\}$ and $B = \{2, 3, 4\}$, then A and B are *subsets* of S, denoted by $A \subset S$ and $B \subset S$ (B is "contained in" S). We denote the fact that 2 is an *element* of A by $2 \in A$. The *union* of A and B is the set consisting of all points that are in either A or B or in both. This is denoted by $A \cup B = \{1, 2, 3, 4\}$. If $C = \{4\}$, then $A \cup C = \{1, 2, 4\}$. The *intersection* of two sets A and B is the set consisting of all points that are in both A and B, and is denoted by $A \cap B$, or merely AB. For the above example $A \cap B = AB = \{2\}$ and $AC = \varnothing$, where $\varnothing$ denotes the null set, or the set consisting of no points.

The *complement* of A, with respect to S, is the set of all points in S that are not in A and is denoted by $\bar{A}$. For the specific sets given above, $\bar{A} = \{3, 4\}$. Two sets are said to be *mutually exclusive*, or *disjoint*, if they have no points in common, as in A and C above.

Venn diagrams can be used to portray effectively the concepts of union, intersection, complement, and disjoint sets, as in Figure 2.1.

FIGURE 2.1
Venn Diagrams of Set Relations

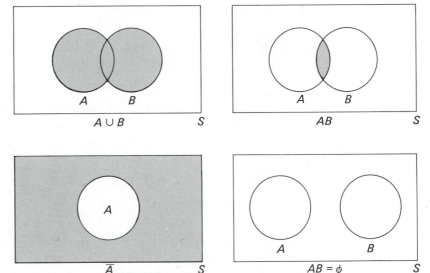

We can easily see from Figure 2.1 that

$$\bar{A} \cup A = S$$

for any set A. Other important relationships among events are the *distributive laws*,

$$A(B \cup C) = AB \cup AC$$

$$A \cup (BC) = (A \cup B)(A \cup C)$$

and *DeMorgan's Laws*,

$$\overline{A \cup B} = \bar{A}\bar{B}$$

$$\overline{AB} = \bar{A} \cup \bar{B}$$

It is important to be able to relate descriptions of sets to their symbolic notation, using the symbols given above, and to list correctly or count the elements in sets of interest. The following example illustrates the point.

EXAMPLE 2.1 _____

Twenty electric motors are pulled from an assembly line and inspected for defects. Eleven of the motors are free of defects, 8 have defects on the exterior finish, and 3 have defects in their assembly and will not run. Let A denote the set of motors having assembly defects and F the set having defects on their finish. Using A and F, write a symbolic notation for

(a)   the set of motors having both types of defects
(b)   the set of motors having at least one type of defect
(c)   the set of motors having no defects
(d)   the set of motors having exactly one type of defect

Then give the number of motors in each set.

Solution   (a)   The motors with both types of defects must be in A and F; thus this event can be written AF. Since only 9 motors have defects, while A contains 3 and F contains 8 motors, 2 motors must be in AF. (See Figure 2.2.)

FIGURE 2.2
Venn Diagram for
Example 2.1
(numbers of motors
shown for each set)

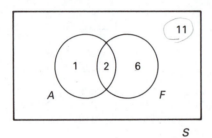

(b)   The motors having at least one type of defect must have either an assembly defect or a finish defect. Hence this set can be written $A \cup F$. Since 11 motors have no defects, 9 must have at least one defect.

(c)   The set of motors having no defects is the complement of the set having at least one defect, and is written $\overline{A \cup F} = \bar{A}\bar{F}$ (by DeMorgan's Law). Clearly 11 motors fall into this set.

(d)   The set of motors having exactly one type of defect must be in A and not in F or in F and not in A. This set can be written $A\bar{F} \cup \bar{A}F$, and 7 motors fall into the set.   ■

EXERCISES ██

2.1   Of 25 microcomputers available in a supply room, 10 have circuit boards for a printer, 5 have circuit boards for a modem, and 13 have neither board. Using P to denote those that have printer boards and M to denote those that have modem boards, symbolically denote the following sets, and give the number of microcomputers in each set.

(a)   Those that have both boards.
(b)   Those that have neither board.
(c)   Those that have printer boards only.
(d)   Those that have exactly one of the boards.

2.2 Five applicants (Jim, Don, Mary, Sue, and Nancy) are available for two identical jobs. A supervisor selects two applicants to fill these jobs.

(a) List all possible ways in which the jobs can be filled. (That is, list all possible selections of two applicants from the five.)

(b) Let A denote the set of selections containing *at least* one male. How many elements are in A?

(c) Let B denote the set of selections containing *exactly* one male. How many elements are in B?

(d) Write the set containing two females in terms of A and B.

(e) List the elements in $\bar{A}$, AB, $A \cup B$, and $\overline{AB}$.

2.3 Use Venn diagrams to verify the distributive laws.

2.4 Use Venn diagrams to verify DeMorgan's Laws.

2.3 A FORMAL DEFINITION OF PROBABILITY

Suppose a regular six-sided die is tossed onto a table and the number on the upper face is observed. This is a probabilistic situation since the number occurring on the upper face cannot be determined in advance. We will analyze the components of this experimental situation and arrive at a definition of probability that will allow us to model mathematically what happens in die tosses, as well as in many similar situations.

First we might toss the die once, or several times, to collect data on possible outcomes. This data-generation phase is called an *experiment*.

DEFINITION 2.1	An **experiment** is the process of making an observation.

Any experiment can result in a number of possible outcomes. A list, or set, or all possible outcomes for an experiment is called a *sample space*.

DEFINITION 2.2	A **sample space** S is a set that includes all possible outcomes for an experiment, listed in a mutually exclusive and exhaustive manner.

"Mutually exclusive" means that the elements of the set do not overlap, and "exhaustive" means that the list contains all possible outcomes.

For the die toss we could write a sample space as

$$S_1 = \{1, 2, 3, 4, 5, 6\}$$

where the integers indicate the possible numbers of dots on the upper face, or as

$$S_2 = \{\text{even, odd}\}$$

Both S_1 and S_2 satisfy Definition 2.2, but S_1 seems the better choice because it gives all the necessary details. S_2 has three possible upper-face outcomes in each listed element, whereas S_1 has only one possible outcome per element.

As another example suppose an inspector is measuring the length of a machined rod. (This measurement process constitutes the experiment.) A sample space could be listed as

$$S_3 = \{1, 2, 3, \ldots 50, 51, 52, \ldots 70, 71, 72, \ldots\}$$

if the length is rounded to the closest integer number of inches. On the other hand an appropriate sample space could be

$$S_4 = \{x \mid x > 0\}$$

read "the set of all real numbers x such that $x > 0$." Whether S_3 or S_4 should be used in a particular problem depends on the nature of the measurement process. If the measurement is to be precise, we need S_4. If only integers are to be used, S_3 will suffice. The point is, then, that sample spaces for a particular experiment are not unique, and must be selected so as to provide all pertinent information for a given situation.

Let us go back to our first example, the toss of a die. Suppose player A can have first turn at a board game if he rolls a six. Therefore, the event, "roll a six" is important to him. Other possible events of interest in the die-tossing experiment are "roll an even number," "roll a number greater than four," and so on.

DEFINITION 2.3 | An **event** is any subset of a sample space.

Definition 2.3 holds as stated for any sample space that has a finite or countable number of elements. Some subsets must be ruled out if a sample space covers a continuum of real numbers, as S_4 given above, but any subset likely to occur in practice can be called an event.

From this definition of events we see that an event is a collection of elements from the sample space. For the die-tossing experiment we can see how this works, for example, by defining

A to be "an even number"

B to be "an odd number"

C to be "a number greater than 4"

E_1 to be "observe a 1"

and, in general,

E_i to be "observe integer i"

Then if $S = \{1, 2, 3, 4, 5, 6\}$,

$A = \{2, 4, 6\}$

$B = \{1, 3, 5\}$

$$C = \{5, 6\}$$

$$E_1 = \{1\}$$

and

$$E_i = \{i\}, \qquad i = 1, 2, 3, 4, 5, 6$$

Sample spaces and events often can be conveniently displayed in Venn diagrams. Some events for the die-tossing experiment are shown in Figure 2.3.

FIGURE 2.3
Venn Diagram for a
Die Toss

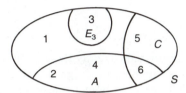

For an experiment we now know how to establish a sample space and to list appropriate events. The next step is to define a probability for these events. We have seen already that the intuitive idea of probability is related to relative frequency of occurrence. A regular die should show an even number, when tossed, about 1/2 of the time and a "3" about 1/6 of the time. So probabilities should be fractions between 0 and 1. One of the integers 1, 2, 3, 4, 5, or 6 must occur every time the die is tossed, and so the total probability associated with the sample space must be 1. In repeated tosses of the die, if a "1" occurs 1/6 of the time, then a "1 or 2" must occur $1/6 + 1/6 = 1/3$ of the time. Since relative frequencies for mutually exclusive events add, then so must the probabilities. These considerations lead to the following definition.

DEFINITION 2.4

Suppose that an experiment has associated with it a sample space S. A **probability** is a numerically valued function that assigns a number $P(A)$ to every event A so that the following axioms hold:

(1) $P(A) \geqslant 0$

(2) $P(S) = 1$

(3) If $A_1, A_2, \ldots,$ is a sequence of mutually exclusive events (i.e., $A_i A_j = \varnothing$ for any $i \neq j$), then

$$P\left(\bigcup_{i=1}^{\infty} A_i \right) = \sum_{i=1}^{\infty} P(A_i)$$

From axiom (3) it follows that, if A and B are mutually exclusive events,

$$P(A \cup B) = P(A) + P(B)$$

This is similar to the addition of relative frequencies in the die-tossing example discussed above.

It is now easy to see that if $A \subset B$, then $P(A) \leqslant P(B)$. To see this, write

$$B = A \cup \bar{A}B$$

so that

$$P(B) = P(A \cup \bar{A}B) = P(A) + P(\bar{A}B)$$

Since $P(\bar{A}B) \geqslant 0$ by axiom (1), it follows that $P(A) \leqslant P(B)$. In particular since $A \subset S$, for any event A, and $P(S) = 1$, then $P(A) \leqslant 1$.

In a similar way we can show that $P(\varnothing) = 0$. Since S and $\varnothing$ are disjoint with $S \cup \varnothing = S$,

$$1 = P(S) = P(S \cup \varnothing) = P(S) + P(\varnothing)$$

The definition of probability tells us only the axioms such a function must obey; it does not tell us what numbers to assign to specific events. The actual assignment of numbers usually comes about from empirical evidence, or from careful thought about the experiment. If a die is balanced, we could toss it a few times to see if the upper faces all seem equally likely to occur. Or we could simply assume this would happen and assign a probability of 1/6 to each of the six elements in S: $P(E_i) = 1/6$, $i = 1, 2, \ldots, 6$. Once we've done this, the model is complete since, by axiom (3), we can now find the probability of any event. For example, for events defined on page 28 and on Figure 2.3,

$$P(A) = P(E_2 \cup E_4 \cup E_6)$$
$$= P(E_2) + P(E_4) + P(E_6)$$
$$= \frac{1}{6} + \frac{1}{6} + \frac{1}{6} = \frac{1}{2}$$

and

$$P(C) = P(E_5 \cup E_6)$$
$$= P(E_5) + P(E_6)$$
$$= \frac{1}{6} + \frac{1}{6} = \frac{1}{3}$$

It is important to remember that Definition 2.4 and the actual assignment of probabilities to events provide a probabilistic model for an experiment. If $P(E_i) = 1/6$ is used in the die-tossing experiment, the model is good or bad depending on how close the long-run relative frequencies for each outcome actually come to these numbers suggested by a theory. If the die is balanced, the model should be good; it tells us what we can expect to happen. If the die is not balanced, then the model is bad and other probabilities should be substituted for the $P(E_i)$. Throughout the remainder of this book we will develop many specific models based on this underlying definition, and discuss practical situations in which they work well. None are perfect, but many are adequate for describing real-world probabilistic phenomena.

EXAMPLE 2.2

A purchasing clerk wants to order supplies from one of three possible vendors, which are numbered 1, 2 and 3. All vendors are equal with respect to quality and price, so the clerk writes each number on a piece of paper, mixes the papers, and blindly selects one number. The order is placed with the vendor whose number is selected. Let E_i denote the event that vendor i is selected ($i = 1, 2, 3$), B the event that vendor 1 or 3 is selected, and C the event that vendor 1 is *not* selected. Find the probabilities of events E_i, B, and C.

Solution

Events E_1, E_2, and E_3 correspond to the elements of S since they represent all the "single possible outcomes." Thus if we assign appropriate probabilities to these events, the probability of any other event is easily found.

Since one number is picked at random from the three available, it should seem intuitively reasonable to assign a probability of 1/3 to each E_i. That is,

$$P(E_1) = P(E_2) = P(E_3) = \frac{1}{3}$$

We find no reason to suspect that one number has a higher chance of being selected than any of the others.

Now

$$B = E_1 \cup E_3$$

and by axiom (3) of Definition 2.4

$$P(B) = P(E_1 \cup E_3) = P(E_1) + P(E_3) = \frac{1}{3} + \frac{1}{3} = \frac{2}{3}$$

Similarly

$$C = E_2 \cup E_3$$

and thus

$$P(C) = P(E_2) + P(E_3) = \frac{2}{3}$$

Note that different probability models could have been selected for the sample space connected with this experiment, but only this model is reasonable under the assumption that the vendors are all equally likely to be selected.

The terms *blindly* and *at random* are interpreted as imposing equal probabilities on the finite number of points in the sample space. ∎

The examples seen so far assign equal probabilities to the elements of a sample space, but this is not always the case. If you have a quarter and a penny in your pocket and pull out the first one you touch, then the quarter may have a higher probability of being chosen because of its larger size.

Often the probabilities that are assigned to events are based on experimental evidence or observational studies that yield relative frequency data on those events of interest. The data provide only approximations to the true probabilities, but these approximations are often quite good, and are usually the only information we have on the events of interest. Example 2.3 illustrates the point.

EXAMPLE 2.3 _____

The data in Figure 2.4 on marital status of the unemployed in the United States is provided by the Bureau of Labor Statistics.

FIGURE 2.4
Marital Status of the
Unemployed

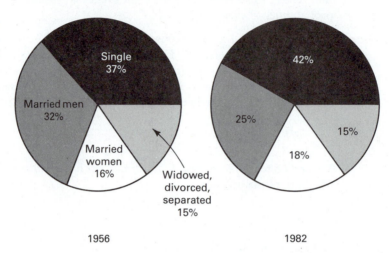

1956   1982

Suppose you had met an unemployed worker in 1956. What is the approximate probability that the unemployed person would have been (a) a married woman? (b) single? (c) married? Answer parts (a), (b), and (c) also for an unemployed worker met in 1982.

Solution   We assume nothing else is known about the unemployed person that you met. (The person is, in effect, randomly chosen from the population of unemployed workers.) Then we see directly from the chart for 1956 that

P(married woman) $= 0.16$

P(single) $= 0.37$

P(married) $= P$(married woman) $+ P$(married man)

$= 0.16 + 0.32 = 0.48$

From the 1982 chart we have

P(married woman) $= 0.18$

P(single) $= 0.42$

P(married) $= 0.18 + 0.25 = 0.43$

Note that we *cannot* answer some potentially interesting questions from this chart. For example we cannot find the probability that the unemployed person is a single woman. ∎

EXERCISES

2.5 A vehicle arriving at an intersection can turn left, turn right, or continue straight ahead. If an experiment consists of observing the movement of one vehicle at this intersection, do the following.
  (a) List the elements of a sample space.
  (b) Attach probabilities to these elements if all possible outcomes are equally likely.
  (c) Find the probability that the vehicle turns, under the probabilistic model of part (b).

2.6 A manufacturing company has two retail outlets. It is known that 30% of the potential customers buy products from outlet I alone, 50% buy from outlet II alone, 10% buy from both I and II, and 10% of the potential customers buy from neither. Let A denote the event that a potential customer, randomly chosen, buys from I and B the event that the customer buys from II. Find the following probabilities.
  (a) $P(A)$ (b) $P(A \cup B)$
  (c) $P(\bar{B})$ (d) $P(AB)$
  (3) $P(A \cup B)$ (f) $P(\bar{A}\bar{B})$
  (g) $P(\overline{A \cup B})$

$33\frac{\%}{}$

2.7 For volunteers coming into a blood center, 1 in 3 have O^+ blood, 1 in 15 have O^-, 1 in 3 have A^+, and 1 in 16 have A^-. What is the probability that the first person who shows up tomorrow to donate blood has

.06 %

  (a) type O^+ blood?   (b) type O blood?
  (c) type A blood?   (d) either type A^+ or O^+ blood?

6%

2.8 Information on modes of transportation for coal leaving the Appalachian region is shown on the following chart. If coal arriving at a certain power plant comes from this region, find the probability that it was transported out of the region

33%

6%

  (a) by truck to rail.
  (b) by water only.
  (c) at least partially by truck.
  (d) at least partially by rail.
  (e) by modes not involving water. (Assume "other" does not involve water.)

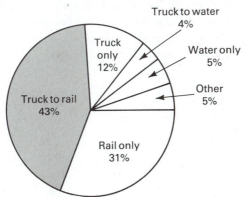

Source: Elmes, G., Transportation Research, *18A, No. 1, 1984, p. 19.*

2.9 Hydraulic assemblies for landing gear coming from an aircraft rework facility are inspected for defects. History shows that 8% have defects in the shafts alone, 6% have defects in the bushings alone, and 2% have defects in both shafts and bushings. If a randomly chosen assembly is to be used on aircraft, find the probability that it has

(a) a bushing defect.

(b) a shaft or bushing defect.

(c) only one of the two types of defects.

(d) no defects in shafts or bushings.

2.4 COUNTING RULES USEFUL IN PROBABILITY

Let's look at the die-tossing experiment from a slightly different perspective. Since there are six outcomes that should be equally likely for a balanced die, then the probability of A, observe an even number, is

$$P(A) = \frac{3}{6} = \frac{\text{number of outcomes favorable to } A}{\text{total number of equally likely outcomes}}$$

This "definition" of probability will work for any experiment resulting in a finite sample space with *equally likely* outcomes. Thus it is important to be able to count the number of possible outcomes for an experiment. The number of outcomes for an experiment can easily become quite large, and counting them is difficult unless one knows a few counting rules. Four such rules are presented as theorems in this section.

Suppose a quality control inspector examines two manufactured items selected from a production line. Item 1 can be defective or nondefective, as can item 2. How many possible outcomes are possible for this experiment? In this case it is easy to list them. Using D_i to denote that the ith item is defective and N_i to denote that the ith item is nondefective, the possible outcomes are

$$D_1D_2 \qquad D_1N_2 \qquad N_1D_2 \qquad N_1N_2$$

These four outcomes could be placed on a two-way table, as in Figure 2.5. This table helps us see that the four outcomes arise from the fact that the first item has two possible outcomes and the second item has two possible outcomes, and hence the experiment of looking at both items has $2 \times 2 = 4$ outcomes. This is an example of the *multiplication rule*, given as Theorem 2.1.

FIGURE 2.5
Possible Outcomes for Inspecting Two Items (D_i denotes that the ith item is defective; N_i denotes that the ith item is nondefective.)

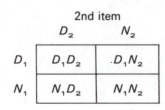

THEOREM 2.1   If the first task of an experiment can result in n_1 possible outcomes and, for each such outcome, the second task can result in n_2 possible outcomes, then there are $n_1 n_2$ possible outcomes for the two tasks together.

The multiplication rule extends to more tasks in a sequence. If, for example, three items were inspected and each could be defective or nondefective, then there would be $2 \times 2 \times 2 = 8$ possible outcomes.

Tree diagrams are also helpful in verifying the multiplication rule and for listing outcomes of experiments. Suppose a firm is deciding where to build two new plants, one in the east and one in the west. Four eastern cities and two western cities are possibilities. Thus there are $n_1 n_2 = 4(2) = 8$ possibilities for locating the two plants. Figure 2.6 shows the listing of these possibilities on a tree diagram.

FIGURE 2.6
Possible Outcomes for Locating Two Plants (A, B, C, D denote eastern cities; E, F denote western cities.)

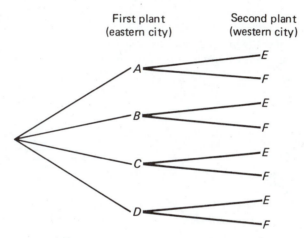

First plant (eastern city)   Second plant (western city)

The multiplication rule (Theorem 2.1) helps only in finding the number of elements in a sample space for an experiment. We must still assign probabilities to these elements to complete our probabilistic model. This is done for the site selection problem in Example 2.4.

EXAMPLE 2.4 _____

Referring to the firm that plans to build two new plants, the eight possible outcomes are shown in Figure 2.6. If all eight choices are equally likely (i.e., one of the pairs of cities is selected at random), find the probability that city E gets selected.

Solution   City E can be selected in four different ways, since there are four possible eastern cities to pair with it. Thus,

$$\{E \text{ gets selected}\} = \{AE\} \cup \{BE\} \cup \{CE\} \cup \{DE\}$$

Each of the 8 outcomes has probability 1/8, since the eight events are assumed to be equally likely. Since these 8 events are mutually exclusive,

$$P\{E \text{ gets selected}\} = P\{AE\} + P\{BE\} + P\{CE\} + P\{DE\}$$

$$= \frac{1}{8} + \frac{1}{8} + \frac{1}{8} + \frac{1}{8} = \frac{1}{2}$$

∎

EXAMPLE 2.5

Five motors (numbered 1 through 5) are available for use, and motor number 2 is defective. Motors 1 and 2 come from supplier I and motors 3, 4, and 5 come from supplier II. Suppose two motors are randomly selected for use on a particular day. Let A denote the event that the defective motor is selected and B the event that at least one motor comes from supplier I. Find $P(A)$ and $P(B)$.

Solution   We can see on the tree diagram in Figure 2.7 that there are 20 possible outcomes for this experiment, which agrees with our calculation using the multiplication rule. That is, there are twenty events of the form $\{1,2\}$, $\{1,3\}$, and so forth. Since the

FIGURE 2.7
Outcomes for
Experiment of
Example 2.5

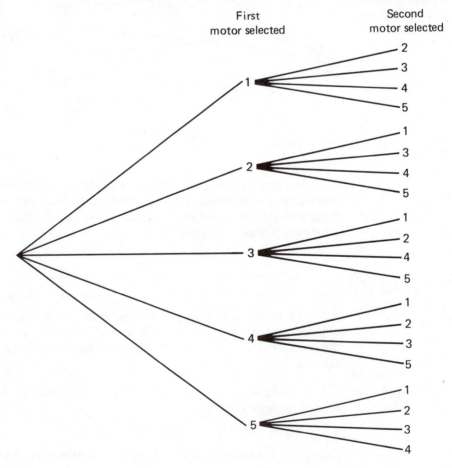

motors are randomly selected, each of the 20 outcomes has probability 1/20. Thus

$$P(A) = P(\{1,2\} \cup \{2,1\} \cup \{2, 3\} \cup \{2,4\} \cup \{2, 5\} \cup \{3, 2\} \cup \{4, 2\} \cup \{5, 2\})$$

$$= \frac{8}{20} = 0.4$$

since the probability of the union is the sum of the probabilities of the events in the union.

The reader can show that B contains 14 of the 20 outcomes and that

$$P(B) = \frac{14}{20} = 0.7$$ ■

The multiplication rule is often used to develop other counting rules. Suppose that from three pilots, a crew of two is to be selected to form a pilot-copilot team. To count the number of ways this can be done, observe that the pilot's seat can be filled in 3 ways and the copilot's in 2 ways (after the pilot is selected), so that there are $3 \cdot 2 = 6$ ways of forming the team. This is an example of a *permutation*, for which a general result is given in Theorem 2.2.

THEOREM 2.2 The number of ordered arrangements, or permutations, of r objects selected from n distinct objects $(r \leqslant n)$ is given by

$$P_r^n = n(n - 1) \cdots (n - r + 1) = \frac{n!}{(n - r)!}$$

Proof The basic idea of a permutation can be thought of as filling r slots in a line, with one object in each slot, by drawing these objects one at a time from a pool of n distinct objects. The first slot can be filled in n ways, but the second in only $(n - 1)$ ways after the first is filled. Thus by the multiplication rule the first two slots can be filled in $n(n - 1)$ ways. Extending this reasoning to r slots, we have the number of ways of filling all r slots is

$$n(n - 1) \cdots (n - r + 1) = \frac{n!}{(n - r)!} = P_r^n$$

Hence the theorem is proved. We illustrate the use of Theorem 2.2 with the following examples.

EXAMPLE 2.6 _____

From among 10 employees, three are to be selected for travel to three out-of-town plants, A, B, and C, one to each plant. Since the plants are in different cities, the order of assigning the employees to the plants is an important consideration. In how many ways can the assignments be made?

Solution   Since order is important, the number of possible distinct assignments is

$$P_3^{10} = \frac{10!}{7!} = 10(9)(8) = 720$$

In other words there are 10 choices for plant A, but then only 9 for plant B and 8 for plant C. This gives a total of $10(9)(8)$ ways of assigning employees to the plants.

■

EXAMPLE 2.7 _____

An assembly operation in a manufacturing plant involves four steps, which can be performed in any order. If the manufacturer wishes to compare experimentally the assembly times for each possible ordering of the steps, how many orderings will the experiment involve?

Solution   The number of orderings is the permutation of $n = 4$ things taken $r = 4$ at a time. (All steps must be accomplished each time.) This turns out to be

$$P_4^4 = \frac{4!}{0!} = 4! = 4 \cdot 3 \cdot 2 \cdot 1 = 24$$

since $0! = 1$ by definition. (In fact $P_r^r = r!$ for any integer r.)   ■

Sometimes order is not important, and we are interested only in the number of subsets of a certain size that can be selected from a given set. The general form of such a *combination* is given in Theorem 2.3.

THEOREM 2.3   The number of distinct subsets, or combinations, of size r that can be selected from n distinct objects ($r \leqslant n$) is given by

$$\binom{n}{r} = \frac{n!}{r!(n-r)!}$$

Proof   The number of *ordered* subsets of size r, selected from n distinct objects, is given by P_r^n. The number of *unordered* subsets of size r is denoted by $\binom{n}{r}$. Since any particular set of r objects can be ordered among themselves in $P_r^r = r!$ ways, it follows that

$$\binom{n}{r} r! = P_r^n$$

or

$$\binom{n}{r} = \frac{1}{r!} P_r^n = \frac{n!}{r!(n-r)!}$$

EXAMPLE 2.8

In Example 2.6 suppose that three employees are to be selected from the ten to go to the same plant. In how many ways can the selection be made?

Solution　Here order is not important, and we merely want to know how many subsets of size $r = 3$ can be selected from $n = 10$ people. The result is

$$\binom{10}{3} = \frac{10!}{3!7!} = \frac{10 \cdot 9 \cdot 8}{1 \cdot 2 \cdot 3} = 120$$ ∎

EXAMPLE 2.9

Refer to Example 2.8. If 2 of the 10 employees are female and 8 are male, what is the probability that exactly one female gets selected among the three?

Solution　We have seen that there are $\binom{10}{3} = 120$ ways to select 3 employees from the 10. Similarly there are $\binom{2}{1} = 2$ ways to select 1 female from the 2 available and $\binom{8}{2} = 28$ ways to select 2 males from the 8 available. If selections are made at random (that is, all subsets of 3 employees are equally likely to be chosen), then the probability of selecting exactly one female is

$$\frac{\binom{2}{1}\binom{8}{2}}{\binom{10}{3}} = \frac{2(28)}{120} = \frac{7}{15}$$ ∎

EXAMPLE 2.10

Five applicants for a job are ranked according to ability with applicant number 1 being best, number 2 second best, and so on. These rankings are unknown to an employer, who simply hires two applicants at random. What is the probability that this employer hires exactly one of the two best applicants?

Solution　The number of possible outcomes for the process of selecting two applicants from five is

$$\binom{5}{2} = \frac{5!}{2!3!} = 10$$

If one of the two best is selected, the selection can be done in

$$\binom{2}{1} = \frac{2!}{1!1!} = 2$$

ways. The other selected applicant must come from among the three lowest-ranking applicants, which can be done in

$$\binom{3}{1} = \frac{3!}{1!2!} = 3$$

ways. Thus the event of interest (hiring one of the two best applicants) can come about in $2 \cdot 6 = 6$ ways. The probability of this event is then $6/10 = 0.6$. ∎

THEOREM 2.4 The number of ways of partitioning n distinct objects into k groups containing n_1, $n_2, \ldots, n_k$ objects, respectively, is

$$\frac{n!}{n_1! n_2! \cdots n_k!}$$

where

$$\sum_{i=1}^{k} n_i = n$$

Proof The partitioning of n objects into k groups can be done by first selecting a subset of size n_1 from the n objects, then selecting a subset of size n_2 from the $n - n_1$ objects that remain, and so on until all groups are filled. The number of ways of doing this is

$$\binom{n}{n_1}\binom{n - n_1}{n_2} \cdots \binom{n - n_1 \cdots \cdots n_{k-1}}{n_k}$$

$$= \frac{n!}{n_1!(n - n_1)!} \cdot \frac{(n - n_1)!}{n_2!(n - n_1 - n_2)!} \cdot \cdots \cdot \frac{(n - n_1 \cdots - n_{k-1})!}{n_k! 0!} = \frac{n!}{n_1! n_2! \cdots n_k!}$$

EXAMPLE 2.11 _____

Suppose that 10 employees are to be divided among three jobs with 3 employees going to job I, 4 to job II, and 3 to job III. In how many ways can the job assignment be made?

Solution This problem involves a partitioning of the $n = 10$ employees into groups of size $n_1 = 3$, $n_2 = 4$, and $n_3 = 3$, and can be accomplished in

$$\frac{n!}{n_1! n_2! n_3!} = \frac{10!}{3! 4! 3!} = \frac{10 \cdot 9 \cdot 8 \cdot 7 \cdot 6 \cdot 5}{3 \cdot 2 \cdot 1 \cdot 3 \cdot 2 \cdot 1} = 4200$$

ways. (Notice the large number of ways this task can be accomplished!) ■

EXAMPLE 2.12 _____

In the setting of Example 2.11 suppose the only three employees of a certain ethnic group all get assigned to job I. What is the probability of this happening under a random assignment of employees to jobs?

Solution We have seen in Example 2.11 that there are 4,200 ways of assigning the ten workers to three jobs. The event of interest assigns three specified employees to job I. It remains to determine how many ways the other seven employees can be assigned to jobs II and III, which is

$$\frac{7!}{4! 3} = \frac{7(6)(5)}{3(2)(1)} = 35$$

Thus the chance of assigning three specific workers all to job I is

$$\frac{35}{4,200} = \frac{1}{120}$$

which is very small, indeed!   ■

EXERCISES

2.10   An experiment consists of observing two vehicles in succession move through the intersection of two streets.
   (a)   List the possible outcomes, assuming each vehicle can go straight, turn right, or turn left.
   (b)   Assuming the outcomes to be equally likely, find the probability that at least one vehicle turns left. (Would this assumption always be reasonable?)
   (c)   Assuming the outcomes to be equally likely, find the probability that at most one vehicle makes a turn.

2.11   A commercial building is designed with two entrances, say I and II. Two customers arrive and enter the building.
   (a)   List the elements of a sample space for this observational experiment.
   (b)   If all elements in (a) are equally likely, find the probability that both customers use door I; that both customers use the same door.

2.12   A corporation has two construction contracts that are to be assigned to one or more of three firms bidding for these contracts. (One firm could receive both contracts.)
   (a)   List the possible outcomes for the assignment of contracts to the firms.
   (b)   If all outcomes are equally likely, find the probability that both contracts go to the same firm.
   (c)   Under the assumptions of (b) find the probability that one specific firm, say firm I, gets at least one contract.

2.13   Among five portable generators produced by an assembly line in one day, there are two defectives. If two generators are selected for sale, find the probability that both will be nondefective. (Assume the two selected for sale are chosen so that every possible sample of size two has the same probability of being selected.)

2.14   Seven applicants have applied for two jobs. How many ways can the jobs be filled if
   (a)   the first person chosen receives a higher salary than the second?
   (b)   there are no differences between the jobs?

2.15   A package of six light bulbs contains two defective bulbs. If three bulbs are selected for use, find the probability that none is defective.

2.16   How many four-digit serial numbers can be formed if no digit is to be repeated within any one number? (The first digit may be a zero.)

2.17   A fleet of eight taxis is to be randomly assigned to three airports A, B, and C, with 2 going to A, 5 to B, and 1 to C.
   (a)   In how many ways can this be done?
   (b)   What is the probability that the specific cab driven by Jones is assigned to airport C?

2.18 Show that $\binom{n}{r} = \binom{n-1}{r-1} + \binom{n-1}{r}$, $1 \leqslant r \leqslant n$.

2.19 Five employees of a firm are ranked from 1 to 5 in their abilities to program a computer. Three of these employees are selected to fill equivalent programming jobs. If all possible choices of three (out of the five) are equally likely, find the probability that
  (a) the employee ranked number 1 is selected.
  (b) the highest-ranked employee among those selected has rank 2 or lower.
  (c) the employees ranked 4 and 5 are selected.

2.20 For a certain style of new automobile the colors blue, white, black, and green are in equal demand. Three successive orders are placed for automobiles of this style. Find the probability that
  (a) one blue, one white, and one green are ordered.
  (b) two blues are ordered.
  (c) at least one black is ordered.
  (d) exactly two of the orders are for the same color.

2.21 A firm is placing three orders for supplies among five different distributors. Each order is randomly assigned to one of the distributors, and a distributor can receive multiple orders. Find the probability that
  (a) all orders go to different distributors.
  (b) all orders go to the same distributor.
  (c) exactly two of the three orders go to one particular distributor.

2.22 An assembly operation for a computer circuit board consists of four operations that can be performed in any order.
  (a) In how many ways can the assembly operation be performed?
  (b) One of the operations involves soldering wire to a microchip. If all possible assembly orderings are equally likely, what is the probability that the soldering comes first or second?

2.23 Nine impact wrenches are to be divided evenly among three assembly lines.
  (a) In how many ways can this be done?
  (b) Two of the wrenches are used and seven are new. What is the probability that a particular line (line *A*) gets both used wrenches?

2.5 CONDITIONAL PROBABILITY AND INDEPENDENCE

Sometimes a sample space for an experiment can be narrowed by extra information available on events in question. For instance a sample space appropriate for measuring weights of all *people* could be narrowed considerably if it is known that only *adults* are to be investigated in the study. This extra information is referred to as *conditional* information. Let's look at a specific example in more detail.

Of 100 students completing an introductory statistics course, 20 were business majors. Ten students received A's in the course, and three of these were business majors. These facts are easily displayed on a Venn diagram, such as Figure 2.8, where *A* represents those students receiving A's and *B* represents business majors.

FIGURE 2.8

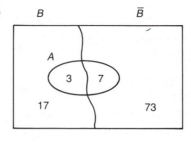

Now for a randomly selected student from this class, $P(A) = 0.1$ and $P(B) = 0.2$. But suppose we know that the randomly selected student is a business major. Then we may want to know the probability that the student received an A, *given* that she is a business major. Among the 20 business majors, 3 received A's. Thus $P(A$ *given* $B)$, written $P(A\,|\,B)$, is 3/20.

From the Venn diagram we see that the conditional (or given) information reduces the effective sample space to just the 20 business majors. Among them, 3 received A's.

Note that

$$P(A\,|\,B) = \frac{3}{20} = \frac{P(AB)}{P(B)} = \frac{3/100}{20/100}$$

This fact motivates Definition 2.5.

DEFINITION 2.5 | If A and B are any two events, then the **conditional probability** of A given B, denoted by $P(A\,|\,B)$, is

$$P(A\,|\,B) = \frac{P(AB)}{P(B)}$$

provided $P(B) > 0$.

EXAMPLE 2.13 _____

From 5 motors, of which one is defective, 2 motors are to be selected at random for use on a particular day. Find the probability that the second motor selected is nondefective, given that the first was nondefective.

Solution   Let N_i denote that the ith motor selected is nondefective. We want $P(N_2\,|\,N_1)$. From Definition 2.5 we have

$$P(N_2\,|\,N_1) = \frac{P(N_1 N_2)}{P(N_1)}$$

Looking at the twenty possible outcomes given in Figure 2.7, we can see that event N_1 contains 16 of these outcomes, and $N_1 N_2$ contains 12. Thus since the twenty outcomes are equally likely,

$$P(N_2 \mid N_1) = \frac{P(N_1 N_2)}{P(N_1)} = \frac{12/20}{16/20} = \frac{12}{16} = \frac{3}{4}$$

Does this answer seem intuitively reasonable?   ■

Conditional probabilities satisfy the three axioms of probability (Definition 2.4), as can be easily shown. First, since $AB \subset B$, then $P(AB) \leqslant P(B)$. Also $P(AB) \geqslant 0$ and $P(B) \geqslant 0$, so that

$$0 \leqslant P(A \mid B) = \frac{P(AB)}{P(B)} \leqslant 1$$

Second,

$$P(S \mid B) = \frac{P(SB)}{P(B)} = \frac{P(B)}{P(B)} = 1$$

Third, if $A_1, A_2, \ldots$ are mutually exclusive events, then so are $A_1 B, A_2 B, \ldots$, and

$$P\left(\bigcup_{i=1}^{\infty} A_i \mid B\right) = \frac{P\left(\left(\bigcup_{i=1}^{\infty} A_i\right) B\right)}{P(B)}$$

$$= \frac{P\left(\bigcup_{i=1}^{\infty} (A_i B)\right)}{P(B)} = \frac{\sum_{i=1}^{\infty} P(A_i B)}{P(B)}$$

$$= \sum_{i=1}^{\infty} \frac{P(A_i B)}{P(B)} = \sum_{i=1}^{\infty} P(A_i \mid B)$$

Conditional probability plays a key role in many practical applications of probability. In these applications important conditional probabilities are often drastically affected by seemingly small changes in the basic information from which the probabilities are derived. The following discussion of a medical application of probability illustrates the point.

A screening test indicates the presence or absence of a disease, and such tests are often used by physicians to detect diseases. Virtually all screening tests, however, have errors associated with their use. On thinking about the possible errors for a moment, it is clear that two different kinds of errors are possible; the test could show a person to have the disease when he or she does not (false positive) or fail to show that a person has the disease when it is present (false negative). Measures of these two types of errors are conditional probabilities called *sensitivity* and *specificity*.

A diagram will help in defining and interpreting these measures where the + indicates presence of the disease under study and the − indicates absence of the disease. The true diagnosis may never be known, but often can be determined by more intensive follow-up tests.

		True Diagnosis		
		+	−	
Test	+	a	b	$a + b$
Result	−	c	d	
		$a + c$	$b + d$	$a + b + c + d = n$

In this scenario n people are tested and $a + b$ are shown by the test to have the disease. Of these, a actually have the disease and b do not have the disease (false positives). Of the $c + d$ who test negative, c actually have the disease (false negatives). Using these labels,

$$\text{sensitivity} = \frac{a}{a + c}$$

the conditional probability of a positive test given that the person has the disease, and

$$\text{specificity} = \frac{d}{b + d}$$

the conditional probability of a negative test given that the person does not have the disease.

Obviously, a good test should have sensitivity and specificity both close to one. If sensitivity is close to one, then c (the number of false negatives) must be small. If specificity is close to one, then b (the number of false positives) must be small. Even when sensitivity and specificity are both close to one, a screening test can produce misleading results if not carefully applied. To see this, we look at one other important measure, the *predictive value* of a test given by

$$\text{predictive value} = \frac{a}{a + b}$$

The predictive value is the conditional probability of the person actually having the disease given that he or she tested positive. Clearly a good test should have a high predictive value, but this is not always possible even for highly sensitive and specific tests. The reason that all three measures cannot always be close to one simultaneously lies in the fact that predictive value is affected by the *prevalence rate* of the disease (that is, the proportion of the population under study that actually has the disease). We illustrate with three numerical situations given as I, II, and III on the diagram.

True Diagnosis

			+	−	
I	Test	+	90	10	100
	Result	−	10	90	
			100	100	200

			+	−	
II	Test	+	90	100	190
	Result	−	10	900	
			100	1000	1100

			+	−	
III	Test	+	90	1000	1090
	Result	−	10	9000	
			100	10,000	10,100

Among the 200 people under study in I, 100 have the disease (a prevalence rate of 50%). The test has sensitivity and specificity each equal to 0.90, and the predictive value is $90/100 = 0.90$. This is a good situation; the test is doing well.

In II the prevalence rate changes to 100/1100, or 9%. Even though the sensitivity and specificity are still 0.90, the predictive value has dropped to $90/190 = 0.47$. In III the prevalence rate is 100/10,000 or about 1%, and the predictive value has dropped farther to 0.08. Thus only 8% of those tested positive actually have the disease, even though the test has high sensitivity and specificity. What does this imply about using screening tests on large populations in which the prevalence rate for the disease being studied is low? An assessment of the answer to this question involves a careful look at conditional probabilities.

If the extra information in knowing that an event B has occurred does not change the probability of A—that is, if $P(A|B) = P(A)$—then events A and B are said to be *independent*. Since

$$P(A|B) = \frac{P(AB)}{P(B)}$$

the condition $P(A|B) = P(A)$ is equivalent to

$$\frac{P(AB)}{P(B)} = P(A)$$

or

$$P(AB) = P(A)P(B)$$

DEFINITION 2.6

Two events A and B are said to be **independent** if

$$P(A|B) = P(A)$$

or

$$P(B|A) = P(B)$$

This is equivalent to stating that

$$P(AB) = P(A)P(B)$$

EXAMPLE 2.14

Suppose that a foreman must select one worker for a special job, from a pool of four available workers, numbered 1, 2, 3, and 4. He selects the worker by mixing the four names and randomly selecting one. Let A denote the event that worker 1 or 2 is selected, B the event that worker 1 or 3 is selected, and C the event that worker 1 is selected. Are A and B independent? Are A and C independent?

Solution   Since the name is selected at random, a reasonable assumption for the probabilistic model is to assign a probability of 1/4 to each individual worker. Then $P(A) = 1/2$, $P(B) = 1/2$, and $P(C) = 1/4$. Since the intersection AB contains only worker 1, $P(AB) = 1/4$. Now $P(AB) = 1/4 = P(A)P(B)$, so A and B are independent. Since AC also contains only worker 1, $P(AC) = 1/4$. But $P(AC) = 1/4 \neq P(A)P(C)$, so A and C *are not* independent. A and C are said to be *dependent* because the fact that C occurs changes the probability that A occurs. ∎

Most situations in which independence issues arise are not like the one portrayed in Example 2.14, in which events are well defined and one merely calculates probabilities to check the definition. Often independence is *assumed* for two events in order to calculate their joint probability. Suppose, for example, A denotes that machine A does not break down today, and B denotes that machine B will not break down today. $P(A)$ and $P(B)$ can be approximated from the repair records of the machines. What is $P(AB)$, the probability that neither machine breaks down today? If we assume independence, $P(AB) = P(A)P(B)$, a straightforward calculation. However, if we do not assume independence, we cannot calculate $P(AB)$ unless we form a model for their dependence structure, or collect data on their joint performance. Is independence a reasonable assumption? It may be if the operation of one machine is not affected by the other, but it may not be if the machines share the same room, power supply, or job foreman. So, you see, independence is often used as a simplifying assumption, and may not hold precisely in all cases for which it is assumed. Remember, probabilistic models are simply models, and they do always precisely mirror reality. This should not be a major concern since all branches of science make simplifying assumptions when developing their models, probabilistic or deterministic.

Now we will show how these definitions aid us in establishing rules for computing probabilities of composite events.

2.6   RULES OF PROBABILITY

We now establish four rules that will help us in calculating probabilities of events. First recall that the complement $\bar{A}$ of an event A is the set of all outcomes in a sample space S that are not in A. Thus $\bar{A}$ and A are mutually exclusive and their union is S. That is

$$\bar{A} \cup A = S$$

It follows that

$$P(\bar{A} \cup A) = P(\bar{A}) + P(A) = P(S) = 1$$

or

$$P(\bar{A}) = 1 - P(A)$$

Thus we have established the following theorem.

THEOREM 2.5   If $\bar{A}$ is the complement of an event A in a sample space S, then

$$P(\bar{A}) = 1 - P(A)$$

EXAMPLE 2.15

A quality control inspector has 10 assembly lines from which to choose products for testing. Each morning of a 5-day week she randomly selects one of the lines to work on for the day. Find the probability that a line is chosen more than once during the week.

Solution   It is easier, here, to think in terms of complements, and first find the probability that no line is chosen more than once. If no line is repeated, 5 different lines must be chosen on successive days, which can be done in

$$P_5^{10} = \frac{10!}{5!} = 10(9)(8)(7)(6)$$

ways. The total number of possible outcomes for the selection of 5 lines without restriction is $(10)^5$ by an extension of the multiplication rule. Thus

$$P(\text{no line is chosen more than once}) = \frac{P_5^{10}}{(10)^5}$$

$$= \frac{10(9)(8)(7)(6)}{(10)^5} = 0.30$$

and

$$P(\text{a line is chosen more than once}) = 1.0 - 0.30 = 0.70$$ ■

Axiom 3 applies to $P(A \cup B)$ if A and B are disjoint. But what happens when A and B are not disjoint? Theorem 2.6 gives the answer.

THEOREM 2.6   If A and B are any two events, then

$$P(A \cup B) = P(A) + P(B) - P(AB)$$

If A and B are mutually exclusive, then

$$P(A \cup B) = P(A) + P(B)$$

Proof   From a Venn diagram for the union of A and B (see Figure 2.1) it is easy to see that

$$A \cup B = A\bar{B} \cup \bar{A}B \cup AB$$

and that the three events on the right-hand side of the equality are mutually exclusive. Hence

$$P(A \cup B) = P(A\bar{B}) + P(\bar{A}B) + P(AB)$$

Now

$$A = A\bar{B} \cup AB$$

and

$$B = \bar{A}B \cup AB$$

so that

$$P(A) = P(A\bar{B}) + P(AB)$$

and

$$P(B) = P(\bar{A}B) + P(AB)$$

It follows that

$$P(A\bar{B}) = P(A) - P(AB)$$

and

$$P(\bar{A}B) = P(B) - P(AB)$$

Substituting into the first equation for $P(A \cup B)$, we have

$$P(A \cup B) = P(A) - P(AB) + P(B) - P(AB) + P(AB)$$
$$= P(A) + P(B) - P(AB)$$

and the proof is complete.

The formula for the probability of the union of k events, $A_1, A_2, \ldots, A_k$, is derived in similar fashion, and is given by

$$P(A_1 \cup A_2 \cup \cdots \cup A_k) = \sum_{i=1}^{k} P(A_i) - \sum_{i<j}\sum P(A_iA_j) + \sum_{i<j<l}\sum\sum P(A_iA_jA_l)$$

$$- \cdots + \cdots - (-1)^k P(A_1A_2 \ldots A_k)$$

The next rule is actually just a rearrangement of the definition of conditional probability, for the case in which a conditional probability may be known and we want to find the probability of an intersection.

THEOREM 2.7   If A and B are any two events, then

$$P(AB) = P(A)P(B|A)$$

$$= P(B)P(A|B)$$

If A and B are independent, then

$$P(AB) = P(A)P(B)$$

We illustrate the use of these three theorems in the following examples.

EXAMPLE 2.16 _____

Records indicate that for the parts coming out of a hydraulic repair shop at an airplane rework facility, 20% will have a shaft defect, 10% will have a bushing defect, and 75% will be defect free. For an item chosen at random from this output, find the probability of

A: The item has at least one type of defect.

B: The item has only a shaft defect.

Solution   The percentages given imply that 5% of the items have both a shaft and a bushing defect. Let D_1 denote the event that an item has a shaft defect, and D_2 the event that it has a bushing defect. Then

$$A = D_1 \cup D_2$$

and

$$P(A) = P(D_1 \cup D_2)$$

$$= P(D_1) + P(D_2) - P(D_1D_2)$$

$$= 0.20 + 0.10 - 0.05$$

$$= 0.25$$

Another possible solution is to observe that the complement of A is the event that an item has no defects. Thus

$$P(A) = 1 - P(\bar{A})$$

$$= 1 - 0.75 = 0.25$$

To find $P(B)$, note that the event D_1 that the item has a shaft defect is the union of the events that it has *only* a shaft defect (B) and the event that it has *both* defects ($D_1 D_2$). That is

$$D_1 = B \cup D_1 D_2$$

where B and $D_1 D_2$ are mutually exclusive. Therefore

$$P(D_1) = P(B) + P(D_1 D_2)$$

or

$$P(B) = P(D_1) - P(D_1 D_2)$$

$$= 0.20 - 0.05 = 0.15$$

You should sketch these events on a Venn diagram and verify the results derived above.   ∎

EXAMPLE 2.17 _____

A section of an electrical circuit has two relays in parallel, as shown in Figure 2.9. The relays operate independently, and when a switch is thrown, each will close properly with probability only 0.8. If the relays are both open, find the probability that current will flow from s to t when the switch is thrown.

FIGURE 2.9
Two Relays in Parallel

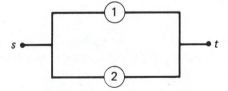

Solution   Let O denote an open relay and C a closed relay. The four outcomes for this experiment are shown by

$$
\begin{array}{ccc}
& \text{Relay} & \text{Relay} \\
& 1 & 2 \\
E_1 = \{(O & , & O)\} \\
E_2 = \{(O & , & C)\} \\
E_3 = \{(C & , & O)\} \\
E_4 = \{(C & , & C)\}
\end{array}
$$

Since the relays operate independently, we can find the probabilities for these outcomes as follows.

$$P(E_1) = P(O)P(O) = (0.2)(0.2) = 0.04$$

$$P(E_2) = P(O)P(C) = (0.2)(0.8) = 0.16$$

$$P(E_1) = P(C)P(O) = (0.8)(0.2) = 0.16$$

$$P(E_1) = P(C)P(C) = (0.8)(0.8) = 0.64$$

if A denotes the event that current will flow from s to t, then

$$A = E_2 \cup E_3 \cup E_4$$

or

$$\bar{A} = E_1$$

(At least one of the relays must close for current to flow.) Thus

$$P(A) = 1 - P(\bar{A})$$

$$= 1 - P(E_1)$$

$$= 1 - 0.04 = 0.96$$

which is the same as $P(E_2) + P(E_3) + P(E_4)$. ■

EXAMPLE 2.18 _____

Three different orders are to be mailed to three suppliers. However, an absent-minded secretary gets the orders mixed up, and just sends them randomly to suppliers. If a match refers to the fact that a supplier gets the correct order, find the probability of

(a)   no matches
(b)   exactly one match

Solution   This problem can be solved by listing outcomes since there are only three orders and suppliers involved, but a more general method of solution will be illustrated. Define the following events

A_1: match for supplier I

A_2: match for supplier II

A_3: match for supplier III

There are $3! = 6$ equally likely ways of randomly sending the orders to suppliers. There are only $2! = 2$ ways of sending the orders to suppliers if one particular supplier is required to have a match. Hence

$$P(A_1) = P(A_2) = P(A_3) = 2/6 = 1/3$$

Similarly it follows that

$$P(A_1 A_2) = P(A_1 A_3) = P(A_2 A_3)$$

$$= P(A_1 A_2 A_3) = 1/6$$

(a) Note that

P(no matches)

$$= 1 - P(\text{at least one match})$$

$$= 1 - P(A_1 \cup A_2 \cup A_3)$$

$$= 1 - [P(A_1) + P(A_2) + P(A_3)$$

$$- P(A_1A_2) - P(A_1A_3)$$

$$- P(A_2A_3) + P(A_1A_2A_3)]$$

$$= 1 - [3(1/3) - 3(1/6) + (1/6)]$$

$$= 1/3$$

(b) We leave it to the reader to show that

P(exactly one match)

$$= P(A_1) + P(A_2) + P(A_3) - 2[P(A_1A_2) + P(A_1A_3) + P(A_2A_3)]$$

$$+ 3P(A_1A_2A_3)$$

$$= 3(1/3) - 2(3)(1/6) + 3(1/6) = 1/2 \qquad \blacksquare$$

The fourth rule we present in this section is based on the notion of a partition of a sample space. Event $B_1, B_2, \ldots, B_k$ are said to partition a sample space S if two conditions exist:

1. $B_iB_j = \varnothing$ for any pair i and j ($\varnothing$ denotes the null set)
2. $B_1 \cup B_2 \cup \cdots \cup B_k = S$

For example, the set of tires in an auto assembly warehouse may be partitioned according to suppliers, or employees of a firm may be partitioned according to level of education. We illustrate a partition for the case $k = 2$ in Figure 2.10. The key idea involving a partition is to observe that an event A (see Figure 2.9) can be written as the union of mutually exclusive events AB_1 and AB_2.

FIGURE 2.10
A Partition of S into B_1 and B_2

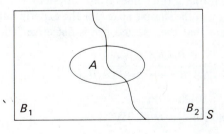

That is

$$A = AB_1 \cup AB_2$$

and thus

$$P(A) = P(AB_1) + P(AB_2)$$

If conditional probabilities $P(A|B_1)$ and $P(A|B_2)$ are known, then $P(A)$ can be found by writing

$$P(A) = P(B_1)P(A|B_1) + P(B_2)P(A|B_2)$$

In problems dealing with partitions it is frequently of interest to find probabilities of the form $P(B_1|A)$, which can be written

$$P(B_1|A) = \frac{P(B_1A)}{P(A)}$$

$$= \frac{P(B_1)P(A|B_1)}{P(B_1)P(A|B_1) + P(B_2)P(A|B_2)}$$

This result is a special case of *Bayes' Rule*, of which Theorem 2.8 is a general statement.

THEOREM 2.8   If $B_1, B_2, \ldots, B_k$ forms a partition of S, and A is any event in S, then

$$P(B_j|A) = \frac{P(B_j)P(A|B_j)}{\sum\limits_{i=1}^{k} P(B_i)P(A|B_i)}$$

The proof is an extension of the results shown above for $k = 2$.

EXAMPLE 2.19 _____

A company buys tires from two suppliers, 1 and 2. Supplier 1 has a record of delivering tires containing 10% defectives, whereas supplier 2 has a defective rate of only 5%. Suppose 40% of the current supply came from supplier 1. If a tire is taken from this supply and observed to be defective, find the probability that it came from supplier 1.

Solution   Let B_i denote the event that a tire comes from supplier i, $i = 1, 2$, and note that B_1 and B_2 form a partition of the sample space for the experiment of selecting one tire. Let A denote the event that the selected tire is defective. Then

$$P(B_1|A) = \frac{P(B_1)P(A|B_1)}{P(B_1)P(A|B_1) + P(B_2)P(A|B_2)}$$

$$= \frac{0.40(0.10)}{0.40(0.10) + (0.60)(0.05)}$$

$$= \frac{0.04}{0.04 + 0.03} = \frac{4}{7}$$

Supplier 1 has a greater probability of being the party supplying the defective tire than does supplier 2.   ■

In practice, approximations for probabilities of events often come from real data. The data may not provide exact probabilities, but the approximations are frequently good enough to provide a basis for clearer insights into the problem at hand.

We illustrate the use of real data in probability calculations with Examples 2.20 and 2.21.

EXAMPLE 2.20

The U.S. Bureau of Labor Statistics provides the data in Table 2.1 as a summary of employment in the United States for 1982.

If an arbitrarily selected U.S. resident was asked, in 1982, to fill out a questionnaire on employment, find the probability that the resident would have been

(a) in the labor force.
(b) employed.
(c) employed and in the labor force.
(d) employed given that he or she was known to be in the labor force.
(e) either not in the labor force or unemployed.

TABLE 2.1
Employment in the U.S., 1982 (figures in millions)

Civilian noninstitutional population	172
Civilian labor force	110
Employed	99
Unemployed	11
Not in the labor force	62

Solution Let L denote the event that the resident is in the labor force and E that he or she is employed.

(a) Since 172 million people comprise the population under study and 110 million are in the labor force,

$$P(L) = \frac{110 \text{ in labor force}}{172 \text{ under study}}$$

(b) Similarly 99 million are employed; thus

$$P(E) = \frac{99 \text{ in force}}{172 \text{ under study}}$$

(c) The employed persons are a subset of those in the labor force; in other words, $EL = E$. Hence

$$P(EL) = P(E) = \frac{99}{172}$$

(d)   Among the 110 million people known to be in the labor force, 99 million are employed. Therefore

$$P(E|L) = \frac{99}{110}$$

Note that this can also be found by using Definition 2.5, as follows:

$$P(E|L) = \frac{P(EL)}{P(L)} = \frac{99/172}{110/172} = \frac{99}{110}$$

(e)   The event that the resident is not in the labor force $\bar{L}$ is mutually exclusive from the event that he or she is unemployed. Therefore

$$P(\bar{L} \cup \bar{E}) = P(\bar{L}) + P(\bar{E})$$

$$= \frac{62}{172} + \frac{11}{172} = \frac{73}{172} \qquad \blacksquare$$

The next example uses a data set that is a little more complicated in structure.

EXAMPLE 2.21

The National Fire Incident Reporting Service gave the information in Table 2.2 on fires reported in 1978. All figures on the table are percents, and the main body of the table shows percentages according to cause. For example, the 22 in the upper left corner shows that 22% of fires in the family homes were caused by heating.

TABLE 2.2
Causes of Fires in Residences, 1978

Causes of Fire	Family Homes	Apartments	Mobile Homes	Hotels/ Motels	Other	All Locations
Heating	22	6	22	8	45	19
Cooking	15	24	13	7	0	16
Incendiary Substance	10	15	7	16	8	11
Smoking	7	18	6	36	19	10
Electrical	8	5	15	7	28	8
Other	38	32	37	26	0	36
All Causes	73	20	3	2	2	

If a residential fire is called into a fire station, find the probability that it is
(a)   a fire caused by heating.
(b)   a fire in a family home.
(c)   a fire caused by heating given that it was in an apartment.
(d)   an apartment fire caused by heating.
(e)   a fire in a family home given that it was caused by heating.

Solution (a) Over all locations 19% of fires are caused by heating. (See the right-hand column of Table 2.2.) Thus

P(heating fire) = 0.19

(b) Over all causes 73% of the fires are in family homes. (See the bottom row of Table 2.2.) Thus

P(family home fire) = 0.73

(c) The second column from the left deals only with apartment fires. Thus 6% of all apartment fires are caused by heating. Note that this is a conditional probability and can be written

P(heating fire|apartment fire) = 0.06

(d) In part (c) we know the fire was in an apartment, and the probability in question was conditional on that information. Here we must find the probability of an intersection between "apartment fire" (say, event *A*) and "heating fire" (say, event *H*). We know *P*(*H*|*A*) directly from the table, so

$$P(AH) = P(A)P(H|A)$$

$$= (0.20)(0.06) = 0.012$$

In other words only 1.2% of all reported fires are caused by heating in apartments.

(e) The figures in the body of Table 2.2 give probabilities of causes, given the location of the fire. We are now asked to find the probability of a location given the cause. This is exactly the situation to which Bayes' Rule (Theorem 2.8) applies. The locations form a partition of the set of all fires into five different groups. We have, then, with obvious shortcuts in wording,

$$P(\text{family}|\text{heating}) = \frac{P(\text{family})P(\text{heating}|\text{family})}{P(\text{heating})}$$

Now *P*(heating) = 0.19 from Table 2.2. However, it might be informative to show how this figure is derived from the other columns of the table. We have

$$P(\text{heating}) = P(\text{family})P(\text{heating}|\text{family})$$

$$+ P(\text{apartment})P(\text{heating}|\text{apartment})$$

$$+ P(\text{mobile})P(\text{heating}|\text{mobile})$$

$$+ P(\text{hotel})P(\text{heating}|\text{hotel}) + P(\text{other})P(\text{heating}|\text{other})$$

$$= (0.73)(0.22) + (0.20)(0.06) + (0.03)(0.22) + (0.02)(0.08)$$

$$+ (0.02)(0.45)$$

$$= 0.19$$

Then

$$P(\text{family}|\text{heating}) = \frac{(0.73)(0.22)}{0.19} = 0.85$$

In other words 85% of heating fires are found in family homes.

EXERCISES

2.24   Vehicles coming into an intersection can turn left or right or go straight ahead. Two vehicles enter an intersection in succession. Find the probability that at least one of the two turns left given that at least one of the two vehicles turns. What assumptions have you made?

2.25   A purchasing office is to assign a contract for computer paper and a contract for microcomputer disks to any one of three firms bidding for these contracts. (Any one firm could receive both contracts.) Find the probability that

     (a)   firm I receives a contract given that both contracts do not go to the same firm.

     (b)   firm I receives both contracts.

     (c)   firm I receives the contract for paper given that it does not receive the contract for disks.

What assumptions have you made?

2.26   The following data give the number of accidental deaths overall, and for three specific causes, for the United States in 1984.

	All Types	Motor Vehicle	Falls	Drowning
All ages	92,911	46,263	11,937	5,388
Under 5	3,652	1,132	114	638
5 to 14	4,198	2,263	68	532
15 to 24	19,801	14,738	399	1,353
25 to 44	25,498	15,036	963	1,549
45 to 64	15,273	6,954	1,624	763
65 to 74	8,424	3,020	1,702	281
75 and over	16,065	3,114	7,067	272
Male	64,053	32,949	6,210	4,420
Female	28,858	13,314	5,727	968

Source: *The World Almanac*, 1988.

You are told that a certain person recently died in an accident. Approximate the probability that

     (a)   it was a motor vehicle accident.

     (b)   it was a motor vehicle accident if you know the person to be male.

     (c)   it was a motor vehicle accident if you know the person to be between 15 and 24 years of age.

     (d)   it was a fall if you know the person to be over age 75.

     (e)   the person was male.

2.27   The data below show the distribution of annual times at work by mode of travel for workers in the central business district of a larger city. The figures are percentages. (The columns should add to 100, but some do not because of rounding.)

Arrival Time	Transit	Drove Alone	Shared Ride with Family Member	Car Pool	All
Before 7:15	17	16	16	19	18
7:15–7:45	35	30	30	42	34
7:45–8:15	32	31	43	35	33
8:15–8:45	10	14	8	2	10
After 8:45	5	11	3	2	6

Source: Hendrickson, C., and E. Plank, *Transportation Research*, 18A, No. 1, 1984.

If a randomly selected worker is asked about his or her travel to work, find the probability that the worker

    (a)   arrives before 7:15 given that the worker drives alone.

    (b)   arrives at or after 7:15 given that the worker drives alone.

    (c)   arrives before 8:15 given that the worker rides in a car pool.

    (d)   Can you find the probability that the worker drives alone, using only these data?

2.28   Table 2.2 shows percentages of fires by cause for certain residential locations. If a fire is reported to be residential, find the probability that it is

    (a)   caused by smoking.

    (b)   in a mobile home.

    (c)   caused by smoking given that it was in a mobile home.

    (d)   in a mobile home given that it was caused by smoking.

2.29   An incoming lot of silicon wafers is to be inspected for defectives by an engineer in a microchip manufacturing plant. In a tray containing twenty wafers assume four are defective. Two wafers are to be randomly selected for inspection. Find the probability that

    (a)   both are nondefective.

    (b)   at least one of the two is nondefective.

    (c)   both are nondefective given that at least one is nondefective.

2.30   In the setting of Exercise 2.29 answer the same three questions if only two among the 20 wafers are assumed to be defective.

2.31   Among five applicants for chemical engineering positions in a firm, two are rated as excellent and the other three are rated as good. A manager randomly chooses two of these applicants to interview. Find the probability that the manager chooses

    (a)   the two rated as excellent.

    (b)   at least one of those rated excellent.

    (c)   the two rated excellent given that one of the chosen two is already known to be excellent.

2.32   Resistors are produced by a certain firm and marketed as 10-ohm resistors. However, the actual ohms of resistance produced by the resistors may vary. It is observed that 5% of the values are below 9.5 ohms and 10% are above 10.5 ohms. If two of these resistors, randomly selected, are used in a system, find the probability that

    (a)   both have actual values between 9.5 and 10.5.

    (b)   at least one has an actual value in excess of 10.5.

2.33 Consider the following segment of an electric circuit with three relays. Current will flow from a to b if there is at least one closed path when the relays are switched to "closed." However, the relays may malfunction. Suppose they close properly only with probability 0.9 when the switch is thrown, and suppose they operate independently of one another. Let A denote the event that current will flow from a to b when the relays are switched to "closed."

   (a) Find $P(A)$.
   (b) Find the probability that relay 1 is closed properly, given that current is known to be flowing from a to b.

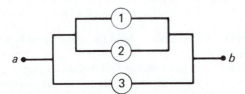

2.34 With relays operating as in Exercise 2.33, compare the probability of current flowing from a to b in the series system shown

with the probability of flow in the parallel system shown.

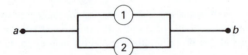

2.35 Electric motors coming off two assembly lines are pooled for storage in a common stockroom, and the room contains an equal number of motors from each line. Motors are periodically sampled from that room and tested. It is known that 10% of the motors from Line I are defective and 15% of the motors from Line II are defective. If a motor is randomly selected from the stockroom and found to be defective, find the probability that it came from Line I.

2.36 Two methods, A and B, are available for teaching a certain industrial skill. The failure rate is 20% for A and 10% for B. However, B is more expensive and hence is used only 30% of the time. (A is used the other 70%.) A worker is taught the skill by one of the methods, but fails to learn it correctly. What is the probability that he was taught by Method A?

2.37 A diagnostic test for a certain disease is said to be 90% accurate in that, if a person has the disease, the test will detect it with probability 0.9. Also, if a person does not have the disease, the test will report that he doesn't have it with probability 0.9. Only 1% of the population has the disease in question. If a person is chosen at random from the population, and the diagnostic test reports him to have the disease, what is the conditional probability that he does, in fact, have the disease? Are you surprised by the size of the answer? Would you call this diagnostic test reliable?

2.38 A proficiency examination for a certain skill was given to 100 employees of a firm. Forty of the employees were male. Sixty of the employees passed the examination, in that they scored above a preset level for satisfactory performance. The breakdown among males and females was as follows.

	Male (*M*)	Female (*F*)
Pass(*P*)	24	36
Fail ($\bar{P}$)	16	24

$\frac{24}{40}$

Suppose an employee is randomly selected from the 100 who took the examination.

    (a) Find the probability that the employee passed given that he was a male. $\frac{24}{60}$
    (b) Find the probability that the employee was male given that he passed.
    (c) Are events *P* and *M* independent?
    (d) Are events *P* and *F* independent?

2.39 By using Venn diagrams or similar arguments, show that, for events *A*, *B*, and *C*,

$$P(A \cup B \cup C) = P(A) + P(B) + P(C) - P(AB) - P(AC) - P(BC) + P(ABC)$$

2.40 By using the definition of conditional probability, show that

$$P(ABC) = P(A)P(B|A)P(C|AB)$$

2.41 There are 23 students in a classroom. What is the probability that at least two of them have the same birthday (day and month)? Assume that the year has 365 days. State your assumptions.

SUPPLEMENTARY EXERCISES

2.42 A coin is tossed four times and the outcome is recorded for each toss.
    (a) List the outcomes for the experiment.
    (b) Let *A* be the event that the experiment yields three heads. List the outcomes in *A*.
    (c) Make a reasonable assignment of probabilities to the outcomes and find *P*(*A*).

2.43 A hydraulic rework shop in a factory turns out seven rebuilt pumps today. Suppose three pumps are still defective. Two of the seven are selected for thorough testing and then classified as defective or nondefective.
    (a) List the outcomes for this experiment.
    (b) Let *A* be the event that the selection includes no defectives. List the outcomes in *A*.
    (c) Assign probabilities to the outcomes and find *P*(*A*).

2.44 The National Maximum Speed Limit (NMSL) of 55 miles per hour has been in force in the U.S. since early 1974. The data below show the percent of vehicles found to travel at various speeds for three types of highways in 1973 (before the NMSL), 1974 (the year the NMSL was put in force), and 1975.

Vehicle Speed (mph)	Rural Interstate						Rural Primary						Rural Secondary					
	Car			Truck			Car			Truck			Car			Truck		
	73	74	75	73	74	75	73	74	75	73	74	75	73	74	75	73	74	75
30–35	0	0	0	0	0	0	0	1	0	1	1	1	4	4	2	6	7	5
35–40	0	0	0	1	1	0	3	2	2	5	5	3	6	9	6	9	11	7
40–45	0	1	1	2	2	2	5	7	5	9	10	7	11	14	10	15	16	13
45–50	2	7	5	5	11	8	13	18	14	20	21	19	19	25	21	22	25	23
50–55	5	24	23	15	29	29	16	29	29	21	30	30	19	23	26	19	21	26
55–60	13	37	41	27	36	40	22	27	32	22	23	27	19	16	22	17	14	19
60–65	21	21	22	25	15	16	19	11	12	14	7	10	11	6	9	7	4	5
65–70	29	7	6	18	5	4	13	3	5	6	2	2	7	2	3	4	2	1
70–75	19	2	2	5	1	1	6	2	1	1	1	1	3	1	1	0	0	1
75–80	7	1	0	2	0	0	1	0	0	0	0	0	1	0	0	1	0	0
80–85	4	0	0	0	0	0	0	0	0	0	0	0	0	0	0	0	0	0

Source: D. B. Kamerud, *Transportation Research*, Vol. 17A, No. 1, p. 61.

(a) Find the probability that a randomly observed car on a rural interstate was traveling less than 55 mph in 1973; less than 55 mph in 1974; less than 55 mph in 1975.

(b) Answer the questions in (a) for a randomly observed truck.

(c) Answer the questions in (a) for a randomly observed car on a rural secondary road.

2.45 An experiment consists of tossing a pair of dice.

    (a) Use the combinatorial theorems to determine the number of outcomes in the sample space S.

    (b) Find the probability that the sum of the numbers appearing on the dice is equal to 7.

2.46 Show that $\binom{3}{0} + \binom{3}{1} + \binom{3}{2} + \binom{3}{3} = 2^3$. Show that in general $\sum_{i=0}^{n}\binom{n}{i} = 2^n$.

2.47 Of the persons arriving at a small airport, 60% fly on major airlines, 30% fly on privately owned airplanes, and 10% fly on commercially owned airplanes not belonging to an airline. Of the persons arriving on major airlines, 50% are traveling for business reasons, while this figure is 60% for those arriving on private planes and 90% for those arriving on other commercially owned planes. For a person randomly selected from a group of arrivals, find the probability that

    (a) the person is traveling on business.

    (b) the person is traveling on business and on a private airplane.

    (c) the person is traveling on business given that he arrived on a commercial airliner.

    (d) the person arrived on a private plane given that he is traveling on business.

2.48 In how many ways can a committee of three be selected from ten people?

2.49 How many different telephone numbers can be formed from a seven-digit number if the first digit cannot be zero?

2.50   A personnel director for a corporation has hired ten new engineers. If three (distinctly different) positions are open at a particular plant, in how many ways can he fill the positions?

2.51   An experimenter wishes to investigate the effect of three variables—pressure, temperature, and type of catalyst—on the yield in a refining process. If the experimenter intends to use three settings each for temperature and pressure and two types of catalysts, how many experimental runs will have to be conducted if he wishes to run all possible combinations of pressure, temperature, and types of catalysts?

2.52   An inspector must perform eight tests on a randomly selected keyboard coming off an assembly line. The sequence in which the tests are conducted is important because the time lost between tests will vary. If an efficiency expert were to study all possible sequences to find the one that required the minimum length of time, how many sequences would be included in his study?

2.53   (a)   Two cards are drawn from a 52-card deck. What is the probability that the draw will yield an ace and a face card in either order?
    (b)   Five cards are drawn from a 52-card deck. What is the probability that all 5 cards will be spades? of the same suit?

2.54   The quarterback on a certain football team completes 60% of his passes. If he tries three passes, assumed independent, in a given quarter, what is the probability that he will complete
    (a)   all three?
    (b)   at least one?
    (c)   at least two?

2.55   Two men each tossing a balanced coin obtain a "match" if both coins are heads or if both coins are tails. If the process is repeated three times,
    (a)   What is the probability of three matches?
    (b)   What is the probability that all six tosses (three for each man) result in "tails"?
    (c)   Coin tossing provides a model for many practical experiments. Suppose that the "coin tosses" represented the answers given by two students for three specific true-false questions on an examination. If the two students gave three matches for answers, would the low probability acquired in part (a) suggest collusion?

2.56   Refer to Exercise 2.55. What is the probability that the pair of coins are tossed four times before a match occurs (that is, the match occurs for the first time on the fourth toss)?

2.57   Suppose that the probability of exposure to the flu during an epidemic is 0.6. Experience has shown that a serum is 80% successful in preventing an inoculated person from acquiring the flu if exposed. A person not inoculated faces a probability of 0.90 of acquiring the flu if exposed. Two persons, one inoculated and one not, are capable of performing a highly specialized task in a business. Assume that they are not at the same location, are not in contact with the same people, and cannot expose each other. What is the probability that at least one will get the flu?

2.58   Two gamblers bet $1 each on the successive tosses of a coin. Each has a bank of $6.
    (a)   What is the probability that they break even after six tosses of the coin?
    (b)   What is the probability that one player, say Jones, wins all the money on the tenth loss of the coin?

2.59 Suppose that the streets of a city are laid out in a grid with streets running north-south and east-west. Consider the following scheme for patrolling an area of 16 blocks by 16 blocks. A patrolman commences walking at the intersection in the center of the area. At the corner of each block, he randomly elects to go north, south, east, or west.

(a) What is the probability that he will reach the boundary of his patrol area by the time he walks the first eight blocks?

(b) What is the probability that he will return to the starting point after walking exactly four blocks?

2.60 Consider two mutually exclusive events, A and B, such that $P(A) > 0$ and $P(B) > 0$. Are A and B independent? Give a proof for your answer.

2.61 An accident victim will die unless he receives in the next 10 minutes an amount of type A Rh^+ blood, which can be supplied by a single donor. It requires 2 minutes to "type" a prospective donor's blood and 2 minutes to complete the transfer of blood. A large number of untyped donors are available and 40% of them have type A Rh^+ blood. What is the probability that the accident victim will be saved if there is only one blood-typing kit available?

2.62 An assembler of electric fans uses motors from two sources. Company A supplies 90% of the motors and company B supplies the other 10%. Suppose it is known that 5% of the motors supplied by company A are defective and 3% of the motors supplied by company B are defective. An assembled fan is found to have a defective motor. What is the probability that this motor was supplied by company B?

2.63 Show that, for three events A, B, and C,

$$P[(A \cup B)|C] = P(A|C) + P(B|C) - P[(A \cap B|C)]$$

2.64 If A and B are independent events, show that A and $\bar{B}$ are also independent.

2.65 Three events A, B, and C are said to be independent if

$$P(AB) = P(A)P(B),$$

$$P(AC) = P(A)P(C),$$

$$P(BC) = P(B)P(C),$$

and

$$P(ABC) = P(A)P(B)P(C)$$

Suppose that a balanced coin is independently tossed two times. Define the following events:

A: head appears on the first toss,

B: head appears on the second toss,

C: both tosses yield the same outcome.

Are A, B, and C independent?

2.66 A line from a to b has midpoint c. A point is chosen at random on the line and marked x (the point x being chosen at random implies that x is equally likely to fall in any subinterval of fixed length l). Find the probability that the line segments ax, bx, and ac can be joined to form a triangle.

2.67 Eight tires of different brands are ranked from 1 to 8 (best to worst) according to mileage performance. If four of these tires are chosen at random by a customer, find the probability that the best tire among those selected by the customer is actually ranked third among the original eight.

2.68 Suppose that n indistinguishable balls are to be arranged in N distinguishable boxes so that each distinguishable arrangement is equally likely. If $n \geqslant N$, show that the probability that no box will be empty is given by

$$\frac{\binom{n-1}{N-1}}{\binom{N+n-1}{N-1}}$$

2.69 Relays in a section of an electrical circuit operate independently, and each one closes properly with probability 0.9 when a switch is thrown. The following two designs, each involving four relays, are presented for a section of a new circuit.

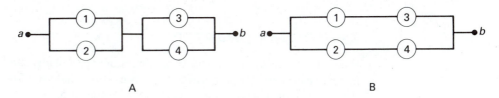

A   B

Which has the higher probability of current flowing from a to b when the switch is thrown?

2.70 A blood test for hepatitis has the following accuracy:

		Test Result	
		+	−
Patient	With Hepatitis	0.90	0.10
	Without Hepatitis	0.01	0.99

The disease rate in the general population is 1 in 10,000.

    (a) What is the probability that a person actually has hepatitis if he gets a positive blood test result?

    (b) A patient is sent for a blood test because he has lost his appetite and has jaundice. The physician knows that this type of patient has probability 1/2 of having hepatitis. If this patient gets a positive result on his blood test, what is the probability that he has hepatitis?

2.71 Show that for any events A and B

    (a) $P(AB) \geqslant P(A) + P(B) - 1$.

    (b) The probability that exactly one of the events occurs is $P(A) + P(B) - 2P(AB)$.

3

Discrete Probability Distributions

ABOUT THIS CHAPTER

Random variables, the numerical outcomes of experiments, can be divided nicely into two categories for practical convenience. Those that arise from count data, like the number of defective items per batch or the number of bacteria per cubic centimeter, are called *discrete* random variables. Probability distributions commonly used to model discrete random variables are discussed in this chapter. Continuous random variables are the subject of Chapter 4.

CONTENTS

3.1 RANDOM VARIABLES AND THEIR PROBABILITY DISTRIBUTIONS

Most of the experiments we encounter generate outcomes that can be interpreted in terms of real numbers, such as heights of children, number of voters favoring a certain candidate, tensile strengths of wires, and numbers of accidents at specified intersections. These numerical outcomes, with values that can change from experiment to experiment, are called *random variables*. We will look at an illustrative example of a random variable before we give a more formal definition.

A section of an electrical circuit has two relays, numbered 1 and 2, operating in parallel. The current will flow when a switch is thrown if either one or both of the relays close. The probability of a relay closing properly is 0.8, which is the same for each relay. The relays operate independently, we assume. Let E_i denote the event the relay i closes properly when the switch is thrown. Then $P(E_i) = 0.8$.

When the switch is thrown, a numerical event of some interest to the operator of this system is X, the number of relays that close properly. Now X can take on only three possible values, since the number of relays that close must be 0, 1, or 2. We can find the probabilities associated with these values of X by relating them to the underlying events E_i. Thus we have

$$P(X = 0) = P(\bar{E}_1 \bar{E}_2)$$
$$= P(\bar{E}_1)P(\bar{E}_2)$$
$$= (0.2)(0.2) = 0.04$$

since $X = 0$ means neither relay closes and the relays operate independently. Similarly

$$P(X = 1) = P(E_1\bar{E}_2 \cup \bar{E}_1 E_2)$$
$$= P(E_1\bar{E}_2) + P(\bar{E}_1 E_2)$$
$$= P(E_1)P(\bar{E}_2) + P(\bar{E}_1)P(E_2)$$
$$= (0.8)(0.2) + (0.2)(0.8)$$
$$= 0.32$$

and

$$P(X = 2) = P(E_1 E_2)$$
$$= P(E_1)P(E_2)$$
$$= (0.8)(0.8) = 0.64$$

The values of X along with their probabilities are more useful for keeping track of the operation of this system than are the underlying events E_i, for it is the *number* of properly closing relays that is the key to whether or not the system will work. The current will flow if X is at least one, and this event has probability

$$P(X \geqslant 1) = P(X = 1 \text{ or } X = 2)$$
$$= P(X = 1) + P(X = 2)$$
$$= 0.32 + 0.64$$
$$= 0.96$$

Note that we have mapped the outcomes of an experiment into a set of three meaningful real numbers, and attached a probability to each. Such situations provide the motivation for Definitions 3.1 and 3.2.

DEFINITION 3.1 | A **random variable** is a real-valued function whose domain is a sample space.

Random variables will be denoted by upper-case letters such as X, Y, and Z. The actual numerical values that a random variable can assume will be denoted by lower-case letters, such as x, y, and z. We can talk about the "the probability that X takes on the value x," $P(X = x)$, denoted by $p(x)$.

In the relay example the random variable X has only three possible values, and it is a relatively simple matter to assign probabilities to these values. Such a random variable is called discrete.

DEFINITION 3.2 | A random variable X is said to be **discrete** if it can take on only a finite number, or a countable infinity, of possible values x.

In this case
(1) $P(X = x) = p(x) \geqslant 0$.
(2) $\Sigma_x P(X = x) = 1$, where the sum is over all possible values x.
The function $p(x)$ is called the **probability function** of X.

The probability function is sometimes called the *probability mass function* of X to denote the idea that a mass of probability is piled up at discrete points.

It is often convenient to list the probabilities for a discrete random variable on a table. With X defined as the number of closed relays in the problem discussed above, the table is as follows:

x	$p(x)$
0	0.04
1	0.32
2	0.64
Total	1.00

This listing is one way of representing the *probability distribution* of X.

Note that the probability function $p(x)$ for any random variable satisfies two properties:

1. $0 \leqslant p(x) \leqslant 1$ for any x.
2. $\Sigma_x p(x) = 1$, where the sum is over all possible values of x.

Functional forms for the probability function $p(x)$ will be given in later sections. We now illustrate another method for arriving at a tabular presentation of a discrete probability distribution.

EXAMPLE 3.1 _____

The output of circuit boards from two assembly lines set up to produce identical boards is mixed into one storage tray. As inspectors examine the boards, it is difficult to determine whether a board comes from line A or line B. A probabilistic assessment of this question is often helpful. Suppose a storage tray contains 10 circuit boards, of which 6 came from line A and 4 from line B. An inspector selects two of these identically appearing boards for inspection. He is interested in X, the number of inspected boards from line A. Find the probability distribution for X.

Solution   The experiment consists of two selections, each of which can result in two outcomes. Let A_i denote the event that the ith board comes from line A and B_i the event that it comes from line B. Then the probability of selecting two boards from line A ($X = 2$) is

$$P(X = 2) = P(A_1 A_2) = P(A \text{ on 1st})P(A \text{ on 2nd}|A \text{ on 1st})$$

The multiplicative law of probability is used, and the probability on the second selection depends upon what happened on the first selection. There are other possibilities for outcomes that will result in other values of X. These outcomes are conveniently listed on the tree in Figure 3.1. The probabilities for the various selections are given on the branches of the tree.

FIGURE 3.1
Outcomes for
Example 3.1

1st Selection	2nd Selection	Probability	Value of X
A ($\frac{6}{10}$)	A ($\frac{5}{9}$)	$\frac{30}{90}$	2
	B ($\frac{4}{9}$)	$\frac{24}{90}$	1
B ($\frac{4}{10}$)	A ($\frac{6}{9}$)	$\frac{24}{90}$	1
	B ($\frac{3}{9}$)	$\frac{12}{90}$	0

It is easily seen that X has three possible outcomes, with probabilities as follows:

x	$p(x)$
0	$\dfrac{12}{90}$
1	$\dfrac{48}{90}$
2	$\dfrac{30}{90}$
Total	1.00

The reader should envision this concept extended to more selections from trays of various structure. ■

We sometimes study the behavior of random variables by looking at their *cumulative probabilities*. That is, for any random variable X we may look at $P(X \leqslant b)$ for any real number b. This is the cumulative probability for X evaluated at b. Thus we can define a function $F(b)$ as

$$F(b) = P(X \leqslant b)$$

This is similar to the study of cumulative frequency plots in Section 1.4.3.

DEFINITION 3.3

The **distribution function** $F(b)$ for a random variable X is defined as

$$F(b) = P(X \leqslant b)$$

If X is discrete,

$$F(b) = \sum_{x=-\infty}^{b} p(x)$$

where $p(x)$ is the probability function.

The distribution function is sometimes called the **cumulative distribution function** (c.d.f.).

The random variable X, denoting the number of relays closing properly (defined at the beginning of this section), has probability distribution given by

$$P(X = 0) = 0.04$$

$$P(X = 1) = 0.32$$

$$P(X = 2) = 0.64$$

Note that

$$P(X \leqslant 1.5) = P(X \leqslant 1.9) = P(X \leqslant 1) = 0.36$$

The distribution function for this random variable then has the form

$$F(b) = \begin{cases} 0 & b < 0 \\ 0.04 & 0 \leqslant b < 1 \\ 0.36 & 1 \leqslant b < 2 \\ 1.00 & 2 \leqslant b \end{cases}$$

This function is graphed in Figure 3.2.

FIGURE 3.2
A Distribution Function for a Discrete Random Variable

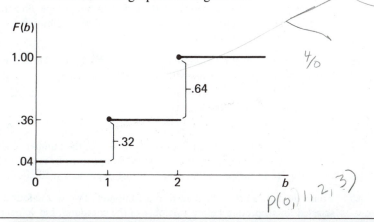

EXERCISES

3.1 Among 10 applicants for an open position 6 are female and 4 are males. Suppose 3 applicants are randomly selected from the applicant pool for final interviews. Find the probability distribution for X, the number of female applicants among the final three.

3.2 The median annual income for household heads in a certain city is $18,000. Four such household heads are randomly selected for an opinion poll.
   (a) Find the probability distribution of X, the number (out of the four) that have annual incomes below $18,000.
   (b) Would you say that it is unusual to see all four below $18,000 in this type of poll? (What is the probability of this event?)

3.3 Wade Boggs of the Boston Red Sox hit .363 in 1987. (He got a hit on 36.3% of his official times at bat.) In a typical game he was up to bat 3 official times. Find the probability distribution for X, the number of hits in a typical game. What assumptions are involved in the answer? Are the assumptions reasonable? Is it unusual for a good hitter to go 0 for 3 in one game?

3.4 A commercial building has two entrances, numbered I and II. Three people enter the building at 9:00 A.M. Let X denote the number that select entrance I. Assuming the people choose entrances independently, find the probability distribution for X. Were any additional assumptions necessary for your answer?

3.5 Table 2.2 on page 56 gives information on causes of residential fires. Suppose four independent residential fires are reported in one day, and let X denote the number, out of the four, that are in family homes.
   (a) Find the probability distribution for X in tabular form.
   (b) Find the probability that at least one of the four fires is in a family home.

3.6 It was observed that 40% of the vehicles crossing a certain toll bridge are commercial trucks. Four vehicles will cross the bridge in the next minute. Find the probability distribution for X, the number of commercial trucks among the four, if the vehicle types are independent of one another.

3.7 Of the people entering a blood bank to donate blood, 1 in 3 have type O^+ blood, and 1 in 15 have type O^- blood. For the next three people entering the blood bank, let X denote the number with O^+ blood and Y the number with O^- blood. Assuming independence among the people, with respect to blood type, find the probability distributions for X and Y. Also find the probability distribution for $X + Y$, the number of people with type O blood.

3.8 Daily sales records for a computer manufacturing firm show that it will sell 0, 1, or 2 mainframe computer systems with probabilities as listed:

Number of sales	0	1	2
Probability	0.7	0.2	0.1

(a) Find the probability distribution for X, the number of sales in a two-day period, assuming that sales are independent from day to day.

(b) Find the probability that at least one sale is made in the two-day period.

3.9 Four microchips are to be placed in a computer. Two of the four chips are randomly selected for inspection before assembly of the computer. Let X denote the number of defective chips found among the two inspected. Find the probability distribution for X if

(a) two of the microchips were defective.

(b) one of the four microchips was defective.

(c) none of the microchips was defective.

3.10 When turned on, each of the three switches in the diagram below works properly with probability 0.9. If a switch is working properly, current can flow through it when it is turned on. Find the probability distribution for Y, the number of closed paths from a to b, when all three switches are turned on.

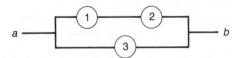

3.2 EXPECTED VALUES OF RANDOM VARIABLES

Since a probability can be thought of as the long-run relative frequency of occurrence for an event, a probability distribution can be interpreted as showing the long-run relative frequency of occurrence for numerical outcomes associated with an experiment. Suppose, for example, that you and a friend are matching balanced coins. Each of you tosses a coin. If the upper faces match, you win $1.00; if they do not match, you lose $1.00 (your friend wins $1.00). The probability of a match is 0.5 and, in the long run, you will win about half of the time. Thus a relative frequency distribution of your winnings should look like Figure 3.3. Note that the negative sign indicates a loss to you.

FIGURE 3.3
Relative Frequency of Winnings

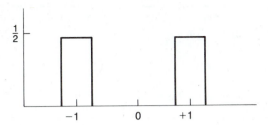

On the average, how much will you win per game over the long run? If Figure 3.3 is a correct display of your winnings, you win -1 half of the time and $+1$ half of the time for an average of

$$(-1)(\tfrac{1}{2}) + (1)(\tfrac{1}{2}) = 0$$

This average is sometimes called your expected winnings per game, or the *expected value* of your winnings. (An expected value of 0 indicates that this is a *fair game*.) The general definition of expected value is given in Definition 3.4. Note that the average is another measure of centrality, similar to the median introduced in Chapter 1.

DEFINITION 3.4

The **expected value** of a discrete random variable X having probability function $p(x)$ is given by[1]

$$E(X) = \sum_x xp(x)$$

(The sum is over all values of x for which $p(x) > 0$.) We sometimes use the notation

$$E(X) = \mu$$

Now payday has arrived and you and your friend up the stakes to \$10 per game of matching coins. You now win -10 or $+10$ with equal probability; your expected winnings per game is

$$(-10)(\tfrac{1}{2}) + (10)(\tfrac{1}{2}) = 0$$

and the game is still fair. The new stakes can be thought of as a function of the old in the sense that if X represents your winnings per game when you were playing for \$1.00, then $10X$ represents your winnings per game when playing for \$10.00. Such functions of random variables arise often, and the extension of the definition of expected value to cover these cases is given in Theorem 3.1.

[1] We assume absolute convergence when the range of X is countable; we talk about an expectation only when it is assumed to exist.

THEOREM 3.1   If X is a discrete random variable with probability distribution $p(x)$ and if $g(x)$ is any real-valued function of X, then

$$E[g(X)] = \sum_x g(x)p(x)$$

(The proof of this theorem will not be given here.)

You and your friend decide to complicate the payoff picture in the coin-matching game by allowing you to win \$1 if the match is tails and \$2 if the match is heads. You still lose \$1 if the coins do not match. Quickly you see that this is not a fair game, since your expected winnings are

$$(-1)\left(\frac{1}{2}\right) + (1)\left(\frac{1}{4}\right) + (2)\left(\frac{1}{4}\right) = 0.25$$

You compensate for this by paying your friend \$1.50 if the coins do not match. Then your expected winnings per game are

$$(-1.5)\left(\frac{1}{2}\right) + (1)\left(\frac{1}{4}\right) + (2)\left(\frac{1}{4}\right) = 0$$

and the game is now fair. What is the difference between this game and the original one in which payoffs were \$1? The difference certainly cannot be explained by the expected value, since both games are fair. You can win more but also lose more with the new payoffs, and the difference between the two games can be partially explained in terms of the variability of your winnings across many games. This increased variability can be seen in Figure 3.4, the relative frequency for your winnings in the new game, which spreads out more than Figure 3.3. Formally, variation is often measured by the *variance* and a related quantity called the *standard deviation*.

FIGURE 3.4
Relative Frequency
of Winnings

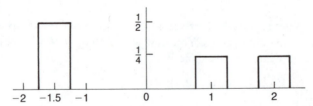

DEFINITION 3.5   The **variance** of a random variable X with expected value μ is given by

$$V(X) = E[(X - \mu)^2]$$

We sometimes use the notation

$$E[(x - \mu)^2] = \sigma^2$$

The smallest value σ^2 can assume is zero, and that would occur if all the probability was at a single point (that is, X takes on a constant value with probability one). The variance will become larger as the points with positive probability spread out more.

Observe that the variance squares the units in which we are measuring. A measure of variation that maintains the original units is the *standard deviation*.

DEFINITION 3.6

The **standard deviation** of a random variable X is the square root of the variance, given by

$$\sigma = \sqrt{\sigma^2} = \sqrt{E[(X - \mu)^2]}$$

Note that the standard deviation is another measure of variability, similar to the interquartile range introduced in Chapter 1.

For the game of Figure 3.3, the variance of your winnings is (with $\mu = 0$)

$$\sigma^2 = E[(X - \mu)^2]$$

$$= (-1)^2 \left(\frac{1}{2}\right) + (1)^2 \left(\frac{1}{2}\right) = 1$$

It follows that $\sigma = 1$ as well. For the game of Figure 3.4 the variance of your winnings is

$$\sigma^2 = (-1.5)^2 \left(\frac{1}{2}\right) + (1)^2 \left(\frac{1}{4}\right) + (1)^2 \left(\frac{1}{4}\right)$$

$$= 2.375$$

and

$$\sigma = 1.54.$$

Which game would you rather play?

The standard deviation can be thought of as the size of a "typical" deviation between an observed outcome and the expected value. For Figure 3.3 each outcome (-1 or $+1$) does deviate precisely one standard deviation from the expected value. For Figure 3.3 the positive values average 1.5 units from the expected value of 0 (as does the negative value), which is approximately one standard deviation.

The mean (expected value) and variance are often used to summarize the information in a large set of data. Table 3.1 gives cancer mortality rates for white males in 67 counties of Florida for the 1970s. These data can be displayed in a histogram as shown in Figure 3.5. To produce this figure, the interval from 155 (a little smaller than the smallest data value) to 265 (a little larger than the largest data value) is divided into 11 convenient intervals of ten units each. The graph shows the midpoints of these intervals. The height of a bar represents the fraction (relative frequency) of data values falling into that interval. If a county is chosen at random

TABLE 3.1
Cancer Mortality Rates (per 100,000) for White Males in Florida (1970s) (Source: U.S. Environmental Protection Agency and National Cancer Institute)

156.5	159.7	167.0	167.8	170.8	171.0	171.4	175.0	177.4	178.4
179.3	179.4	179.8	181.3	181.4	181.4	185.1	185.2	185.7	185.8
186.3	189.4	190.3	190.4	191.2	191.7	191.8	192.4	192.9	193.1
193.3	194.3	195.0	195.0	195.1	195.4	197.2	200.0	200.1	200.2
201.0	202.7	203.4	203.9	204.4	207.8	208.0	209.2	210.2	213.5
213.8	214.1	215.2	219.4	219.7	222.0	224.2	224.3	230.4	233.4
234.1	235.1	235.2	237.1	240.4	241.7	256.8			

FIGURE 3.5
Relative Frequency Histogram of Data in Table 3.1

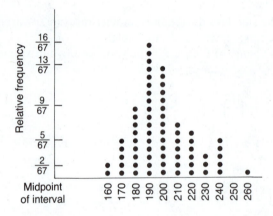

(so that each county has a probability of selection equal to 1/67), the chance that its mortality rate X exceeds 225 is 9/67, the sum of the relative frequencies in the last three bars on the right. The mean value of X is given by

$$\mu = \sum xp(x)$$

$$= \sum x\left(\frac{1}{67}\right) = 199.34$$

which is simply the average of the 67 measurements. Similarly

$$\sigma = \sqrt{\sum(x - \mu)^2 p(x)}$$

$$= \sqrt{\sum(x - \mu)^2\left(\frac{1}{67}\right)} = 21.56$$

An interpretation of these numbers is that a "typical" county in Florida had a mortality rate of about 200 ± 22 for the 1970s.

If we look at the same measurement for another state, the picture changes. Table 3.2 shows cancer mortality rates in 93 counties of Nebraska for the 1970s, and Figure 3.6 gives a histogram for these data. For a randomly selected county

TABLE 3.2
Cancer Mortality Rates (per 100,000) for White Males in Nebraska (1970s)

37.2	96.1	114.4	116.9	121.1	129.6	135.1	136.6	144.4	146.1
146.7	147.3	147.5	148.1	148.2	149.6	150.4	150.6	151.8	152.7
153.6	153.9	154.6	156.1	156.6	157.2	157.9	158.5	159.8	161.1
161.3	161.9	162.4	162.6	162.6	163.0	163.2	163.9	164.9	165.8
166.8	167.1	169.0	171.4	171.5	172.3	174.1	175.6	176.1	177.4
178.9	179.3	179.5	179.7	179.8	180.2	181.9	182.8	183.1	183.6
184.8	184.9	188.0	188.2	188.3	189.5	189.7	191.4	191.6	191.6
192.3	192.5	192.8	194.4	194.6	197.1	197.6	197.6	198.7	200.7
203.2	203.5	211.1	222.0	223.3	230.2	240.0	387.1	398.3	440.7
577.9	622.5	692.1							

FIGURE 3.6
Relative Frequency
Histogram of Data
in Table 3.2

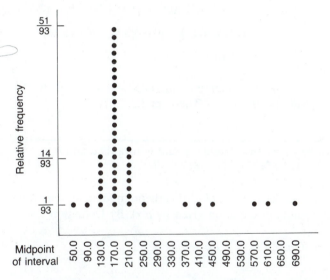

mortality rate from Nebraska, $\mu = 192.47$ and $\sigma = 95.16$. Note that Figures 3.5 and 3.6 have approximately the same mean, but differ considerably in standard deviation, a measure of variability (or spread) in the data. It would not be surprising to see a mortality rate around 300 in Nebraska, but this never seems to occur in Florida (at least, it didn't occur in the 1970s).

We now provide other examples and extensions of these basic results.

EXAMPLE 3.2

The manager of a stockroom in a factory knows from his study of records that the daily demand (number of times used) for a certain tool has the following probability distribution:

Demand	0	1	2
Probability	0.1	0.5	0.4

(That is, 50% of the daily records show that the tool was used one time.) If X denotes the daily demand, find $E(X)$ and $V(X)$.

Solution From Definition 3.4 we see that

$$E(X) = \sum_x xp(x)$$

$$= 0(0.1) + 1(0.5) + 2(0.4) = 1.3$$

The tool is used an average of 1.3 times per day.

From Definition 3.5 we see that

$$V(X) = E[(X - \mu)^2]$$

$$= \sum_x (x - \mu)^2 p(x)$$

$$= (0 - 1.3)^2(0.1) + (1 - 1.3)^2(0.5) + (2 - 1.3)^2(0.4)$$

$$= (1.69)(0.1) + (0.09)(0.5) + (0.49)(0.4)$$

$$= 0.410$$

Our work in manipulating expected values is greatly facilitated by making use of the two results of Theorem 3.2.

THEOREM 3.2 For any random variable X and constants a and b,

(i) $E(aX + b) = aE(X) + b$.

(ii) $V(aX + b) = a^2 V(X)$.

We sketch a proof of this theorem for a discrete random variable X having probability distribution given by $p(x)$. By Definition 3.4

$$E(aX + b) = \sum_x (ax + b)p(x)$$

$$= \sum_x [(ax)p(x) + bp(x)]$$

$$= \sum_x axp(x) + \sum_x bp(x)$$

$$= a \sum_x xp(x) + b \sum_x p(x)$$

$$= aE(X) + b$$

(Note that $\sum_x p(x)$ must equal unity.) Also by Definition 3.5

$$V(aX + b) = E[(aX + b) - E(aX + b)]^2$$

$$= E[aX + b - (aE(X) + b)]^2$$

$$= E[aX - aE(X)]^2$$

$$= E[a^2(X - E(X))^2]$$
$$= a^2 E[X - E(X)]^2$$
$$= a^2 V(X)$$

We illustrate the use of these results in the following example.

EXAMPLE 3.3

In Example 3.2 suppose it costs the factory $10 each time the tool is used. Find the mean and variance of the daily costs for use of this tool.

Solution   Recall that the X of Example 3.2 is the daily demand. The daily cost of using this tool is then $10X$. We have, by Theorem 3.2,

$$E(10X) = 10E(X) = 10(1.3)$$
$$= 13$$

The factory should budget $13 per day to cover the cost of using the tool.
Also by Theorem 3.2

$$V(10X) = (10)^2 V(X) = 100(0.410)$$
$$= 41$$

(We will make use of this value in a later example.)   ■

Theorem 3.2 leads us to a more efficient computational formula for a variance, as given in Theorem 3.3.

THEOREM 3.3   If X is a random variable with mean μ, then
$$V(X) = E(X^2) - \mu^2$$

Proof   The proof is as follows. Starting with the definition of variance, we have
$$V(X) = E[(X - \mu)^2] = E(X^2 - 2X\mu + \mu^2)$$
$$= E(X^2) - E(2X\mu) + E(\mu^2)$$
$$= E(X^2) - 2\mu E(X) + \mu^2$$
$$= E(X^2) - 2\mu^2 + \mu^2$$
$$= E(X^2) - \mu^2$$

EXAMPLE 3.4

Use the result of Theorem 3.3 to compute the variance of X as given in Example 3.2.

Solution   In Example 3.2 X has probability distribution given by

x	0	1	2
$p(x)$	0.1	0.5	0.4

and we saw that $E(X) = 1.3$. Now

$$E(X^2) = \sum_x x^2 p(x)$$

$$= (0)^2(0.1) + (1)^2(0.5) + (2)^2(0.4)$$

$$= 0 + 0.5 + 1.6$$

$$= 2.1$$

By Theorem 3.3

$$V(X) = E(X^2) - \mu^2$$

$$= 2.1 - (1.3)^2 = 0.41$$

■

We have computed means and variances for a number of probability distributions, and argued that these two quantities give us some useful information on the center and spread of the probability mass. Now suppose we know only the mean and variance for a probability distribution. Can we say anything specific about probabilities for certain intervals? The answer is "yes," and two useful results on the relationship between mean, standard deviation, and relative frequency will now be discussed.

The first result applies only to mound-shaped, reasonably symmetric distributions such as the one pictured in Figure 3.5. In such cases approximately 68% of the observations will lie within one standard deviation of the mean, and approximately 95% of the observations will lie within two standard deviations of the mean. For the data on which Figure 3.5 is based, $\mu = 199.34$ and $\sigma = 21.56$. The interval

$$(\mu - \sigma, \mu + \sigma)$$

or

(177.78, 220.90)

contains $46/67 = 0.686$ or approximately 68% of the observations. The interval

$$(\mu - 2\sigma, \mu + 2\sigma)$$

or

(156.22, 242.46)

contains $66/67 = 0.985$ or a little more than 95% of the observations.

Contrasting Figure 3.5 to Figure 3.6, we see that Figure 3.6 is not nearly as symmetric, having a long tail to the right. For the data of Figure 3.6, $\mu = 192.47$ and $\sigma = 95.16$. The interval $(\mu - \sigma, \mu + \sigma)$ contains about 91% of the measurements, and the interval $(\mu - 2\sigma, \mu + 2\sigma)$ contains about 93%. So the 68% − 95%

rule does not work well in this case. If we have only a mean and standard deviation for a probability distribution and know nothing about the shape of the distribution, we must be careful in applying the 68%–95% rule. It may provide a poor approximation.

The second result is a theorem relating μ, σ, and relative frequencies (probabilities) that gives very rough approximations but works in all cases.

THEOREM 3.4 *Tchebysheff's Theorem.*   Let X be a random variable with mean μ and variance σ^2. Then for any positive k

$$P(|X - \mu| < k\sigma) \geq 1 - \frac{1}{k^2}$$

The proof of Theorem 3.4 begins with the definition of $V(X)$ and then makes substitutions in the sum defining this quantity. Now

$$V(X) = \sigma^2 = \sum_{-\infty}^{\infty} (x - \mu)^2 p(x)$$

$$= \sum_{-\infty}^{\mu - k\sigma} (x - \mu)^2 p(x) + \sum_{\mu - k\sigma}^{\mu + k\sigma} (x - \mu)^2 p(x) + \sum_{\mu + k\sigma}^{\infty} (x - \mu)^2 p(x)$$

The first sum stops at the largest value of x smaller than $\mu - k\sigma$, and the third sum begins at the smallest value of x larger than $\mu + k\sigma$; the middle sum collects the remaining terms. Observe that the middle sum is always nonnegative, and for both of the outside sums

$$(x - \mu)^2 \geq k^2 \sigma^2$$

Eliminating the middle sum and substituting for $(x - \mu)^2$ in the other two yields

$$\sigma^2 \geq \sum_{-\infty}^{\mu - k\sigma} k^2 \sigma^2 p(x) + \sum_{\mu + k\sigma}^{\infty} k^2 \sigma^2 p(x)$$

or

$$\sigma^2 \geq k^2 \sigma^2 \left[\sum_{-\infty}^{\mu - k\sigma} p(x) + \sum_{\mu + k\sigma}^{\infty} p(x) \right]$$

or

$$\sigma^2 \geq k^2 \sigma^2 P(|X - \mu| \geq k\sigma)$$

It follows that

$$P(|X - \mu| \geq k\sigma) \leq \frac{1}{k^2}$$

or

$$P(|X - \mu| < k\sigma) \geq 1 - \frac{1}{k^2}$$

The inequality in the statement of the theorem is equivalent to

$$P(\mu - k\sigma < X < \mu + k\sigma) \geqslant 1 - \frac{1}{k^2}$$

To interpret this result, let $k = 2$, for example. Then the interval $\mu - 2\sigma$ to $\mu + 2\sigma$ must contain at least $1 - 1/k^2 = 1 - 1/4 = 3/4$ of the probability mass for the random variable.

We give more specific illustrations in the following two examples.

EXAMPLE 3.5
_____

The daily production of electric motors at a certain factory averaged 120 with a standard deviation of 10.

(a)   What fraction of days will have a production level between 100 and 140?

(b)   Find the shortest interval certain to contain at least 90% of the daily production levels.

Solution   (a)   The interval 100 to 140 is $\mu - 2\sigma$ to $\mu + 2\sigma$, with $\mu = 120$ and $\sigma = 10$. Thus $k = 2$,

$$1 - \frac{1}{k^2} = 1 - \frac{1}{4} = \frac{3}{4}$$

and at least 75% of the days will have total production in this interval. (This could be closer to 95% if the daily production figures show a mound-shaped, symmetric relative frequency distribution.)

(b)   To find k, we must set $(1 - 1/k^2)$ equal to 0.9 and solve for k. That is,

$$1 - \frac{1}{k^2} = 0.9$$

$$\frac{1}{k^2} = 0.1$$

$$k^2 = 10$$

or

$$k = \sqrt{10} = 3.16$$

The interval

$$\mu - 3.16\sigma \text{ to } \mu + 3.16\sigma$$

or

$$120 - 3.16(10) \text{ to } 120 + 3.16(10)$$

or

$$88.4 \text{ to } 151.6$$

will then contain at least 90% of the daily production levels.   ∎

EXAMPLE 3.6

The daily cost for use of a certain tool has a mean of $13 and a variance of 41. (See Example 3.3.) How often will this cost exceed $30?

Solution   First we must find the distance between the mean and 30, in terms of the standard deviation of the distribution of costs. We have

$$\frac{30 - \mu}{\sqrt{\sigma^2}} = \frac{30 - 13}{\sqrt{41}} = \frac{17}{6.4} = 2.66$$

or 30 is 2.66 standard deviations above the mean. Letting $k = 2.66$ in Theorem 3.4, we have that the interval

$$\mu - 2.66\sigma \text{ to } \mu + 2.66\sigma$$

or

$$13 - 2.66(6.4) \text{ to } 13 + 2.66(6.4)$$

or

$$-4 \text{ to } 30$$

must contain at least

$$1 - \frac{1}{k^2} = 1 - \frac{1}{(2.66)^2} = 1 - 0.14 = 0.86$$

of the probability. Since the daily cost cannot be negative, at most 0.14 of the probability mass can exceed 30. Thus the cost cannot exceed $30 more than 14% of the time.   ∎

EXERCISES

3.11   You are to pay $2.00 to play a game consisting of drawing one ticket at random from a box of numbered tickets. You win the amount (in dollars) of the number on the ticket you draw. Two boxes are available with numbered tickets as shown below:

I | 0, 1, 2 | II | 0, 0, 0, 1, 4 |

(a)   Find the expected value and variance of your *net* gain per play with box I.
(b)   Repeat part (a) for box II.
(c)   Given that you have decided to play, which box would you choose and why?

3.12   College students in Florida are required to pass a College Level Academic Skills Test (CLAST) before completing their college work. The scores for 60 students on the mathematics portion of this test are shown below:

```
225  264  272  272  279  282  282  287  287  287  289
292  292  295  295  295  295  300  300  303  303  303
306  306  306  310  313  313  313  313  313  313  317
317  317  320  320  320  324  324  324  324  324  324
324  324  324  329  334  334  334  334  346  346  353
363  363  375  375  375
```

    (a)   If one student's score is randomly selected to represent these 60, find μ and σ.

    (b)   Do you think the 68%–95% rule will work well to interpret these and similar sets of scores?

3.13 Daily sales records for a computer manufacturing firm show that it will sell 0, 1, or 2 mainframe computer systems with probabilities as listed:

Number of Sales	0	1	2
Probability	0.7	0.2	0.1

Find the expected value, variance, and standard deviation for daily sales.

3.14 Approximately 10% of the glass bottles coming off a production line have serious defects in the glass. If two bottles are randomly selected for inspection, find the expected value and variance of the number of inspected bottles with serious defects.

3.15 Two construction contracts are to be randomly assigned to one or more of three firms— I, II, and III. A firm may receive more than one contract. If each contract has a potential profit of $90,000, find the expected potential profit for firm I. Also find the expected potential profit for firms I and II together.

3.16 If two balanced coins are tossed, what are the expected value and variance of the number of heads observed?

3.17 The number of breakdowns for a university computer system is closely monitored by the director of the computing center, since it is critical to the efficient operation of the center. The number averages 4 per week, with a standard deviation of 0.8 per week.

    (a)   Find an interval that must include at least 90% of the weekly figures on number of breakdowns.

    (b)   The center director promises that the number of breakdowns will rarely exceed 8 in a one-week period. Is the director safe in this claim? Why?

3.18 Keeping an adequate supply of spare parts on hand is an important function of the parts department of a large electronics firm. The monthly demand for microcomputer printer boards was studied for some months and found to average 28 with a standard deviation of 4. How many printer boards should be stocked at the beginning of each month to ensure that the demand will exceed the supply with probability less than 0.10?

3.19 An important feature of golf cart batteries is the number of minutes they will perform before needing to be recharged. A certain manufacturer advertises batteries that will run, under a 75-amp discharge test, for an average of 100 minutes, with a standard deviation of 5 minutes.

    (a)   Find an interval that must contain at least 90% of the performance periods for batteries of this type.

    (b)   Would you expect many batteries to die out in less than 80 minutes? Why?

3.20 Costs of equipment maintenance are an important part of a firm's budget. Each visit by a field representative to check out a malfunction in a word processing system costs $40. The word processing system is expected to malfunction approximately 5 times per month, and the standard deviation of the number of malfunctions is 2.

    (a)   Find the expected value and standard deviation of the monthly cost of visits by the field representative.

    (b)   How much should the firm budget per month so the cost of these visits will be covered at least 75% of the time?

At this point it may seem like every problem has its own unique probability distribution, and we must start from the basics to construct such a distribution each time a new problem comes up. Fortunately that is not the case. Certain basic probability distributions can be developed that will serve as models for a large number of practical problems. In the remainder of this chapter we consider some fundamental discrete distributions, looking at the theoretical assumptions that underlie these distributions as well as the means, variances, and applications of the distributions.

3.3 THE BERNOULLI DISTRIBUTION

Consider the inspection of a single item taken from an assembly line. Suppose a 0 is recorded if the item is nondefective, and a 1 is recorded if the item is defective. If X is the random variable denoting the condition of the inspected item, then $X = 0$ with probability $(1 - p)$ and $X = 1$ with probability p, where p denotes the probability of observing a defective item. The probability distribution of X is then given by

$$p(x) = p^x(1 - p)^{1 - x} \qquad x = 0,1$$

where $p(x)$ denotes the probability that $X = x$. Such a random variable is said to have a *Bernoulli distribution* or to represent the outcome of a single Bernoulli trial. Any random variable denoting the presence or absence of a certain condition in an observed phenomenon will possess a distribution of this type. Frequently one of the outcomes is termed "success" and the other "failure." Note that $p(x) \geqslant 0$ and $\Sigma\, p(x) = 1$.

Suppose that we repeatedly observe items of this type, and record a value of X for each item observed. What is the average of X we would expect to see? By Definition 3.4 we have that the expected value of X is given by

$$E(X) = \sum_x xp(x)$$

$$= 0p(0) + 1p(1)$$

$$= 0(1 - p) + 1(p) = p$$

Thus if 10% of the items are defective, we expect to observe an *average* of 0.1 defective per item inspected. (In other words we would expect to see one defective for every ten items inspected.)

For the Bernoulli random variable X the variance (see Theorem 3.3) is

$$V(X) = E(X^2) - [E(X)]^2$$

$$= \sum_x x^2 p(x) - p^2$$

$$= 0(1 - p) + 1(p) - p^2$$

$$= p - p^2 = p(1 - p)$$

The Bernoulli random variable will be used as a building block to form other probability distributions, such as the binomial distribution of Section 3.4.

The properties of the Bernoulli distribution are summarized here.

THE BERNOULLI DISTRIBUTION

$$p(x) = p^x(1 - p)^{1-x} \qquad x = 0, 1 \qquad \text{for } 0 \leqslant p \leqslant 1$$

$$E(X) = p \qquad\qquad V(X) = p(1 - p)$$

3.4 THE BINOMIAL DISTRIBUTION

Instead of inspecting a single item, as we do with the Bernoulli random variable, suppose we now independently inspect n items and record values for $X_1, X_2, \ldots,$ X_n, where $X_i = 1$ if the ith inspected item is defective and $X_i = 0$ otherwise. We have, in fact, observed a sequence of n independent Bernoulli random variables. One especially interesting function of $X_1, \ldots, X_n$ is the sum

$$Y = \sum_{i=1}^{n} X_i$$

which denotes the number of defectives among the n sampled items.

We can easily find the probability distribution for Y under the assumption that $P(X_i = 1) = p$, where p remains constant over all trials. For simplicity let's look at the specific case of $n = 3$. The random variable Y can then take on four possible values—0, 1, 2, and 3. For Y to be 0, all three X_i's must be 0. Thus

$$P(Y = 0) = P(X_1 = 0, X_2 = 0, X_3 = 0)$$

$$= P(X_1 = 0)P(X_2 = 0)P(X_3 = 0)$$

$$= (1 - p)^3$$

Now if $Y = 1$, then exactly one of the X_i's is 1 and the other two are 0. Thus

$$P(Y = 1) = P[(X_1 = 1, X_2 = 0, X_3 = 0) \cup (X_1 = 0, X_2 = 1, X_3 = 0)$$

$$\cup (X_1 = 0, X_2 = 0, X_3 = 1)]$$

$$= P(X_1 = 1, X_2 = 0, X_3 = 0) + P(X_1 = 0, X_2 = 1, X_3 = 0)$$

$$+ P(X_1 = 0, X_2 = 0, X_3 = 1)$$

(since the three possibilities are mutually exclusive)

$$= P(X_1 = 1)P(X_2 = 0)P(X_3 = 0)$$

$$+ P(X_1 = 0)P(X_2 = 1)P(X_3 = 0)$$

$$+ P(X_1 = 0)P(X_2 = 0)P(X_3 = 1)$$

$$= p(1 - p)^2 + p(1 - p)^2 + p(1 - p)^2$$

$$= 3p(1 - p)^2$$

For $Y = 2$ two X_i's must be 1 and one must be 0, which can also occur in three mutually exclusive ways. Hence

$$P(Y = 2) = P(X_1 = 1, X_2 = 1, X_3 = 0) + P(X_1 = 1, X_2 = 0, X_3 = 1)$$

$$+ P(X_1 = 0, X_2 = 1, X_3 = 1)$$

$$= 3p^2(1 - p)$$

The event $Y = 3$ can occur only if all X_i's are 1, so

$$P(Y = 3) = P(X_1 = 1, X_2 = 1, X_3 = 1)$$

$$= P(X_1 = 1)P(X_2 = 1)P(X_3 = 1)$$

$$= p^3$$

Notice that the coefficient in each of the expressions for $P(Y = y)$ is the number of ways of selecting y positions, in the sequence, in which to place 1s. Since there are three positions in the sequence, this number amounts to $\binom{3}{y}$. Thus we can write

$$P(Y = y) = \binom{3}{y} p^y(1 - p)^{3 - y} \qquad y = 0, 1, 2, 3, \text{ when } n = 3$$

For general n the probability that Y takes on a specific value, say y, is given by the term $p^y(1 - p)^{n - y}$ multiplied by the number of possible outcomes resulting in exactly y defectives being observed. This number is the number of ways of selecting y positions for defectives in the n possible positions of the sequence, and is given by

$$\binom{n}{y} = \frac{n!}{y!(n - y)!}$$

where $n! = n(n - 1) \cdots 1$ and $0! = 1$. Thus in general

$$P(Y = y) = p(y) = \binom{n}{y} p^y(1 - p)^{n - y} \qquad y = 0, 1, \ldots, n$$

The probability distribution given above is referred to as the *binomial* distribution.

To summarize, a random variable Y possesses a binomial distribution if five conditions exist:
1. The experiment consists of a fixed number n of identical trials.
2. Each trial can result in one of only two possible outcomes, called "success" or "failure."
3. The probability of "success" p is constant from trial to trial.
4. The trials are independent.
5. Y is defined to be the number of successes among the n trials.

Many experimental situations result in a random variable that can be adequately modeled by the binomial distribution. In addition to counts of the number of defectives in a sample of n items, examples may include counts of the number of employees favoring a certain retirement policy out of n employees interviewed, the number of pistons in an eight-cylinder engine that are misfiring, and the number of electronic systems sold this week out of the n that are manufactured.

A general formula for $p(x)$ identifies a family of distributions indexed by certain constants called *parameters*. Thus n and p are the parameters of the binomial distribution.

To see how choices for the parameters affect the probability function $p(x)$ for the binomial distribution, Figures 3.7 through 3.11 show graphs of $p(x)$ for various choices of n and p. For $n = 20$ and $p = 0.1$ (Figure 3.7) the graph of $p(x)$ has a peak around $x = 1$ and $x = 2$ and then drops off sharply. For $p = 0.3$ and the same n (Figure 3.8) the graph becomes more symmetric with a peak at $x = 6$. With $p = 0.5$ (Figure 3.9) the graph is perfectly symmetric around $x = 10$, and with $p = 0.8$ (Figure 3.10) the graph is mound shaped but not quite symmetric around $x = 16$. To see what happens when n is increased, we observe that for $n = 50$ and $p = 0.3$ (Figure 3.11) the graph is a little more symmetric than it was in the case $n = 20$ and $p = 0.3$. Also Figure 3.11 centers at $x = 15$ rather than $x = 6$.

FIGURE 3.7
Binomial Probability
Function; $n = 20$,
$p = 0.1$

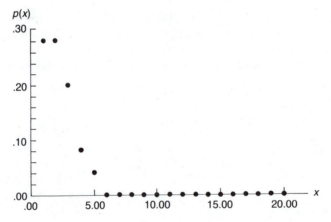

FIGURE 3.8
Binomial Probability
Function; $n = 20$,
$p = 0.3$

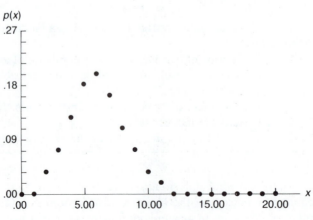

FIGURE 3.9
Binomial Probability Function; n = 20, p = 0.5

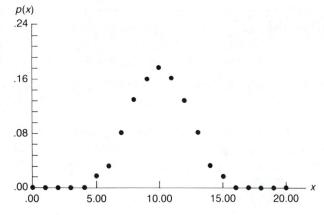

FIGURE 3.10
Binomial Probability Function; n = 20, p = 0.8

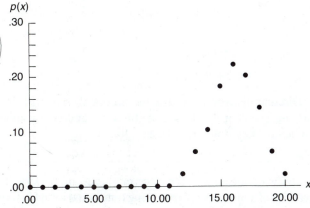

FIGURE 3.11
Binomial Probability Function; n = 50, p = 0.3

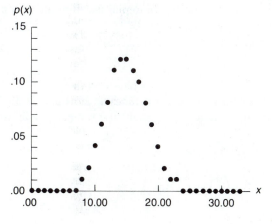

EXAMPLE 3.7

Suppose that a large lot of fuses contains 10% defectives. If four fuses are randomly sampled from the lot, find the probability that exactly one fuse is defective. Find the probability that at least one fuse in the sample of four is defective.

Solution   We assume that the four trials are independent and that the probability of observing a defective is the same (0.1) for each trial. This would be approximately true if the lot is indeed *large*. (If the lot contained only a few fuses, removal of one fuse would substantially change the probability of observing a defective on the second draw.) Thus the binomial distribution provides a reasonable model for this experiment, and we have, with Y denoting the number of defectives,

$$p(1) = \binom{4}{1}(0.1)^1(0.9)^3 = 0.2916$$

To find $P(Y \geq 1)$, observe that

$$P(Y \geq 1) = 1 - P(Y = 0) = 1 - p(0)$$

$$= 1 - \binom{4}{0}(0.1)^0(0.9)^4$$

$$= 1 - (0.9)^4$$

$$= 0.3439$$

Discrete distributions, like the binomial, can arise in situations where the underlying problem involves a continuous random variable that is not discrete. The following example provides an illustration.

EXAMPLE 3.8

In a study of lifelengths for a certain type of battery it was found that the probability of a lifelength X exceeding four hours is 0.135. If three such batteries are in use in independently operating systems, find the probability that only one of the batteries lasts four hours or more.

Solution   Letting Y denote the number of batteries lasting four hours or more, we can reasonably assume Y to have a binomial distribution with $p = 0.135$. Hence

$$P(Y = 1) = p(1) = \binom{3}{1}(0.135)^1(0.865)^2$$

$$= 0.303$$

There are numerous ways of finding $E(Y)$ and $V(Y)$ for a binomially distributed random variable Y. We might use the basic definition and compute

$$E(Y) = \sum_x yp(y)$$

$$= \sum_{y=0}^{n} y\binom{n}{y}p^y(1 - p)^{n-y}$$

but direct evaluation of this expression is a little tricky. Another approach is to make use of the results on linear functions of random variables. We will show in Chapter 5 that, since the binomial Y arose as a sum of independent Bernoulli random variables, $X_1, \ldots, X_n$,

$$E(Y) = E\left[\sum_{i=1}^{n} X_i\right] = \sum_{i=1}^{n} E(X_i)$$

$$= \sum_{i=1}^{n} p = np$$

and

$$V(Y) = \sum_{i=1}^{n} V(X_i) = \sum_{i=1}^{n} p(1-p) = np(1-p)$$

EXAMPLE 3.9

Refer to Example 3.7. Suppose that the four fuses sampled from the lot were shipped to a customer before being tested, on a guarantee basis. Assume that the cost of making the shipment good is given by $C = 3Y^2$, where Y denotes the number of defectives in the shipment of four. Find the expected repair cost.

Solution   We know that

$$E(C) = E(3Y^2) = 3E(Y^2)$$

and it now remains to find $E(Y^2)$. We have from Theorem 3.3 that

$$V(Y) = E(Y-\mu)^2 = E(Y^2) - \mu^2$$

Since $V(Y) = np(1-p)$ and $\mu = E(Y) = np$, we see that

$$E(Y^2) = V(Y) + \mu^2$$
$$= np(1-p) + (np)^2$$

For Example 3.7, $p = 0.1$ and $n = 4$, and hence

$$E(C) = 3E(Y^2) = 3[np(1-p) + (np)^2]$$
$$= 3[(4)(0.1)(0.9) + (4)^2(0.1)^2]$$
$$= 1.56$$

If the costs were originally in dollars, we could expect to pay an average of $1.56 in repair costs for each shipment of four fuses.   ■

Table 2 of the Appendix gives cumulative binomial probabilities for selected values of n and p. The entries in the table are values of

$$\sum_{y=0}^{a} p(y) = \sum_{y=0}^{a} \binom{n}{y} p^y (1-p)^{n-y}$$

The following example illustrates the use of Table 2.

EXAMPLE 3.10 _____

An industrial firm supplies 10 manufacturing plants with a certain chemical. The probability that any one firm calls in an order on a given day is 0.2, and this is the same for all 10 plants. Find the probability that, on the given day, the number of plants calling in orders is (a) at most 3, (b) at least 3, (c) exactly 3.

Solution   Let Y denote the number of plants calling in orders on the day in question. If the plants order independently, then Y can be modeled to have a binomial distribution with $p = 0.2$.

(a) We then have

$$P(Y \leqslant 3) = \sum_{y=0}^{3} p(y)$$

$$= \sum_{y=0}^{3} \binom{10}{y} (0.2)^y (0.8)^{10-y}$$

$$= 0.879$$

from Table 2(b).

(b) Note that

$$P(Y \geqslant 3) = 1 - P(Y \leqslant 2)$$

$$= 1 - \sum_{y=0}^{2} \binom{10}{y} (0.2)^y (0.8)^{10-y}$$

$$= 1 - 0.678 = 0.322$$

(c) Observe that

$$P(Y = 3) = P(Y \leqslant 3) - P(Y \leqslant 2)$$

$$= 0.879 - 0.678 = 0.201$$

from results just established.   ■

The examples used to this point have specified n and p to calculate probabilities or expected values. Sometimes, however, it is necessary to choose n so as to achieve a specified probability. Example 3.11 illustrates the point.

EXAMPLE 3.11 _____

The guidance system for a rocket operates correctly with probability p when called upon. Independent but identical backup systems are installed in the rocket so that the probability that at least one system will operate correctly when called upon is no less than 0.99. Let n denote the number of guidance systems in the rocket. How large must n be to achieve the specified probability of at least one guidance system operating if (a) $p = 0.9$? (b) $p = 0.8$?

Solution   Let Y denote the number of correctly operating systems. If the systems are identical and independent, Y has a binomial distribution. Thus

$$P(Y \geqslant 1) = 1 - P(Y = 0)$$

$$= 1 - \binom{n}{0} p^0 (1 - p)^n$$

$$= 1 - (1 - p)^n$$

The conditions specify that n must be such that $P(Y \geqslant 1) = 0.99$ or more.

(a) When $p = 0.9$,

$$P(Y \geqslant 1) = 1 - (1 - 0.9)^n \geqslant 0.99$$

results in

$$1 - (0.1)^n \geqslant 0.99$$

or

$$(0.1)^n \leqslant 1 - 0.99 = 0.01$$

so $n = 2$. That is, installing two guidance systems will satisfy the specifications.

(b) When $p = 0.8$,

$$P(Y \geqslant 1) = 1 - (1 - 0.8)^n \geqslant 0.99$$

results in

$$(0.2)^n \leqslant 0.01$$

Now $(0.2)^2 = 0.04$, and $(0.2)^3 = 0.008$, and we must go to $n = 3$ systems so that

$$P(Y \geqslant 1) = 1 - (0.2)^3 = 0.992 > 0.99$$

[*Note*: We cannot achieve the 0.99 probability exactly since Y can assume only integer values.] ■

We now move on to a discussion of other discrete random variables, but the binomial distribution, summarized here, will be used frequently throughout the text.

THE BINOMIAL DISTRIBUTION

$$p(y) = \binom{n}{y} p^y (1 - p)^{n - y} \qquad y = 0, 1, \ldots, n \qquad \text{for } 0 \leqslant p \leqslant 1$$

$$E(Y) = np \qquad\qquad V(Y) = np(1 - p)$$

EXERCISES

3.21 Let X denote a random variable having a binomial distribution with $p = 0.2$ and $n = 4$. Find

(a) $P(X = 2)$

(b) $P(X \geqslant 2)$

(c) $P(X \leqslant 2)$

(d) $E(X)$

(e) $V(X)$

3.22   Let X denote a random variable having a binomial distribution with $p = 0.4$ and $n = 20$. Use Table 2 of the Appendix to evaluate

(a) $P(X \leqslant 6)$ (b) $P(X \geqslant 12)$

(c) $P(X = 8)$

3.23   A machine that fills boxes of cereal underfills a certain proportion p. If 25 boxes are randomly selected from the output of this machine, find the probability that no more than two are underfilled when

(a) $p = 0.1$ (b) $p = 0.2$

3.24   In testing the lethal concentration of a chemical found in polluted water it is found that a certain concentration will kill 20% of the fish that are subjected to it for 24 hours. If 20 fish are placed in a tank containing this concentration of chemical, find the probability that after 24 hours,

(a)   exactly 14 survive.   (b)   at least 10 survive.

(c)   at most 16 survive.

3.25   Refer to Exercise 3.24.

(a)   Find the number expected to survive, out of 20.

(b)   Find the variance of the number of survivors, out of 20.

3.26   Among persons donating blood to a clinic 80% have Rh^+ (that is, have the Rhesus factor present in their blood). Five people donate blood at the clinic on a particular day.

(a)   Find the probability that at least one of the 5 does not have the Rh factor.

(b)   Find the probability that at most 4 of the 5 have Rh^+ blood.

3.27   The clinic needs 5 Rh^+ donors on a certain day. How many people must donate blood to have the probability of at least 5 Rh^+ donors over 0.90?

3.28   The *U.S. Statistical Abstract* reports that the median family income in the U.S. for 1985 was $27,735. Among four randomly selected families find the probability that

(a)   all four had incomes above $27,735 in 1985.

(b)   one of the four had an income below $27,735 in 1985.

3.29   According to an article by B. E. Sullivan in *Transportation Research* (18A, No. 2, 1984, p. 119), 55% of U.S. corporations say that one of the most important factors in locating a corporate headquarters is the "quality of life" for the employees. If five firms are contacted by the governor of Florida concerning possible relocation to that state, find the probability that at least three say that "quality of life" is an important factor in their decision. What assumptions have you made in finding this answer?

3.30   Goranson and Hall (*Aeronautical Journal*, November 1980, pp. 279–280) explain that the probability of detecting a crack in an airplane wing is the product of p_1, the probability of inspecting a plane with a wing crack; p_2, the probability of inspecting the detail in which the crack is located; and p_3, the probability of detecting the damage.

(a)   What assumptions justify the multiplication of these probabilities?

(b)   Suppose $p_1 = 0.9$, $p_2 = 0.8$, and $p_3 = 0.5$ for a certain fleet of planes. If three planes are inspected from this fleet, find the probability that a wing crack will be detected in at least one of them.

3.31   A missile protection system consists of n radar sets operating independently, each with probability 0.9 of detecting an aircraft entering a specified zone. (All radar sets cover the same zone.) If an airplane enters the zone, find the probability that it will be detected if

(a) $n = 2$ (b) $n = 4$

3.32   Refer to Exercise 3.31. How large must n be if it is desired to have a probability of 0.99 of detecting an aircraft entering the zone?

3.33   A complex electronic system is built with a certain number of backup components in its subsystems. One subsystem has four identical components, each with probability 0.2 of failing in less than 1,000 hours. The subsystem will operate if any two or more of the four components are operating. Assuming that the components operate independently, find the probability that
    (a)   exactly two of the four components last longer than 1,000 hours.
    (b)   the subsystem operates longer than 1,000 hours.

3.34   An oil exploration firm is to drill ten wells, with each well having probability 0.1 of successfully producing oil. It costs the firm $10,000 to drill each well. A successful well will bring in oil worth $500,000.
    (a)   Find the firm's expected gain from the ten wells.
    (b)   Find the standard deviation of the firm's gain.

3.35   A firm sells four items randomly selected from a large lot known to contain 10% defectives. Let Y denote the number of defectives among the four sold. The purchaser of the item will return the defectives for repair, and the repair cost is given by

$$C = 3Y^2 + Y + 2$$

Find the expected repair cost.

3.36   From a large lot of new tires, n are to be sampled by a potential buyer, and the number of defectives X is to be observed. If at least one defective is observed in the sample of n, the entire lot is to be rejected by the potential buyer. Find n so that the probability of detecting at least one defective is approximately 0.90 if
    (a)   10% of the lot is defective.
    (b)   5% of the lot is defective.

3.37   Ten motors are packaged for sale in a certain warehouse. The motors sell for $100 each, but a "double-your-money-back" guarantee is in effect for any defectives the purchaser might receive. Find the expected net gain for the seller if the probability of any one motor being defective is 0.08. (Assume the quality of any one motor is independent of that of the others.)

3.5   THE GEOMETRIC DISTRIBUTION

Suppose a series of test firings of a rocket engine can be represented by a sequence of independent Bernoulli random variables with $X_i = 1$ if the ith trial result in a successful firing and $X_i = 0$ otherwise. Assume that the probability of a successful firing is constant for the trials, and let this probability be denoted by p. For this problem we might be interested in the number of the trial on which the first successful firing occurs. If Y denotes the number of the trial on which the first success occurs, then

$$P(Y = y) = p(y) = P(X_1 = 0, X_2 = 0, \ldots, X_{y-1} = 0, X_y = 1)$$
$$= P(X_1 = 0)P(X_2 = 0) \cdots P(X_{y-1} = 0)P(X_y = 1)$$
$$= (1 - p)^{y-1}p \qquad y = 1, 2, \ldots$$

because of the independence of the trials. This formula is referred to as the *geometric probability distribution*. Note that this random variable can take on a countably infinite number of possible values.

In addition to the rocket-firing example just given, other situations result in a random variable whose probability can be modeled by a geometric distribution: the number of customers contacted before the first sale is made, the number of years a dam is in service before it overflows, and the number of automobiles going through a radar check before the first speeder is detected.

The following example illustrates the use of the geometric distribution.

EXAMPLE 3.12 _____

A recruiting firm finds that 30% of the applicants for a certain industrial job have advanced training in computer programming. Applicants are interviewed sequentially and are selected at random from the pool. Find the probability that the first applicant having advanced training in programming is found on the fifth interview.

Solution The probability of finding a suitably trained applicant will remain relatively constant from trail to trail if the pool of applicants is reasonably large. It then makes sense to define Y as the number of the trial on which the first applicant having advanced training in programming is found, and to model Y as having a geometric distribution. Thus

$$P(Y = 5) = p(5) = (0.7)^4(0.3)$$

$$= 0.072 \qquad \blacksquare$$

From the basic definition

$$E(Y) = \sum_y yp(y) = \sum_{y=1}^{\infty} yp(1-p)^{y-1}$$

$$= p \sum_{y=1}^{\infty} y(1-p)^{y-1}$$

$$= p[1 + 2(1-p) + 3(1-p)^2 + \cdots]$$

The infinite series can be split up into a triangular array of series as follows:

$$E(Y) = p[1 \quad +(1-p) \quad +(1-p)^2 \quad + \cdots$$
$$+(1-p) \quad +(1-p)^2 \quad + \cdots$$
$$+(1-p)^2 \quad + \cdots$$
$$+ \cdots]$$

Each line on the right side is an infinite, decreasing geometric progression with common ratio $(1-p)$. Thus the first line inside the bracket sums to $1/p$, the second to $(1-p)/p$, the third to $(1-p)^2/p$, and so on.[2] On accumulating these totals, we

_____

[2]Recall that

$$a + ax + ax^2 + ax^3 + \cdots = \frac{a}{1-x}$$

if $|x| < 1$.

then have

$$E(Y) = p\left[\frac{1}{p} + \frac{1-p}{p} + \frac{(1-p)^2}{p} + \cdots\right]$$

$$= 1 + (1-p) + (1-p)^2 + \cdots$$

$$= \frac{1}{1-(1-p)} = \frac{1}{p}$$

This answer for $E(Y)$ should seem intuitively realistic. For example if 10% of a certain lot of items are defective, and if an inspector looks at randomly selected items one at a time, then he should expect to wait until the tenth trial to see the first defective.

The variance of the geometric distribution will be derived in Chapter 5. The result, however, is

$$V(Y) = \frac{1-p}{p^2}$$

EXAMPLE 3.13

Refer to Example 3.12 and let Y denote the number of the trial on which the first applicant having advanced training in computer programming is found. Suppose that the first applicant with advanced training is offered the position, and the applicant accepts. If each interview costs $30.00, find the expected value and variance of the total cost of interviewing until the job is filled. Within what interval would this cost be expected to fall?

Solution   Since Y is the number of the trial on which the interviewing process ends, the total cost of interviewing is $C = 30Y$. Now

$$E(C) = 30E(Y) = 30\left(\frac{1}{p}\right)$$

$$= 30\left(\frac{1}{0.3}\right) = 100$$

and

$$V(C) = (30)^2 V(Y) = \frac{900(1-p)}{p^2}$$

$$= \frac{900(0.7)}{(0.3)^2}$$

$$= 7000$$

The standard deviation of C is then $\sqrt{V(C)} = \sqrt{7000} = 83.67$. Tchebysheff's Theorem (see Section 3.2) says that C will lie within two standard deviations of its mean at least 75% of the time. Thus it is quite likely that C will be between

$$100 - 2(83.67) \text{ and } 100 + 2(83.67)$$

or

$$-67.34 \text{ and } 267.34$$

Since the lower bound is negative, that end of the interval is meaningless. However, we can still say that it is quite likely that the total cost of such an interviewing process will be less than \$267.34.   ■

THE GEOMETRIC DISTRIBUTION ——————————

$$p(y) = p(1 - p)^{y-1} \qquad y = 1, 2, \ldots \qquad \text{for } 0 < p < 1$$

$$E(Y) = \frac{1}{p} \qquad\qquad V(Y) = \frac{1 - p}{p^2}$$

3.6 THE NEGATIVE BINOMIAL DISTRIBUTION

In Section 3.5 we saw that the geometric distribution models the probabilistic behavior of the number of the trial on which the *first success* occurs in a sequence of independent Bernoulli trials. But what if we were interested in the number of the trial for the second success, or third success, or, in general, the *r*th success? The distribution governing the probabilistic behavior in these cases is called the *negative binomial distribution.*

Let Y denote the number of the trial on which the *r*th success occurs in a sequence of independent Bernoulli trials with p denoting the common probability of "success." We can derive the distribution of Y from known facts. Now

$P(Y = y) = P[(1\text{st}(y - 1)$ trials contain $(r - 1)$ successes and yth trial is a success]

$\qquad = P[(1\text{st}(y - 1)$ trials contain $(r - 1)$ successes]$P[y$th trial is a success]

Since the trials are independent, the joint probability can be written as a product of probabilities. The first probability statement is identical to that resulting in a binomial model, and hence

$$P(Y = y) = p(y) = \binom{y - 1}{r - 1} p^{r-1}(1 - p)^{y-r} \times p$$

$$= \binom{y - 1}{r - 1} p^r(1 - p)^{y-r} \qquad y = r, r + 1, \ldots$$

EXAMPLE 3.14 ——————————————————————————

As in Example 3.12, 30% of the applicants for a certain position have advanced training in computer programming. Suppose that three jobs requiring advanced programming training are open. Find the probability that the third qualified applicant is found on the fifth interview, if the applicants are interviewed sequentially and at random.

Solution Again we assume independent trials with 0.3 being the probability of finding a qualified candidate on any one trial. Let Y denote the number of the trial on which the third qualified candidate is found. Then Y can reasonably be assumed to have a negative binomial distribution, so

$$P(Y = 5) = p(5) = \binom{4}{2}(0.3)^3(0.7)^2$$

$$= 6(0.3)^3(0.7)^2$$

$$= 0.079 \qquad \blacksquare$$

The expected value, or mean, and the variance for the negative binomially distributed Y are found easily by analogy with the geometric distribution. Recall that Y denotes the number of the trial on which the rth success occurs. Let W_1 denote the number of the trial on which the first success occurs, W_2 the number of trials between the first success and the second success, including the trial of the second success, W_3 the number of trials between the second success and the third success, and so forth. The results of the trials are then as diagramed next (F standing for failure and S for success).

$$\underbrace{FF \cdots FS}_{W_1} \ \underbrace{F \cdots FS}_{W_2} \ \underbrace{F \cdots FS}_{W_3}$$

It is easy to observe that $Y = \sum_{i=1}^{r} W_i$, where the W_i's are independent and each has a geometric distribution. Thus by results to be derived in Chapter 5,

$$E(Y) = \sum_{i=1}^{r} E(W_i) = \sum_{i=1}^{r} \left(\frac{1}{p}\right) = \frac{r}{p}$$

and

$$V(Y) = \sum_{i=1}^{r} V(W_i) = \sum_{i=1}^{r} \frac{1-p}{p^2} = \frac{r(1-p)}{p^2}$$

EXAMPLE 3.15

A large stockpile of used pumps contains 20% that are unusable and need repair. A repairman is sent to the stockpile with 3 repair kits. He selects pumps at random and tests them one at a time. If a pump works, he goes on to the next one. If a pump doesn't work, he uses one of his repair kits on it. Suppose that it takes 10 minutes to test a pump if it works, and 30 minutes to test and repair a pump that doesn't work. Find the expected value and variance of the total time it takes the repairman to use up his 3 kits.

Solution Letting Y denote the number of the trial on which the 3rd defective pump is found, we see that Y has a negative binomial distribution with $p = 0.2$. The total time T taken to use up the three repair kits is

$$T = 10(Y - 3) + 3(30) = 10Y + 3(20)$$

(Each test takes 10 minutes, but the repairing takes 20 extra minutes.)
It follows that

$$E(T) = 10E(Y) + 3(20)$$

$$= 10\left(\frac{3}{0.2}\right) + 3(20)$$

$$= 150 + 60 = 210$$

and

$$V(T) = (10)^2 V(Y)$$

$$= (10)^2 \left[\frac{3(0.8)}{(0.2)^2}\right]$$

$$= 100[60]$$

$$= 6000$$

Thus the total time to use up the kits has an expected value of 210 minutes, with a standard deviation of $\sqrt{6000} = 77.46$ minutes. ■

The negative binomial distribution is used to model a wide variety of phenomena, from number of defects per square yard in fabrics to numbers of individuals in an insect population after many generations.

THE NEGATIVE BINOMIAL DISTRIBUTION

$$p(y) = \binom{y-1}{r-1} p^r (1-p)^{y-r} \qquad y = r, r+1, \ldots \qquad \text{for } 0 < p < 1$$

$$E(Y) = \frac{r}{p} \qquad\qquad V(Y) = \frac{r(1-p)}{p^2}$$

EXERCISES

3.38   Let Y denote a random variable having a geometric distribution, with probability of success on any trial denoted by p.
(a)   Find $P(Y \geqslant 2)$ if $p = 0.1$.
(b)   Find $P(Y > 4 \mid Y > 2)$ for general p. Compare with the unconditional probability $P(Y > 2)$.

3.39   Let Y denote a negative binomial random variable with $p = 0.4$. Find $P(Y \geqslant 4)$ if
(a) $r = 2$ (b) $r = 4$

3.40   Suppose that 10% of the engines manufactured on a certain assembly line are defective. If engines are randomly selected one at a time and tested, find the probability that the first nondefective engine is found on the second trial.

3.41   Refer to Exercise 3.40. Find the probability that the third nondefective engine is found
  (a)   on the fifth trial.
  (b)   on or before the fifth trial.

3.42   Refer to Exercise 3.40. Given that the first two engines are defective, find the probability that at least two more engines must be tested before the first nondefective is found.

3.43   Refer to Exercise 3.40. Find the mean and variance of the number of the trial on which
  (a)   the first nondefective engine is found.
  (b)   the third nondefective engine is found.

3.44   The employees of a firm that manufactures insulation are being tested for indications of asbestos in their lungs. The firm is requested to send three employees who have positive indications of asbestos to a medical center for further testing. If 40% of the employees have positive indications of asbestos in their lungs, find the probability that ten employees must be tested to find three positives.

3.45   Refer to Exercise 3.44. If each test costs $20, find the expected value and variance of the total cost of conducting the tests to locate three positives. Do you think it is highly likely that the cost of completing these tests would exceed $350?

3.46   If one-third of the persons donating blood at a clinic have O^+ blood, find the probability that the
  (a)   first O^+ donor is the fourth donor of the day.
  (b)   second O^+ donor is the fourth donor of the day.

3.47   A geological study indicates that an exploratory oil well drilled in a certain region should strike oil with probability 0.2. Find the probability that the
  (a)   first strike of oil comes on the third well drilled.
  (b)   third strike of oil comes on the fifth well drilled.
What assumptions are necessary for your answers to be correct?

3.48   In the setting of Exercise 3.47 suppose a company wants to set up three producing wells. Find the expected value and variance of the number of wells that must be drilled to find the three successful ones.

3.49   A large lot of tires contains 10% defectives. Four are to be chosen to place on a car.
  (a)   Find the probability that six tires must be selected from the lot to get four good ones.
  (b)   Find the expected value and variance of the number of selections that must be made to get four good tires.

3.50   The telephone lines coming into an airline reservation office are all occupied about 60% of the time.
  (a)   If you are calling this office, what is the probability that you complete your call on the first try? second try? third try?
  (b)   If you and a friend both must complete separate calls to this reservation office, what is the probability that it takes a total of four tries for the two of you?

3.51 An appliance comes in two colors, white and brown, which are in equal demand. A certain dealer in these appliances has three of each color in stock, although this is not known to the customers. Customers arrive and independently order these appliances. Find the probability that
  (a) the third white is ordered by the fifth customer.
  (b) the third brown is ordered by the fifth customer.
  (c) all of the whites are ordered before any of the browns.
  (d) all of the whites are ordered before all of the browns.

3.7 THE POISSON DISTRIBUTION

A number of probability distributions come about through limiting arguments applied to other distributions. One such very useful distribution is called the *Poisson*.

Consider the development of a probabilistic model for the number of accidents occurring at a particular highway intersection in a period of one week. We can think of the time interval as being split up into n subintervals such that

$$P(\text{one accident in a subinterval}) = p$$

$$P(\text{no accidents in a subinterval}) = 1 - p$$

Note that we are assuming the same value of p holds for all subintervals, and the probability of more than one accident in any one subinterval is zero. If the occurrence of accidents can be regarded as independent from subinterval to subinterval, then the total number of accidents in the time period (which equals the total number of subintervals containing one accident) will have a binomial distribution.

Although there is no unique way to choose the subintervals and we therefore know neither n nor p, it seems reasonable to assume that as n increases, p should decrease. Thus we want to look at the limit of the binomial probability distribution as $n \to \infty$ and $p \to 0$.

To simplify matters, we take the limit under the restriction that the mean, np in the binomial case, remains constant at a value we will call λ.

Now with $np = \lambda$ or $p = \lambda/n$ we have

$$\lim_{n \to \infty} \binom{n}{y}\left(\frac{\lambda}{n}\right)^y\left(1 - \frac{\lambda}{n}\right)^{n-y}$$

$$= \lim_{n \to \infty} \frac{\lambda^y}{y!}\left(1 - \frac{\lambda}{n}\right)^n \frac{n(n-1)\cdots(n-y+1)}{n^y}\left(1 - \frac{\lambda}{n}\right)^{-y}$$

$$= \frac{\lambda^y}{y!}\lim_{n \to \infty}\left(1 - \frac{\lambda}{n}\right)^n\left(1 - \frac{\lambda}{n}\right)^{-y}\left(1 - \frac{1}{n}\right)\left(1 - \frac{2}{n}\right)\cdots\left(1 - \frac{y-1}{n}\right)$$

Noting that

$$\lim_{n \to \infty}\left(1 - \frac{\lambda}{n}\right)^n = e^{-\lambda}$$

and that all other terms involving n tend to unity, we have the limiting distribution

$$p(y) = \frac{\lambda^y}{y!}\, e^{-\lambda} \qquad y = 0, 1, 2, \ldots$$

Recall that λ denotes the mean number of occurrences in one time period (a week for the example under consideration), and hence if t nonoverlapping time periods were considered, the mean would be λt.

The above distribution, called the *Poisson with parameter λ*, can be used to model counts in areas or volumes, as well as in time. For example we may use this distribution to model the number of flaws in a square yard of textile, the number of bacteria colonies in a cubic centimeter of water, or the number of times a machine fails in the course of a workday.

To see how various choices of λ affect the shape of the Poisson probability function, we show graphs of this function for $\lambda = 1$ (Figure 3.12) and $\lambda = 5$ (Figure 3.13). Note that the probability function is very asymmetric for $\lambda = 1$, but is mound shaped and quite symmetric when λ gets as large as 5. We illustrate the use of the Poisson distribution in the following example.

FIGURE 3.12
Poisson Probability Function; $\lambda = 1$

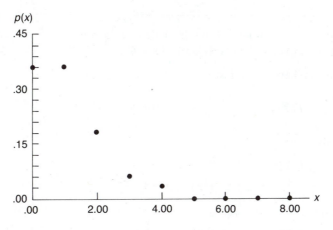

FIGURE 3.13
Poisson Probability Function; $\lambda = 5$

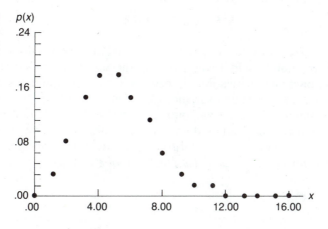

EXAMPLE 3.16

For a certain manufacturing industry the number of industrial accidents averages three per week. Find the probability that no accidents will occur in a given week.

Solution   If accidents tend to occur independently of one another, and if they occur at a constant rate over time, the Poisson model provides an adequate representation of the probabilities. Thus

$$p(0) = \frac{3^0}{0!} e^{-3} = e^{-3} = 0.05$$ ∎

Table 3 of the Appendix gives values for cumulative Poisson probabilities of the form

$$\sum_{y=0}^{a} e^{-\lambda} \frac{\lambda^y}{y!}$$

The following example illustrates the use of Table 3.

EXAMPLE 3.17

Refer to Example 3.16 and let Y denote the number of accidents in the given week. Find $P(Y \leqslant 4)$, $P(Y \geqslant 4)$, and $P(Y = 4)$.

Solution   From Table 3 we have

$$P(Y \leqslant 4) = \sum_{y=0}^{4} \frac{(3)^y}{y!} e^{-3} = 0.815$$

Also

$$P(Y \geqslant 4) = 1 - P(Y \leqslant 3)$$
$$= 1 - 0.647 = 0.353$$

and

$$P(Y = 4) = P(Y \leqslant 4) - P(Y \leqslant 3)$$
$$= 0.815 - 0.647 = 0.168$$ ∎

We can intuitively determine what the mean and variance of a Poisson distribution should be by recalling the mean and variance of a binomial distribution and the relationship between the two distributions. A binomial distribution has mean np and variance $np(1 - p) = np - (np)p$. Now if n gets large and p remains at $np = \lambda$, the variance $np - (np)p = \lambda - \lambda p$ should tend toward λ. In fact the Poisson distribution does have both mean and variance equal to λ.

The mean of the Poisson distribution is easily derived formally if one remembers the *Taylor series* expansion of e^x, namely

$$e^x = 1 + x + \frac{x^2}{2!} + \frac{x^3}{3!} + \cdots$$

Then

$$E(Y) = \sum_y yp(y) = \sum_{y=0}^{\infty} y \frac{\lambda^y}{y!} e^{-\lambda} = \sum_{y=1}^{\infty} y \frac{\lambda^y}{y!} e^{-\lambda}$$

$$= \lambda e^{-\lambda} \sum_{y=1}^{\infty} \frac{\lambda^{y-1}}{(y-1)!}$$

$$= \lambda e^{-\lambda} \left(1 + \lambda + \frac{\lambda^2}{2!} + \frac{\lambda^3}{3!} + \cdots \right)$$

$$= \lambda e^{-\lambda} e^{\lambda} = \lambda$$

The formal derivation of the fact that

$$V(Y) = \lambda$$

is left as an exercise for the interested reader. [*Hint*: First find $E(Y^2) = E[Y(Y-1)] + E(Y).$]

EXAMPLE 3.18

The manager of an industrial plant is planning to buy a new machine of either type A or type B. For each day's operation the number of repairs X that machine A requires is a Poisson random variable with mean $0.10t$, where t denotes the time (in hours) of daily operation. The number of daily repairs Y for machine B is Poisson with mean $0.12t$. The daily cost of operating A is $C_A(t) = 10t + 30X^2$; for B it is $C_B(t) = 8t + 30Y^2$. Assume that the repairs take negligible time and each night the machines are to be cleaned, so they operate like new machines at the start of each day. Which machine minimizes the expected daily cost if a day consists of (a) ten hours? (b) twenty hours?

Solution The expected cost for A is

$$E[C_A(t)] = 10t + 30E(X^2)$$

$$= 10t + 30[V(X) + (E(X))^2]$$

$$= 10t + 30[0.10t + 0.01t^2]$$

$$= 13t + 0.3t^2$$

Similarly

$$E[C_B(t)] = 8t + 30E(Y^2)$$

$$= 8t + 30[0.12t + 0.0144t^2]$$

$$= 11.6t + 0.432t^2$$

For part (a)

$$E[C_A(10)] = 13(10) + 0.3(10)^2 = 160$$

and

$$E[C_B(10)] = 11.6(10) + 0.432(10)^2 = 159.2$$

which results in the choice of machine B.

For part (b)

$$E[C_A(20)] = 380$$

and

$$E[C_B(20)] = 404.8$$

which results in the choice of machine A. In conclusion, B is more economical for short time periods because of its smaller hourly operating cost. However for long time periods A is more economical because it tends to be repaired less frequently. ■

THE POISSON DISTRIBUTION

$$p(y) = \frac{\lambda^y}{y!} e^{-\lambda} \qquad y = 0, 1, 2, \ldots$$

$$E(Y) = \lambda \qquad V(Y) = \lambda$$

EXERCISES

3.52 Let Y denote a random variable having a Poisson distribution with mean $\lambda = 2$. Find
    (a) $P(Y = 4)$ (b) $P(Y \geqslant 4)$
    (c) $P(Y < 4)$ (d) $P(Y \geqslant 4 \mid Y \geqslant 2)$

3.53 The number of telephone calls coming into the central switchboard of an office building averages 4 per minute.
    (a) Find the probability that no calls will arrive in a given one-minute period.
    (b) Find the probability that at least two calls will arrive in a given one-minute period.
    (c) Find the probability that at least two calls will arrive in a given two-minute period.

3.54 The quality of computer disks is measured by sending the disks through a certifier that counts the number of missing pulses. A certain brand of computer disks has averaged 0.1 missing pulse per disk.
    (a) Find the probability that the next inspected disk will have no missing pulse.
    (b) Find the probability that the next inspected disk will have more than one missing pulse.
    (c) Find the probability that neither of the next two inspected disks will contain any missing pulse.

3.55 The National Maximum Speed Limit (NMSL) of 55 miles per hour has been in force in the United States since early 1974. The benefits of this law have been studied by D. B. Kamerud (*Transportation Research*, 17A, No. 1, 1983, pp. 51–64), who reports that the fatality rate for interstate highways with the NMSL in 1975 is approximately 16 per 10^9 vehicle miles.

   (a) Find the probability of at most 15 fatalities occurring in 10^9 vehicle miles.

   (b) Find the probability of at least 20 fatalities occurring in 10^9 vehicle miles.

(Assume that fatalities per vehicle mile follows a Poisson distribution.)

3.56 In the article cited in Exercise 3.55 the projected fatality rate for 1975 if the NMSL had not been in effect would have been 25 per 10^9 vehicle miles. Under these conditions

   (a) find the probability of at most 15 fatalities occurring in 10^9 vehicle miles.

   (b) find the probability of at least 20 fatalities occurring in 10^9 vehicle miles.

   (c) compare the answers in parts (a) and (b) to those in Exercise 3.55.

3.57 In a timesharing computer system the number of teleport inquiries averages 0.2 per millisecond and follows a Poisson distribution.

   (a) Find the probability that no inquiries are made during the next millisecond.

   (b) Find the probability that no inquiries are made during the next three milliseconds.

3.58 Rebuilt ignition systems leave an aircraft rework facility at the rate of 3 per hour on the average. The assembly line needs four ignition systems in the next hour. What is the probability that they will be available?

3.59 Customer arrivals at a checkout counter in a department store have a Poisson distribution with an average of 8 per hour. For a given hour find the probability that

   (a) exactly 8 customers arrive.

   (b) no more than 3 customers arrive.

   (c) at least 2 customers arrive.

3.60 Refer to Exercise 3.59. If it takes approximately 10 minutes to service each customer, find the mean and variance of the total service time connected to the customer arrivals for one hour. (Assume that an unlimited number of servers are available, so that no customer has to wait for service.) Is it highly likely that total service time would exceed 200 minutes?

3.61 Refer to Exercise 3.59. Find the probability that exactly two customers arrive in the two-hour period of time

   (a) between 2:00 P.M. and 4:00 P.M. (one continuous two-hour period).

   (b) between 1:00 P.M. and 2:00 P.M. and between 3:00 P.M. and 4:00 P.M. (two separate one-hour periods for a total of two hours).

3.62 The number of imperfections in the weave of a certain textile has a Poisson distribution with a mean of 4 per square yard.

   (a) Find the probability that a one-square-yard sample will contain at least one imperfection.

   (b) Find the probability that a three-square-yard sample will contain at least one imperfection.

3.63 Refer to Exercise 3.62. The cost of repairing the imperfections in the weave is $10 per imperfection. Find the mean and standard deviation of the repair costs for an eight-square-yard bolt of the textile in question.

3.64 The number of bacteria colonies of a certain type in samples of polluted water has a Poisson distribution with a mean of 2 per cubic centimeter.

2 COLONIES

(a) If four one-cubic-centimeter samples are independently selected from this water, find the probability that at least one sample will contain one or more bacteria colonies.

(b) How many one-cubic-centimeter samples should be selected to have a probability of approximately 0.95 of seeing at least one bacteria colony?

3.65 Let Y have a Poisson distribution with mean λ. Find $E[Y(Y-1)]$ and use the result to show that $V(Y) = \lambda$.

3.66 A food manufacturer uses an extruder (a machine that produces bite-sized foods such as cookies and many snack foods) that produces revenue for the firm at the rate of $200 per hour when in operation. However, the extruder breaks down on the average of two times for every ten hours of operation. If Y denotes the number of breakdowns during the time of operation, the revenue generated by the machine is given by

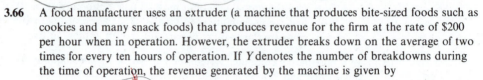

$$R = 200t - 50Y^2$$

where t denotes hours of operation. The extruder is shut down for routine maintenance on a regular schedule, and operates like a new machine after this maintenance. Find the optimal maintenance interval t_0 so the expected revenue is maximized between shutdowns.

3.67 The number of cars entering a parking lot is a random variable having a Poisson distribution with a mean of 4 per hour. The lot holds only 12 cars.

(a) Find the probability that the lot fills up in the first hour. (Assume all cars stay in the lot longer than one hour.)

(b) Find the probability that fewer than 12 cars arrive during an eight-hour day.

3.8 THE HYPERGEOMETRIC DISTRIBUTION

The distributions already discussed in this chapter have as their basic building block a series of *independent* Bernoulli trials. The examples, such as sampling from large lots, depict situations in which the trials of the experiment do generate, for all practical purposes, independent outcomes.

Suppose that we have a relatively small lot consisting of N items, of which k are defective. If two items are sampled sequentially, then the outcome for the second draw is very much influenced by what happened on the first draw, provided that the first item drawn remains out of the lot. A new distribution must be developed to handle this situation involving *dependent* trials.

In general suppose a lot consists of N items, of which k are of one type (called successes) and $N - k$ are of another type (called failures). Suppose that n items are sampled randomly and sequentially from the lot, with none of the sampled items being replaced. (This is called *sampling without replacement*.) Let $X_i = 1$ if the ith draw results in a success and $X_i = 0$ otherwise, $i = 1, \ldots, n$, and let Y denote the total number of successes among the n sampled items. To develop the probability distribution for Y, let us start by looking at a special case for $Y = y$. One way for y successes to occur is to have

$$X_1 = 1, X_2 = 1, \ldots, X_y = 1, X_{y+1} = 0, \ldots, X_n = 0$$

We know that

$$P(X_1 = 1, X_2 = 2) = P(X_1 = 1)P(X_2 = 1 | X_1 = 1)$$

and this result can be extended to give

$$P(X_1 = 1, X_2 = 1, \ldots, X_y = 1, X_{y+1} = 0, \ldots, X_n = 0)$$
$$= P(X_1 = 1)P(X_2 = 1 | X_1 = 1)P(X_3 = 1 | X_2 = 1, X_1 = 1) \cdots$$
$$P(X_n = 0 | X_{n-1} = 0, \ldots, X_{y+1} = 0, X_y = 1, \ldots, X_1 = 1)$$

Now

$$P(X_1 = 1) = \frac{k}{N}$$

if the item is randomly selected, and similarly

$$P(X_2 = 1 | X_1 = 1) = \frac{k-1}{N-1}$$

since, at this point, one of the k successes has been removed. Using this idea repeatedly, we see that

$$P(X_1 = 1, \ldots, X_y = 1, X_{y+1} = 0, \ldots, X_n = 0)$$
$$= \left(\frac{k}{N}\right)\left(\frac{k-1}{N-1}\right) \cdots \left(\frac{k-y+1}{N-y+1}\right)$$
$$\times \left(\frac{N-k}{N-y}\right) \cdots \left(\frac{N-k-n+y+1}{N-n+1}\right)$$

provided $y \leq k$. A more compact way to write the above expression is to employ factorials, arriving at the formula

$$\frac{\dfrac{k!}{(k-y)!} \times \dfrac{(N-k)!}{(N-k-n+y)!}}{\dfrac{N!}{(N-n)!}}$$

(The reader can check the equivalence of the two expressions.)

Any specified arrangement of y successes and $(n-y)$ failures will have the same probability as the one derived above for all successes followed by all failures; the terms will merely be rearranged. Thus to find $P(Y = y)$, we need only to count how many of these arrangements are possible. Just as in the binomial case, the number of such arrangements is $\binom{n}{y}$. Hence we have

$$P(Y = y) = \binom{n}{y} \frac{\dfrac{k!}{(k-y)!} \times \dfrac{(N-k)!}{(N-k-n+y)!}}{\dfrac{N!}{(N-n)!}} = \frac{\dbinom{k}{y}\dbinom{N-k}{n-y}}{\dbinom{N}{n}}$$

Of course, $0 \leq y \leq k \leq N$, $0 \leq y \leq n \leq N$.

This formula is referred to as the *hypergeometric probability distribution*. Note that it arises from a situation quite similar to the binomial, except that the trials are *dependent*.

Experiments that result in a random variable possessing a hypergeometric distribution usually involve counting the number of "successes" in a sample taken from a small lot. Examples could include counting the number of males that show up on a committee of five randomly selected from among twenty employees or counting the number of Brand A alarm systems sold in three sales from a warehouse containing two Brand A and four Brand B systems.

EXAMPLE 3.19 _____

A personnel director selects two employees for a certain job from a group of six employees, of which one is female and five are male. Find the probability that the female is selected for one of the jobs.

Solution   If the selections are made at random, and if Y denotes the number of females selected, then the hypergeometric distribution would provide a good model for the behavior of Y. Hence

$$P(Y = 1) = p(1) = \frac{\binom{1}{1}\binom{5}{1}}{\binom{6}{2}} = \frac{1 \times 5}{15} = \frac{1}{3}$$

Here $N = 6$, $k = 1$, $n = 2$, and $y = 1$.

It might be instructive to see this calculation from basic principles, letting $X_1 = 1$ if the ith draw results in the female, and $X_i = 0$ otherwise. Then

$$P(Y = 1) = P(X_1 = 1, X_2 = 0) + P(X_1 = 0, X_2 = 1)$$

$$= P(X_1 = 1)P(X_2 = 0|X_1 = 1) + P(X_1 = 0)P(X_2 = 1|X_1 = 0)$$

$$= \left(\frac{1}{6}\right)\left(\frac{5}{5}\right) + \left(\frac{5}{6}\right)\left(\frac{1}{5}\right) = \frac{1}{3}$$ ■

The techniques needed to derive the mean and variance of the hypergeometric distribution will be given in Chapter 5. The results are

$$E(Y) = n\left(\frac{k}{N}\right)$$

$$V(Y) = n\left(\frac{k}{N}\right)\left(1 - \frac{k}{N}\right)\left(\frac{N - n}{N - 1}\right)$$

Observe that, since the probability of selecting a success on one draw is k/N, the mean of the hypergeometric distribution has the same form as the mean of the binomial distribution. Also the variance of the hypergeometric is like the variance of the binomial, multiplied by $(N - n)/(N - 1)$, a correction factor for dependent samples.

EXAMPLE 3.20

In an assembly line production of industrial robots, gear box assemblies can be installed in one minute each if holes have been properly drilled in the boxes, and in ten minutes each if holes must be redrilled. Twenty gear boxes are in stock, and it is assumed that two will have improperly drilled holes. Five gear boxes must be selected from the 20 available for installation in the next five robots in line.

(a) Find the probability that all five gear boxes will fit properly.
(b) Find the expected value, variance, and standard deviation of the time it takes to install these five gear boxes.

Solution

(a) In this problem $N = 20$, and the number of nonconforming boxes is assumed to be $k = 2$, according to the manufacturer's usual standards. Let Y denote the number of nonconforming boxes (number with improperly drilled holes) in the sample of 5. Then

$$P(Y = 0) = \frac{\binom{2}{0}\binom{18}{5}}{\binom{20}{5}}$$

$$= \frac{(1)(8{,}568)}{15{,}504} = 0.55$$

(b) The total time T taken to install the boxes (in minutes) is

$$T = 10Y + (5 - Y)$$

$$= 9Y + 5$$

since each of Y nonconforming boxes takes 10 minutes to install, and the others take only one minute. To find $E(T)$ and $V(T)$, we first need $E(Y)$ and $V(Y)$.

$$E(Y) = n\left(\frac{k}{N}\right) = 5\left(\frac{2}{20}\right) = 0.5$$

and

$$V(Y) = n\left(\frac{k}{N}\right)\left(1 - \frac{k}{N}\right)\left(\frac{N - n}{N - 1}\right)$$

$$= 5(0.1)(1 - 0.1)\left(\frac{20 - 5}{20 - 1}\right)$$

$$= 0.355$$

It follows that

$$E(T) = 9E(Y) + 5$$

$$= 9(0.5) + 5 = 9.5$$

and

$$V(T) = (9)^2 V(Y)$$

$$= 81(0.355) = 28.755$$

Thus installation time should average 9.5 minutes, with a standard deviation of $\sqrt{28.755} = 5.4$ minutes.   ∎

THE HYPERGEOMETRIC DISTRIBUTION

$$p(y) = \frac{\binom{k}{y}\binom{N-k}{n-y}}{\binom{N}{n}} \qquad y = 0, 1, \ldots, k \text{ with } \binom{b}{a} = 0 \text{ if } a > b$$

$$E(Y) = n\left(\frac{k}{N}\right) \qquad\qquad V(Y) = n\left(\frac{k}{N}\right)\left(1 - \frac{k}{N}\right)\left(\frac{N-n}{N-1}\right)$$

EXERCISES

3.68   From a box containing 4 white and 3 red balls, 2 balls are selected at random without replacement. Find the probability that
  (a)   exactly one white ball is selected.
  (b)   at least one white ball is selected.
  (c)   two white balls are selected given that at least one white ball is selected.
  (d)   the second ball drawn is white.

3.69   A warehouse contains 10 printing machines, 4 of which are defective. A company randomly selects five of the machines for purchase. What is the probability that all five of the machines are nondefective?

3.70   Refer to Exercise 3.69. The company purchasing the machines returns the defective ones for repair. If it costs $50 to repair each machine, find the mean and variance of the total repair cost. In what interval would you expect the repair costs on these five machines to lie? (Use Tchebysheff's Theorem.)

3.71   A corporation has a pool of 6 firms, 4 of which are local, from which they can purchase certain supplies. If 3 firms are randomly selected without replacement, find the probability that
  (a)   at least one selected firm is not local.
  (b)   all three selected firms are local.

3.72   A foreman has 10 employees from whom he must select 4 to perform a certain undesirable task. Among the 10 employees 3 belong to a minority ethnic group. The foreman selected all three minority employees (plus one other) to perform the undesirable task. The minority group then protested to the union steward that they were discriminated against by the foreman. The foreman claimed that the selection was completely at random. What do you think?

3.73   Specifications call for a type of thermistor to test out at between 9,000 and 10,000 ohms at 25°C. From 10 thermistors available 3 are to be selected for use. Let Y denote the number among the 3 that do not conform to specifications. Find the probability distribution for Y (tabular form) if
   (a)   the 10 contain 2 thermistors not conforming to specifications.
   (b)   the 10 contain 4 thermistors not conforming to specifications.

3.74   Used photocopying machines are returned to the supplier, cleaned, and then sent back out on lease agreements. Major repairs are not made and, as a result, some customers receive malfunctioning machines. Among 8 used photocopiers in supply today, 3 are malfunctioning. A customer wants to lease 4 of these machines immediately. Hence 4 machines are quickly selected and sent out, with no further checking. Find the probability that the customer receives
   (a)   no malfunctioning machines.
   (b)   at least one malfunctioning machine.
   (c)   three malfunctioning machines.

3.75   An eight-cylinder automobile engine has two misfiring spark plugs. If all four plugs are removed from one side of the engine, what is the probability that the two misfiring ones are among them?

3.76   The "worst-case" requirements are defined in the design objectives for a brand of computer terminal. A quick preliminary test indicates that 4 out of a lot of 10 such terminals failed the "worst-case" requirements. Five of the 10 are randomly selected for further testing. Let Y denote the number, among the 5, that failed the preliminary test. Find
   (a) $P(Y \geqslant 1)$ (b) $P(Y \geqslant 3)$
   (c) $P(Y \geqslant 4)$ (d) $P(Y \geqslant 5)$

3.77   An auditor checking the accounting practices of a firm samples 3 accounts from an accounts receivable list of 8. Find the probability that the auditor sees at least one past due account if there are
   (a)   2 such accounts among the 8.
   (b)   4 such accounts among the 8.
   (c)   7 such accounts among the 8.

3.78   A group of 6 software packages available to solve a linear programming problem have been ranked from 1 to 6 (best to worst). An engineering firm selects 2 of these packages for purchase without looking at the ratings. Let Y denote the number of packages purchased by the firm that are ranked 3, 4, 5, or 6. Show the probability distribution for Y in tabular form.

3.79   Lot acceptance sampling procedures for an electronics manufacturing firm call for sampling n items from a lot of N items, and accepting the lot if $Y \leqslant c$, where Y is the number of nonconforming items in the sample. For an incoming lot of 20 printer covers, 5 are to be sampled. Find the probability of accepting the lot if $c = 1$ and the actual number of nonconforming covers in the lot is

   (a)   0   (b)   1   (c)   2   (d)   3   (e)   4

3.80   In the setting and terminology of Exercise 3.79, answer the same questions if $c = 2$.

3.81 Two assembly lines (I and II) have the same rate of defectives in their production of voltage regulators. Five regulators are sampled from each line and tested. Among the total of ten tested regulators there were 4 defectives. Find the probability that exactly 2 of the defectives came from line I.

*3.9 THE MOMENT-GENERATING FUNCTION

We saw in earlier sections that if $g(Y)$ is a function of a random variable Y with probability distribution given by $p(y)$, then

$$E[g(Y)] = \sum_y g(y)p(y)$$

A special function with many theoretical uses in probability theory is the expected value of e^{tY}, for a random variable Y, and this expected value is called the *moment-generating function* (mgf). We denote mgf's by $M(t)$, and thus

$$M(t) = E(e^{tY})$$

The expected values of powers of a random variable are often called *moments*. Thus $E(Y)$ is the first moment and $E(Y^2)$ the second moment of Y. One use for the moment-generating function is that it does, in fact, generate moments of Y. When $M(t)$ exists, it is differentiable in a neighborhood of the origin $t = 0$, and derivatives may be taken inside the expectation. Thus

$$M^{(1)}(t) = \frac{dM(t)}{dt} = \frac{d}{dt} E[e^{tY}]$$

$$= E\left[\frac{d}{dt} e^{tY}\right] = E[Ye^{tY}]$$

Now if we set $t = 0$, we have

$$M^{(1)}(0) = E(Y)$$

Going on to the second derivative,

$$M^{(2)}(t) = E(Y^2 e^{tY})$$

and

$$M^{(2)}(0) = E(Y^2)$$

In general

$$M^{(k)}(0) = E(Y^k)$$

It is often easier to evaluate $M(t)$ and its derivatives than to find the moments of the random variable directly. Other theoretical uses of the mgf will be seen in later chapters.

*Optional section.

EXAMPLE 3.21

Evaluate the moment-generating function for the geometric distribution and use it to find the mean and variance of this distribution.

Solution   We have, for the geometric random variable Y,

$$M(t) = E(e^{tY}) = \sum_{y=1}^{\infty} e^{ty} p(1-p)^{y-1}$$

$$= pe^t \sum_{y=1}^{\infty} (1-p)^{y-1}(e^t)^{y-1}$$

$$= pe^t \sum_{y=1}^{\infty} [(1-p)e^t]^{y-1}$$

$$= pe^t \{1 + [(1-p)e^t] + [(1-p)e^t]^2 + \cdots$$

$$= pe^t \left[\frac{1}{1-(1-p)e^t} \right]$$

since the series is geometric with common ratio $(1-p)e^t$.
   To evaluate the mean, we have

$$M^{(1)}(t) = \frac{[1-(1-p)e^t]pe^t - pe^t[-(1-p)e^t]}{[1-(1-p)e^t]^2}$$

$$= \frac{pe^t}{[1-(1-p)e^t]^2}$$

and

$$M^{(1)}(0) = \frac{p}{[1-(1-p)]^2} = \frac{1}{p}$$

To evaluate the variance, we first need

$$E(Y^2) = M^{(2)}(0)$$

Now

$$M^{(2)}(t) = \frac{[1-(1-p)e^t]^2 pe^t - pe^t\{2[1-(1-p)e^t](-1)(1-p)e^t\}}{[1-(1-p)e^t]^4}$$

and

$$M^{(2)}(0) = \frac{p^3 + 2p^2(1-p)}{p^4} = \frac{p+2(1-p)}{p^2}$$

Hence

$$V(Y) = E(Y^2) - [E(Y)]^2$$

$$= \frac{p+2(1-p)}{p^2} - \frac{1}{p^2} = \frac{1-p}{p^2}$$

■

Moment-generating functions have some very important properties that make them extremely useful in finding expected values and determining the probability distributions of random variables. These properties will be discussed in detail in Chapters 4 and 5, but one such property is given in Exercise 3.85.

EXERCISES

3.82 Find the moment-generating function for the Bernoulli random variable.

3.83 Show that the moment-generating function for the binomial random variable is given by

$$M(t) = [pe^t + (1 - p)]^n$$

Use this result to derive the mean and variance for the binomial distribution.

3.84 Show that the moment-generating function for the Poisson random variable with mean λ is given by

$$M(t) = e^{\lambda(e^t - 1)}$$

Use this result to derive the mean and variance for the Poisson distribution.

3.85 If X is a random variable with moment-generating function $M(t)$, and Y is a function of X given by $Y = aX + b$, show that the moment-generating function for Y is $e^{tb}M(at)$.

3.86 Use the result of Exercise 3.85 to show that

$$E(Y) = aE(X) + b$$

and

$$V(Y) = a^2 V(X)$$

---

*3.10   ACTIVITIES FOR THE COMPUTER

Computers lend themselves nicely for use in the area of probability. Not only can computers be used to calculate probabilities, but one can use computers to simulate random variables from specified probability distributions. A *simulation* can be thought of as an experiment performed on the computer. Simulation is used to analyze problems that are both theoretical and applied. A simulated model attempts to copy the behavior of a situation under consideration. Some practical applications of simulation might be to model inventory control problems, queuing systems, production lines, medical systems, and flight patterns of major jets. Simulation can also be used to determine the behavior of a complicated random variable whose precise probability distribution function is difficult to evaluate mathematically.

Generating observations from a probability distribution is based upon random numbers on [0, 1]. A random number R_i on the interval [0, 1] is one that is selected upon the condition that each number between 0 and 1 has the same probability of being selected. A sequence of numbers that appears to follow a certain pattern or trend would not be considered random. Most computer languages have built-in random generators that will give a number on $[0, 1]$. If a built-in generator is not available, algorithms are available for setting up one.

Given that a random number, R_i, on $[0, 1]$ can be generated, a brief description will now be given for generating discrete random variables for the distributions discussed in this chapter.

Bernoulli.   Let p = probability of success. If $R_i \leqslant p$, then $X_i = 1$. Otherwise $X_i = 0$.

Binomial.   A binomial random variable X_i can be expressed as the sum of n independent Bernoulli random variables, Y_j; that is, $X_i = \Sigma \, Y_j, j = 1, \ldots, n$. Thus to simulate X_i with parameters n and p, simulate n Bernoulli random variables as stated previously. X_i = sum of the n Bernoulli variables.

Geometric.   Let X_i = the number of trials necessary for the first success with p = probability of success. $X_i = m$, where m = the number of R_i's generated until the condition $R_i \leqslant p$ is met.

Negative Binomial.   Let X_i = the number of trials necessary until the rth success with p = probability of success. A negative binomial random variable X_i can be expressed as the sum of r independent geometric random variables Y_j; that is, $X_i = \Sigma \, Y_j, j = 1, \ldots, r$. Thus to simulate X_i with parameter p, simulate r geometric random variables as stated previously. X_i = sum of the r geometric variables.

Poisson.   Generating Poisson random variables will be discussed at the end of Chapter 4.

Let's consider some simple examples of possible uses for simulating some discrete random variables. Suppose that n_1 items are to be inspected from one production line and n_2 items are to be inspected from another production line. Let p_1 = probability of a defective from line 1 and p_2 = probability of a defective from line 2. Let X be a binomial random variable with parameters n_1 and p_1. Let Y be a binomial random variable with parameters n_2 and p_2. A variable of interest is W, which is the total number of defective items observed in both production lines. Let $W = X + Y$. Unless $p_1 = p_2$, the distribution of W will not be binomial. To see how the distribution of W will behave, a simulation could be performed. Useful information could be obtained from the simulation by looking at a histogram of the W_i's generated and also considering the values of the sample mean and sample variance. Let's consider the following random variables X and Y: X is binomial with $n_1 = 7, p_1 = 0.2$ and Y is binomial with $n_1 = 8, p_2 = 0.6$. Defining $W = X + Y$, a simulation was performed that produced the histogram shown in Figure 3.14.

The sample mean was 6.2 with a sample standard deviation of 1.76. In Chapter 5 we will be able to show that these values are very close to the expected value of $\mu_w = 6.2$ and $\sigma_w = 1.74$. (This calculation will be possible after linear

FIGURE 3.14
Simulating Sums of
Binomial Random
Variables

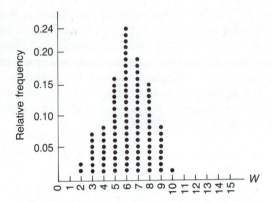

functions of random variables are discussed.) From the histogram, the probability that the total number of defective items is at least 9 is given by 0.09.

Another example of interest might be the coupon-collector problem, which incorporates the geometric distribution.

Suppose that there are *n* distinct colors of coupons. Each time an individual obtains a coupon, we assume that it is equally likely to be any one of the *n* colors and the selection of the coupon is independent of a previously obtained coupon. Suppose the situation exists that one can redeem a set of coupons for a prize if each possible color coupon is represented in the set. We define the random variable $X =$ the number of necessary coupons to be selected in order that one has a complete set of each color coupon. Questions of interest might be

1. What is the expected number of coupons needed in order to obtain this complete set; that is, what is $E(X)$?
2. What is the standard deviation of X?
3. What is the probability that one must select at most x coupons to obtain this complete set?

Instead of answering the above questions by deriving the distribution function of X, one might try simulation. Two simulations are presented below. The first histogram, Figure 3.15, represents a simulation where *n*, the number of different color coupons, is equal to 5. The sample mean was computed to be 11.06, with a

FIGURE 3.15
Simulating Sums of
Geometric Random
Variables

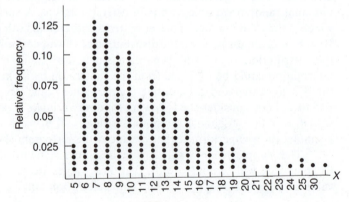

sample standard deviation of 4.65. Suppose one is interested in finding $P\{X \leqslant 10\}$. Using the results of the simulations, the relative frequency probability is given to be 0.555. It might also be noted that from this simulation, the largest number of coupons needed to obtain the complete set was 31.

The second histogram, Figure 3.16, represents a simulation where $n = 10$. The sample mean was computed to be 29.07 with a sample standard deviation of 10.16. The $P\{X \leqslant 20\} = 0.22$ from the simulated values. In this simulation the largest number of coupons necessary to obtain the complete set was 71.

FIGURE 3.16
Simulating Sums of Geometric Random Variables

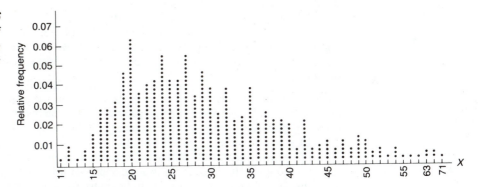

SUPPLEMENTARY EXERCISES

3.87 Construct probability histograms for the binomial probability distribution for $n = 5$, $p = 0.1, 0.5,$ and 0.9. (Table 2 of the Appendix will reduce the amount of calculation.) Note the symmetry for $p = 0.5$ and the direction of skewness for $p = 0.1$ and 0.9.

3.88 Use Table 2 of the Appendix to construct a probability histogram for the binomial probability distribution for $n = 20$ and $p = 0.5$. Note that almost all the probability falls in the interval $5 \leqslant y \leqslant 15$.

3.89 The probability that a single radar set will detect an enemy plane is 0.9. If we have five radar sets, what is the probability that exactly four sets will detect the plane? (Assume that the sets operate independently of each other.) At least one set?

3.90 Suppose that the four engines of a commercial aircraft were arranged to operate independently and that the probability of in-flight failure of a single engine is 0.01. What is the probability that, on a given flight,
    (a) No failures are observed?
    (b) No more than one failure is observed?

3.91 Sampling for defectives from large lots of manufactured product yields a number of defectives Y that follows a binomial probability distribution. A sampling plan consists in specifying the number of items to be included in a sample n and an acceptance number a. The lot is accepted if $Y \leqslant a$ and rejected if $Y > a$. Let p denote the proportion of defectives in the lot. For $n = 5$ and $a = 0$, calculate the probability of lot acceptance if

    (a) $p = 0$ (b) $p = 0.1$ (c) $p = 0.3$
    (d) $p = 0.5$ (e) $p = 1.0$

A graph showing the probability of lot acceptance as a function of lot fraction defective is called the *operating characteristic curve* for the sample plan. Construct this curve for the plan $n = 5$, $a = 0$. Note that a sampling plan is an example of statistical inference. Accepting or rejecting a lot based on information contained in the sample is equivalent to concluding that the lot is either "good" or "bad," respectively. "Good" implies that a low fraction is defective and that the lot is therefore suitable for shipment.

3.92   Refer to Exercise 3.91. Use Table 2 of the Appendix to construct the operating characteristic curve for a sampling plan with

    (a) $n = 10, a = 0$ (b) $n = 10, a = 1$ (c) $n = 10, a = 2$

For each, calculate P(lot acceptance) for $p = 0, 0.05, 0.1, 0.3, 0.5$, and 1.0. Our intuition suggests that sampling plan (a) would be much less likely to accept bad lots than plans (b) and (c). A visual comparison of the operating characteristic curves will confirm this supposition.

3.93   A quality-control engineer wishes to study the alternative sampling plans $n = 5, a = 1$ and $n = 25, a = 5$. On a sheet of graph paper construct the operating characteristic curves for both plans; make use of acceptance probabilities at $p = 0.05$, $p = 0.10$, $p = 0.20$, $p = 0.30$, and $p = 0.40$ in each case.

    (a)   If you were a seller producing lots with fraction defective ranging from $p = 0$ to $p = 0.10$, which of the two sampling plans would you prefer?

    (b)   If you were a buyer wishing to be protected against accepting lots with fraction defective exceeding $p = 0.30$, which of the two sampling plans would you prefer?

3.94   For a certain section of a pine forest the number of diseased trees per acre Y has a Poisson distribution with mean $\lambda = 10$. The diseased trees are sprayed with an insecticide at a cost of $3.00 per tree, plus a fixed overhead cost for equipment rental of $50.00. Letting C denote the total spraying cost for a randomly selected acre, find the expected value and standard deviation for C. Within what interval would you expect C to lie with probability at least 0.75?

3.95   In checking river water samples for bacteria, water is placed in a culture medium so that certain bacteria colonies can grow if those bacteria are present. The number of colonies per dish averages 12 for water samples from a certain river.

    (a)   Find the probability that the next dish observed will have at least 10 colonies.

    (b)   Find the mean and standard deviation of the number of colonies per dish.

    (c)   Without calculating exact Poisson probabilities, find an interval in which at least 75% of the colony count measurements should lie.

3.96   The number of vehicles passing a specified point on a highway averages 10 per minute.

    (a)   Find the probability that at least 15 vehicles pass this point in the next minute.

    (b)   Find the probability that at least 15 vehicles pass this point in the next two minutes.

3.97   A production line often produces a variable number N of items each day. Suppose each item produced has the same probability p of not conforming to manufacturing standards. If N has a Poisson distribution with mean λ, then the number of nonconforming items in one day's production Y has a Poisson distribution with mean λp.

The average number of resistors produced by a facility in one day has a Poisson distribution with mean 100. Typically, 5% of the resistors produced do not meet specifications.

(a) Find the expected number of resistors not meeting specifications for a given day.

(b) Find the probability that all resistors will meet the specifications on a given day.

(c) Find the probability that more than 5 resistors fail to meet specifications on a given day.

3.98 A certain type of bacteria cell divides at a constant rate λ over time. (That is, the probability that a cell will divide in a small interval of time t is approximately λt.) Given that a population starts out at time zero with k cells of this type, and cell divisions are independent of one another, the size of the population at time t, $Y(t)$, has the probability distribution

$$P[Y(t) = n] = \binom{N-1}{k-1} e^{-\lambda k t}(1 - e^{-\lambda t})^{n-k}$$

(a) Find the expected value of $Y(t)$ in terms of λ and t.

(b) If, for a certain type of bacteria cell, $\lambda = 0.1$ per second, and the population starts out with 2 cells at time zero, find the expected population size after 5 seconds.

3.99 The probability that any one vehicle will turn left at a particular intersection is 0.2. The left-turn lane at this intersection has room for three vehicles. If five vehicles arrive at this intersection while the light is red, find the probability that the left-turn lane will hold all of the vehicles that want to turn left.

3.100 Refer to Exercise 3.99. Find the probability that six cars must arrive at the intersection while the light is red to fill up the left-turn lane.

3.101 For any probability function $p(y)$, $\Sigma \, y p(y) = 1$ if the sum is taken over all possible values y that the random variable in question can assume. Show that this is true for

(a) the binomial distribution.

(b) the geometric distribution.

(c) the Poisson distribution.

3.102 The supply office for a large construction firm has three welding units of Brand A in stock. If a welding unit is requested, the probability is 0.7 that the request will be for this particular brand. On a typical day five requests for welding units come to the office. Find the probability that all three Brand A units will be in use on that day.

3.103 Refer to Exercise 3.102. If the supply office also stocks three welding units that are not Brand A, find the probability that exactly one of these units will be left immediately after the third Brand A unit is requested.

3.104 The probability of a customer arrival at a grocery service counter in any 1 second is equal to 0.1. Assume that customers arrive in a random stream and hence that the arrival at any 1 second is independent of any other.

(a) Find the probability that the first arrival will occur during the third 1-second interval.

(b) Find the probability that the first arrival will not occur until at least the third 1-second interval.

3.105 Sixty percent of a population of consumers is reputed to prefer Brand A toothpaste. If a group of consumers are interviewed, what is the probability that exactly five people have to be interviewed before encountering a consumer who prefers Brand A? At least five people?

3.106 The mean number of automobiles entering a mountain tunnel per 2-minute period is one. An excessive number of cars entering the tunnel during a brief period of time produces a hazardous situation.
    (a) Find the probability that the n number of autos entering the tunnel during a 2-minute period exceeds three.
    (b) Assume that the tunnel is observed during ten 2-minute intervals, thus giving ten independent observations, $Y_1, Y_2, \ldots, Y_{10}$, on a Poisson random variable. Find the probability that $Y > 3$ during at least one of the ten 2-minute intervals.

3.107 Suppose that 10% of a brand of microcomputers will fail before their guarantee has expired. If 1000 computers are sold this month, find the expected value and variance of Y, the number that have not failed during the guarantee period. Within what limit would Y be expected to fall? [*Hint*: Use Tchebysheff's Theorem.]

3.108 (a) Consider a binomial experiment for $n = 20$, $p = 0.05$. Use Table 2 of the Appendix to calculate the binomial probabilities for $Y = 0, 1, 2, 3, 4$.
    (b) Calculate the same probabilities using the Poisson approximation with $\lambda = np$. Compare.

3.109 The manufacturer of a low-calorie dairy drink wishes to compare the taste appeal of a new formula (B) with that of the standard formula (A). Each of four judges is given three glasses in random order, two containing formula A and the other containing formula B. Each judge is asked to state which glass he most enjoyed. Suppose that the two formulas are equally attractive. Let Y be the number of judges stating a preference for the new formula.
    (a) Find the probability function for Y.
    (b) What is the probability that at least three of the four judges state a preference for the new formula?
    (c) Find the expected value of Y.
    (d) Find the variance of Y.

3.110 Show that the hypergeometric probability function approaches the binomial in the limit as $N \to \infty$ and $p = r/N$ remains constant. That is, show that

$$\lim_{N \to \infty} \frac{\binom{r}{y}\binom{N-r}{n-y}}{\binom{N}{n}} = \binom{n}{y} p^y q^{n-y}$$

for $p = r/N$ constant and $q = 1 - p$.

3.111 A lot of $N = 100$ industrial products contains 40 defectives. Let Y be the number of defectives in a random sample of size 20. Find $p(10)$ by using
    (a) the hypergeometric probability distribution.
    (b) the binomial probability distribution.
Is N large enough so that the binomial probability function is a good approximation to the hypergeometric probability function?

3.112　For simplicity let us assume that there are two kinds of drivers. The safe drivers, which are 70 percent of the population, have a probability of 0.1 of causing an accident in a year. The rest of the population are accident makers, who have a probability of 0.5 of causing an accident in a year. The insurance premium is $400 times one's probability of causing an accident in the following year. A new subscriber has an accident during the first year. What should be his insurance premium for the next year?

3.113　A merchant stocks a certain perishable item. He knows that on any given day he will have a demand for either 2, 3, or 4 of these items with probabilities 0.1, 0.4, and 0.5, respectively. He buys the items for $1.00 each and sells them for $1.20 each. If any are left at the end of the day, they represent a total loss. How many items should the merchant stock to maximize his expected daily profit?

3.114　It is known that 5% of a population have disease A, which can be discovered by a blood test. Suppose that N (a large number) people are to be tested. This can be done in two ways.
1. Each person is tested separately.
2. The blood samples of k people are pooled together and analyzed. (Assume that $N = nk$, with n an integer.) If the test is negative, all of them are healthy (that is, just this one test is needed). If the test is positive, each of the k persons must be tested separately (that is, a total of $k + 1$ tests are needed).
    (a)　For fixed k what is the expected number of tests needed in method (2)?
    (b)　Find the k that will minimize the expected number of tests in method (2).
    (c)　How many tests does part (b) save in comparison with part (a)?

3.115　Four possible winning numbers for a lottery—AB-4536, NH-7812, SQ-7855, and ZY-3221—are given to you. You will win a prize if one of your numbers matches with one of the winning numbers. You are told that there is one first prize of $100,000, two second prizes of $50,000 each, and 10 third prizes of $1,000 each. The only thing you need to do is to mail the coupon back. No purchase is required. From the structure of the numbers you have received, it is obvious that the entire list consists of all the permutations of two alphabets followed with four digits. Is the coupon worth mailing back for $0.25 postage?

4

Continuous Probability Distributions

ABOUT THIS CHAPTER

Random variables that are not discrete, like measurements of lifelength or weight, can often be classified as *continuous*. This is the second category of random variables we will discuss. Probability distributions commonly used to model continuous random variables are presented in this chapter.

CONTENTS

4.1 CONTINUOUS RANDOM VARIABLES AND THEIR PROBABILITY DISTRIBUTIONS

All of the random variables discussed in Chapter 3 were discrete; each could assume only a finite number or countable infinity of values. But many of the random variables seen in practice have more than a countable collection of possible values. Weights of patients coming into a clinic may be anywhere from, say, 80 to 300 pounds. Diameters of machined rods from a certain industrial process may be anywhere from 1.2 to 1.5 centimeters. Proportions of impurities in ore samples may run from 0.10 to 0.80. These random variables can take on any value in an interval of real numbers. That is not to say that every value in the interval can be found in sample data if one looks long enough; one may never observe a patient weighing exactly 172.38 pounds. That is to say that no one value can be ruled out as a possible observation; one could possibly have a patient weighing 172.38 pounds, and so this number must be considered in the set of possible outcomes. Since random variables of this type have a continuum of possible values, they are called *continuous random variables*. Probability distributions for continuous random variables will now be developed, with the basic ideas being presented in the context of an experiment on lifelengths.

An experimenter is measuring the lifelength X of a transistor. In this case there is an infinite number of possible values that X can assume. We cannot assign a positive probability to each possible outcome of the experiment because, no matter how small we might make the individual probabilities, they would sum to a value greater than one when accumulated over the entire sample space. We can, however, assign positive probabilities to *intervals* of real numbers in a manner consistent with the axioms of probability. To introduce the basic ideas involved here, let us consider a specific example in some detail.

Suppose that we have conducted an experiment to measure the lifelengths of 50 batteries of a certain type, selected from a larger population of such batteries. The experiment is completed and the observed lifelengths are as given in Table 4.1.

TABLE 4.1
Lifelengths of Batteries (in hundreds of hours)

0.406	0.685	4.778	1.725	8.223
2.343	1.401	1.507	0.294	2.230
0.538	0.234	4.025	3.323	2.920
5.088	1.458	1.064	0.774	0.761
5.587	0.517	3.246	2.330	1.064
2.563	0.511	2.782	6.426	0.836
0.023	0.225	1.514	3.214	3.810
3.334	2.325	0.333	7.514	0.968
3.491	2.921	1.624	0.334	4.490
1.267	1.702	2.634	1.849	0.186

The relative frequency histogram for these data (Figure 4.1) shows clearly that most of the lifelengths are near zero, and the frequency drops off rather smoothly as we look at larger lifelengths. Here 32% of the 50 observations fall into the first subinterval and another 22% fall into the second. There is a decline in frequency as we proceed across the subintervals, until the last subinterval (8 to 9) contains a single observation.

FIGURE 4.1
Relative Frequency Histogram of Data from Table 2.3

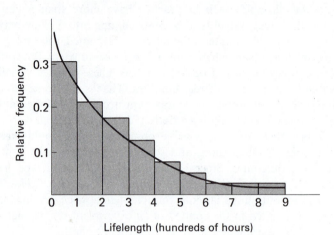

This sample relative frequency histogram not only allows us to picture how the sample behaves, but it also gives us some insight into a possible probabilistic model for the random variable X. The histogram of Figure 4.1 looks like it could be approximated quite closely by a negative exponential curve. The particular function

$$f(x) = \frac{1}{2}e^{-x/2} \qquad x > 0$$

is sketched through the histogram in Figure 4.1 and seems to fit reasonably well. Thus we could take this function as a mathematical model for the behavior of the random variable X. Note that the area under $f(x)$ equals one. If we want to use a battery of this type in the future, we might want to know the probability that it will last longer than four hundred hours. This probability can be approximated by the area under the curve to the right of the value 4; that is, by

$$\int_{4}^{\infty} \frac{1}{2}e^{-x/2}\, dx = 0.135$$

Note that this figure is quite close to the observed sample fraction of lifetimes that exceed 4, namely, $(8/50) = 0.16$. One might suggest that, since the sample fraction 0.16 is available, we do not really need the model. But there are other questions for which the model would give more satisfactory answers than could otherwise be obtained. For example suppose we are interested in the probability that X is greater than 9. Then the model suggests the answer

$$\int_9^\infty \frac{1}{2} e^{-x/2} \, dx = 0.011$$

whereas the sample shows no observations in excess of 9. These are quite simple examples, and we will see many examples of more involved questions for which a model is quite essential.

Why did we choose the exponential function as a model here? Wouldn't some others do just as well? The choice of a model is a fundamental problem, and we will spend considerable time in later sections delving into theoretical and practical reasons for these choices. For this early discussion we will merely suggest some models that look like they might do the job.

The function $f(x)$, which models the relative frequency behavior of X, is called the *probability density function*.

DEFINITION 4.1	A random variable X is said to be **continuous** if it can take on the infinite number of possible values associated with intervals of real numbers, and there is a function $f(x)$, called the **probability density function**, such that (1) $f(x) \geqslant 0$ for all x. (2) $\int_{-\infty}^{\infty} f(x)\,dx = 1$. (3) $P(a \leqslant X \leqslant b) = \int_a^b f(x)\,dx$.

Note that for a continuous random variable X,

$$P(X = a) = \int_a^a f(x) \, dx = 0$$

for any specific value a. The fact that we must assign zero probability to any specific value should not disturb us, since there is an infinite number of possible values that X can assume. For example, out of all the possible values that the lifelength of a transistor can take on, what is the probability that the transistor we are using will last exactly 497.392 hours? Assigning probability zero to this event does not rule out 497.392 as a possible lifelength, but it does say that the chance of observing this particular lifelength is extremely small.

EXAMPLE 4.1

Refer to the random variable X of the lifelength example, which has associated with it a probability density function of the form

$$f(x) = \begin{cases} \frac{1}{2} e^{-x/2} & x > 0 \\ 0 & \text{elsewhere} \end{cases}$$

Find the probability that the lifelength of a particular battery of this type is less than 200 or greater than 400 hours.

Solution   Let A denote the event that X is less than 2 and B the event that X is greater than 4. Then since A and B are mutually exclusive,

$$P(A \cup B) = P(A) + P(B)$$

$$= \int_0^2 \frac{1}{2} e^{-x/2} \, dx + \int_4^\infty \frac{1}{2} e^{-x/2} \, dx$$

$$= (1 - e^{-1}) + (e^{-2})$$

$$= 1 - 0.368 + 0.135$$

$$= 0.767 \qquad \blacksquare$$

EXAMPLE 4.2
_____

Refer to Example 4.1. Find the probability that a battery of this type lasts more than 300 hours given that it has already been in use for more than 200 hours.

Solution   We are interested in $P(X > 3 | X > 2)$ and, by the definition of conditional probability, we have

$$P(X > 3 | X > 2) = \frac{P(X > 3)}{P(X > 2)}$$

since the intersection of the events $(X > 3)$ and $(X > 2)$ is the event $(X > 3)$. Now

$$\frac{P(X > 3)}{P(X > 2)} = \frac{\int_3^\infty \frac{1}{2} e^{-x/2} \, dx}{\int_2^\infty \frac{1}{2} e^{-x/2} \, dx} = \frac{e^{-3/2}}{e^{-1}} = e^{-1/2} = 0.606 \qquad \blacksquare$$

Sometimes it is convenient to look at cumulative probabilities of the form $P(X \leqslant b)$. To do this, we can make use of the distribution function.

DEFINITION 4.2

The **distribution function** for a random variable X is defined as

$$F(b) = P(X \leqslant b).$$

If X is continuous with probability density function $f(x)$, then

$$F(b) = \int_{-\infty}^b f(x) \, dx.$$

Note that $F'(x) = f(x)$.

In the lifelength-of-batteries example, X has a probability density function given by

$$f(x) = \begin{cases} \frac{1}{2} e^{-x/2} & x > 0 \\ 0 & \text{elsewhere} \end{cases}$$

Thus

$$F(b) = P(X \leqslant b) = \int_0^b \frac{1}{2} e^{-x/2} \, dx$$

$$= -e^{-x/2}\big|_0^b$$

$$= 1 - e^{b/2} \qquad b > 0$$

$$= 0 \qquad\qquad b \leqslant 0$$

The function is shown graphically in Figure 4.2.

FIGURE 4.2
*A Distribution
Function for a
Continuous
Random Variable*

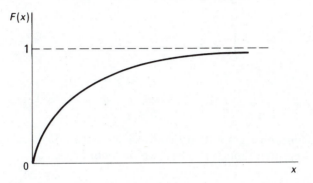

EXAMPLE 4.3

*not
continuous
distribution*

A supplier of kerosene has a 200-gallon tank filled at the beginning of each week. His weekly demands show a relative frequency behavior that increases steadily up to 100 gallons, and then levels off between 100 and 200 gallons. Letting X denote weekly demand in hundreds of gallons, suppose the relative frequencies for demand are modeled adequately by

$$f(x) = \begin{cases} = 0 & x < 0 \\ = x & 0 \leqslant x \leqslant 1 \\ = 1/2 & 1 < x \leqslant 2 \\ = 0 & x > 2 \end{cases}$$

FIGURE 4.3
f(x) for Example 4.3

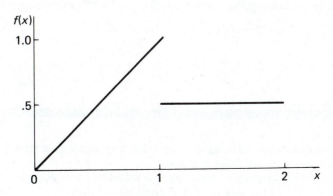

This function has the graphical form shown in Figure 4.3. Find $F(b)$ for this random variable. Use $F(b)$ to find the probability that demand will exceed 150 gallons on a given week.

Solution   From the definition

$$F(b) = \int_{-\infty}^{b} f(x)\, dx$$

$$= 0 \qquad\qquad\qquad b < 0$$

$$= \int_{-\infty}^{b} x\, dx = \frac{b^2}{2} \qquad 0 \leqslant b \leqslant 1$$

$$= \frac{1}{2} + \int_{1}^{b} \frac{1}{2} dx$$

$$= \frac{1}{2} + \frac{b-1}{2} = \frac{b}{2} \qquad 1 < b \leqslant 2$$

$$= 1 \qquad\qquad\qquad b > 2$$

This function is graphed in Figure 4.4. Note that $F(b)$ is continuous over the whole real line, even though $f(b)$ has two discontinuities.

FIGURE 4.4
F(b) for Example 4.3

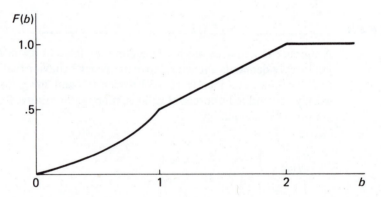

The probability that demand will exceed 150 gallons is given by

$$P(X > 1.5) = 1 - P(X \leqslant 1.5) = 1 - F(1.5)$$

$$= 1 - \frac{1.5}{2} = 0.25$$

■

EXERCISES

4.1   For each of the following situations, define an appropriate random variable and state whether it is continuous or discrete.

    (a)   An environmental engineer is looking at ten field plots to determine whether or not they contain a certain type of insect.

discrete *specific*

more than one sample   *not Poisson no upper limit you don't*   *no*

(b) A quality-control technician samples a continuously produced fabric in square-yard sections, and counts the number of defects he observes for each sampled section.

(c) A metallurgist counts the number of grains seen in a cross-sectional sample of aluminum.

(d) The metallurgist of (c) measures the area proportion covered by grains of a certain size, rather than simply counting them.

4.2 Suppose a random variable X has a probability density function given by

$$f(x) = \begin{cases} kx(1-x) & 0 \leqslant x \leqslant 1 \\ 0 & \text{elsewhere} \end{cases}$$

(a) Find the value of k that makes this a probability density function.
(b) Find $P(0.4 \leqslant X \leqslant 1)$.
(c) Find $P(X \leqslant 0.4 | X \leqslant 0.8)$.
(d) Find $F(b) = P(X \leqslant b)$ and sketch the graph of this function.

4.3 The effectiveness of solar-energy heating units depends upon the amount of radiation available from the sun. For a typical October, daily total solar radiation in Tampa, Florida, approximately follows the probability density function given below (units are hundreds of calories):

$$f(x) = \begin{cases} \frac{3}{32}(x-2)(6-x) & 2 \leqslant x \leqslant 6 \\ 0 & \text{elsewhere} \end{cases}$$

(a) Find the probability that solar radiation will exceed 300 calories on a typical October day.
(b) What amount of solar radiation is exceeded on exactly 50% of the October days according to this model?

4.4 An accounting firm that does not have its own computing facilities rents time from a consulting company. The firm must plan its computing budget carefully and hence has studied the weekly use of CPU time quite thoroughly. The weekly use of CPU time approximately follows the probability density function given by (measurements in hours)

$$f(x) = \begin{cases} \frac{3}{64}x^2(4-x) & 0 \leqslant x \leqslant 4 \\ 0 & \text{elsewhere} \end{cases}$$

(a) Find the distribution function $F(x)$ for weekly CPU time X.
(b) Find the probability that CPU time used by the firm will exceed 2 hours for a selected week.
(c) The current budget of the firm covers only 3 hours of CPU time per week. How often will the budgeted figure be exceeded?
(d) How much CPU time should be budgeted per week if this figure is to be exceeded with probability only 0.10?

4.5 The pH, a measure of the acidity of water, is important in studies of acid rain. For a certain Florida lake, baseline measurements on acidity are made so any changes caused by acid rain can be noted. The pH of water samples from the lake is a random variable X with probability density function

$$f(x) = \begin{cases} \frac{3}{8}(7-x)^2 & 5 \leqslant x \leqslant 7 \\ 0 & \text{elsewhere} \end{cases}$$

(a) Sketch the curve of $f(x)$.

(b) Find the distribution function $F(x)$ for X.

(c) Find the probability that the pH will be less than 6 for a water sample from this lake.

(d) Find the probability that the pH of a water sample from this lake will be less than 5.5 given that it is known to be less than 6.

4.6 The "on" temperature of a thermostatically controlled switch for an air conditioning system is set at 60°, but the actual temperature X at which the switch turns on is a random variable having probability density function

$$f(x) = \begin{cases} \frac{1}{2} & 59 \leqslant x \leqslant 61 \\ 0 & \text{elsewhere} \end{cases}$$

(a) Find the probability that it takes a temperature in excess of 60° to turn the switch on.

(b) If two such switches are used independently, find the probability that they both require a temperature in excess of 60° to turn on.

4.7 The proportion of time, during a forty-hour work week, that an industrial robot was in operation was measured for a large number of weeks, and the measurements can be modeled by the probability density function

$$f(x) = \begin{cases} 2x & 0 \leqslant x \leqslant 1 \\ 0 & \text{elsewhere} \end{cases}$$

If X denotes the proportion of time this robot will be in operation during a coming week, find the following:

(a) $P(X > 1/2)$ (b) $P(X > 1/2 | X > 1/4)$

(c) $P(X > 1/4 | X > 1/2)$

(d) Find $F(x)$ and graph this function. Is $F(x)$ continuous?

4.8 The proportion of impurities X in certain copper ore samples is a random variable having probability density function

$$f(x) = \begin{cases} 12x^2(1 - x) & 0 \leqslant x \leqslant 1 \\ 0 & \text{elsewhere} \end{cases}$$

If four such samples are independently selected, find the probability that

(a) exactly one has a proportion of impurities exceeding 0.5.

(b) at least one has a proportion of impurities exceeding 0.5.

4.2 EXPECTED VALUES OF CONTINUOUS RANDOM VARIABLES

As in the discrete case we often want to summarize the information contained in a probability distribution by calculating expected values for the random variable and certain functions of the random variable.

DEFINITION 4.3 | The **expected value** of a continuous random variable X having probability density function $f(x)$ is given by[1]

$$E(x) = \int_{-\infty}^{\infty} xf(x)\,dx$$

For functions of random variables we have the following theorem.

THEOREM 4.1   If X is a continuous random variable with probability distribution $f(x)$ and if $g(x)$ is any real-valued function of X, then

$$E[g(X)] = \int_{-\infty}^{+\infty} g(x)f(x)\,dx$$

The proof of Theorem 4.1 will not be given here.

The definitions of variance and standard deviation and the properties given in Theorems 3.2 and 3.3 hold for the continuous case as well.

We illustrate the expectations of continuous random variables in the following examples.

EXAMPLE 4.4

For a lathe in a machine shop let X denote the percentage of time out of a 40-hour work week that the lathe is actually in use. Suppose X has a probability density function given by

$$f(x) = \begin{cases} 3x^2 & 0 \leqslant x \leqslant 1 \\ 0 & \text{elsewhere} \end{cases}$$

Find the mean and variance of X.

Solution   From Definition 4.3 we have

$$E(X) = \int_{-\infty}^{\infty} xf(x)\,dx$$

$$= \int_{0}^{1} x(3x^2)\,dx$$

$$= \int_{0}^{1} 3x^3\,dx$$

$$= 3\left[\frac{x^4}{4}\right]_{0}^{1} = \frac{3}{4} = 0.75$$

Thus on the average the lathe is in use 75% of the time.

[1]We assume absolute convergence of the integrals.

To compute $V(X)$, we first find $E(X^2)$ by

$$E(X^2) = \int_{-\infty}^{\infty} x^2 f(x)\, dx$$

$$= \int_0^1 x^2 (3x^2)\, dx$$

$$= \int_0^1 3x^4\, dx$$

$$= 3\left[\frac{x^5}{5}\right]_0^1 = \frac{3}{5} = 0.60$$

Then

$$V(X) = E(X^2) - \mu^2$$

$$= 0.60 - (0.75)^2$$

$$= 0.60 - 0.5625 = 0.0375 \qquad \blacksquare$$

EXAMPLE 4.5 _____

The weekly demand X for kerosene at a certain supply station has a density function given by

$$f(x) = \begin{cases} x & 0 \leqslant x \leqslant 1 \\ 1/2 & 1 < x \leqslant 2 \\ 0 & \text{elsewhere} \end{cases}$$

Find the expected weekly demand.

Solution Using Definition 4.3 to find $E(X)$, we must now carefully observe that $f(x)$ has different nonzero forms over two disjoint regions. Thus

$$E(X) = \int_{-\infty}^{\infty} xf(x)\, dx$$

$$= \int_0^1 x(x)\, dx + \int_1^2 x(1/2)\, dx$$

$$= \int_0^1 x^2\, dx + \int_1^2 x\, dx$$

$$= \left[\frac{x^3}{3}\right]_0^1 + \frac{1}{2}\left[\frac{x^2}{2}\right]_1^2$$

$$= \frac{1}{3} + \frac{1}{2}\left[2 - \frac{1}{2}\right]$$

$$= \frac{1}{3} + \frac{3}{4} = \frac{13}{12} = 1.08$$

That is, the expected weekly demand is for 108 gallons. $\qquad \blacksquare$

Tchebysheff's Theorem (Theorem 3.4) holds for continuous random variables, just as it does for discrete. Thus if X is continuous with mean μ and standard deviation σ, then

$$P(|X - \mu| < k\sigma) \geqslant 1 - \frac{1}{k^2}$$

for any positive number k. We illustrate the use of this result in the next example.

EXAMPLE 4.6

The weekly amount Y spent for chemicals in a certain firm has a mean of $445 and a variance of $236. Within what interval would these weekly costs for chemicals be expected to lie at least 75% of the time?

Solution To find an interval guaranteed to contain at least 75% of the probability mass for Y, we get

$$1 - \frac{1}{k^2} = 0.75$$

which gives

$$\frac{1}{k^2} = 0.25$$

$$k^2 = \frac{1}{0.25} = 4$$

or

$$k = 2$$

Thus the interval $\mu - 2\sigma$ to $\mu + 2\sigma$ will contain at least 75% of the probability. This interval is given by

$$445 - 2\sqrt{236} \quad \text{to} \quad 445 + 2\sqrt{236}$$
$$445 - 30.62 \quad \text{to} \quad 445 + 30.72,$$

or

$$414.28 \quad \text{to} \quad 475.72 \quad \blacksquare$$

EXERCISES

4.9 The temperature X at which a thermostatically controlled switch turns on has probability density function

$$f(x) = \begin{cases} \frac{1}{2} & 59 \leqslant x \leqslant 61 \\ 0 & \text{elsewhere} \end{cases}$$

Find $E(X)$ and $V(X)$.

4.10 The proportion of time X that an industrial robot is in operation during a 40-hour week is a random variable with probability density function

$$f(x) = \begin{cases} 2x & 0 \leqslant x \leqslant 1 \\ 0 & \text{elsewhere} \end{cases}$$

(a) Find $E(X)$ and $V(X)$.

(b) For the robot under study the profit Y for a week is given by

$$Y = 200X - 60$$

Find $E(Y)$ and $V(Y)$.

(c) Find an interval in which the profit should lie for at least 75% of the weeks that the robot is in use. [*Hint*: Use Tchebysheff's Theorem.]

4.11 Daily total solar radiation for a certain location in Florida in October has probability density function

$$f(x) = \begin{cases} \frac{3}{32}(x - 2)(6 - x) & 2 \leqslant x \leqslant 6 \\ 0 & \text{elsewhere} \end{cases}$$

with measurements in hundreds of calories. Find the expected daily solar radiation for October.

4.12 Weekly CPU time used by an accounting firm has probability density function (measured in hours)

$$f(x) = \begin{cases} \frac{3}{64}x^2(4 - x) & 0 \leqslant x \leqslant 4 \\ 0 & \text{elsewhere} \end{cases}$$

(a) Find the expected value and variance of weekly CPU time.

(b) The CPU time costs the firm $200 per hour. Find the expected value and variance of the weekly cost for CPU time.

(c) Will the weekly cost exceed $600 very often? Why?

4.13 The pH of water samples from a specific lake is a random variable X with probability density function

$$f(x) = \begin{cases} \frac{3}{8}(7 - x)^2 & 5 \leqslant x \leqslant 7 \\ 0 & \text{elsewhere} \end{cases}$$

(a) Find $E(X)$ and $V(X)$.

(b) Find an interval shorter than (5, 7) in which at least 3/4 of the pH measurements must lie.

(c) Would you expect to see a pH measurement below 5.5 very often? Why?

4.14 A retail grocer has a daily demand X for a certain food sold by the pound, such that X (measured in hundreds of pounds) has probability density function

$$f(x) = \begin{cases} 3x^2 & 0 \leqslant x \leqslant 1 \\ 0 & \text{elsewhere} \end{cases}$$

(He cannot stock over 100 pounds.) The grocer wants to order $100k$ pounds of food on a certain day. He buys the food at 6 cents per pound and sells it at 10 cents per pound. What value of k will maximize his expected daily profit?

4.3 THE UNIFORM DISTRIBUTION

We now move from a general discussion of continuous random variables to discussions of specific models found useful in practice. Consider an experiment that consists of observing events occurring in a certain time frame, such as buses arriving at a bus stop or telephone calls coming into a switchboard. Suppose we know that one such event has occurred in the time interval (a, b). (A bus arrived between 8:00 and 8:10.) It may then be of interest to place a probability distribution on the actual time of occurrence of the event under observation, which we will denote by X. A very simple model assumes that X is equally likely to lie in any small subinterval, say of length d, no matter where that subinterval lies within (a, b). This assumption leads to the *uniform* probability distribution, which has probability density function given by

$$f(x) = \frac{1}{b-a} \qquad a \leqslant x \leqslant b$$

$$= 0 \qquad \text{elsewhere}$$

This density function is graphed in Figure 4.5.

FIGURE 4.5
The Uniform Probability Density Function

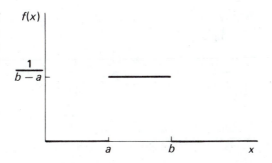

The distribution function for a uniformly distributed X is given by

$$F(x) = \int_a^x \frac{1}{b-a}\, dx = \frac{x-a}{b-a} \qquad a \leqslant x \leqslant b$$

If we consider a subinterval $(c, c + d)$ contained entirely within (a, b), we have

$$P(c \leqslant X \leqslant c + d) = F(c + d) - F(c)$$

$$= \frac{(c+d) - a}{b-a} - \frac{c-a}{b-a}$$

$$= \frac{d}{b-a}$$

Note that this probability does not depend on the location c but only on the length d of the subinterval.

A relationship exists between the uniform distribution and the Poisson distribution, which was introduced in Section 3.7. Suppose the number of events occurring in an interval, say $(0, t)$, has a Poisson distribution. If exactly one of these events is known to have occurred in the interval (a, b), with $a \geqslant 0$ and $b \leqslant t$, then the conditional probability distribution of the actual time of occurrence for this event is uniform over (a, b). (The proof of this fact will not be given here.)

Paralleling the material presented in Chapter 3, we now look at the mean and variance of the uniform distribution. From Definition 4.3 we see that

$$E(X) = \int_{-\infty}^{\infty} xf(x)\, dx = \int_{a}^{b} x\left(\frac{1}{b-a}\right) dx$$

$$= \left(\frac{1}{b-a}\right)\left(\frac{b^2 - a^2}{2}\right) = \frac{a+b}{2}$$

It is intuitively reasonable that the mean value of a uniformly distributed random variable should lie at the midpoint of the interval.

Recalling from Theorem 3.3 that $V(X) = E(X - \mu)^2 = E(X^2) - \mu^2$, we have, for the uniform case,

$$E(X^2) = \int_{-\infty}^{\infty} x^2 f(x)\, dx$$

$$= \int_{a}^{b} x^2 \left(\frac{1}{b-a}\right) dx$$

$$= \left(\frac{1}{b-a}\right)\left(\frac{b^3 - a^3}{3}\right) = \frac{b^2 + ab + a^2}{3}$$

Thus

$$V(X) = \frac{b^2 + ab + a^2}{3} - \left(\frac{a+b}{2}\right)^2$$

$$= \frac{1}{12}\left[4(b^2 + ab + a^2) - 3(a+b)^2\right]$$

$$= \frac{1}{12}(b - a)^2$$

This result may not be intuitive, but we do see that the variance depends only upon the length of the interval (a, b).

EXAMPLE 4.7 _____

The failure of a circuit board causes a computing system to be shut down until a new board is delivered. Delivery time X is uniformly distributed over the interval one to five days. The cost C of this failure and shutdown consists of a fixed cost c_0 for the new part and a cost that increases proportional to X^2, so that

$$C = c_0 + c_1 X^2$$

(a) Find the probability that the delivery time is two or more days.

(b) Find the expected cost of a single component failure in terms of c_0 and c_1.

Solution   (a) The delivery time X is uniformly distributed from one to five days, which gives

$$f(x) = \begin{cases} \frac{1}{4} & 1 \leq x \leq 5 \\ 0 & \text{elsewhere} \end{cases}$$

Thus

$$P(X \geq 2) = \int_2^5 \left(\frac{1}{4}\right) dx$$

$$= \frac{1}{4}(5 - 2) = \frac{3}{4}$$

(b) We know that

$$E(C) = c_0 + c_1 E(X^2)$$

so it remains to find $E(X^2)$. This could be found directly from the definition or by using the variance and the fact that

$$E(X^2) = V(X) + \mu^2$$

Using the latter approach,

$$E(X^2) = \frac{(b - a)^2}{12} + \left(\frac{a + b}{2}\right)^2$$

$$= \frac{(5 - 1)^2}{12} + \left(\frac{1 + 5}{2}\right)^2 = \frac{31}{3}$$

Thus

$$E(C) = c_0 + c_1 \left(\frac{31}{3}\right)$$

Constant function

We now summarize the properties of the uniform distribution.

THE UNIFORM DISTRIBUTION

always integral constant

$$f(x) = \frac{1}{b - a} \qquad a \leq x \leq b$$

$$= 0 \qquad \text{elsewhere}$$

$$E(X) = \frac{a + b}{2} \qquad V(X) = \frac{(b - a)^2}{12}$$

EXERCISES

4.15 Suppose X has a uniform distribution over the interval (a, b).
  (a) Find $F(x)$.
  (b) Find $P(X > c)$ for some point c between a and b.
  (c) If $a \leqslant c \leqslant d \leqslant b$, find $P(X > d \mid X > c)$.

4.16 Upon studying low bids for shipping contracts, a microcomputer manufacturing firm finds that intrastate contracts have low bids that are uniformly distributed between 20 and 25, in units of thousands of dollars. Find the probability that the low bid on the next intrastate shipping contract is
  (a) below $22,000   (b) in excess of $24,000
  (c) Find the average cost of low bids on contracts of this type.

4.17 If a point is *randomly* located in an interval (a, b) and X denotes its distance from a, then X will be assumed to have a uniform distribution over (a, b).
  A plant efficiency expert randomly picks a spot along a 500-foot assembly line from which to observe work habits. Find the probability that he is
  (a) within 25 feet of the end of the line.
  (b) within 25 feet of the beginning of the line.
  (c) closer to the beginning than to the end of the line.

4.18 A bomb is to be dropped along a mile-long line that stretches across a practice target. The target center is at the midpoint of the line. The target will be destroyed if the bomb falls within a tenth of a mile to either side of the center. Find the probability that the target is destroyed if the bomb falls randomly along the line.

4.19 A telephone call arrived at a switchboard at a random time within a one-minute interval. The switchboard was fully busy for 15 seconds into this one-minute period. Find the probability that the call arrived when the switchboard was not fully occupied.

4.20 Beginning at 12:00 midnight a computer center is up for one hour and down for two hours on a regular cycle. A person who doesn't know the schedule dials the center at a random time between 12:00 midnight and 5:00 A.M. What is the probability that the center will be operating when he dials in?

4.21 The number of defective circuit boards among those coming out of a soldering machine follows a Poisson distribution. On a particular 8-hour day one defective board is found.
  (a) Find the probability that it was produced the first hour of operation of that day.
  (b) Find the probability that it was produced during the last hour of operation for that day.
  (c) Given that no defective boards were seen during the first four hours of operation, find the probability that the defective board was produced during the fifth hour.

4.22 In determining the range of an acoustic source by triangulation, the time at which the spherical wave front arrives at a receiving sensor must be measured accurately. According to an article by Perruzzi and Hilliard (*Journal of the Acoustical Society of America*, 75(1), 1984, pp. 197–201), measurement errors in these times can be modeled as having uniform distributions. Suppose measurement errors are uniformly distributed from -0.05 to $+0.05$ microsecond.

      (a)   Find the probability that a particular arrival time measurement will be in error by less than 0.01 microsecond.

      (b)   Find the mean and variance of these measurement errors.

4.23   In the setting of Exercise 4.22 suppose the measurement errors are uniformly distributed from −0.02 to +0.05 microsecond.

      (a)   Find the probability that a particular arrival time measurement will be in error by less than 0.01 microsecond.

      (b)   Find the mean and variance of these measurement errors.

4.24   According to Y. Zimmels (*AIChE Journal*, 29(4), 1983, pp. 669–676), the sizes of particles used in sedimentation experiments often have uniform distributions. It is important to study both the mean and variance of particle sizes since, in sedimentation with mixtures of various-sized particles, the larger particles hinder the movements of the smaller ones.

          Suppose spherical particles have diameters uniformly distributed between 0.01 and 0.05 centimeter. Find the mean and variance of the *volumes* of these particles. (Recall that the volume of a sphere is $\frac{4}{3}\pi r^3$.)

4.25   Arrivals of customers at a certain checkout counter follow a Poisson distribution. It is known that during a given 30-minute period one customer arrived at the counter. Find the probability that he arrived during the last five minutes of the 30-minute period.

4.26   A customer's arrival at a counter is uniformly distributed over a 30-minute period. Find the conditional probability that the customer arrived during the last 5 minutes of the 30-minute period given that there were no arrivals during the first 10 minutes of the period.

4.27   In tests of stopping distances for automobiles, those automobiles traveling at 30 miles per hour before the brakes are applied tend to travel distances that appear to be uniformly distributed between two points *a* and *b*. Find the probability that one of these automobiles

      (a)   stops closer to *a* than to *b*.

      (b)   stops so that the distance to *a* is more than three times the distance to *b*.

4.28   Suppose three automobiles are used in a test of the type discussed in Exercise 4.27. Find the probability that exactly one of the three travels past the midpoint between *a* and *b*.

4.29   The cycle time for trucks hauling concrete to a highway construction site is uniformly distributed over the interval 50 to 70 minutes.

      (a)   Find the expected value and variance for these cycle times.

      (b)   How many trucks would you expect to have to schedule to this job so that a truck load of concrete can be dumped at the site every 15 minutes?

4.4   THE EXPONENTIAL DISTRIBUTION

The lifelength data of Section 4.1 displayed a probabilistic behavior that was not uniform, but rather one in which the probability over intervals of constant length decreased as the intervals moved further and further to the right. We saw that an exponential curve seemed to fit this data rather well, and now discuss the exponential probability distribution in more detail. In general the exponential density function is given by

$$f(x) = \frac{1}{\theta} e^{-x/\theta} \qquad x \geqslant 0$$

$$= 0 \qquad \text{elsewhere}$$

where the parameter θ is a constant that determines the rate at which the curve decreases.

An exponential density function with $\theta = 2$ was sketched in Figure 4.1 and, in general, the exponential functions have the form shown in Figure 4.6.

FIGURE 4.6
An Exponential
Probability Density
Function

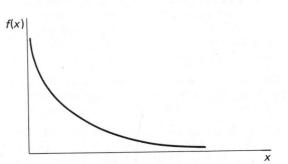

Many random variables occurring in engineering and the sciences can be appropriately modeled as having exponential distributions. Figure 4.7 shows two examples of relative frequency distributions for times between arrivals (interarrival times) of vehicles at a fixed point on a one-directional roadway. Both of these relative frequency histograms can be modeled quite nicely by exponential functions. Note that the higher traffic density causes shorter interarrival times to be more frequent.

Finding expected values for the exponential distribution is simplified by an understanding of a certain type of integral called a *gamma* (Γ) *function*. The function $\Gamma(\alpha)$, for $\alpha \geqslant 1$, is defined by

$$\Gamma(\alpha) = \int_0^\infty x^{\alpha-1} e^{-x} \, dx$$

Integration by parts can be used to show that $\Gamma(\alpha + 1) = \alpha\Gamma(\alpha)$. It follows that $\Gamma(n) = (n - 1)!$ for any positive integer n. The integral

$$\int_0^\infty x^{\alpha-1} e^{-x/\beta} \, dx$$

for positive constants α and β can be evaluated by making the transformation $y = x/\beta$, or $x = \beta y$, $dx = \beta \, dy$. We have then

$$\int_0^\infty (\beta y)^{\alpha-1} e^{-y} \beta \, dy = \beta^\alpha \int_0^\infty y^{\alpha-1} e^{-y} \, dy = \beta^\alpha \Gamma(\alpha)$$

It is useful to note that $\Gamma(1/2) = \sqrt{\pi}$.

FIGURE 4.7
Interarrival Times of Vehicles on a One-Directional Roadway

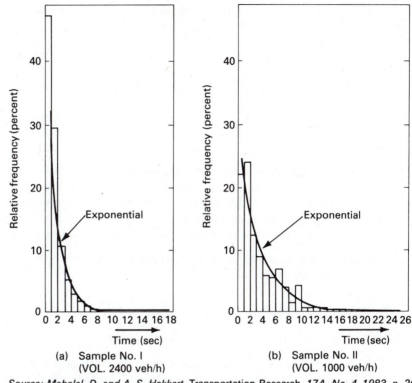

(a) Sample No. I
(VOL. 2400 veh/h)

(b) Sample No. II
(VOL. 1000 veh/h)

Source: Mahalel, D. and A. S. Hakkert, Transportation Research, *17A, No. 4, 1983, p. 267.*

Using the above result, we see that, for the exponential distribution,

$$E(X) = \int_{-\infty}^{\infty} xf(x)\,dx = \int_{0}^{\infty} x\left(\frac{1}{\theta}\right) e^{-x/\theta}\,dx$$

$$= \frac{1}{\theta} \int_{0}^{\infty} xe^{-x/\theta}\,dx$$

$$= \frac{1}{\theta}\, \Gamma(2)\theta^2 = \theta$$

Thus the parameter θ is actually the mean of the distribution.

To evaluate the variance of the exponential distribution, first we can find

$$E(X^2) = \int_{0}^{\infty} x^2 \left(\frac{1}{\theta}\right) e^{-x/\theta}\,dx$$

$$= \frac{1}{\theta}\, \Gamma(3)\theta^3 = 2\theta^2$$

It follows that

$$V(X) = E(X^2) - \mu^2$$

$$= 2\theta^2 - \theta^2 = \theta^2$$

and θ becomes the standard deviation as well as the mean.

The distribution function for the exponential case has a simple form, seen to be

$$F(t) = 0 \qquad\qquad \text{for } t < 0$$

$$F(t) = P(X \leq t) = \int_0^t \frac{1}{\theta} e^{-x/\theta}\, dx$$

$$= -e^{-x/\theta}|_0^t = \boxed{1 - e^{-t/\theta}} \qquad\qquad \text{for } t \geq 0$$

EXAMPLE 4.8

A sugar refinery has three processing plants, all receiving raw sugar in bulk. The amount of sugar that one plant can process in one day can be modeled as having an exponential distribution with a mean of 4 (measurements in tons), for each of the three plants. If the plants operate independently, find the probability that exactly two of the three plants process more than 4 tons on a given day.

Solution The probability that any given plant processes more than 4 tons is, with X denoting the amount used,

$$P(X > 4) = \int_4^\infty f(x)\, dx = \int_4^\infty \frac{1}{4} e^{-x/4}\, dx$$

$$= -e^{-x/4}|_4^\infty = e^{-1} = 0.37$$

[*Note:* $P(X > 4) \neq 0.5$ even though the mean of X is 4.]
Knowledge of the distribution function could allow us to evaluate this immediately as

$$P(X > 4) = 1 - P(X \leq 4) = 1 - (1 - e^{-4/4})$$

$$= e^{-1}$$

Assuming the three plants operate independently, the problem is to find the probability of two successes out of three tries, where 0.37 denotes the probability of success. This is a binomial problem, and the solution is

$$P(\text{exactly 2 use more than 4 tons}) = \binom{3}{2}(0.37)^2(0.63)$$

$$= 3(0.37)^2(0.63)$$

$$= 0.26 \qquad\blacksquare$$

EXAMPLE 4.9

Consider a particular plant in Example 4.8. How much raw sugar should be stocked for that plant each day so that the chance of running out of product is only 0.05?

Solution Let a denote the amount to be stocked. Since the amount to be used X has an exponential distribution, we have

$$P(X > a) = \int_a^\infty \frac{1}{4} e^{-x/4} \, dx = e^{-a/4}$$

We want to choose a so that

$$P(X > a) = e^{-a/4} = 0.05$$

and solving this equation yields

$$a = 11.98 \qquad\qquad \blacksquare$$

As in the uniform case, there is a relationship between the exponential distribution and the Poisson distribution. Suppose events are occurring in time according to a Poisson distribution with a rate of λ events per hour. Thus in t hours the number of events, say Y, will have a Poisson distribution with mean value λt. Suppose we start at time zero and ask the question "How long do I have to wait to see the first event occur?" Let X denote the length of time until this first event. Then

$$P(X > t) = P[Y = 0 \text{ on the interval } (0, t)]$$
$$= (\lambda t)^0 e^{-\lambda t}/0 = e^{-\lambda t}$$

and

$$P(X \leqslant t) = 1 - P(X > t) = 1 - e^{-\lambda t}$$

We see that $P(X \leqslant t) = F(t)$, the distribution function for X, has the form of an exponential distribution function with $\lambda = (1/\theta)$. Upon differentiating, we see that the probability density function of X is given by

$$f(t) = \frac{dF(t)}{dt} = \frac{d(1 - e^{-\lambda t})}{dt}$$
$$= \lambda e^{-\lambda t}$$
$$= \frac{1}{\theta} e^{-t/\theta} \qquad t > 0$$

and X has an exponential distribution. Actually we need not start at time zero for it can be shown that the waiting time from the occurrence of any one event until the occurrence of the next event will have an exponential distribution for events occurring according to a Poisson distribution.

Besides probability density and distribution functions, another function is of use in studying properties of continuous distributions, especially in working with lifelength data. Suppose X denotes the lifelength of a component with density function $f(x)$ and distribution function $F(x)$. The *failure rate function* $r(t)$ is defined as

$$r(t) = \frac{f(t)}{1 - F(t)} \qquad t > 0, F(t) < 1$$

For an intuitive look at what $r(t)$ is measuring, suppose dt denotes a very small interval around the point t. Then $f(t) \, dt$ is approximately the probability that X

takes on a value in $(t, t + dt)$. Also $1 - F(t) = P(X > t)$. Thus

$$r(t)\, dt = \frac{f(t)\, dt}{1 - F(t)}$$

$$\approx P[X \in (t,\ t + dt) | X > t]$$

In other words $r(t)\, dt$ represents the probability of failure during the time interval $(t, t + dt)$ given that the component has survived up to time t.

For the exponential case

$$r(t) = \frac{f(t)}{1 - F(t)} = \frac{\frac{1}{\theta} e^{-t/\theta}}{e^{-t/\theta}} = \frac{1}{\theta}$$

or X has a *constant* failure rate. It is unlikely that many individual components have a constant failure rate over time (most fail more frequently as they age), but it may be true of some systems that undergo regular preventive maintenance.

We have summarized the properties of the exponential distribution.

THE EXPONENTIAL DISTRIBUTION

$$f(x) = \frac{1}{\theta} e^{-x/\theta} \qquad x > 0$$

$$= 0 \qquad\qquad \text{elsewhere}$$

$$E(X) = \theta \qquad\qquad V(X) = \theta^2$$

EXERCISES

4.30 Suppose Y has an exponential density function with mean θ. Show that $P(Y > a + b | Y > a) = P(Y > b)$. This is referred to as the "memoryless" property of the exponential distribution.

4.31 The magnitudes of earthquakes recorded in a region of North America can be modeled by an exponential distribution with mean 2.4 as measured on the Richter scale. Find the probability that the next earthquake to strike this region will
   (a)   exceed 3.0 on the Richter scale.
   (b)   fall between 2.0 and 3.0 on the Richter scale.

4.32 Refer to Exercise 4.31. Out of the next ten earthquakes to strike this region, find the probability that at least one will exceed 5.0 on the Richter scale.

4.33 A pumping station operator observes that the demand for water at a certain hour of the day can be modeled as an exponential random variable with a mean of 100 cfs (cubic feet per second).
   (a)   Find the probability that the demand will exceed 200 cfs on a randomly selected day.

Handwritten notes in top margin: $P(x > b) = .01$

Handwritten notes in left margin: rate growth + decay, meaning waiting time, $\sigma = \frac{1}{2}\mu$, larger

(b) What is the maximum water-producing capacity that the station should keep on line for this hour so that the demand will exceed this production capacity with a probability of only 0.01?

4.34 Suppose customers arrive at a certain checkout counter at the rate of two every minute.

    (a) Find the mean and variance of the waiting time between successive customer arrivals.

    (b) If a clerk takes 3 minutes to serve the first customer arriving at the counter, what is the probability that at least one more customer is waiting when the service of the first customer is completed?

4.35 The length of time X to complete a certain key task in house construction is an exponentially distributed random variable with a mean of 10 hours. The cost C of completing this task is related to the square of the time to completion by the formula

$$C = 100 + 40X + 3X^2$$

    (a) Find the expected value and variance of C.

    (b) Would you expect C to exceed 2000 very often?

4.36 The interaccident times (times between accidents) for all fatal accidents on scheduled American domestic passenger airflights, 1948–1961, were found to follow an exponential distribution with a mean of approximately 44 days (Pyke, *Journal of the Royal Statistical Society*, (B), 27, 1968, p. 426).

    (a) If one of those accidents occurred on July 1, find the probability that another one occurred in that same month.

    (b) Find the variance of the interaccident times.

    (c) What does the information given above suggest about the clumping of airline accidents?

4.37 The lifelengths of automobile tires of a certain brand, under average driving conditions, are found to follow an exponential distribution with mean 30 (in thousands of miles). Find the probability that one of these tires bought today will last

    (a) over 30,000 miles.

    (b) over 30,000 miles given that it already has gone 15,000 miles.

4.38 The dial-up connections from remote terminals come into a computing center at the rate of 4 per minute. The calls follow a Poisson distribution. If a call arrives at the beginning of a one-minute period, find the probability that a second call will not arrive in the next 20 seconds.

4.39 The breakdowns of an industrial robot follow a Poisson distribution with an average of 0.5 breakdown per eight-hour work day. If this robot is placed in service at the beginning of the day, find the probability that

    (a) it will not break down during the day.

    (b) it will work for at least four hours without breaking down.

    (c) Does what happened the day before have any effect on your answers above? Why?

4.40 One-hour carbon monoxide concentrations in air samples from a large city are found to have an exponential distribution with a mean of 3.6 ppm (J. Zamurs, *Air Pollution Control Association Journal*, 34(6), 1984, p. 637).

    (a) Find the probability that a concentration will exceed 9 ppm.

    (b) A traffic control strategy reduced the mean to 2.5 ppm. Now find the probability that a concentration will exceed 9 ppm.

4.41 The weekly rainfall totals for a section of the midwestern United States follow an exponential distribution with a mean of 1.6 inches.

    (a) Find the probability that a weekly rainfall total in this section will exceed 2 inches.

    (b) Find the probability that the weekly rainfall totals will not exceed 2 inches in either of the next two weeks.

4.42 The service times at teller windows in a bank were found to follow an exponential distribution with a mean of 3.2 minutes. A customer arrives at a window at 4:00 P.M.

    (a) Find the probability that he will still be there at 4:02 P.M.

    (b) Find the probability that he will still be there at 4:04 given that he was there at 4:02.

4.43 In deciding how many customer service representatives to hire and in planning their schedules, it is important for a firm marketing electronic typewriters to study repair times for the machines. Such a study revealed that repair times have approximately an exponential distribution with a mean of 22 minutes.

    (a) Find the probability that a repair time will last less than 10 minutes.

    (b) The charge for typewriter repairs is $50 for each half hour, or part thereof. What is the probability that a repair job will result in a charge of $100?

    (c) In planning schedules, how much time should be allowed for each repair so that the chance of any one repair time exceeding this allowed time is only 0.10?

4.44 Explosive devices used in a mining operation cause nearly circular craters to form in a rocky surface. The radii of these craters are exponentially distributed with a mean of 10 feet. Find the mean and variance of the area covered by such a crater.

4.5 THE GAMMA DISTRIBUTION

Many sets of data, of course, will not have relative frequency curves with the smooth decreasing trend found in the exponential model. It is perhaps more common to see distributions that have low probabilities for intervals close to zero, with the probability increasing for a while as the interval moves to the right (in the positive direction) and then decreasing as the interval moves to the extreme positive side. That is, the relative frequency curves appear as in Figure 4.8. In the case of electronic components, for example, few will have very short lifelengths, many will have something close to an average lifelength, and very few will have extraordinarily long lifelengths.

FIGURE 4.8
A Common Relative Frequency Curve

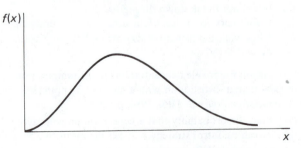

A class of functions that serve as good models for this type of behavior is the *gamma* class. The gamma probability density function is given by

$$f(x) = \frac{1}{\Gamma(\alpha)\beta^\alpha} x^{\alpha-1}e^{-x/\beta} \qquad x \geq 0$$

$$= 0 \qquad\qquad\qquad \text{elsewhere}$$

where α and β are parameters that determine the specific shape of the curve. Note immediately that the gamma density reduces to the exponential when $\alpha = 1$. The parameters α and β must be positive but need not be integers. The symbol $\Gamma(\alpha)$ is defined by

$$\Gamma(\alpha) = \int_0^\infty x^{\alpha-1}e^{-x}\,dx$$

Since we have already seen that

$$\int_0^\infty x^{\alpha-1}e^{-x/\beta}\,dx = \beta^\alpha\Gamma(\alpha)$$

it follows that the gamma density function will integrate to one.

Some typical gamma densities are shown in Figure 4.9.

FIGURE 4.9
The Gamma Density
Function, $\beta = 1$

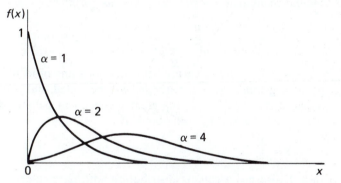

An example of a real data set that closely follows a gamma distribution is shown in Figure 4.10. The data are six-week summer rainfall totals for Ames, Iowa. Notice that many totals are from two to eight inches, but occasionally a rainfall total goes well beyond eight inches. Of course, no rainfall measurements can be negative.

The derivation of expectations here is very similar to the exponential case of Section 4.4. We have

$$E(X) = \int_{-\infty}^\infty xf(x)\,dx = \int_0^\infty x\,\frac{1}{\Gamma(\alpha)\beta^\alpha} x^{\alpha-1}e^{-x/\beta}\,dx$$

$$= \frac{1}{\Gamma(\alpha)\beta^\alpha}\int_0^\infty x^\alpha e^{-x/\beta}\,dx$$

$$= \frac{1}{\Gamma(\alpha)\beta^\alpha}\,\Gamma(\alpha+1)\beta^{\alpha+1} = \alpha\beta$$

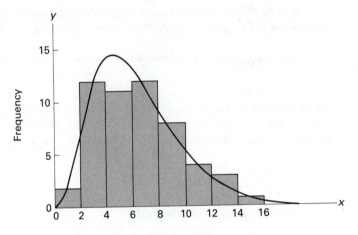

FIGURE 4.10
Summer Rainfall
(6-week totals) for
Ames, Iowa

Rainfall (inches)

Source: G. L. Barger and H. C. S. Thom, Agronomy Journal, 41, 1949, p. 521.

Similar manipulations yield $E(X^2) = \alpha(\alpha + 1)\beta^2$ and hence

$$V(X) = E(X^2) - \mu^2$$

$$= \alpha(\alpha + 1)\beta^2 - \alpha^2\beta^2 = \alpha\beta^2$$

A simple and often used property of sums of identically distributed, independent gamma random variables will be stated, but not proved, at this point. Suppose $X_1, X_2, \ldots, X_n$ represent independent gamma random variables with parameters α and β as used above. If

$$Y = \sum_{i=1}^{n} X_i$$

then Y also has a gamma distribution with parameters $n\alpha$ and β. Thus one can immediately see that

$$E(Y) = n\alpha\beta$$

and

$$V(Y) = n\alpha\beta^2$$

EXAMPLE 4.10 _____

A certain electronic system having lifelength X_1 with an exponential distribution and mean 400 hours is supported by an identical backup system with lifelength X_2. The backup system takes over immediately when the primary system fails. If the systems operate independently, find the probability distribution and expected value for the total lifelength of the primary and backup system.

Solution   Letting Y denote the total lifelength, we have $Y = X_1 + X_2$, where X_1 and X_2 are independent exponential random variables, each with mean $\beta = 400$. By the results stated above, Y will then have a gamma distribution with $\alpha = 2$ and $\beta = 400$, that is,

$$f_Y(y) = \frac{1}{\Gamma(2)(400)^2}\, ye^{-y/400} \qquad y > 0$$

The mean value is given by

$$E(Y) = \alpha\beta = 2(400) = 800$$

which is intuitively reasonable. ∎

EXAMPLE 4.11

Suppose that the length of time Y to conduct a periodic maintenance check (from previous experience) on a dictating machine follows a gamma-type distribution with $\alpha = 3$ and $\beta = 2$ (minutes). Suppose that a new repairman requires 20 minutes to check a machine. Does it appear that his time to perform a maintenance check disagrees with prior experience?

Solution   The mean and variance for the length of maintenance times (prior experience) are

$$\mu = \alpha\beta \qquad \text{and} \qquad \sigma^2 = \alpha\beta^2$$

Then for our example

$$\mu = \alpha\beta = (3)(2) = 6 \qquad \sigma^2 = \alpha\beta^2 = (3)(2)^2 = 12 \qquad \sigma = \sqrt{12} = 3.46$$

and the observed deviation $(Y - \mu)$ is $20 - 6 = 14$ minutes.

For our example $y = 20$ minutes exceeds the mean $\mu = 6$ by $k = 14/3.46$ standard deviations. Then from Tchebysheff's Theorem

$$P(|Y - \mu| \geqslant k\sigma) \leqslant \frac{1}{k^2}$$

or

$$P(|Y - 6| \geqslant 14) \leqslant \frac{1}{k^2} = \frac{(3.46)^2}{(14)^2} = 0.06$$

Note that this probability is based on the assumption that the distribution of maintenance times has not changed from prior experience. Then observing that $P(Y \geqslant 20 \text{ minutes})$ is small, we must conclude that either our new maintenance man has generated a lengthy maintenance time that occurs with low probability or he is somewhat slower than his predecessors. Noting the low probability for $P(Y \geqslant 20)$, we would be inclined to favor the latter view. ∎

The failure rate function $r(t)$ for the gamma case is not easily displayed since $F(t)$ does not have a simple closed form. However for $\alpha > 1$ this function will increase but is always bounded above by $1/\beta$. A typical form is shown in Figure 4.11. The properties of the gamma distribution are summarized in the box on page 152.

FIGURE 4.11
The Failure Rate
Function for the
Gamma Distribution
(α > 1)

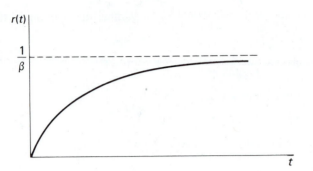

THE GAMMA DISTRIBUTION _____

$$f(x) = \frac{1}{\Gamma(\alpha)\beta^\alpha} x^{\alpha-1} e^{-x/\beta} \qquad x > 0$$

$$= 0 \qquad\qquad \text{elsewhere}$$

$$E(X) = \alpha\beta \qquad\qquad V(X) = \alpha\beta^2$$

EXERCISES

4.45 Four-week summer rainfall totals in a certain section of the midwestern United States have a relative frequency histogram that appears to fit closely to a gamma distribution with $\alpha = 1.6$ and $\beta = 2.0$.
  (a) Find the mean and variance of this distribution of four-week rainfall totals.
  (b) Find an interval that will include the rainfall total for a selected four-week period with probability at least 0.75.

4.46 Annual incomes for engineers in a certain industry have approximately a gamma distribution with $\alpha = 600$ and $\beta = 50$.
  (a) Find the mean and variance of these incomes.
  (b) Would you expect to find many engineers in this industry with an annual income exceeding $35,000?

4.47 The weekly downtime Y (in hours) for a certain industrial machine has approximately a gamma distribution with $\alpha = 3$ and $\beta = 2$. The loss, in dollars, to the industrial operation as a result of this downtime is given by

$$L = 30Y + 2Y^2$$

  (a) Find the expected value and variance of L.
  (b) Find an interval that will contain L on approximately 89% of the weeks that the machine is in use.

4.48 Customers arrive at a checkout counter according to a Poisson process with a rate of 2 per minute. Find the mean, variance, and probability density function of the waiting time

between the opening of the counter and
- (a) the arrival of the second customer.
- (b) the arrival of the third customer.

4.49 Suppose two houses are to be built and each will involve the completion of a certain key task. The task has an exponentially distributed time to completion with a mean of 10 hours. Assuming the completion times are independent for the two houses, find the expected value and variance of
- (a) the total time to complete both tasks.
- (b) the average time to complete the two tasks.

4.50 The total sustained load on the concrete footing of a planned building is the sum of the dead load plus the occupancy load. Suppose the dead load X_1 has a gamma distribution with $\alpha_1 = 50$ and $\beta_1 = 2$ while the occupancy load X_2 has a gamma distribution with $\alpha_2 = 20$ and $\beta_2 = 2$. (Units are in kips, or thousands of pounds.)
- (a) Find the mean, variance, and probability density function of the total sustained load on the footing.
- (b) Find a value for the sustained load that should be exceeded only with probability less than 1/16.

4.51 A forty-year history of maximum river flows for a certain small river in the United States shows a relative frequency histogram that can be modeled by a gamma density function with $\alpha = 1.6$ and $\beta = 150$. (Measurements are in cubic feet per second.)
- (a) Find the mean and standard deviation of the annual maximum river flows.
- (b) Within what interval will the maximum annual flow be contained with probability at least 8/9?

4.52 The time intervals between dial-up connections to a computer center from remote terminals are exponentially distributed with a mean of 15 seconds. Find the mean, variance, and probability distribution of the waiting time from the opening of the computer center until the fourth dial-up connection from a remote terminal.

4.53 If service times at a teller window of a bank are exponentially distributed with a mean of 3.2 minutes, find the probability distribution, mean, and variance of the time taken to serve three waiting customers.

4.54 The response times at an online terminal have approximately a gamma distribution with a mean of 4 seconds and a variance of 8. Write the probability density function for these response times.

4.6 THE NORMAL DISTRIBUTION

Perhaps the most widely used of all the continuous probability distributions is the one referred to as the *normal distribution*. The normal probability density function has the familiar symmetric "bell" shape as indicated in Figure 4.12. The curve is centered at the mean value μ and its spread is, of course, measured by the variance σ^2. These two parameters, μ and σ^2, completely determine the shape and location of the normal density function, whose functional form is given by

$$f(x) = \frac{1}{\sqrt{2\pi}\sigma} e^{-(x-\mu)^2/2\sigma^2} \qquad -\infty < x < \infty$$

FIGURE 4.12
A Normal Density
Function

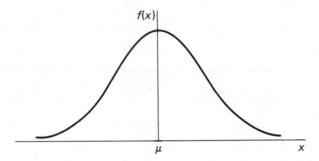

The basic reason that the normal distribution works well as a model for many different types of measurements generated in real experiments will be discussed in some detail in Chapter 7. For now we simply say that any time responses tend to be averages of independent quantities, the normal distribution will quite likely provide a reasonably good model for their relative frequency behavior. Many naturally occurring measurements tend to have relative frequency distributions closely resembling the normal curve, probably because nature tends to "average out" the effects of the many variables that relate to a particular response. For example, heights of adult American males tend to have a distribution that shows many measurements clumped closely about a mean height, with relatively few very short or very tall males in the population. In other words the relative frequency distribution is close to normal.

In contrast, lifelengths of biological organisms or electronic components tend to have relative frequency distributions that are not normal, or close to normal. This often results from the fact that lifelength measurements are a product of "extreme" behavior, not "average" behavior. A component may fail because of one extremely hard shock rather than the average effect of many shocks. Thus the normal distribution is not often used to model lifelengths, and consequently we will not discuss the failure rate function for this distribution.

A naturally occurring example of the normal distribution is seen in Michelson's measures of the speed of light. A histogram of these measurements is given in Figure 4.13. The distribution is not perfectly symmetrical but still exhibits an approximately normal shape.

A very important property of the normal distribution, which will be proved in Section 4.9, is that any linear function of a normally distributed random variable is also normally distributed. That is, if X has a normal distribution with mean μ and variance σ^2, and $Y = aX + b$ for constants a and b, then Y is also normally distributed. It is easily seen that

$$E(Y) = a\mu + b \qquad V(Y) = a^2\sigma^2$$

Suppose Z has a normal distribution with $\mu = 0$ and $\sigma = 1$. This random variable Z is said to have a *standard normal distribution*. Direct integration will show that $E(Z) = 0$ and $V(Z) = 1$. We have

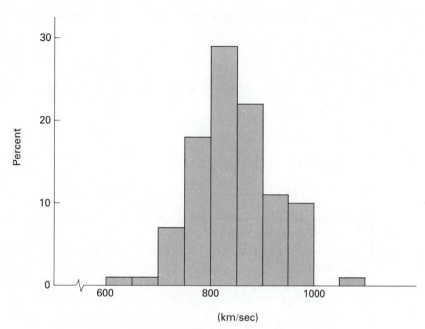

FIGURE 4.13
Michelson's 100 Measures of the Speed of Light in Air (−299,000)

Source: Astronomical Papers, *1881, p. 231.*

$$E(Z) = \int_{-\infty}^{\infty} z \frac{1}{\sqrt{2\pi}} e^{-z^2/2} \, dz$$

$$= \frac{1}{\sqrt{2\pi}} \int_{-\infty}^{\infty} e^{-z^2/2} z \, dz$$

$$= \frac{1}{\sqrt{2\pi}} [-e^{-z^2/2}]_{-\infty}^{\infty} = 0$$

Similarly

$$E(Z^2) = \int_{-\infty}^{\infty} \frac{1}{\sqrt{2\pi}} z^2 e^{-z^2/2} \, dz$$

$$= \frac{1}{\sqrt{2\pi}} (2) \int_{0}^{\infty} z^2 e^{-z^2/2} \, dz$$

On making the transformation $u = z^2$, the integral becomes

$$\frac{1}{\sqrt{2\pi}} \int_{0}^{\infty} u^{1/2} e^{-u/2} \, du = \frac{1}{\sqrt{2\pi}} \Gamma\left(\frac{3}{2}\right)(2)^{3/2}$$

$$= \frac{1}{\sqrt{2\pi}} (2)^{3/2} \left(\frac{1}{2}\right)\Gamma\left(\frac{1}{2}\right) = 1$$

since $\Gamma(\frac{1}{2}) = \sqrt{\pi}$. Therefore $E(Z) = 0$ and $V(Z) = E(Z^2) - \mu^2 = E(Z^2) = 1$.

For any normally distributed random variable X, with parameters μ and σ^2,

$$Z = \frac{X - \mu}{\sigma}$$

will have a standard normal distribution. Then

$$X = Z\sigma + \mu$$

$$E(X) = \sigma E(Z) - \mu = \mu$$

and

$$V(X) = \sigma^2 V(Z) = \sigma^2$$

This shows that the parameters μ and σ^2 do, indeed, measure the mean and variance of the distribution.

Since any normally distributed random variable can be transformed to the standard normal, probabilities can be evaluated for any normal distribution simply by having a table of standard normal integrals available. Such a table is given in Table 4 of the Appendix. Table 4 gives numerical values for

$$P(0 \leqslant Z \leqslant z) = \int_0^z \frac{1}{\sqrt{2\pi}} e^{-x^2/2} \, dx$$

Values of the integral are given for z between 0.00 and 3.09.

We will now show how to use Table 4 to find $P(-0.5 \leqslant Z \leqslant 1.5)$ for a standard normal variable Z. Figure 4.14 will aid in visualizing the necessary areas.

FIGURE 4.14
A Standard Normal Density Function

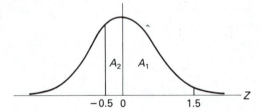

We first must write the probability in terms of intervals to the left and right of zero, the mean of the distribution. This produces

$$P(-0.5 \leqslant Z \leqslant 1.5) = P(0 \leqslant Z \leqslant 1.5) + P(-0.5 \leqslant Z \leqslant 0)$$

Now $P(0 \leqslant Z \leqslant 1.5) = A_1$ in Figure 4.14 and is found by looking up $z = 1.5$ in Table 4. The result is $A_1 = 0.4332$. Similarly $P(-0.5 \leqslant Z \leqslant 0) = A_2$ in Figure 4.14 and is found by looking up $z = 0.5$ in Table 4. Areas under the standard normal curve for negative z values are equal to those for corresponding positive z values since the curve is symmetric around zero. We find $A_2 = 0.1915$. It follows that

$$P(-0.5 \leqslant Z \leqslant 1.5) = A_1 + A_2 = 0.4332 + 0.1915 = 0.6247$$

EXAMPLE 4.12 _____

If Z denotes a standard normal variable, find

(a) $P(Z \leqslant 1)$　　　　　　　　　　　　　(b) $P(Z > 1)$

(c) $P(Z < -1.5)$　　　　　　　　　　　　(d) $P(-1.5 \leqslant Z \leqslant 0.5)$

Also find a value of z, say z_0, such that $P(0 \leqslant Z \leqslant z_0) = 0.49$.

Solution   This example provides practice in using Table 4. We see that

(a) $P(Z \leqslant 1) = P(Z \leqslant 0) + P(0 \leqslant Z \leqslant 1)$

$$= 0.5 + 0.3413 = 0.8413$$

.49

(b) $P(Z > 1) = 0.5 - P(0 \leqslant Z \leqslant 1)$

$$= 0.5 - 0.3413 = 0.1587$$

(c) $P(Z < -1.5) = P(Z > 1.5)$

$$= 0.5 - P(0 \leqslant Z \leqslant 1.5)$$

$$= 0.5 - 0.4332 = 0.0668$$

(d) $P(-1.5 \leqslant Z \leqslant 0.5) = P(-1.5 \leqslant Z \leqslant 0) + P(0 \leqslant Z \leqslant 0.5)$

$$= P(0 \leqslant Z \leqslant 1.5) + P(0 \leqslant Z \leqslant 0.5)$$

$$= 0.4332 + 0.1915$$

$$= 0.6247$$

To find the value of z_0, we must look for the given probability of 0.49 on the area side of Table 4. The closest we can come is at 0.4901, which corresponds to a z value of 2.33. Hence $z_0 = 2.33$.　■

The next example illustrates how the standardization works to allow Table 4 to be used for any normally distributed random variable.

EXAMPLE 4.13 _____

A firm that manufactures and bottles apple juice has a machine that automatically fills 16-ounce bottles. There is, however, some variation in the ounces of liquid dispensed into each bottle by the machine. Over a long period of time the average amount dispensed into the bottles was 16 ounces, but there is a standard deviation of one ounce in these measurements. If the ounces of fill per bottle can be assumed to be normally distributed, find the probability that the machine will dispense more than 17 ounces of liquid in any one bottle.

Solution   Let X denote the ounces of liquid dispensed into one bottle by the filling machine. Then X is assumed to be normally distributed with mean 16 and standard deviation 1. Hence

$$P(X > 17) = P\left(\frac{X - \mu}{\sigma} > \frac{17 - \mu}{\sigma}\right) = P\left(Z > \frac{17 - 16}{1}\right) = P(Z > 1) = 0.1587$$

The answer is found from Table 4 since $Z = (X - \mu)/\sigma$ has a *standard* normal distribution.　■

EXAMPLE 4.14 _____

Suppose that another machine, similar to the one of Example 4.13, operates so that ounces of fill have a mean equal to the dial setting for "amount of liquid," but have a standard deviation of 1.2 ounces. Find the proper setting for the dial so that 17-ounce bottles will overflow only 5% of the time. Assume that the amounts dispensed have a normal distribution.

Solution   Letting X denote the amount of liquid dispensed, we are now looking for μ value such that

$$P(X > 17) = 0.05$$

Now

$$P(X > 17) = P\left(\frac{X - \mu}{\sigma} > \frac{17 - \mu}{\sigma}\right)$$

$$= P\left(Z > \frac{17 - \mu}{1.2}\right)$$

From Table 4 we know that if

$$P(Z > z_0) = 0.05$$

then $z_0 = 1.645$. Thus it must be that

$$\frac{17 - \mu}{1.2} = 1.645$$

and

$$\mu = 17 - 1.2(1.645) = 15.026 \qquad\blacksquare$$

As mentioned earlier, we will make much more use of the normal distribution in later chapters. The properties of the normal distribution are summarized here.

THE NORMAL DISTRIBUTION _____

$$f(x) = \frac{1}{\sqrt{2\pi}\sigma} e^{-(x-\mu)^2/2\sigma^2} \qquad -\infty < x < \infty$$

$$E(X) = \mu \qquad V(X) = \sigma^2$$

EXERCISES ▐▬▬▬▬▬▬▬▬▬▬▬▬▬▬▬▬▬▬▬▬▬▬▬▬▬▬

4.55   Use Table 4 of the Appendix to find the following probabilities for a standard normal random variable Z.

    (a) $P(0 \leqslant Z \leqslant 1.2)$ (b) $P(-0.9 \leqslant Z \leqslant 0)$

     (c) $P(0.3 \leqslant Z \leqslant 1.56)$ (d) $P(-0.2 \leqslant Z \leqslant 0.2)$
     (e) $P(-2.00 \leqslant Z \leqslant -1.56)$

4.56   For a standard normal random variable Z, use Table 4 of the Appendix to find a number z_0 such that
     (a) $P(Z \leqslant z_0) = 0.5$ (b) $P(Z \leqslant z_0) = 0.8749$
     (c) $P(Z \geqslant z_0) = 0.117$ (d) $P(Z \geqslant z_0) = 0.617$
     (e) $P(-z_0 \leqslant Z \leqslant z_0) = 0.90$ (f) $P(-z_0 \leqslant Z \leqslant z_0) = 0.95$

4.57   The weekly amount spent for maintenance and repairs in a certain company has approximately a normal distribution with a mean of $400 and a standard deviation of $20. If $450 is budgeted to cover repairs for next week, what is the probability that the actual costs will exceed the budgeted amount?

4.58   In the setting of Exercise 4.57 how much should be budgeted weekly for maintenance and repairs so that the budgeted amount will be exceeded with probability only 0.1?

4.59   A machining operation produces steel shafts having diameters that are normally distributed with a mean of 1.005 inches and a standard deviation of 0.01 inch. Specifications call for diameters to fall within the interval 1.00 ± 0.02 inches. What percentage of the output of this operation will fail to meet specifications?

4.60   Refer to Exercise 4.59. What should be the mean diameter of the shafts produced to minimize the fraction not meeting specifications?

4.61   Wires manufactured for use in a certain computer system are specified to have resistances between 0.12 and 0.14 ohm. The actual measured resistances of the wires produced by Company A have a normal probability distribution with a mean of 0.13 ohm and a standard deviation of 0.005 ohm.
     (a)   What is the probability that a randomly selected wire from Company A's production will meet the specifications?
     (b)   If four such wires are used in the system and all are selected from Company A, what is the probability that all four will meet the specifications?

4.62   At a temperature of 25°C the resistances of a type of thermistor are normally distributed with a mean of 10,000 ohms and a standard deviation of 4000 ohms. The thermistors are to be sorted, with those having resistances between 8,000 and 15,000 ohms being shipped to a vendor. What fraction of these thermistors will be shipped?

4.63   A vehicle driver gauges the relative speed of a preceding vehicle by the speed with which the image of the width of that vehicle varies. This speed is proportional to the speed X of variation of the angle at which the eye subtends this width. According to P. Ferrani and others (*Transportation Research*, 18A, a, 1984, pp. 50, 51), a study of many drivers revealed X to be normally distributed with a mean of zero and a standard deviation of 10 (10^{-4} radian per second). What fraction of these measurements is more than 5 units away from zero? What fraction is more than 10 units away from zero?

4.64   A type of capacitor has resistances that vary according to a normal distribution with a mean of 800 megohms and a standard deviation of 200 megohms. (See W. Nelson, *Industrial Quality Control*, 1967, pp. 261–268, for a more thorough discussion.) A certain application specifies capacitors with resistances between 900 and 1000 megohms.
     (a)   What proportion of these capacitors will meet this specification?
     (b)   If two capacitors are randomly chosen from a lot of capacitors of this type, what is the probability that both will satisfy the specification?

4.65 Sick leave time used by employees of a firm in one month has approximately a normal distribution with a mean of 200 hours and a variance of 400.
(a) Find the probability that total sick leave for next month will be less than 150 hours.
(b) In planning schedules for next month, how much time should be budgeted for sick leave if that amount is to be exceeded with a probability of only 0.10?

4.66 The times of first failure of a unit of a brand of ink jet printers are approximately normally distributed with a mean of 1500 hours and a standard deviation of 200 hours.
(a) What fraction of these printers will fail before 1000 hours?
(b) What should be the guarantee time for these printers if the manufacturer wants only 5% to fail within the guarantee period?

4.67 A machine for filling cereal boxes has a standard deviation of 1 ounce on ounces of fill per box. What setting of the mean ounces of fill per box will allow 16-ounce boxes to overflow only 1% of the time? Assume that the ounces of fill per box are normally distributed.

4.68 Refer to Exercise 4.67. Suppose the standard deviation σ is not known but can be fixed at certain levels by carefully adjusting the machine. What is the largest value of σ that will allow the actual value dispensed to be within one ounce of the mean with probability at least 0.95?

4.7 THE BETA DISTRIBUTION

Except for the uniform distribution of Section 4.3 the continuous distributions discussed thus far are defined as nonzero functions over an infinite interval. It is of value to have at our disposal another class of distributions that can be used to model phenomena constrained to a finite interval of possible values. One such class, the beta distribution, is very useful for modeling the probabilistic behavior of certain random variables, such as proportions, constrained to fall in the interval $(0, 1)$. The beta distribution has the functional form

$$f(x) = \frac{\Gamma(\alpha + \beta)}{\Gamma(\alpha)\Gamma(\beta)} x^{\alpha - 1}(1 - x)^{\beta - 1} \qquad 0 < x < 1$$

$$= 0 \qquad\qquad\qquad\qquad \text{elsewhere}$$

where α and β are positive constants. The constant term in $f(x)$ is necessary so that

$$\int_0^1 f(x)\, dx = 1$$

In other words

$$\int_0^1 x^{\alpha - 1}(1 - x)^{\beta - 1}\, dx = \frac{\Gamma(\alpha)\Gamma(\beta)}{\Gamma(\alpha + \beta)}$$

for positive α and β. This is a handy result to keep in mind. The graphs of some common beta density functions are shown in Figure 4.15.

One measurement of interest in the process of sintering copper is the proportion of the volume that is solid, as opposed to the proportion made up of

FIGURE 4.15
*The Beta Density
Function*

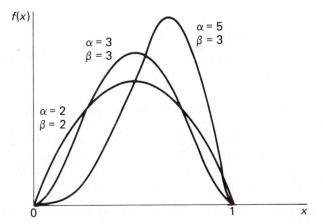

voids. (The proportion due to voids is sometimes called the *porosity* of the solid.) Figure 4.16 shows a relative frequency histogram of proportions of solid copper in samples from a sintering process. This distribution could be modeled with a beta distribution having a large α and small β.

FIGURE 4.16
*Solid Mass in
Sintered Linde
Copper*

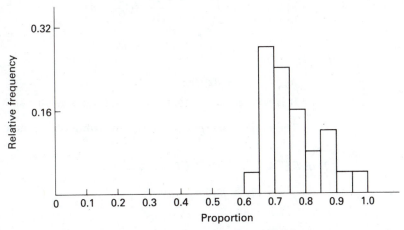

Source: Department of Materials Science, University of Florida

The expected value of a beta random variable is easily found since

$$E(X) = \int_0^1 x \frac{\Gamma(\alpha + \beta)}{\Gamma(\alpha)\Gamma(\beta)} x^{\alpha - 1}(1 - x)^{\beta - 1} dx$$

$$= \frac{\Gamma(\alpha + \beta)}{\Gamma(\alpha)\Gamma(\beta)} \int_0^1 x^{\alpha}(1 - x)^{\beta - 1} dx$$

$$= \frac{\Gamma(\alpha + \beta)}{\Gamma(\alpha)\Gamma(\beta)} \times \frac{\Gamma(\alpha + 1)\Gamma(\beta)}{\Gamma(\alpha + \beta + 1)}$$

$$= \frac{\alpha}{\alpha + \beta}$$

[Recall that $\Gamma(n + 1) = n\Gamma(n)$.] Similar manipulations reveal that

$$V(X) = \frac{\alpha\beta}{(\alpha + \beta)^2(\alpha + \beta + 1)}$$

We illustrate the use of this density function in an example.

EXAMPLE 4.15

A gasoline wholesale distributor has bulk storage tanks holding a fixed supply. The tanks are filled every Monday. Of interest to the wholesaler is the proportion of this supply that is sold during the week. Over many weeks this proportion has been observed to be modeled fairly well by a beta distribution with $\alpha = 4$ and $\beta = 2$. Find the expected value of this proportion. Is it highly likely that the wholesaler will sell at least 90% of his stock in a given week?

Solution By the results given above, with X denoting the proportion of the total supply sold in a given week,

$$E(X) = \frac{\alpha}{\alpha + \beta} = \frac{4}{6} = \frac{2}{3}$$

For the second part we are interested in

$$P(X > 0.9) = \int_{0.9}^{1} \frac{\Gamma(4 + 2)}{\Gamma(4)\Gamma(2)} x^3(1 - x)dx$$

$$= 20 \int_{0.9}^{1} (x^3 - x^4)dx$$

$$= 20(0.004) = 0.08$$

It is not very likely that 90% of the stock will be sold in a given week. ■

The basic properties of the beta distribution are summarized here.

THE BETA DISTRIBUTION

$$f(x) = \frac{\Gamma(\alpha + \beta)}{\Gamma(\alpha)\Gamma(\beta)} x^{\alpha - 1}(1 - x)^{\beta - 1} \qquad 0 < x < 1$$

$$= 0 \qquad\qquad\qquad\qquad \text{elsewhere}$$

$$E(X) = \frac{\alpha}{\alpha + \beta} \qquad\qquad V(X) = \frac{\alpha\beta}{(\alpha + \beta)^2(\alpha + \beta + 1)}$$

EXERCISES

4.69 Suppose X has a probability density function given by

$$f(x) = \begin{cases} kx^3(1 - x)^2 & 0 \leq x \leq 1 \\ 0 & \text{elsewhere} \end{cases}$$

(a) Find the value of k that makes this a probability density function.

(b) Find $E(X)$ and $V(X)$.

4.70 If X has a beta distribution with parameters α and β, show that

$$V(X) = \frac{\alpha\beta}{(\alpha + \beta)^2(\alpha + \beta + 1)}$$

4.71 During an eight-hour shift, the proportion of time X that a sheet-metal stamping machine is down for maintenance or repairs has a beta distribution with $\alpha = 1$ and $\beta = 2$. That is,

$$f(x) = \begin{cases} 2(1 - x) & 0 \leqslant x \leqslant 1 \\ 0 & \text{elsewhere} \end{cases}$$

The cost (in hundreds of dollars) of this downtime, due to lost production and cost of maintenance and repair, is given by

$$C = 10 + 20X + 4X^2$$

(a) Find the mean and variance of C.

(b) Find an interval in which C will lie with probability at least 0.75.

4.72 The percentage of impurities per batch in a certain type of industrial chemical is a random variable X having the probability density function

$$f(x) = \begin{cases} 12x^2(1 - x) & 0 \leqslant x \leqslant 1 \\ 0 & \text{elsewhere} \end{cases}$$

(a) Suppose a batch with more than 40% impurities cannot be sold. What is the probability that a randomly selected batch will not be allowed to be sold?

(b) Suppose the dollar value of each batch is given by

$$V = 5 - 0.5X$$

Find the expected value and variance of V.

4.73 To study the disposal of pollutants emerging from a power plant, the prevailing wind direction was measured for a large number of days. The direction is measured on a scale of $0°$ to $360°$, but by dividing each daily direction by 360, the measurements can be rescaled to the interval $(0, 1)$. These rescaled measurements X are found to follow a beta distribution with $\alpha = 4$ and $\beta = 2$. Find $E(X)$. To what angle does this mean correspond?

4.74 Errors in measuring the arrival time of a wave front from an acoustic source can sometimes be modeled by a beta distribution. (See J. J. Perruzzi and E. J. Hilliard, *Journal of the Acoustical Society of America*, 75(1), 1984, p. 197.)

Suppose these errors have a beta distribution with $\alpha = 1$ and $\beta = 2$, with measurements in microseconds.

(a) Find the probability that such a measurement error will be less than 0.5 microsecond.

(b) Find the mean and standard deviation of these error measurements.

4.75 In blending fine and coarse powders prior to copper sintering, proper blending is necessary for uniformity in the finished product. One way to check the blending is to select many small samples of the blended powders and measure the weight fractions of the fine particles. These measurements should be relatively constant if good blending has been achieved.

    (a)   Suppose the weight fractions have a beta distribution with $\alpha = \beta = 3$. Find their mean and variance.

    (b)   Repeat (a) for $\alpha = \beta = 2$.

    (c)   Repeat (a) for $\alpha = \beta = 1$.

    (d)   Which of the three cases, (a), (b), (c), would exemplify the best blending?

4.76   The proportion of pure iron in certain ore samples has a beta distribution with $\alpha = 3$ and $\beta = 1$.

    (a)   Find the probability that one of these samples will have more than 50% pure iron.

    (b)   Find the probability that two out of three samples will have less than 30% pure iron.

4.8   THE WEIBULL DISTRIBUTION

We have suggested that the gamma distribution can often serve as a probabilistic model for lifetimes of components or systems. However, the failure rate function for the gamma distribution has an upper bound that limits its applicability to real systems. For this and other reasons other distributions often provide better models for lifelength data. One such distribution is the *Weibull*, which will be explored in this section.

    A Weibull density function has the form

$$f(x) = \frac{\gamma}{\theta} x^{\gamma - 1} e^{-x^{\gamma}/\theta} \qquad x > 0$$

$$= 0 \qquad\qquad\qquad \text{elsewhere}$$

for positive parameters θ and γ. For $\gamma = 1$ this becomes an exponential density. For $\gamma > 1$ the functions look something like the gamma functions of Section 4.5, but have somewhat different mathematical properties. We can integrate directly to see that

$$F(x) = \int_0^x \frac{\gamma}{\theta} t^{\gamma - 1} e^{-t^{\gamma}/\theta} \, dt$$

$$= -e^{-t^{\gamma}/\theta}\big|_0^1 = 1 - e^{-x^{\gamma}/\theta} \qquad x > 0$$

    A convenient way to look at properties of the Weibull density is to use the transformation $Y = X^{\gamma}$. Then

$$F_Y(y) = P(Y \leqslant y) = P(X^{\gamma} \leqslant y) = P(X \leqslant y^{1/\gamma})$$

$$= F_X(y^{1/\gamma}) = 1 - e^{-(y^{1/\gamma})^{\gamma}/\theta}$$

$$= 1 - e^{-y/\theta} \qquad y > 0$$

Hence

$$f_Y(y) = \frac{dF_Y(y)}{dy} = \frac{1}{\theta} e^{-y/\theta} \qquad y > 0$$

and Y has the familiar exponential density.

If we want to find $E(X)$ for an X having the Weibull distribution, then

$$E(X) = E(Y^{1/\gamma}) = \int_0^\infty y^{1/\gamma} \frac{1}{\theta} e^{-y/\theta} dy$$

$$= \frac{1}{\theta} \int_0^\infty y^{1/\gamma} e^{-y/\theta} dy$$

$$= \frac{1}{\theta} \Gamma\left(1 + \frac{1}{\gamma}\right) \theta^{(1 + 1/\gamma)} = \theta^{1/\gamma} \Gamma\left(1 + \frac{1}{\gamma}\right)$$

The above result follows from recognizing the integral to be of the gamma type.

If we let $\gamma = 2$ in the Weibull density, we see that $Y = X^2$ has an exponential distribution. To reverse the idea outlined above, if we start with an exponentially distributed random variable Y, then the square root of Y will have a Weibull distribution with $\gamma = 2$. We can illustrate empirically by taking the square roots of the data from an exponential distribution, given in Table 4.1. These square roots are given in Table 4.2.

TABLE 4.2
Square Roots of the Lifelengths of Table 4.1

0.637	0.828	2.186	1.313	2.868
1.531	1.184	1.228	0.542	1.493
0.733	0.484	2.006	1.823	1.709
2.256	1.207	1.032	0.880	0.872
2.364	0.719	1.802	1.526	1.032
1.601	0.715	1.668	2.535	0.914
0.152	0.474	1.230	1.793	1.952
1.826	1.525	0.577	2.741	0.984
1.868	1.709	1.274	0.578	2.119
1.126	1.305	1.623	1.360	0.431

A relative frequency histogram for these data is given in Figure 4.17. Notice that the exponential form has now disappeared and that the curve given by the Weibull density with $\gamma = 2$, $\theta = 2$ (seen in Figure 4.18) is a much more plausible model for these observations.

The Weibull distribution is commonly used as a model for lifelengths because of the properties of its failure rate function. In this case

$$r(t) = \frac{f(t)}{1 - F(t)} = \frac{\dfrac{\gamma}{\theta} t^{\gamma-1} e^{-t^\gamma/\theta}}{e^{-t^\gamma/\theta}} = \frac{\gamma}{\theta} t^{\gamma-1}$$

which, for $\gamma > 1$, is a monotonically increasing function with no upper bound. By appropriate choice of γ this failure rate function can be made to model many sets of lifelength data seen in practice.

We can illustrate the use of the Weibull distribution in the following example.

FIGURE 4.17
*Relative Frequency
Histogram for the
Data of Table 4.2*

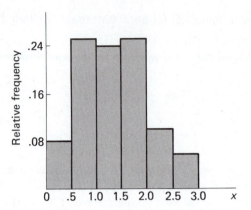

FIGURE 4.18
*The Weibull Density
Function $\gamma = 2$,
$\theta = 2$*

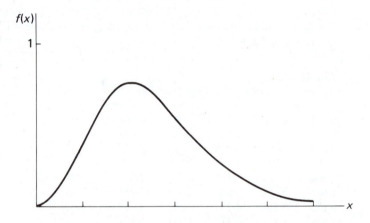

EXAMPLE 4.16

The length of service time during which a certain type of thermistor produces resistances within its specifications has been observed to follow a Weibull distribution with $\theta = 50$ and $\gamma = 2$ (measurements in thousands of hours).

(a) Find the probability that one of these thermistors, to be installed in a system today, will function properly for over 10,000 hours.

(b) Find the expected lifelength for thermistors of this type.

Solution

(a) The Weibull distribution has a closed-form expression for $F(x)$. Thus if X represents the lifelength of the thermistor in question,

$$P(X > 10) = 1 - F(10)$$

$$= 1 - [1 - e^{-(10)^2/50}]$$

$$= e^{-(10)^2/50} = e^{-2} = 0.14$$

since $\theta = 50$ and $\gamma = 2$.

(b)   We know from above that

$$E(X) = \theta^{1/\gamma}\Gamma\left(1 + \frac{1}{\gamma}\right)$$

$$= (50)^{1/2}\Gamma\left(\frac{3}{2}\right)$$

$$= (50)^{1/2}\,\frac{1}{2}\,\Gamma\left(\frac{1}{2}\right)$$

$$= (50)^{1/2}\left(\frac{1}{2}\right)\sqrt{\pi} = 6.27$$

Thus the average service time for these thermistors is 6270 hours.   ■

THE WEIBULL DISTRIBUTION

$$f(x) = \frac{\gamma}{\theta}\,x^{\gamma-1}e^{-x^\gamma/\theta} \qquad x > 0$$

$$= 0 \qquad\qquad\qquad \text{elsewhere}$$

$$E(X) = \theta^{1/\gamma}\Gamma\left(1 + \frac{1}{\gamma}\right)$$

$$V(X) = \theta^{2/\gamma}\left\{\Gamma\left(1 + \frac{2}{\gamma}\right) - \left[\Gamma\left(1 + \frac{1}{\gamma}\right)\right]^2\right\}$$

EXERCISES

4.77   Fatigue life, in hundreds of hours, for a certain type of bearings has approximately a Weibull distribution with $\gamma = 2$ and $\theta = 4$.
  (a)   Find the probability that a bearing of this type fails in less than 200 hours.
  (b)   Find the expected value of the fatigue life for these bearings.

4.78   The maximum flood levels, in millions of cubic feet per second, for a certain U.S. river have a Weibull distribution with $\gamma = 1.5$ and $\theta = 0.6$. (See Cohen, Whitten, and Ding, *Journal of Quality Technology*, 16(3), 1984, p. 165, for more details.) Find the probability that the maximum flood level for next year
  (a)   will exceed 0.5.
  (b)   will be less than 0.8.

4.79   The time necessary to achieve proper blending of copper powders before sintering was found to have a Weibull distribution with $\gamma = 1.1$ and $\theta = 2$ (measurements in minutes). Find the probability that a proper blending takes less than 2 minutes.

4.80   The ultimate tensile strength of steel wire used to wrap concrete pipe was found to have a Weibull distribution with $\gamma = 1.2$ and $\theta = 270$ (measurements in thousands of pounds).

Pressure in the pipe at a certain point may require an ultimate tensile strength of at least 300,000 pounds. What is the probability that a wire will possess this required strength?

4.81 The yield strengths of certain steel beams have a Weibull distribution with $\gamma = 2$ and $\theta = 3600$ (measurements in pounds per square inch). Two such beams are used in a construction project, which calls for yield strengths in excess of 70,000 psi. Find the probability that both beams meet the specifications for the project.

4.82 The pressure, in thousand psi, exerted on the tank of a steam boiler has a Weibull distribution with $\gamma = 1.8$ and $\theta = 1.5$. The tank is built to withstand pressures of 2000 psi. Find the probability that this limit will be exceeded.

4.83 Resistors being used in the construction of an aircraft guidance system have lifelengths that follow a Weibull distribution with $\gamma = 2$ and $\theta = 10$, with measurements in thousands of hours.
    (a) Find the probability that a randomly selected resistor of this type has a lifelength that exceeds 5000 hours.
    (b) If three resistors of this type are operating independently, find the probability that exactly one of the three burns out prior to 5000 hours of use.

4.84 Refer to Exercise 4.83. Find the mean and variance of the lifelength of a resistor of the type discussed.

4.85 Maximum wind-gust velocities in summer thunderstorms were found to follow a Weibull distribution with $\gamma = 2$ and $\theta = 400$ (measurements in feet per second). Engineers designing structures in the areas in which these thunderstorms are found are interested in finding a gust velocity that will be exceeded only with probability 0.01. Find such a value.

4.86 The velocities of gas particles can be modeled by the Maxwell distribution, with probability density function given by

$$f(v) = 4\pi \left(\frac{m}{2\pi KT}\right)^{3/2} v^2 e^{-v^2(m/2KT)} \qquad v > 0$$

where m is the mass of the particle, K is Boltzmann's constant, and T is the absolute temperature.
    (a) Find the mean velocity of these particles.
    (b) The kinetic energy of a particle is given by $(1/2)mV^2$. Find the mean kinetic energy for a particle.

*4.9 MOMENT-GENERATING FUNCTIONS FOR CONTINUOUS RANDOM VARIABLES

As in the case of discrete distributions, the moment-generating functions of continuous random variables help in finding expected values and in identifying certain properties of probability distributions. We now present a short discussion of moment-generating functions for continuous random variables.

*Optional section.

The moment-generating function of a continuous random variable X with probability density function $f(x)$ is given by

$$M(t) = E(e^{tX}) = \int_{-\infty}^{\infty} e^{tx} f(x)\, dx$$

when the integral exists.

For the exponential distribution this becomes

$$M(t) = \int_{0}^{\infty} e^{tx} \frac{1}{\theta} e^{-x/\theta}\, dx$$

$$= \frac{1}{\theta} \int_{0}^{\infty} e^{-x(1/\theta - t)}\, dx$$

$$= \frac{1}{\theta} \int_{0}^{\infty} e^{-x(1 - \theta t)/\theta}\, dx$$

$$= \frac{1}{\theta} \Gamma(1)\left(\frac{\theta}{1 - \theta t}\right) = (1 - \theta t)^{-1}$$

Since $M(t)$ needs to exist only in a neighborhood of zero, t can be small enough to make $1 - \theta t$ positive.

We can now use $M(t)$ to find $E(X)$, since

$$M'(0) = E(X)$$

and for the exponential distribution we have

$$E(X) = M'(0) = [-(1 - \theta t)^{-2}(-\theta)]_{t=0}$$

$$= \theta$$

Similarly we could find $E(X^2)$ and then $V(X)$ by using the moment-generating function.

An argument analogous to the one used for the exponential distribution will show that, for the gamma distribution,

$$M(t) = (1 - \beta t)^{-\alpha}$$

From this we can see that, if X has a gamma distribution,

$$E(X) = M'(0) = [-\alpha(1 - \beta t)^{-\alpha-1}(-\beta)]_{t=0}$$

$$= \alpha\beta$$

$$E(X^2) = M^{(2)}(0) = [\alpha\beta(-\alpha - 1)(1 - \beta t)^{-\alpha-2}(-\beta)]_{t=0}$$

$$= \alpha(\alpha + 1)\beta^2$$

$$E(X^3) = M^{(3)}(0) = [\alpha(\alpha + 1)\beta^2(-\alpha - 2)(1 - \beta t)^{-\alpha-3}(-\beta)]_{t=0}$$

$$= \alpha(\alpha + 1)(\alpha + 2)\beta^3$$

and so on.

EXAMPLE 4.17 _____

The force h exerted by a mass m moving at velocity v is

$$h = \frac{mv^2}{2}$$

Consider a device that fires a serrated nail into concrete at a mean velocity of 500 feet per second, where V, the random velocity, possesses a density function

$$f(v) = \frac{v^3 e^{-v/b}}{b^4 \Gamma(4)} \qquad b = 500, \, v \geqslant 0$$

If each nail possesses mass m, find the expected force exerted by a nail.

Solution
$$E(H) = E\left(\frac{mV^2}{2}\right) = \frac{m}{2} E(V^2)$$

The density function for V is a gamma-type function with $\alpha = 4, \beta = 500$. Therefore

$$E(V^2) = \alpha(\alpha + 1)\beta^2$$
$$= (4)(5)(500)^2$$
$$= 5,000,000$$

and

$$E(H) = \frac{m}{2} E(V^2)$$

$$= \frac{m}{2} (5,000,000) = 2,500,000m \qquad \blacksquare$$

Two important properties of moment-generating functions are as follows:

1. If a random variable X has moment-generating function $M_X(t)$, then $Y = aX + b$, for constants a and b, has moment-generating function

$$M_Y(t) = e^{tb} M_X(at)$$

2. Moment-generating functions are unique. That is, if two random variables have the same moment-generating function, then they have the same probability distribution. If they have different moment-generating functions, they have different distributions.

Property 1 is easily shown (see Exercise 3.85). The proof of property 2 is beyond the scope of this textbook, but we make use of it in identifying distributions, as will be demonstrated below.

To see something of the usefulness of these properties, suppose X has a gamma distribution with parameters α and β. Then

$$M_X(t) = (1 - \beta t)^{-\alpha}$$

as shown above. Now let

$$Y = aX + b$$

for constants a and b. From property 1 above,

$$M_Y(t) = e^{tb}[1 - \beta(at)]^{-\alpha}$$
$$= e^{tb}[1 - (a\beta)t]^{-\alpha}$$

Since this moment-generating function is *not* of the gamma form (it has an extra multiplier of e^{tb}), Y will *not* have a gamma distribution (from property 2). If, however, $b = 0$ and $Y = aX$, then Y will have a gamma distribution with parameters α and $a\beta$.

The normal random variable is so commonly used that we should investigate its moment-generating function. We do so by first finding the moment-generating function of $X - \mu$, where X is normally distributed with mean μ and variance σ^2. Now

$$E(e^{t(X - \mu)}) = \int_{-\infty}^{\infty} e^{t(x - \mu)} \frac{1}{\sigma\sqrt{2\pi}} e^{-(x - \mu)^2/2\sigma^2} \, dx$$

Letting $y = x - \mu$, the integral becomes

$$\frac{1}{\sigma\sqrt{2\pi}} \int_{-\infty}^{\infty} e^{ty - y^2/2\sigma^2} = \frac{1}{\sigma\sqrt{2\pi}} \int_{-\infty}^{\infty} \exp\left[-\frac{1}{2\sigma^2}(y^2 - 2\sigma^2 ty)\right] dy$$

Upon the completing of the square in the exponent, we have

$$-\frac{1}{2\sigma^2}(y^2 - 2\sigma^2 ty) = -\frac{1}{2\sigma^2}(y^2 - 2\sigma^2 ty + \sigma^4 t^2) + \frac{1}{2}t^2\sigma^2$$

$$= -\frac{1}{2\sigma^2}(y - \sigma^2 t)^2 + \frac{1}{2}t^2\sigma^2$$

so that the integral becomes

$$e^{t^2\sigma^2/2} \frac{1}{\sigma\sqrt{2\pi}} \int_{-\infty}^{\infty} \exp\left[-\frac{1}{2\sigma^2}(y - \sigma^2 t)^2\right] dy = e^{t^2\sigma^2/2}$$

since the remaining integrand forms a normal probability density that integrates to unity.

If $Y = X - \mu$ has a moment-generating function given by

$$M_Y(t) = e^{t^2\sigma^2/2}$$

then

$$X = Y + \mu$$

has a moment-generating function given by

$$M_X(t) = e^{t\mu} M_Y(t)$$
$$= e^{t\mu} e^{t^2\sigma^2/2}$$
$$= e^{t\mu + t^2\sigma^2/2}$$

by property 1 stated above.

For this same normal random variable X, let

$$Z = \frac{X - \mu}{\sigma} = \frac{1}{\sigma} X - \frac{\mu}{\sigma}$$

Then by property 1

$$M_Z(t) = e^{-\mu t/\sigma} e^{\mu t/\sigma + t^2 \sigma^2/2\sigma^2}$$

$$= e^{t^2/2}$$

The normal generating function of Z has the form of a moment-generating function for a normal random variable with mean zero and variance one. Thus by property 2, Z must have that distribution.

EXERCISES

4.87 Show that a gamma distribution with parameters α and β has moment-generating function

$$M(t) = (1 - \beta t)^{-\alpha}$$

4.88 Using the moment-generating function for the exponential distribution with mean θ, find $E(X^2)$. Use this result to show that $V(X) = \theta^2$.

4.89 Let Z denote a standard normal random variable. Find the moment-generating function of Z directly from the definition.

4.90 Let Z denote a standard normal random variable. Find the moment-generating function of Z^2. What does the uniqueness property of the moment-generating function tell you about the distribution of Z^2?

---

*4.10 ACTIVITIES FOR THE COMPUTER

When generating random variables from a continuous distribution with a cumulative distribution function (CDF) $F(x)$, the inverse transformation method—or inverse CDF method—may be considered. Using this method, the three-step procedure is as follows:

1. Generate R_i, which is uniform over the interval $(0, 1)$. This is denoted as $R_i \sim U(0, 1)$.
2. Let $R_i = F(X)$, where $F(x)$ is the CDF for the distribution of x's.
3. Evaluate (2) for X, which gives $X = F^{-1}(R)$.

Uniform. The probability density function (pdf) for the uniform is defined as $f(x) = 1/(b - a)$, $a \leqslant x \leqslant b$. The CDF is given by $F(x) = (x - a)/(b - a)$. Using the inverse CDF method to generate uniform random numbers x_i, let $R_i = (x_i - a)/(b - a)$. Solving for x_i, we obtain $x_i = a + (b - a)R_i$.

Exponential. The pdf for the exponential is defined as $f(x) = (1/\theta)e^{-x/\theta}$, $x \geqslant 0$. The CDF is given by $F(x) = 1 - e^{-x/\theta}$. Using the inverse CDF method, one can generate exponential random numbers x_i by letting $R_i = 1 - e^{-x_i/\theta}$. Solving for x_i, $e^{-x_i/\theta} = 1 - R_i$. Taking the ln on both sides, we obtain $x_i = -\theta \ln(1 - R_i)$. Since $(1 - R_i) \sim U(0, 1)$, we let $x_i = -\theta \ln R_i$.

Gamma. If X_i is a random variable from the gamma distribution with parameters (α, β) and α is an integer n, then X_i is the *convolution* (sum) of n exponential random variables y_j with parameter β. Thus, to simulate a gamma random variable X_i, generate n exponential random variables as stated previously. Then let $X_i = \Sigma Y_j$, where $j = 1, 2, \ldots, n$.

Normal. In attempting to apply the inverse CDF method to generate normal random variables, a problem arises in that there is no explicit closed-form expression for the CDF of X, $F(x)$, or the inverse function, $F^{-1}(R)$. Methods of approximation have been developed that work well in simulating normal random variables, and one of these will be discussed in Chapter 6.

Let us now consider a situation involving exponential random variables. In engineering, one is often concerned with a *parallel system*. A system is defined as a parallel system if it functions when at least one of the components of the system is working. Let's consider a parallel system that has 2 components. We define the random variable $X = $ time until failure for component 1 and the random variable $Y = $ time until failure for component 2. X and Y are independent exponential random variables with the mean for $X = \theta_x$ and the mean for $Y = \theta_y$. A variable of interest is the maximum of X and Y; that is, let $W = \max\{X, Y\}$. By simulation we can see how the distribution of W will behave. Figure 4.19 gives the results of three simulations for this situation.

FIGURE 4.19
Simulations of the Distribution of Maximum Exponential Random Variables in a Parallel System

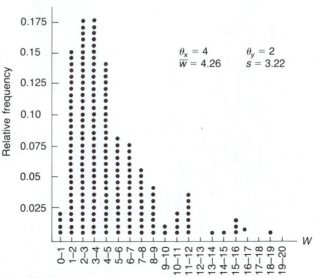

(a) Simulation 1

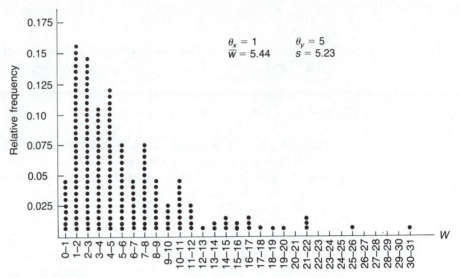

(b) Simulation 2

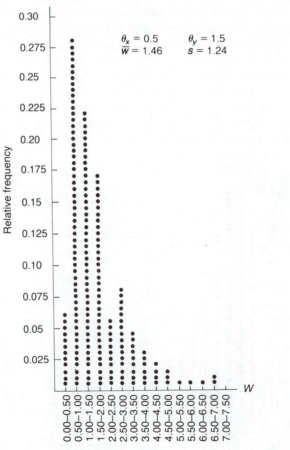

(c) Simulation 3

Note that the distribution of W does *not* look like an exponential distribution. The maximum has relatively low probability of being close to zero, but could be quite large compared to most values of X or Y.

SUPPLEMENTARY EXERCISES

4.91   Let Y possess a density function

$$f(y) = \begin{cases} cy & 0 \leqslant y \leqslant 2 \\ 0 & \text{elsewhere} \end{cases}$$

(a)   Find c.
(b)   Find $F(y)$.
(c)   Graph $f(y)$ and $F(y)$.
(d)   Use $F(y)$ in part (b) to find $P(1 \leqslant Y \leqslant 2)$.
(e)   Use the geometric figure for $f(y)$ to calculate $P(1 \leqslant Y \leqslant 2)$.

4.92   Let Y have the density function given by

$$f(y) = \begin{cases} cy^2 + y & 0 \leqslant y \leqslant 2 \\ 0 & \text{elsewhere} \end{cases}$$

(a)   Find c.
(b)   Find $F(y)$.
(c)   Graph $f(y)$ and $F(y)$.
(d)   Use $F(y)$ in part (b) to find $F(-1)$, $F(0)$, and $F(1)$.
(e)   Find $P(0 \leqslant Y \leqslant 0.5)$.
(f)   Find the mean and variance of Y.

4.93   Let Y have the density function given by

$$f(y) = \begin{cases} 0.2 & -1 < y \leqslant 0 \\ 0.2 + cy & 0 < y \leqslant 1 \\ 0 & \text{elsewhere} \end{cases}$$

Answer parts (a) through (f), Exercise 4.92.

4.94   The grade-point averages of a large population of college students are approximately normally distributed with mean equal to 2.4 and standard deviation equal to 0.5. What fraction of the students will possess a grade-point average in excess of 3.0?

4.95   Refer to Exercise 4.94. If students possessing a grade-point average equal to or less than 1.9 are dropped from college, what percentage of the students will be dropped?

4.96   Refer to Exercise 4.94. Suppose that three students are randomly selected from the student body. What is the probability that all three will possess a grade-point average in excess of 3.0?

4.97   A machine operation produces bearings with a diameter that is normally distributed with mean and standard deviation equal to 3.005 and 0.001, respectively. Customer specifications require the bearing diameters to lie in the interval 3.000 ± 0.0020. Those outside the interval are considered scrap and must be remachined or used as stock for smaller bearings. With the existing machine setting, what fraction of total production will be scrap?

4.98 Refer to Exercise 4.97. Suppose that five bearings are drawn from production. What is the probability that at least one will be defective?

4.99 Let Y have density function

$$f(y) = \begin{cases} cye^{-2y} & 0 \leqslant y \leqslant \infty \\ 0 & \text{elsewhere} \end{cases}$$

(a) Give the mean and variance for Y.
(b) Give the moment-generating function for Y.
(c) Find the value of c.

4.100 Find $E(X^k)$ for the beta random variable. Then find the mean and variance for the beta random variable.

4.101 The yield force of a steel reinforcing bar of a certain type is found to be normally distributed with a mean of 8500 pounds and a standard deviation of 80 pounds. If three such bars are to be used on a certain project, find the probability that all three will have yield forces in excess of 8700 pounds.

4.102 The lifetime X of a certain electronic component is a random variable with density function

$$f(x) = \begin{cases} (1/100)e^{-x/100} & x > 0 \\ 0 & \text{elsewhere} \end{cases}$$

Three of these components operate independently in a piece of equipment. The equipment fails if at least two of the components fail. Find the probability that the equipment operates for at least 200 hours without failure.

4.103 An engineer has observed that the gap times between vehicles passing a certain point on a highway have an exponential distribution with a mean of ten seconds.
(a) Find the probability that the next gap observed will be no longer than one minute.
(b) Find the probability density function for the sum of the next four gap times to be observed.
What assumptions are necessary for this answer to be correct?

4.104 The proportion of time, per day, that all checkout counters in a supermarket are busy is a random variable X having probability density function

$$f(x) = \begin{cases} kx^2(1 - x)^4 & 0 \leqslant x \leqslant 1 \\ 0 & \text{elsewhere} \end{cases}$$

(a) Find the value of k that makes this a probability density function.
(b) Find the mean and variance of X.

4.105 If the lifelength X for a certain type of battery has a Weibull distribution with $\gamma = 2$ and $\theta = 3$ (with measurements in years), find the probability that the battery lasts less than 4 years given that it is now 2 years old.

4.106 The time (in hours) it takes a manager to interview an applicant has an exponential distribution with $\theta = 1/2$. Three applicants arrive at 8:00 A.M. and interviews begin. A fourth applicant arrives at 8:45 A.M. What is the probability that he has to wait before seeing the manager?

4.107 The weekly repair cost Y for a certain machine has a probability density function given by

$$f(y) = \begin{cases} 3(1-y)^2 & 0 < y < 1 \\ 0 & \text{elsewhere} \end{cases}, \quad .$$

with measurements in hundreds of dollars. How much money should be budgeted each week for repair cost so that the actual cost will exceed the budgeted amount only 10 percent of the time?

4.108 A builder of houses has to order some supplies that have a waiting time for delivery Y uniformly distributed over the interval one to four days. Since he can get by without them for two days, the cost of the delay is a fixed $100 for any waiting time up to two days. However after two days the cost of the delay is $100 plus $20 per day for any time beyond two days. That is, if the waiting time is 3.5 days, the cost of the delay is $100 + \$29(1.5) = \$130$. Find the expected value of the builder's cost due to waiting for supplies.

4.109 There is a relationship between incomplete gamma integrals and sums of Poisson probabilities given by

$$\frac{1}{\Gamma(\alpha)} \int_{\lambda}^{\infty} y^{\alpha-1} e^{-y} dy = \sum_{y=0}^{\alpha-1} \frac{\lambda^y e^{-\lambda}}{y!}$$

for integer values of α.

If Y has a gamma distribution with $\alpha = 2$ and $\beta = 1$, find $P(Y > 1)$ by using the above equality and Table 3 of the Appendix.

4.110 The weekly downtime in hours for a certain production line has a gamma distribution with $\alpha = 3$ and $\beta = 2$. Find the probability that the downtime for a given week will not exceed 10 hours.

4.111 Suppose that plants of a certain species are randomly dispensed over a region, with a mean density of λ plants per unit area. That is, the number of plants in a region of area A has a Poisson distribution with mean λA. For a randomly selected plant in this region let R denote the distance to the nearest neighboring plant.
  (a) Find the probability density function for R. [*Hint:* Note that $P(R > r)$ is the same as the probability of seeing no plants in a circle of radius r.]
  (b) Find $E(R)$.

4.112 A random variable X is said to have a *lognormal* distribution if $Y = \ln(X)$ has a normal distribution. (The symbol "ln" denotes natural logarithm.) In this case X must be nonnegative. The shape of the lognormal probability density function is similar to that of the gamma distribution, with long tails to the right. The equation of the lognormal density function is

$$f(x) = \frac{1}{\sqrt{2\pi}\sigma x} e^{-(\ln(x)-\mu)^2/2\sigma^2} \qquad x > 0$$

$$= 0 \qquad\qquad\qquad\qquad \text{elsewhere}$$

Since $\ln(x)$ is a monotonic function of x,

$$P(X \leqslant x) = P[\ln(X) \leqslant \ln(x)] = P[Y \leqslant \ln(x)]$$

where Y has a normal distribution with mean μ and variance σ^2. Thus probabilities in the lognormal case can be found by transforming them to the normal case.

If X has a lognormal distribution with $\mu = 4$ and $\sigma^2 = 1$, find

(a) $P(X \leqslant 4)$ (b) $P(X > 8)$

4.113 If X has a lognormal distribution with parameters μ and σ^2, then it can be shown that

$$E(X) = e^{\mu + \sigma^2/2}$$

and

$$V(X) = e^{2\mu + \sigma^2}(e^{\sigma^2} - 1)$$

The grains comprising polycrystalline metals tend to have weights that follow a lognormal distribution. For a certain type of aluminum, gram weights have a lognormal distribution with $\mu = 3$ and $\sigma = 4$ (in units of 10^{-2} gram).

(a) Find the mean and variance of the grain weights.

(b) Find an interval in which at least 75% of the grain weights should lie. [*Hint:* use Tchebysheff's Theorem.]

(c) Find the probability that a randomly chosen grain weighs less than the mean gram weight.

*4.114 Let Y denote a random variable with probability density function given by

$$f(y) = (1/2)e^{-|y|} \qquad -\infty < y < \infty$$

Find the moment-generating function of Y and use it to find $E(Y)$.

5

Multivariate Probability Distributions

ABOUT THIS CHAPTER

In most experimental investigations multiple types of observations are made on the same experimental units. Tensile and torsion properties are both measured on the same sample of wire, or amount of cement and porosity are both measured on the same sample of concrete. This chapter provides techniques for modeling the joint behavior of two or more random variables being studied together.

CONTENTS

5.1 BIVARIATE AND MARGINAL PROBABILITY DISTRIBUTIONS

Chapters 3 and 4 dealt with experiments that produced a single numerical response, or random variable, of interest. We discussed, for example, the lifelengths X of a type of battery or the strength Y of a steel casing. Often, however, we want to study the joint behavior of two random variables, such as the joint behavior of lifelength *and* casing strength for these batteries. Perhaps in such a study we can identify a region in which there is a combination of lifelength and casing strength that will be optimal in terms of balancing cost of manufacturing with customer satisfaction. To proceed with such a study, we must know how to handle joint probability distributions. When only two random variables are involved, these joint distributions are called *bivariate distributions*. We will discuss the bivariate case in some detail; extensions to more than two variables follow along similar lines.

Other situations in which bivariate probability distributions are important come to mind easily. The physician studies the joint behavior of pulse and amount of exercise, or blood pressure and weight. The educator studies the joint behavior of grades and time devoted to study, or the interrelationship of pretest and posttest scores. The economist studies the joint behavior of business volume and profits. In fact most real problems we come across will have more than one underlying random variable of interest.

To simplify the basic ideas of joint distributions, we will consider a bivariate example in some detail. Suppose two light bulbs are to be drawn sequentially from a box containing four good bulbs and one defective bulb. The drawings are made without replacement; that is, the first one is not put back into the box before the second one is drawn. Let X_1 denote the number of defectives observed on the first draw and X_2 the number of defectives observed on the second draw. Note that X_1 and X_2 can take on only the values zero and one. How might we calculate the probability of the event ($X_1 = 0$ and $X_2 = 0$)? There are five choices for the bulb drawn first, and then only four choices for the bulb drawn second. Thus there are $5 \times 4 = 20$ possible ordered outcomes for the drawing of two bulbs. Since the draws were supposedly made at random, a reasonable probabilistic model for this experiment would be one that assigns equal probability of 1/20 to each of the twenty possible outcomes. Now how many of these 20 outcomes are in the event $X_1 = 0$, $X_2 = 0$? There are four good bulbs in the box, any one of which could occur on the first draw. After that one good bulb is removed, three are left for the second draw. Thus there are $4 \times 3 = 12$ outcomes that result in no defectives on either draw. The probability of the event in question is now apparent:

$$P(X_1 = 0, X_2 = 0) = \frac{12}{20} = 0.6$$

After similar calculations for the other possible values of X_1 and X_2, we can arrange the probabilities as in Table 5.1

TABLE 5.1
Joint Distribution of X_1 and X_2

X_2 \ X_1	0	1	Row Totals (marginal probabilities for X_2)
0	0.6	0.2	0.8
1	0.2	0	0.2
Column Totals (marginal probabilities for X_1)	0.8	0.2	1.0

The entries in Table 5.1 represent the joint probability distribution of X_1 and X_2. Sometimes we write

$$P(X_1 = x_1, X_2 = x_2) = p(x_1, x_2)$$

and call $p(x_1, x_2)$ the *joint probability function* of (X_1, X_2). From these probabilities we can extract the probabilities that represent the behavior of X_1, or X_2, by itself. Note that

$$P(X_1 = 0) = P(X_1 = 0, X_2 = 0) + P(X_1 = 0, X_2 = 1)$$

$$= 0.6 + 0.2 = 0.8$$

This entry (0.8) is found on Table 5.1 as the total of the entries in the first column (under $X_1 = 0$). In similar fashion one finds $P(X_1 = 1) = 0.2$, the total of the second-column entries. These values, 0.8 and 0.2, are placed conveniently on the margin of the table and hence are termed the *marginal probabilities* of X_1. The marginal probability distribution for X_2 is found by looking at the row totals, which show $P(X_2 = 0) = 0.8$ and $P(X_2 = 1) = 0.2$. The distributions considered in previous chapters could be thought of as marginal probability distributions.

In addition to joint and marginal probabilities we may be interested in conditional probabilities, such as the probability that the second bulb turns out to be good given that we have found the first one to be good, or $P(X_2 = 0 | X_1 = 0)$. Looking back at our probabilistic model for this experiment, we see that, after it is known that the first bulb is good, there are only three good bulbs among the four remaining. Thus we should have

$$P(X_2 = 0 | X_1 = 0) = \frac{3}{4}$$

The definition of conditional probability gives

$$P(X_2 = 0 | X_1 = 0) = \frac{P(X_1 = 0, X_2 = 0)}{P(X_1 = 0)}$$

$$= \frac{0.6}{0.8} = \frac{3}{4}$$

so the result agrees with our intuitive assessment above.

The bivariate discrete case is summarized in Definition 5.1.

DEFINITION 5.1	Let X_1 and X_2 be discrete random variables. The **joint probability distribution** of X_1 and X_2 is given by $$p(x_1, x_2) = P(X_1 = x_1, X_2 = x_2)$$ defined for all real numbers x_1 and x_2. The function $p(x_1, x_2)$ is called the **joint probability function** of X_1 and X_2. The **marginal probability functions** of X_1 and X_2, respectively, are given by $$p_1(x_1) = \sum_{x_2} p(x_1, x_2)$$ and $$p_2(x_2) = \sum_{x_1} p(x_1, x_2)$$

EXAMPLE 5.1 ───

There are three checkout counters in operation at a local supermarket. Two customers arrive at the counters at different times, when the counters are serving no other customers. It is assumed that the customers then choose a checkout station at random, and independently of one another. Let X_1 denote the number of times counter A is selected and X_2 the number of times counter B is selected by the two customers. Find the joint probability distribution of X_1 and X_2. Find the probability that one of the customers visits counter B given that one of the customers is known to have visited counter A.

Solution   For convenience let us introduce X_3, defined as the number of customers visiting C. Now the event $(X_1 = 0$ and $X_2 = 0)$ is equivalent to the event $(X_1 = 0, X_2 = 0,$ and $X_3 = 2)$. It follows that

$$P(X_1 = 0, X_2 = 0) = P(X_1 = 0, X_2 = 0, X_3 = 2)$$

$$= P(\text{customer I selects counter } C \text{ and customer II selects counter } C)$$

$$= P(\text{customer I selects counter } C) \times P(\text{customer II selects counter } C)$$

$$= \frac{1}{3} \times \frac{1}{3} = \frac{1}{9}$$

since customers' choices are independent and each customer makes a random selection from among the three available counters. The joint probability distribution is shown in Table 5.2.

It is slightly more complicated to calculate $P(X_1 = 1, X_2 = 0) = P(X_1 = 1, X_2 = 0, X_3 = 1)$. In this event customer I could select counter A and customer II could select counter C, or I could select C and II could select A. Thus

TABLE 5.2
Joint Distribution of X_1 and X_2 for Example 5.1

X_2 \ X_1	0	1	2	Marginal Probabilities for X_2
0	1/9	2/9	1/9	4/9
1	2/9	2/9	0	4/9
2	1/9	0	0	1/9
Marginal Probabilities for X_1	4/9	4/9	1/9	1

$$P(X_1 = 1, X_2 = 0) = P(\text{I selects } A)P(\text{II selects } C)$$
$$+ P(\text{I selects } C)P(\text{II selects } A)$$
$$= \frac{1}{3} \times \frac{1}{3} + \frac{1}{3} \times \frac{1}{3} = \frac{2}{9}$$

Similar arguments will allow one to derive the results in Table 5.1. The second statement asks for

$$P(X_2 = 1 | X_1 = 1) = \frac{P(X_1 = 1, X_2 = 1)}{P(X_1 = 1)}$$

$$\frac{2/9}{4/9} = \frac{1}{2}$$

Does this answer 1/2 agree with your intuition?   ■

As we move to the continuous case, let us first quickly review the situation in one dimension. If $f(x)$ denotes the probability density function of a random variable X, then $f(x)$ represents a relative frequency curve, and probabilities such as $P(a \leqslant X \leqslant b)$ are represented as areas under this curve. That is

$$P(a \leqslant X \leqslant b) = \int_a^b f(x) \, dx$$

Now suppose we are interested in the joint behavior of two continuous random variables, say X_1 and X_2, where X_1 and X_2 might, for example, represent the amounts of two different hydrocarbons found in an air sample taken for a pollution study. The relative frequency behavior of these two random variables can be modeled by a bivariate function, $f(x_1, x_2)$, which forms a probability, or relative frequency, surface in three dimensions. Figure 5.1 shows such a surface. The probability that X_1 lies in some interval and X_2 lies in another interval is then represented as a volume under this surface. Thus

$$P(a_1 \leqslant X_1 \leqslant a_2, b_1 \leqslant X_2 \leqslant b_2) = \int_{b_1}^{b_2} \int_{a_1}^{a_2} f(x_1, x_2) \, dx_1 \, dx_2$$

FIGURE 5.1
A Bivariate Density Function

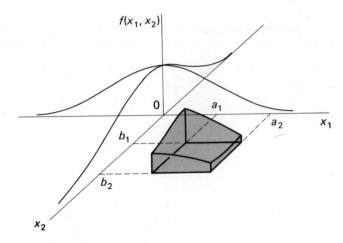

Note that the above integral simply gives the volume under the surface and over the shaded region in Figure 5.1.

 We will illustrate the actual computations involved in such a bivariate problem with a very simple example.

EXAMPLE 5.2

A certain process for producing an industrial chemical yields a product containing two predominant types of impurities. For a certain volume of sample from this process, let X_1 denote the proportion of impurities in the sample and let X_2 denote the proportion of type I impurity among all impurities found. Suppose the joint distribution of X_1 and X_2, after investigation of many such samples, can be adequately modeled by the following function:

$$f(x_1, x_2) = \begin{cases} 2(1 - x_1) & 0 \leqslant x_1 \leqslant 1, 0 \leqslant x_2 \leqslant 1 \\ 0 & \text{elsewhere} \end{cases}$$

This function graphs as the surface given in Figure 5.2. Calculate the probability that X_1 is less than 0.5 and that X_2 is between 0.4 and 0.7.

FIGURE 5.2
Probability Density Function for Example 5.2

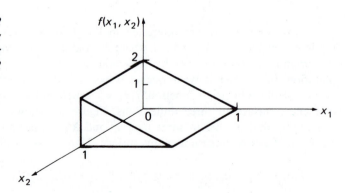

Solution   From the preceding discussion we see that

$$P(0 \leqslant x_1 \leqslant 0.5, 0.4 \leqslant x_2 \leqslant 0.7) = \int_{0.4}^{0.7} \int_0^{0.5} 2(1 - x_1) \, dx_1 \, dx_2$$

$$= \int_{0.4}^{0.7} [-(1 - x_1)^2]_0^{0.5} \, dx_2$$

$$= \int_{0.4}^{0.7} (0.75) \, dx_2$$

$$= 0.75(0.7 - 0.4) = 0.75(0.3) = 0.225$$

Thus the fraction of such samples having less than 50% impurities and a relative proportion of type I impurities between 40% and 70% is 0.225.   ■

Just as the univariate, or marginal, probabilities were computed by summing over rows or columns in the discrete case, the univariate density function for X_1 in the continuous case can be found by integrating ("summing") over values of X_2. Thus the *marginal density function of X_1—$f_1(x_1)$*—is given by

$$f_1(x_1) = \int_{-\infty}^{\infty} f(x_1, x_2) \, dx_2$$

Similarly the marginal density function of X_2—$f_2(x_2)$—is given by

$$f_2(x_2) = \int_{-\infty}^{\infty} f(x_1, x_2) \, dx_1$$

EXAMPLE 5.3 _____

For Example 5.2 find the marginal probability density functions for X_1 and X_2.

Solution   Let's first try to visualize what the answers should look like, before going through the integration. To find $f_1(x_1)$, we are to accumulate all probabilities in the x_2 direction. Look at Figure 5.2 and think of collapsing the wedge-shaped figure back onto the $(x_1, f_1(x_1, x_2))$ plane. Then much more probability mass will build up toward the zero point of the x_1-axis than toward the unity point. In other words the function $f_1(x_1)$ should be high at zero and low at one. Formally

$$f_1(x_1) = \int_{-\infty}^{\infty} f(x_1, x_2) dx_2$$

$$= \int_0^1 2(1 - x_1) \, dx_2$$

$$= 2(1 - x_1) \qquad 0 < x_1 \leqslant 1$$

The function graphs as in Figure 5.3. Note that our conjecture is correct. Thinking of how $f_2(x_2)$ should look geometrically, suppose the wedge of Figure 5.2 is forced back onto the $(x_2, f(x_1, x_2))$ plane. Then the probability mass should

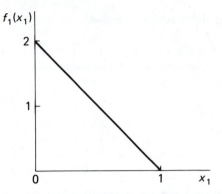

FIGURE 5.3
Probability Density
Function $f_1(x_1)$ for
Example 5.3

accumulate equally all along the $(0, 1)$ interval on the x_2-axis. Mathematically

$$f_2(x_2) = \int_{-\infty}^{\infty} f(x_1, x_2) \, dx_1$$

$$= \int_0^1 2(1 - x_1) \, dx_1$$

$$= [-(1 - x_1)]_0^1 \qquad 0 < x_2 \leqslant 1$$

$$= 1 \qquad 0 \leqslant x_2 \leqslant 1$$

and again our conjecture is verified, as seen in Figure 5.4.

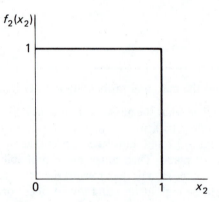

FIGURE 5.4
Probability Density
Function $f_2(x_2)$ for
Example 5.3

The bivariate continuous case is summarized in Definition 5.2.

DEFINITION 5.2   Let X_1 and X_2 be continuous random variables. The **joint probability distribution function** of X_1 and X_2, if it exists, is given by a nonnegative function $f(x_1, x_2)$, which is such that

$$P(a_1 \leqslant X_1 \leqslant a_2, b_1 \leqslant X_2 \leqslant b_2) = \int_{b_1}^{b_2} \int_{a_1}^{a_2} f(x_1, x_2) \, dx_1 \, dx_2$$

The **marginal probability density functions** of X_1 and X_2, respectively, are given by

$$f_1(x_1) = \int_{-\infty}^{\infty} f(x_1, x_2)\, dx_2$$

and

$$f_2(x_2) = \int_{-\infty}^{\infty} f(x_1, x_2)\, dx_1$$

As another (and somewhat more complicated) example, consider the following.

EXAMPLE 5.4

Gasoline is to be stocked in a bulk tank once each week and then sold to customers. Let X_1 denote the proportion of the tank that is stocked on a particular week and let X_2 denote the proportion of the tank that is sold in that same week. Due to limited supplies X_1 is not fixed in advance, but varies from week to week. Suppose a study of many weeks shows the joint relative frequency behavior of X_1 and X_2 to be such that the following density function provides an adequate model:

$$f(x_1, x_2) = 3x_1 \qquad 0 \leqslant x_2 \leqslant x_1 \leqslant 1$$
$$= 0 \qquad \text{elsewhere}$$

Note that X_2 must always be less than or equal to X_1. This density function is graphed in Figure 5.5. Find the probability that X_2 will be between 0.2 and 0.4 for a given week.

FIGURE 5.5
The Joint Density Function for Example 5.4

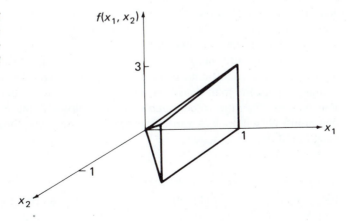

Solution   The question refers to the marginal behavior of X_2. Thus it is necessary to find

$$f_2(x_2) = \int_{-\infty}^{\infty} f(x_1, x_2)\, dx_1$$

$$= \int_{x_2}^{1} 3x_1\, dx_1 = \frac{3}{2}\, x_1^2\big|_{x_2}^{1}$$

$$= \frac{3}{2}(1 - x_2^2) \qquad 0 < x_2 \leqslant 1$$

It follows directly that

$$P(0.2 \leqslant X_2 \leqslant 0.4) = \int_{0.2}^{0.4} \frac{3}{2}(1 - x_2^2)\, dx_2$$

$$= \frac{3}{2}\left(x_2 - \frac{x_2^3}{3}\right)\Big|_{0.2}^{0.4}$$

$$= \frac{3}{2}\left\{\left[0.4 - \frac{(0.4)^3}{3}\right] - \left[0.2 - \frac{(0.2)^3}{3}\right]\right\}$$

$$= 0.272$$

Note the marginal density of X_2 graphs as a function that is high at $x_2 = 0$ and then tends to zero as x_2 tends to one. Does this agree with your intuition after looking at Figure 5.5?   ■

5.2 CONDITIONAL PROBABILITY DISTRIBUTIONS

Recall that in the bivariate *discrete* case the conditional probabilities for X_1 for a given X_2 were found by fixing attention on the particular row in which $X_2 = x_2$ and then looking at the relative probabilities within that row. That is, the individual cell probabilities were divided by the marginal total for that row in order to obtain conditional probabilities.

In the bivariate *continuous* case the form of the probability density function representing the conditional behavior of X_1 for a given value of X_2 is found by slicing through the joint density in the x_1 direction at the particular value of X_2. This function then has to be weighted by the marginal density function for X_2 at that point. We will look at a specific example before giving the general definition of conditional density functions.

EXAMPLE 5.5

Refer to the joint density function of Example 5.4. Find the conditional probability that X_2 is less than 0.2 given that X_1 was known to be 0.5.

Solution   Slicing through $f(x_1, x_2)$ in the x_2 direction at $x_1 = 0.5$ yields

$$f(0.5, x_2) = 3(0.5) = 1.5 \qquad 0 \leqslant x_2 \leqslant 0.5$$

Thus the conditional behavior of X_2 for a given X_1 of 0.5 is constant over the interval $(0, 0.5)$. The marginal value of $f(x_1)$ at $x_1 = 0.5$ is obtained as follows:

$$f_1(x_1) = \int_{-\infty}^{+\infty} f(x_1, x_2) \, dx_2$$

$$= \int_0^{x_1} 3x_1 \, dx_2$$

$$= 3x_1^2 \qquad 0 \leqslant x_1 \leqslant 1$$

$$f_1(0.5) = 3(0.5)^2 = 0.75$$

Upon dividing, we see that the conditional behavior of X_2 for a given X_1 of 0.5 is represented by the function

$$f(x_2 | x_1 = 0.5) = \frac{f(0.5, x_2)}{f_1(0.5)}$$

$$= \frac{1.5}{0.75} = 2 \qquad 0 < x_2 < 0.5$$

This function of x_2 has all the properties of a probability density function, and is the conditional density function for X_2 at $X_1 = 0.5$. Then

$$P(X_2 < 0.2 | X_1 = 0.5) = \int_0^{0.2} f(x_2 | x_1 = 0.5) \, dx_2$$

$$= \int_0^{0.2} 2 \, dx_2 = 0.2(2) = 0.4$$

That is, among all weeks in which the tank was half full immediately after stocking, sales amounted to less than 20% of the tank 40% of the time. ∎

We see that the above manipulations used to obtain conditional density functions in the continuous case are analogous to those used to obtain conditional probabilities in the discrete case, except that integrals are used in place of sums. The formal definition of a conditional density function is given below.

DEFINITION 5.3 | Let X_1 and X_2 be jointly continuous random variables with joint probability density function $f(x_1, x_2)$ and marginal densities $f_1(x_1)$ and $f_2(x_2)$, respectively. Then, the **conditional probability density function** of X_1 given $X_2 = x_2$ is given by

$$f(x_1 | x_2) = \begin{cases} \dfrac{f(x_1, x_2)}{f_2(x_2)} & f_2(x_2) > 0 \\[2mm] 0 & \text{elsewhere} \end{cases}$$

and the conditional probability density function of X_2 given $X_1 = x_1$ is given by

$$f(x_2 | x_1) = \begin{cases} \dfrac{f(x_1, x_2)}{f_1(x_1)} & f_1(x_1) > 0 \\ 0 & \text{elsewhere} \end{cases}$$

EXAMPLE 5.6

A soft-drink machine has a random amount Y_2 in supply at the beginning of a given day and dispenses a random amount Y_1 during the day (with measurements in gallons). It is not resupplied during the day, hence $Y_1 \leqslant Y_2$. It has been observed that Y_1 and Y_2 have joint density

$$f(y_1, y_2) = \begin{cases} 1/2 & 0 \leqslant y_1 \leqslant y_2, \, 0 \leqslant y_2 \leqslant 2 \\ 0 & \text{elsewhere} \end{cases}$$

That is, the points (y_1, y_2) are uniformly distributed over the triangle with the given boundaries. Find the conditional density of Y_1 given $Y_2 = y_2$. Evaluate the probability that less than 1/2 gallon is sold, given that the machine contains 1 gallon at the start of the day.

Solution   The marginal density of Y_2 is given by

$$f_2(y_2) = \int_{-\infty}^{\infty} f(y_1, y_2) \, dy_1$$

$$= \begin{cases} \displaystyle\int_0^{y_2} (1/2) \, dy_1 = (1/2) y_2 & 0 \leqslant y_2 \leqslant 2 \\ 0 & \text{elsewhere} \end{cases}$$

By Definition 5.3

$$f_1(y_1 | y_2) = \frac{f(y_1, y_2)}{f_2(y_2)}$$

$$= \begin{cases} \dfrac{1/2}{(1/2) y_2} = \dfrac{1}{y_2} & 0 \leqslant y_1 \leqslant y_2 \leqslant 2 \\ 0 & \text{elsewhere} \end{cases}$$

The probability of interest is

$$P(Y_1 < 1/2 \,|\, Y_2 = 1) = \int_{-\infty}^{1/2} f(y_1 | y_2 = 1) \, dy_1$$

$$= \int_0^{1/2} (1) \, dy_1 = 1/2$$

Note that if the machine had contained 2 gallons at the start of the day, then

$$P\left(Y_1 \leqslant \frac{1}{2}\middle| Y_2 = 2\right) = \int_0^{1/2} \left(\frac{1}{2}\right) dy_1$$

$$= \frac{1}{4}$$

Thus the amount sold is highly dependent upon the amount in supply. ∎

5.3 INDEPENDENT RANDOM VARIABLES

Before defining independent random variables, we will recall once again that two events A and B are independent if $P(AB) = P(A)P(B)$. Somewhat analogously, two discrete random variables are independent if

$$P(X_1 = x_1, X_2 = x_2) = P(X_1 = x_1)P(X_2 = x_2)$$

for all real numbers x_1 and x_2. A similar idea carries over to the continuous case.

DEFINITION 5.4

Discrete random variables X_1 and X_2 are said to be independent if

$$P(X_1 = x_1, X_2 = x_2) = P(X_1 = x_1)P(X_2 = x_2)$$

for all real numbers x_1 and x_2.
    Continuous random variables X_1 and X_2 are said to be independent if

$$f(x_1, x_2) = f_1(x_1)f_2(x_2)$$

for all real numbers x_1 and x_2.

The concepts of joint probability density functions and independence extend immediately to n random variables, where n is any finite positive integer. The n random variables $X_1, X_2, \ldots, X_n$ are said to be *independent* if their joint density function $f(x_1, \ldots, x_n)$ is given by

$$f(x_1, \ldots, x_n) = f_1(x_1)f_2(x_2) \cdots f_n(x_n)$$

for all real numbers $x_1, x_2, \ldots, x_n$.

EXAMPLE 5.7 ─────────────────────────────────────

Show that the random variables having the joint distribution of Table 5.1 are not independent.

Solution   It is necessary to check only one entry in the table. We see that $P(X_1 = 0, X_2 = 0) = 0.6$, whereas $P(X_1 = 0) = 0.8$ and $P(X_2 = 0) = 0.8$. Since

$$P(X_1 = 0, X_2 = 0) \neq P(X_1 = 0)P(X_2 = 0)$$

the random variables cannot be independent. ∎

EXAMPLE 5.8
_____

Show that the random variables in Example 5.2, page 184, are independent.

Solution   Here

$$f(x_1, x_2) = 2(1 - x_1) \qquad 0 \leqslant x_1 \leqslant 1, 0 \leqslant x_2 \leqslant 1$$

$$= 0 \qquad \qquad \text{elsewhere}$$

We saw in Example 5.3 that

$$f_1(x_1) = 2(1 - x_1) \qquad 0 \leqslant x_1 \leqslant 1$$

and

$$f_2(x_2) = 1 \qquad \qquad 0 \leqslant x_2 \leqslant 1$$

Thus $f(x_1, x_2) = f_1(x_1)f_2(x_2)$ for all real numbers x_1 and x_2, and X_1 and X_2 are independent random variables ∎

EXERCISES

5.1 Two construction contracts are to be randomly assigned to one or more of three firms. Numbering the firms I, II, and III, let X_1 be the number of contracts assigned to firm I and X_2 the number assigned to firm II. (A firm may receive more than one contract.)
   (a) Find the joint probability distribution for X_1 and X_2.
   (b) Find the marginal probability distribution for X_1.
   (c) Find $P(X_1 = 1 | X_2 = 1)$.

5.2 A radioactive particle is randomly located in a square area with sides one unit in length. Let X_1 and X_2 denote the coordinates of the particle. Since the particle is equally likely to fall in any subarea of a fixed size, a reasonable model for (X_1, X_2) is given by

$$f(x_1, x_2) = \begin{cases} 1 & 0 \leqslant x_1 \leqslant 1, 0 \leqslant x_2 \leqslant 1 \\ 0 & \text{elsewhere} \end{cases}$$

   (a) Sketch the probability density surface.
   (b) Find $P(X_1 \leqslant 0.2, X_2 \leqslant 0.4)$.
   (c) Find $P(0.1 \leqslant X_1 \leqslant 0.3, X_2 > 0.4)$.

5.3 A group of 9 executives of a certain firm contain 4 who are married, 3 who are single, and 2 who are divorced. Three of the executives are to be selected for promotion. Let X_1 denote the number of married executives and X_2 the number of single executives among the three selected for promotion. Assuming that the three are randomly selected from the nine available, find the joint probability distribution for X_1 and X_2.

5.4 An environmental engineer measures the amount (by weight) of particulate pollution in air samples (of a certain volume) collected over the smokestack of a coal-operated power plant. Let X_1 denote the amount of pollutant per sample when a certain cleaning device on the stack is not operating, and X_2 the amount of pollutant per sample when the cleaning device is operating, under similar environmental conditions. It is observed that

X_1 is always greater than $2X_2$, and the relative frequency behavior of (X_1, X_2) can be modeled by

$$f(x_1, x_2) = \begin{cases} k & 0 \leqslant x_1 \leqslant 2,\, 0 \leqslant x_2 \leqslant 1,\, 2x_2 \leqslant x_1 \\ 0 & \text{elsewhere} \end{cases}$$

(That is, X_1 and X_2 are randomly distributed over the region inside the triangle bounded by $x_1 = 2$, $x_2 = 0$, and $2x_2 = x_1$.)
- (a) Find the value of k that makes this a probability density function.
- (b) Find $P(X_1 \geqslant 3X_2)$. (That is, find the probability that the cleaning device will reduce the amount of pollutant by one third or more.)

5.5 Refer to Exercise 5.2.
- (a) Find the marginal density function for X_1.
- (b) Find $P(X_1 \leqslant 0.5)$.
- (c) Are X_1 and X_2 independent?

5.6 Refer to Exercise 5.4.
- (a) Find the marginal density function of X_2.
- (b) Find $P(X_2 \leqslant 0.4)$.
- (c) Are X_1 and X_2 independent?
- (d) Find $P(X_2 \leqslant 1/4 \,|\, X_1 = 1)$.

5.7 Let X_1 and X_2 denote the proportion of two different chemicals found in a sample mixture of chemicals used as an insecticide. Suppose X_1 and X_2 have joint probability density given by

$$f(x_1, x_2) = \begin{cases} 2 & 0 \leqslant x_1 \leqslant 1,\, 0 \leqslant x_2 \leqslant 1,\, 0 \leqslant x_1 + x_2 \leqslant 1 \\ 0 & \text{elsewhere} \end{cases}$$

(Note that $X_1 + X_2$ must be at most unity since the random variables denote proportions within the same sample.)
- (a) Find $P(X_1 \leqslant 3/4,\, X_2 \leqslant 3/4)$.
- (b) Find $P(X_1 \leqslant 1/2,\, X_2 \leqslant 1/2)$.
- (c) Find $P(X_1 \leqslant 1/2 \,|\, X_2 \leqslant 1/2)$.

5.8 Refer to Exercise 5.7.
- (a) Find the marginal density functions for X_1 and X_2.
- (b) Are X_1 and X_2 independent?
- (c) Find $P(X_1 > 1/2 \,|\, X_2 = 1/4)$.

5.9 Let X_1 and X_2 denote the proportions of time, out of one work week, that employees I and II, respectively, actually spend on performing their assigned tasks. The joint relative frequency behavior of X_1 and X_2 is modeled by the probability density function

$$f(x_1, x_2) = \begin{cases} x_1 + x_2 & 0 \leqslant x_1 \leqslant 1,\, 0 \leqslant x_2 \leqslant 1 \\ 0 & \text{elsewhere} \end{cases}$$

- (a) Find $P(X_1 < 1/2,\, X_2 > 1/4)$.
- (b) Find $P(X_1 + X_2 \leqslant 1)$.
- (c) Are X_1 and X_2 independent?

5.10 Refer to Exercise 5.9. Find the probability that employee I spends more than 75% of the week on his assigned task, given that employee II spends exactly 50% of the work week on his assigned task.

5.11 An electronic surveillance system has one of each of two different types of components in joint operations. Letting X_1 and X_2 denote the random lifelengths of the components of type I and type II, respectively, the joint probability density function is given by

$$f(x_1, x_2) = \begin{cases} (1/8)x_1 e^{-(x_1+x_2)/2} & x_1 > 0, x_2 > 0 \\ 0 & \text{elsewhere} \end{cases}$$

(Measurements are in hundreds of hours.)
   (a) Are X_1 and X_2 independent?
   (b) Find $P(X_1 > 1, X_2 > 1)$.

5.12 A bus arrives at a bus stop at a randomly selected time within a one-hour period. A passenger arrives at the bus stop at a randomly selected time within the same hour. The passenger will wait for the bus up to 1/4 of an hour. What is the probability that the passenger will catch the bus? [*Hint*: Let X_1 denote the bus arrival time and X_2 the passenger arrival time. If these arrivals are independent, then

$$f(x_1, x_2) = \begin{cases} 1 & 0 \leqslant x_1 \leqslant 1, 0 \leqslant x_2 \leqslant 1 \\ 0 & \text{elsewhere} \end{cases}$$

Now find $P(X_2 \leqslant X_1 \leqslant X_2 + 1/4)$.]

5.13 Two friends are to meet at a library. Each arrives at independently and randomly selected times within a fixed one-hour period. Each agrees to wait no more than 10 minutes for the other. Find the probability that they meet.

5.14 Two quality control inspectors each interrupt a production line at randomly, but independently, selected times within a given day (of 8 hours). Find the probability that the two interruptions will be more than 4 hours apart.

5.15 Two telephone calls come into a switchboard at random times in a fixed one-hour period. If the calls are made independently of each other,
   (a) find the probability that both are made in the first half hour.
   (b) find the probability that the two calls are within 5 minutes of each other.

5.16 A bombing target is in the center of a circle with radius one mile. A bomb falls at a randomly selected point inside that circle. If the bomb destroys everything within 1/2 mile of its landing point, what is the probability that it destroys the target?

5.4 EXPECTED VALUES OF FUNCTIONS OF RANDOM VARIABLES

As we have seen, we often encounter problems that involve more than one random variable. We may be interested in the lifelengths of five different electronic components within the same system or three different strength test measurements on the same section of cable. We will now discuss how to find expected values of functions of more than one random variable.

Definition 5.5 gives the basic result for finding expected values in the bivariate case. The definition can, of course, be generalized to higher dimensions.

DEFINITION 5.5	Suppose the discrete random variables (X_1, X_2) have joint probability function given by $p(x_1, x_2)$. If $g(X_1, X_2)$ is any real-valued function of (X_1, X_2), then $$E[g(X_1, X_2)] = \sum_{x_1}\sum_{x_2} g(x_1, x_2)p(x_1, x_2)$$ The sum is over all values of (x_1, x_2) for which $p(x_1, x_2) > 0$. If (X_1, X_2) are continuous random variables with probability density function $f(x_1, x_2)$, then $$E[g(X_1, X_2)] = \int_{-\infty}^{\infty} \int_{-\infty}^{\infty} g(x_1, x_2)f(x_1, x_2)\,dx_1\,dx_2$$

If X_1 and X_2 are independent, then it follows easily from Definition 5.5 that

$$E[g(X_1)h(X_2)] = E[g(X_1)]E[h(X_2)]$$

A function of two variables that is commonly of interest in probabilistic and statistical problems is the *covariance*.

DEFINITION 5.6	The **covariance** between two random variables X_1 and X_2 is given by $$\text{Cov}(X_1, X_2) = E[(X_1 - \mu)(X_2 - \mu)]$$ where $$\mu_1 = E(X_1) \text{ and } \mu_2 = E(X_2)$$

The covariance helps us to assess the relationship between two variables, in the following sense. If X_2 tends to be large at the same time X_1 is large, and if X_2 tends to be small when X_1 is small, then X_1 and X_2 will have a *positive covariance*. If, on the other hand, X_2 tends to be small when X_1 is large and large when X_1 is small, then the variables will have a *negative covariance*.

The next theorem gives an easier computational form for the covariance.

THEOREM 5.1   If X_1 has mean μ_1 and X_2 has mean μ_2, then

$$\text{Cov}(X_1, X_2) = E(X_1 X_2) - \mu_1\mu_2$$

If X_1 and X_2 are independent random variables, then $E(X_1, X_2) = E(X_1)E(X_2)$. Using Theorem 5.1, it is then clear that independence between X_1 and X_2 implies that $\text{Cov}(X_1, X_2) = 0$. The converse is not necessarily true. That is, zero covariance does not imply that the variables are independent. To see this, look at the following joint distribution:

$$X_1$$

		-1	0	1	
	-1	1/8	1/8	1/8	3/8
X_2	0	1/8	0	1/8	2/8
	1	1/8	1/8	1/8	3/8
		3/8	2/8	3/8	1

Clearly $E(X_1) = E(X_2) = 0$, and $E(X_1 X_2) = 0$ as well. Therefore $\text{Cov}(X_1, X_2) = 0$. On the other hand

$$P(X_1 = -1, X_2 = -1) = 1/8 \neq P(X_1 = -1)P(X_2 = -1)$$

and so the random variables are dependent.

We will now calculate the covariance in two examples.

EXAMPLE 5.9

A firm that sells word processing systems keeps track of the number of customers who call on any one day and the number of orders placed on any one day. Let X_1 denote the number of calls, X_2 the number of orders placed, and $p(x_1, x_2)$ the joint probability function for (X_1, X_2); records indicate that

$$p(0, 0) = 0.04 \qquad p(2, 0) = 0.20$$

$$p(1, 0) = 0.16 \qquad p(2, 1) = 0.30$$

$$p(1, 1) = 0.10 \qquad p(2, 2) = 0.20$$

That is, for any given day the probability of, say, 2 calls and 1 order is 0.30. Find $\text{Cov}(X_1, X_2)$.

Solution   Theorem 5.1 suggests that we first find $E(X_1, X_2)$, which is

$$E(X_1 X_2) = \sum_{x_1} \sum_{x_2} x_1 x_2 p(x_1, x_2)$$

$$= (0 \times 0)p(0, 0) + (1 \times 0)p(1, 0) + (1 \times 1)p(1, 1) + (2 \times 0)p(2, 0)$$

$$+ (2 \times 1)p(2, 1) + (2 \times 2)p(2, 2)$$

$$= 0(0.4) + 0(0.16) + 1(0.10) + 0(0.20) + 2(0.30) + 4(0.20)$$

$$= 1.50$$

Now we must find $E(X_1) = \mu_1$ and $E(X_2) = \mu_2$. The marginal distributions of X_1 and X_2 are given on the following charts:

x_1	$p(x_1)$
0	0.04
1	0.26
2	0.70

x_2	$p(x_2)$
0	0.40
1	0.40
2	0.20

It follows that

$$\mu_1 = 1(0.26) + 2(0.70) = 1.66$$

and

$$\mu_2 = 1(0.40) + 2(0.20) = 0.80$$

Thus

$$Cov(X_1, X_2) = E(X_1 X_2) - \mu_1 \mu_2$$
$$= 1.50 - 1.66(0.80)$$
$$= 1.50 - 1.328 = 0.172$$

Does the positive covariance agree with your intuition?   ■

When a problem involves n random variables $X_1, X_2, \ldots, X_n$, we are often interested in studying linear combinations of those variables. For example, if the random variables measure the quarterly incomes for n plants in a corporation, we may want to look at their sum or average. If X_1 represents the monthly cost of servicing defective plants before a new quality control system was installed, and X_2 denotes that cost after the system was placed in operation, then we may want to study $X_1 - X_2$. Theorem 5.2 gives a general result on the mean and variance of a linear combination of random variables.

THEOREM 5.2   Let $Y_1, \ldots, Y_n$ and $X_1, \ldots, X_m$ be random variables with $E(Y_i) = \mu_i$ and $E(X_i) = \xi_i$. Define

$$U_1 = \sum_{i=1}^{n} a_i Y_i, \qquad U_2 = \sum_{j=1}^{m} b_j X_i$$

for constants $a_1, \ldots, a_n, b_1, \ldots, b_m$. Then

(a) $E(U_1) = \sum_{i=1}^{n} a_i \mu_i$

(b) $V(U_1) = \sum_{i=1}^{n} a_i^2 V(Y_i) + 2 \sum\sum_{i<j} a_i a_j Cov(Y_i, Y_j)$

where the double sum is over all pairs (i, j) with $i < j$ and

(c) $Cov(U_1, U_2) = \sum_{i=1}^{n} \sum_{j=1}^{m} a_i b_j Cov(Y_i, X_j)$

Proof   Part (a) follows directly from the definition of expected value and properties of sums or integrals. To prove part (b), we appeal to the definition of variance and write

$$V(U_1) = E[U_1 - E(U_1)]^2$$
$$= E\left[\sum_{i=1}^{n} a_i Y_i - \sum_{i=1}^{n} a_i \mu_i \right]^2$$

$$= E\left[\sum_{i=1}^{n} a_i(Y_i - \mu_i)\right]^2$$

$$= E\left[\sum_{i=1}^{n} a_i^2(Y_i - \mu_i)^2 + \sum\sum_{i \neq j} a_i a_j(Y_i - \mu_i)(Y_j - \mu_j)\right]$$

$$= \sum_{i=1}^{n} a_i^2 E(y_i - \mu_i)^2 + \sum\sum_{i \neq j} a_i a_j E[(Y_i - \mu_i)(Y_j - \mu_j)]$$

By definition of variance and covariance, we then have

$$V(U_1) = \sum_{i=1}^{n} a_i^2 V(Y_i) + \sum\sum_{i \neq j} a_i a_j \text{Cov}(Y_i, Y_j)$$

Note that $\text{Cov}(Y_i, Y_j) = \text{Cov}(Y_j, Y_i)$, and hence we can write

$$V(U_1) = \sum_{i=1}^{n} a_i^2 V(Y_i) + 2 \sum\sum_{i < j} \text{Cov}(Y_i, Y_j)$$

Part (c) is obtained by similar steps. We have

$$\text{Cov}(U_1, U_2) = E\{[U_1 - E(U_1)][U_2 - E(U_2)]\}$$

$$= E\left[\left(\sum_{i=1}^{n} a_i Y_i - \sum_{i=1}^{n} a_i \mu_i\right)\left(\sum_{j=1}^{m} b_j X_j - \sum_{j=1}^{m} b_j \xi_j\right)\right]$$

$$= E\left\{\left[\sum_{i=1}^{n} a_i(Y_i - \mu_i)\right]\left[\sum_{j=1}^{m} b_j(X_j - \xi_j)\right]\right\}$$

$$= E\left[\sum_{i=1}^{n}\sum_{j=1}^{m} a_i b_j(Y_i - \mu_i)(X_j - \xi_j)\right]$$

$$= \sum_{i=1}^{n}\sum_{j=1}^{m} a_i b_j E(Y_i - \mu_i)(X_j - \xi_j)$$

$$= \sum_{i=1}^{n}\sum_{j=1}^{m} a_i b_j \text{Cov}(Y_i, X_j)$$

On observing that $\text{Cov}(Y_i, Y_j) = V(Y_i)$, we can see that part (b) is a special case of part (c).

EXAMPLE 5.10

Let $Y_1, Y_2, \ldots, Y_n$ be independent random variables with $E(Y_i) = \mu$ and $V(Y_i) = \sigma^2$. (These variables may denote the outcomes on n independent trials of an experiment.) Defining

$$\bar{Y} = \frac{1}{n}\sum_{i=1}^{n} Y_i$$

show that $E(\bar{Y}) = \mu$ and $V(\bar{Y}) = \sigma^2/n$.

Solution Note that $\bar{Y}$ is a linear function with all constants a_i equal to $1/n$. That is,

$$\bar{Y} = \left(\frac{1}{n}\right) Y_1 + \cdots + \left(\frac{1}{n}\right) Y_n.$$

By Theorem 5.2, part (a),

$$E(\bar{Y}) = \sum_{i=1}^{n} a_i \mu = \mu \sum_{i=1}^{n} a_i$$

$$= \mu \sum_{i=1}^{n} \frac{1}{n} = \frac{n\mu}{n} = \mu$$

By Theorem 5.2, part (b),

$$V(\bar{Y}) = \sum_{i=1}^{n} a_i^2 V(Y_i) + 2 \sum\sum_{i<j} a_{ij} \text{Cov}(Y_i, Y_j)$$

but the covariance terms are all zero since the random variables are independent. Thus

$$V(\bar{Y}) = \sum_{i=1}^{n} \left(\frac{1}{n}\right)^2 \sigma^2 = \frac{n\sigma^2}{n^2} = \frac{\sigma^2}{n} \qquad \blacksquare$$

EXAMPLE 5.11

With X_1 denoting the amount of gasoline stocked in a bulk tank at the beginning of a week and X_2 the amount sold during the week, $Y = X_1 - X_2$ represents the amount left over at the end of the week. Find the mean and variance of Y if the joint density function of (X_1, X_2) is given by

$$f(x_1, x_2) = \begin{cases} 3x_1 & 0 \leqslant x_2 \leqslant x_1 \leqslant 1 \\ 0 & \text{elsewhere} \end{cases}$$

Solution We must first find the means and variances of X_1 and X_2. The marginal density of X_1 is found to be

$$f_1(x_1) = \begin{cases} 3x_1^2 & 0 \leqslant x_1 \leqslant 1 \\ 0 & \text{elsewhere} \end{cases}$$

Thus

$$E(X_1) = \int_0^1 x_1(3x_1^2)dx = 3\left[\frac{x_1^4}{4}\right]_0^1 = \frac{3}{4}$$

Note that $E(X_1)$ can be found directly from the joint density function by

$$E(X_1) = \int_0^1 \int_0^{x_1} x_1(3x_1) dx_2 \, dx$$

$$= \int_0^1 3x_1^2 [x_2]_0^{x_1} \, dx_1$$

$$= \int_0^1 3x_1^3 \, dx_1 = \frac{3}{4} [x_1^4]_0^1 = \frac{3}{4}$$

The marginal density of X_2 is found to be

$$f_2(x_2) = \begin{cases} \frac{3}{2}(1 - x_2^2) & 0 \leqslant x_2 \leqslant 1 \\ 0 & \text{elsewhere} \end{cases}$$

Thus

$$E(X_2) = \int_0^1 (x_2) \frac{3}{2} (1 - x_2^2) \, dx_2$$

$$= \frac{3}{2} \int_0^1 (x_2 - x_2^3) \, dx_2 = \frac{3}{2} \left\{ \left[\frac{x_2^2}{2} \right]_0^1 - \left[\frac{x_2^4}{4} \right]_0^1 \right\} = \frac{3}{2} \left\{ \frac{1}{2} - \frac{1}{4} \right\} = \frac{3}{8}$$

By similar arguments it follows that

$$E(X_1^2) = 3/5$$

$$V(X_1) = 3/5 - (3/4)^2 = 0.375$$

$$E(X_2^2) = 1/5$$

and

$$V(X_2) = 1/5 - (3/8)^2 = 0.0594$$

The next step is to find $\text{Cov}(X_1, X_2)$. Now

$$E(X_1 X_2) = \int_0^1 \int_0^{x_1} (x_1 x_2) 3 x_1 \, dx_2 \, dx_1$$

$$= 3 \int_0^1 \int_0^{x_1} x_1^2 x_2 \, dx_2 \, dx_1$$

$$= 3 \int_0^1 x_1^2 \left[\frac{x_2^2}{2} \right]_0^{x_1} dx_1$$

$$= \frac{3}{2} \int_0^1 x_1^4 \, dx_1$$

$$= \frac{3}{2} \left[\frac{x_1^5}{5} \right]_0^1 = \frac{3}{10}$$

and

$$\text{Cov}(X_1, X_2) = E(X_1 X_2) - \mu_1 \mu_2$$

$$= \frac{3}{10} - \left(\frac{3}{4} \right)\left(\frac{3}{8} \right) = 0.0188$$

From Theorem 5.2

$$E(Y) = E(X_1) - E(X_2)$$

$$= \frac{3}{4} - \frac{3}{8} = \frac{3}{8} = 0.375$$

and

$$V(Y) = V(X_1) + V(X_2) + 2(1)(-1)\mathrm{Cov}(X_1, X_2)$$

$$= 0.0375 + 0.0594 - 2(0.0188)$$

$$= 0.0593 \qquad \blacksquare$$

When the random variables in use are independent, the variance of a linear function simplifies since the covariances become zero.

EXAMPLE 5.12 _____

A firm purchases two types of industrial chemicals. The amount of type I chemical X_1 purchased per week has $E(X_1) = 40$ gallons with $V(X_1) = 4$. The amount of type II chemical X_2 purchased has $E(X_2) = 65$ gallons with $V(X_2) = 8$. Type I costs \$3 per gallon while type II costs \$5 per gallon. Find the mean and variance of the total weekly amount spent for these types of chemicals, assuming X_1 and X_2 are independent.

Solution   The dollar amount spent per week is given by

$$Y = 3X_1 + 5X_2$$

From Theorem 5.2

$$E(Y) = 3E(X_1) + 5E(X_2)$$

$$= 3(40) + 5(65)$$

$$= 445$$

and

$$V(Y) = (3)^2 V(X_1) + (5)^2 V(X_2)$$

$$= 9(4) + 25(8)$$

$$= 236$$

The firm can expect to spend \$445 per week on chemicals.   ■

EXAMPLE 5.13 _____

Suppose an urn contains r white balls and $(N - r)$ black balls. A random sample of n balls is drawn without replacement and Y, the number of white balls in the sample, is observed. From Chapter 3 we know that Y has a hypergeometric probability distribution. Find the mean and variance of Y.

Solution   We first observe some characteristics of sampling without replacement. Suppose that the sampling is done sequentially and we observe outcomes for $X_1, X_2, \ldots, X_n$, where

$$X_i = \begin{cases} 1 & \text{if the } i\text{th draw results in a white ball} \\ 0 & \text{otherwise} \end{cases}$$

Unquestionably $P(X_1 = 1) = r/N$. But it is also true that $P(X_2 = 1) = r/N$ since

$$P(X_2 = 1) = P(X_1 = 1, X_2 = 1) + P(X_1 = 0, X_2 = 1)$$

$$= P(X_1 = 1)P(X_2 = 1|X_1 = 1) + P(X_1 = 0)P(X_2 = 1|X_1 = 0)$$

$$= \frac{r}{N} \frac{r-1}{N-1} + \frac{N-r}{N} \frac{r}{N-1}$$

$$= \frac{r(N-1)}{N(N-1)} = \frac{r}{N}$$

The same is true for X_k; that is,

$$P(X_k = 1) = \frac{r}{N} \qquad k = 1, \ldots, n$$

Thus the probability of drawing a white ball on any draw, given no knowledge of the outcomes on previous draws, is r/N.

In a similar way it can be shown that

$$P(X_j = 1, X_k = 1) = \frac{r(r-1)}{N(N-1)} \qquad j \neq k$$

Now observe that $Y = \sum_{i=1}^{n} X_i$ and hence

$$E(Y) = \sum_{i=1}^{n} E(X_i) = n\left(\frac{r}{N}\right)$$

To find $V(Y)$, we need $V(X_i)$ and $\text{Cov}(X_i, X_j)$. Since X_i is 1 with probability r/N and 0 with probability $1 - (r/N)$, it follows that

$$V(X_i) = \frac{r}{N}\left(1 - \frac{r}{N}\right)$$

Also

$$\text{Cov}(X_i, X_j) = E(X_i X_j) - E(X_i)E(X_j)$$

$$= \frac{r(r-1)}{N(N-1)} - \left(\frac{r}{N}\right)^2$$

$$= -\frac{r}{N}\left(1 - \frac{r}{N}\right)\frac{1}{N-1}$$

since $X_i X_j = 1$ if and only if $X_i = 1$ and $X_j = 1$. From Theorem 5.2 we have that

$$V(Y) = \sum_{i=1}^{n} V(X_i) + 2 \sum\sum_{i<j} \text{Cov}(X_i, X_j)$$

$$= n\frac{r}{N}\left(1 - \frac{r}{N}\right) + 2 \sum\sum_{i<j}\left[-\frac{r}{N}\left(1 - \frac{r}{N}\right)\frac{1}{N-1}\right]$$

$$= n\frac{r}{N}\left(1 - \frac{r}{N}\right) - n(n-1)\frac{r}{N}\left(1 - \frac{r}{N}\right)\frac{1}{N-1}$$

since there are $n(n-1)/2$ terms in the double summation. A little algebra yields

$$V(Y) = n\frac{r}{N}\left(1 - \frac{r}{N}\right)\frac{N-n}{N-1}$$ ∎

An appreciation for the usefulness of Theorem 5.2 can be gained by trying to find the expected value and variance for the hypergeometric random variable by proceeding directly from the definition of an expectation. The necessary summations are exceedingly difficult to obtain.

EXERCISES

5.17 Table 5.1 on page 181 shows the joint probability distribution of the number of defectives observed on the first draw (X_1) and the second draw (X_2) from a box containing 4 good bulbs and 1 defective bulb.
  - (a) Find $E(X_1)$, $V(X_1)$, $E(X_2)$, and $V(X_2)$.
  - (b) Find $\text{Cov}(X_1, X_2)$.
  - (c) The variable $Y = X_1 + X_2$ denotes the total number of defectives observed on the two draws. Find $E(Y)$ and $V(Y)$.

5.18 In a study of particulate pollution in air samples over a smokestack, X_1 represents the amount of pollutant per sample when a cleaning device is not operating and X_2 the amount when the cleaning device is operating. Assume that (X_1, X_2) has joint probability density function

$$f(x_1, x_2) = \begin{cases} 1 & 0 \leqslant x_1 \leqslant 2, 0 \leqslant x_2 \leqslant 1, 2x_2 \leqslant x_1 \\ 0 & \text{elsewhere} \end{cases}$$

The random variable $Y = X_1 - X_2$ represents the amount by which the weight of pollutant can be reduced by using the cleaning device.
  - (a) Find $E(Y)$ and $V(Y)$.
  - (b) Find an interval in which values of Y should lie at least 75% of the time.

5.19 The proportions X_1 and X_2 of two chemicals found in samples of an insecticide have joint probability density function

$$f(x_1, x_2) = \begin{cases} 2 & 0 \leqslant x_1 \leqslant 1, 0 \leqslant x_2 \leqslant 1, 0 \leqslant x_1 + x_2 \leqslant 1 \\ 0 & \text{elsewhere} \end{cases}$$

The random variable $Y = X_1 + X_2$ denotes the proportion of the insecticides due to both chemicals combined.
  - (a) Find $E(Y)$ and $V(Y)$.
  - (b) Find an interval in which values of Y should lie for at least 50% of the samples of insecticide.

5.20 For a sheet-metal stamping machine in a certain factory the time between failures X_1 has a mean (MTBF) of 56 hours and a variance of 16. The repair time X_2 has a mean (MTTR) of 5 hours and a variance of 4.
  - (a) If X_1 and X_2 are independent, find the expected value and variance of $Y = X_1 + X_2$, which represents one operation/repair cycle.
  - (b) Would you expect an operation/repair cycle to last more than 75 hours? Why?

5.21 A particular fast-food outlet is interested in the joint behavior of the random variables Y_1, defined as the total time between a customer's arrival at the store and his leaving the service window, and Y_2, the time that the customer waits in line before reaching the service window. Since Y_1 contains the time a customer waits in line, we must have $Y_1 \geqslant Y_2$. The relative frequency distribution of observed values of Y_1 and Y_2 can be modeled by the probability density function

$$f(y_1, y_2) = \begin{cases} e^{-y_1} & 0 \leqslant y_2 \leqslant y_1 < \infty \\ 0 & \text{elsewhere} \end{cases}$$

(a) Find $P(Y_1 < 2, Y_2 > 1)$.
(b) Find $P(Y_1 \geqslant 2Y_2)$.
(c) Find $P(Y_1 - Y_2 \geqslant 1)$. [*Note:* $Y_1 - Y_2$ denotes the time spent at the service window.]
(d) Find the marginal density functions for Y_1 and Y_2.

5.22 Refer to Exercise 5.21. If a customer's total waiting time plus service time is known to be more than two minutes, find the probability that he waited less than one minute to be served.

5.23 Refer to Exercise 5.21. The random variable $Y_1 - Y_2$ represents the time spent at the service window.
(a) Find $E(Y_1 - Y_2)$.
(b) Find $V(Y_1 - Y_2)$.
(c) Is it highly likely that a customer would spend more than two minutes at the service window?

5.24 Refer to Exercise 5.21. Suppose a customer spends a length of time y_1 at the store. Find the probability that this customer spends less than half of that time at the service window.

5.25 Prove Theorem 5.1.

5.5 THE MULTINOMIAL DISTRIBUTION

Suppose that an experiment consists of n independent trials, much like the binomial case, but that each trial can result in any one of k possible outcomes. For example, a customer checking out of a grocery store may choose any one of k checkout counters. Now suppose the probability that a particular trial results in outcome i is denoted by p_i, $i = 1, \ldots, k$, and p_i remains constant from trial to trial. Let Y_i, $i = 1, \ldots, k$, denote the number of the n trials resulting in outcome i. In developing a formula for $P(Y_1 = y_1, \ldots, Y_k = y_k)$, we first call attention to the fact that, because of independence of trials, the probability of having y_1 outcomes of type 1 through y_k outcomes of type k in a particular order will be

$$p_1^{y_1} p_2^{y_2} \cdots p_k^{y_k}$$

It remains only to count the number of such orderings, and this is the number of ways of partitioning the n trials into y_1 type 1 outcomes, y_2 type 2 outcomes, and so on through y_k type k outcomes, or

$$\frac{n!}{y_1! y_2! \cdots y_k!}$$

where

$$\sum_{i=1}^{k} y_i = n$$

Hence

$$P(Y_1 = y_1, \ldots, Y_k = y_k) = \frac{n!}{y_1! y_2! \cdots y_k!} p_1^{y_1} p_2^{y_2} \cdots p_k^{y_k}$$

This is called the *multinomial probability distribution*. Note that if $k = 2$, we are back into the binomial case.

The following example illustrates the computations.

EXAMPLE 5.14

Items under inspection are subject to two types of defects. About 70% of the items in a large lot are judged to be defect free, whereas 20% have a type A defect alone and 10% have a type B defect alone. (None have both types of defects.) If six of these items are randomly selected from the lot, find the probability that three have no defects, one has a type A defect, and two have type B defects.

Solution   If we can assume that the outcomes are independent from trial to trial (item to item in our sample), which they nearly would be in a large lot, then the multinomial distribution provides a useful model. Letting Y_1, Y_2, and Y_3 denote the number of trials resulting in zero, type A, and type B defectives, respectively, we have $p_1 = 0.7$, $p_2 = 0.2$, and $p_3 = 0.1$. It follows that

$$P(Y_1 = 3, Y_2 = 1, Y_3 = 2) = \frac{6!}{3! 1! 2!} (0.7)^3 (0.2)(0.1)^2$$

$$= 0.041 \qquad \blacksquare$$

EXAMPLE 5.15

Find $E(Y_i)$ and $V(Y_i)$ for the multinomial probability distribution.

Solution   We are concerned with the marginal distribution of Y_i, the number of trials falling in cell i. Imagine all the cells, excluding cell i, combined into a single large cell. Hence every trial will result in cell i or not with probabilities p_i and $1 - p_i$, respectively, and Y_i possesses a binomial marginal probability distribution. Consequently

$$E(Y_i) = np_i$$

$$V(Y_i) = np_i q_i \qquad \text{where } q_i = 1 - p_i$$

[*Note*: The same results can be obtained by setting up the expectations and evaluating. For example,

$$E(Y_i) = \sum_{y_1} \sum_{y_2} \cdots \sum_{y_k} y_1 \frac{n!}{y_1! y_2! \cdots y_k!} p_1^{y_1} p_2^{y_2} \cdots p_k^{y_k}$$

Since we have already derived the expected value and variance of Y_i, we leave the tedious summation of this expectation to the interested reader.]   ∎

EXAMPLE 5.16
_____

If $Y_1, \ldots, Y_k$ have the multinomial distribution given in Example 5.15, find $\text{Cov}(Y_s, Y_t)$, $s \neq t$.

Solution   Thinking of the multinomial experiment as a sequence of n independent trials, we define

$$U_i = \begin{cases} 1 & \text{if trial } i \text{ results in class } s \\ 0 & \text{otherwise} \end{cases}$$

and

$$W_i = \begin{cases} 1 & \text{if trial } i \text{ results in class } t \\ 0 & \text{otherwise} \end{cases}$$

Then

$$Y_s = \sum_{i=1}^{n} U_i \quad \text{and} \quad Y_t = \sum_{j=1}^{n} W_j$$

To evaluate $\text{Cov}(Y_s, Y_t)$, we need the following results:

$$E(U_i) = p_s$$
$$E(W_j) = p_t$$
$$\text{Cov}(U_i, W_j) = 0 \quad \text{if } i \neq j \text{ since the trials are independent}$$

and

$$\text{Cov}(U_i, W_i) = E(U_i W_i) - E(U_i)E(W_i)$$
$$= 0 - p_s p_t$$

since $U_i W_i$ always equals zero. From Theorem 5.2 we then have

$$\text{Cov}(Y_s, Y_t) = \sum_{i=1}^{n} \sum_{j=1}^{n} \text{Cov}(U_i, W_j)$$

$$= \sum_{i=1}^{n} \text{Cov}(U_i, W_i) + \sum \sum_{i \neq j} \text{Cov}(U_i, W_j)$$

$$= \sum_{i=1}^{n} (-p_s p_t) + 0$$

$$= -n p_s p_t$$

Note that the covariance is negative, which is to be expected since a large number of outcomes in cell s would force the number in cell t to be small.   ■

MULTINOMIAL EXPERIMENT

1. The experiment consists of n identical trials.
2. The outcome of each trial falls into one of k classes or cells.
3. The probability that the outcome of a single trial will fall in a particular cell, say cell i, is $p_i(i = 1, 2, \ldots, k)$, and remains the same from trial to trial. Note that $p_1 + p_2 + p_3 + \cdots + p_k = 1$.
4. The trials are independent.
5. The random variables of interest are Y_1, Y_2, $\ldots$, Y_k, where $Y_i(i = 1, 2, \ldots, k)$ is equal to the number of trials in which the outcome falls in cell i. Note that $Y_1 + Y_2 + Y_3 + \cdots + Y_k = n$.

THE MULTINOMIAL DISTRIBUTION

$$P(Y_1 = y_1, \ldots, Y_K = y_k) = \frac{n!}{y_1! y_2! \cdots y_k!} \, p_1^{y_1} p_2^{y_2} \cdots p_k^{y_k}$$

where $\sum_{i=1}^{k} y_i = n$ and $\sum_{i=1}^{k} p_i = 1$

$$E(Y_i) = np_i \qquad V(Y_i) = np_i(1 - p_i) \qquad i = 1, \ldots, k$$

$$\mathrm{Cov}(Y_i, Y_j) = -np_i p_j \qquad\qquad i \neq j$$

EXERCISES

5.26 The National Fire Incident Reporting Service says that, among residential fires, approximately 73% are in family homes, 20% are in apartments, and the other 7% are in other types of dwellings. If 4 fires are independently reported in one day, find the probability that 2 are in family homes, 1 is in an apartment, and 1 is in another type of dwelling.

5.27 The typical cost of damages for a fire in a family home is $20,000, whereas the typical cost in an apartment fire is $10,000, and in other dwellings only $2,000. Using the information in Exercise 5.26, find the expected total damage cost for four independently reported fires.

5.28 Wing cracks, in the inspection of commercial aircrafts, are reported as nonexistent, detectable, or critical. The history of a certain fleet shows that 70% of the planes inspected have no wing cracks, 25% have detectable wing cracks, and 5% have critical

wing cracks. For the next 5 planes inspected find the probability that
  (a)   1 has a critical crack, 2 have detectable cracks, and 2 have no cracks.
  (b)   at least 1 critical crack is observed.

5.29   With the recent emphasis upon solar energy, solar radiation has been carefully monitored at various sites in Florida. For typical July days in Tampa, 30% have total radiation of at most 5 calories, 60% have total radiation of at most 6 calories, and 100% have total radiation of at most 8 calories. A solar collector for a hot water system is to be run for 6 days. Find the probability that 3 of the days produce no more than 5 calories, 1 of the days produces between 5 and 6 calories, and 2 of the days produce between 6 and 8 calories. What assumptions are you making in order for your answer to be correct?

5.30   The U.S. Bureau of Labor Statistics reports that, as of 1981, approximately 21% of the adult population under age 65 are between 18 and 24 years of age, 28% are between 25 and 34, 19% are between 35 and 44, and 32% are between 45 and 64. An automobile manufacturer wants to obtain opinions on a new design from five randomly chosen adults from the group. Of the five so selected, find the approximate probability that two are between 18 and 24, two are between 25 and 44, and one is between 45 and 64.

5.31   Customers leaving a subway station can exit through any one of three gates. Assuming that any one customer is equally likely to select any one of the three gates, find the probability that, among four customers,
  (a)   2 select gate A, 1 selects gate B, and 1 selects gate C.
  (b)   all four select the same gate.
  (c)   all three gates are used.

5.32   Among a large number of applicants for a certain position, 60% have only a high school education, 30% have some college training, and 10% have completed a college degree. If 5 applicants are selected to be interviewed, find the probability that at least one will have completed a college degree. What assumptions are necessary for your answer to be valid?

5.33   In a large lot of manufactured items 10% contain exactly one defect, and 5% contain more than one defect. If 10 items are randomly selected from this lot for sale, the repair costs total

$$Y_1 + 3Y_2$$

where Y_1 denotes the number among the ten having one defect, and Y_2 the number with two or more defects. Find the expected value of the repair costs. Find the variance of the repair costs.

5.34   Refer to Exercise 5.33. If Y denotes the number of items containing at least one defect, among the ten sampled items, find the probability that
  (a) Y is exactly 2.   (b) Y is at least 1.

5.35   Vehicles arriving at an intersection can turn right or left, or continue straight ahead. In a study of traffic patterns at this intersection over a long period of time, engineers have noted that 40% of the vehicles turn left, 25% turn right, and the remainder continue straight ahead.
  (a)   For the next five cars entering this intersection, find the probability that one turns left, one turns right, and three continue straight ahead.

(b) For the next five cars entering the intersection, find the probability that at least one turns right.

(c) If 100 cars enter the intersection in a day, find the expected value and variance of the number turning left. What assumptions are necessary for your answer to be valid?

*5.6 MORE ON THE MOMENT-GENERATING FUNCTION

The use of moment-generating functions for identifying distributions of random variables is particularly useful when working with sums of independent random variables. For example suppose X_1 and X_2 are independent, exponential random variables, each with mean θ, and $Y = X_1 + X_2$. The moment-generating function of Y is given by

$$M_Y(t) = E(e^{tY}) = E(e^{t(X_1 + X_2)})$$
$$= E(e^{tX_1} e^{tX_2})$$
$$= E(e^{tX_1})E(e^{tX_2})$$
$$= M_{X_1}(t)M_{X_2}(t)$$

since X_1 and X_2 are independent. From Chapter 4

$$M_{X_1}(t) = M_{X_2}(t) = (1 - \theta t)^{-1}$$

and so

$$M_Y(t) = (1 - \theta t)^{-2}$$

Upon recognizing the form of this moment-generating function, we can immediately conclude that Y has a gamma distribution with $\alpha = 2$ and $\beta = \theta$. This result can be generalized to the sum of an independent gamma random variables with common scale parameter β.

EXAMPLE 5.17

Let X_1 denote the number of vehicles passing a particular point on the eastbound lane of a highway in one hour. Suppose the Poisson distribution with mean λ_1 is a reasonable model for X_1. Now let X_2 denote the number of vehicles passing a point on the westbound lane of the same highway in one hour. Suppose X_2 has a Poisson distribution with mean λ_2. Of interest is $Y = X_1 + X_2$, the total traffic count in both lanes, per hour. Find the probability distribution for Y if X_1 and X_2 are assumed independent.

*Optional section.

Solution It is known from Chapter 4 that

$$M_{X_1}(t) = e^{\lambda_1(e^t - 1)}$$

and

$$M_{X_2}(t) = e^{\lambda_2(e^t - 1)}$$

By the property of moment-generating functions seen above,

$$\begin{aligned} M_Y(t) &= M_{X_1}(t)M_{X_2}(t) \\ &= e^{\lambda_1(e^t - 1)}e^{\lambda_2(e^t - 1)} \\ &= e^{(\lambda_1 + \lambda_2)(e^t - 1)} \end{aligned}$$

Now the moment-generating function for Y has the form of a Poisson moment-generating function with mean $(\lambda_1 + \lambda_2)$. Thus by the uniqueness property Y must have a Poisson distribution with mean $\lambda_1 + \lambda_2$.

The fact that we can add independent Poisson random variables and still retain the Poisson properties is important in many applications. ■

EXERCISES

5.36 Find the moment-generating function for the negative binomial random variable. Use it to derive the mean and variance of that distribution.

5.37 There are two entrances to a parking lot. Cars arrive at entrance I according to a Poisson distribution with an average of three per hour, and at entrance II according to a Poisson distribution with an average of four per hour. Find the probability that exactly three cars arrive at the parking lot in a given hour.

5.38 Let X_1 and X_2 denote independent normally distributed random variables, not necessarily having the same mean or variance. Show that, for any constants a and b, $Y = aX_1 + bX_2$ is normally distributed.

5.39 Resistors of a certain type have resistances that are normally distributed with a mean of 100 ohms and a standard deviation of 10 ohms. Two such resistors are connected in series, which causes the total resistance in the circuit to be the sum of the individual resistances. Find the probability that the total resistance
(a) exceeds 220 ohms.
(b) is less than 190 ohms.

5.40 A certain type of elevator has a maximum weight capacity X_1, which is normally distributed with a mean and standard deviation of 5000 and 300 pounds, respectively. For a certain building equipped with this type of elevator, the elevator loading X_2 is a normally distributed random variable with a mean and standard deviation of 4000 and 400 pounds, respectively. For any given time that the elevator is in use, find the probability that it will be overloaded, assuming X_1 and X_2 are independent.

5.7 CONDITIONAL EXPECTATIONS

Section 5.2 contains a discussion of conditional probability functions and conditional density functions, which we shall now relate to *conditional expectations*. Conditional expectations are defined in the same manner as univariate expectations except that the conditional density is used in place of the marginal density function.

DEFINITION 5.7	If X_1 and X_2 are any two random variables, the **conditional expectation** of X_1 given that $X_2 = x_2$ is defined to be $$E(X_1 \mid X_2 = x_2) = \int_{-\infty}^{\infty} x_1 f(x_1 \mid x_2)\, dx_1$$ if X_1 and X_2 are jointly continuous and $$E(X_1 \mid X_2 = x_2) = \sum_{y_1} x_1 p(x_1 \mid x_2)$$ if X_1 and X_2 are jointly discrete.

EXAMPLE 5.18

Refer to the Y_1 and Y_2 of Example 5.6, page 190, which we shall relabel as X_1 and X_2.

$$f(x_1, x_2) = \begin{cases} 1/2 & 0 \leqslant x_1 \leqslant x_2,\ 0 \leqslant x_2 \leqslant 2 \\ 0 & \text{elsewhere} \end{cases}$$

Find the conditional expectation of amount of sales X_1 given that $X_2 = 1$.

Solution In Example 5.6 we found that

$$f(x_1 \mid x_2) = \begin{cases} 1/x_2 & 0 \leqslant x_1 \leqslant x_2 \leqslant 2 \\ 0 & \text{elsewhere} \end{cases}$$

Thus from Definition 5.7

$$E(X_1 \mid X_2 = 1) = \int_{-\infty}^{\infty} x_1 f(x_1 \mid x_2)\, dx_1$$

$$= \int_0^1 x_1(1)\, dx_1$$

$$= \frac{x_1^2}{2}\Big|_0^1 = \frac{1}{2}$$

That is, if the soft-drink machine contains 1 gallon at the start of the day, the expected sales for that day is 1/2 gallon. ∎

The conditional expectation of X_1 given $X_2 = x_2$ is a function of x_2. If we now let X_2 range over all its possible values, we can think of the conditional expectation as a function of the random variable X_2 and hence we can find the expected value of the conditional expectation. The result of this type of iterated expectation is given in Theorem 5.3.

THEOREM 5.3   Let X_1 and X_2 denote random variables. Then

$$E(X_1) = E[E(X_1 | X_2)]$$

where, on the right-hand side, the inside expectation is with respect to the conditional distribution of X_1 given X_2, and the outside expectation is with respect to the distribution of X_2.

Proof   Let X_1 and X_2 have joint density function $f(x_1, x_2)$ and marginal densities $f_1(x_1)$ and $f_2(x_2)$, respectively. Then

$$E(X_1) = \int_{-\infty}^{\infty} x_1 f_1(x_2) \, dx$$

$$= \int_{-\infty}^{\infty} \int_{-\infty}^{\infty} x_1 f(x_1, x_2) \, dx_1, \, dx_2$$

$$= \int_{-\infty}^{\infty} \int_{-\infty}^{\infty} x_1 f_1(x_1 | x_2) f_2(x_2) \, dx_1 \, dx_2$$

$$= \int_{-\infty}^{\infty} \left[\int_{-\infty}^{\infty} x_1 f_1(x_1 | x_2) \, dx_1 \right] f_2(x_2) \, dx_2$$

$$= \int_{-\infty}^{\infty} E(X_1 | X_2 = x_2) f_2(x_2) \, dx_2$$

$$= E[E(X_1 | X_2)]$$

The proof is similar for the discrete case.

EXAMPLE 5.19

A quality-control plan for an assembly line involves sampling $n = 10$ finished items per day and counting Y, the number of defectives. If p denotes the probability of observing a defective, then Y has a binomial distribution if the number of items produced by the line is large. But p varies from day to day and is assumed to have a uniform distribution on the interval 0 to 1/4. Find the expected value of Y for any given day.

Solution   From Theorem 5.3 we know that

$$E(Y) = E[E(Y | p)]$$

For a given p, Y has binomial distribution and hence

$$E(Y|p) = np$$

Thus

$$E(Y) = E(np) = nE(p)$$

$$= n \int_0^{1/4} 4p\, dp$$

$$= n(1/8)$$

and for $n = 10$,

$$E(Y) = 10/8 = 5/4$$

This inspection policy should average 5/4 defectives per day, in the long run. The calculations could be checked by actually finding the unconditional distribution of Y and computing $E(Y)$ directly. ∎

5.8 COMPOUNDING AND ITS APPLICATIONS

The univariate probability distributions of Chapters 3 and 4 depend on one or more parameters; once the parameters are known, the distributions are completely specified. However, these parameters are frequently unknown and, as in Example 5.19, may sometimes be regarded as random quantities. Assigning distributions to these parameters and then finding the marginal distributions of the original random variable is known as *compounding*. This process has theoretical as well as practical uses, as we illustrate next.

EXAMPLE 5.20

Suppose that Y denotes the number of bacteria per cubic centimeter in a certain liquid and that, for a given location, Y has a Poisson distribution with mean λ. Also assume λ varies from location to location and, for a location chosen at random, λ has a gamma distribution with parameters α and β, where α is a positive integer. Find the probability distribution for the bacteria count Y at a randomly selected location.

Solution Since λ is random, the Poisson assumption applies to the conditional distribution of Y for fixed λ. Thus

$$p(y|\lambda) = \frac{\lambda^y e^{-\lambda}}{y!} \qquad y = 0, 1, 2, \ldots$$

Also

$$f(\lambda) = \frac{1}{\Gamma(\alpha)\beta^\alpha} \lambda^{\alpha-1} e^{-\lambda/\beta} \qquad \lambda > 0$$

$$= 0 \qquad\qquad\qquad \text{elsewhere}$$

Then the joint distribution of λ and Y is given by

$$g(y, \lambda) = p(y|\lambda)f(\lambda)$$

$$= \frac{1}{y!\Gamma(\alpha)\beta^\alpha} \lambda^{y+\alpha-1} e^{-\lambda[1+(1/\beta)]}$$

The marginal distribution of Y is found by integrating over λ and yields

$$p(y) = \frac{1}{y!\Gamma(\alpha)\beta^\alpha} \int_0^\infty \lambda^{y+\alpha-1} e^{\lambda[1+(1/\beta)]} dx$$

$$= \frac{1}{y!\Gamma(\alpha)\beta^\alpha} \Gamma(y+\alpha) \left(1 + \frac{1}{\beta}\right)^{-(y+\alpha)}$$

Since α is integer,

$$p(y) = \frac{(y+\alpha-1)!}{(\alpha-1)!y!} \left(\frac{1}{\beta}\right)^\alpha \left(\frac{\beta}{1+\beta}\right)^{y+\alpha}$$

$$= \binom{y+\alpha-1}{\alpha-1} \left(\frac{1}{1+\beta}\right)^\alpha \left(\frac{\beta}{1+\beta}\right)^y$$

If we let $y + \alpha = n$ and $1/(1 + \beta) = p$, then $p(y)$ has the form of a negative binomial distribution. Hence the negative binomial distribution is a reasonable model for counts in which the mean count may be random. ∎

EXAMPLE 5.21 _____

Suppose that a customer arrives at a checkout counter in a store just as the counter is opening. A random number N of customers will be ahead of him since some customers may arrive early. Suppose that this number has the probability distribution

$$p(n) = P(N = n) = pq^n \qquad n = 0, 1, 2, \ldots$$

where $0 < p < 1$ and $q = 1 - p$. (This is a form of the geometric distribution.)

Customer service times are assumed to be independent and identically distributed exponential random variables with mean θ. Find the expected waiting time for this customer to complete his checkout.

Solution   For a given value of n the waiting time W is the sum of $n + 1$ independent exponential random variables and thus has a gamma distribution with $\alpha = n + 1$ and $\beta = \theta$. That is,

$$f(w|n) = \frac{1}{\Gamma(n+1)\theta^{n+1}} w^n e^{-w/\theta}$$

Hence

$$f(w, n) = \frac{p}{\Gamma(n+1)\theta^{n+1}} (qw)^n e^{-w/\theta}$$

and $f(w) = \dfrac{p}{\theta} e^{-w/\theta} \displaystyle\sum_{n=0}^{\infty} \left(\dfrac{qw}{\theta}\right)^n \dfrac{1}{n!}$

$\qquad\qquad = \dfrac{p}{\theta} e^{-w/\theta} e^{qw/\theta}$

$\qquad\qquad = \dfrac{p}{\theta} e^{-(w/\theta)(1-q)}$

$\qquad\qquad = \dfrac{p}{\theta} e^{-w(p/\theta)}$

The waiting time W is still exponential, but the mean is (θ/p).

SUPPLEMENTARY EXERCISES

5.41 Let X_1 and X_2 have the joint probability density function given by

$$f(x_1, x_2) = \begin{cases} Kx_1, x_2 & 0 \leqslant x_1 \leqslant 1, 0 \leqslant x_2 \leqslant 1 \\ 0 & \text{elsewhere} \end{cases}$$

(a) Find the value of K that makes this a probability density function.
(b) Find the marginal densities of X_1 and X_2.
(c) Find the joint distribution for X_1 and X_2.
(d) Find the probability $P(X_1 < 1/2, X_2 < 3/4)$.
(e) Find the probability $P(X_1 \leqslant 1/2 | X_2 > 3/4)$.

5.42 Let X_1 and X_2 have the joint density function given by

$$f(x_1, x_2) = \begin{cases} 3x_1 & 0 \leqslant x_2 \leqslant x_1 \leqslant 1 \\ 0 & \text{elsewhere} \end{cases}$$

(a) Find the marginal density functions of X_1 and X_2.
(b) Find $P(X_1 \leqslant 3/4, X_2 \leqslant 1/2)$.
(c) Find $P(X_1 \leqslant 1/2 | X_2 \geqslant 3/4)$.

5.43 From a group consisting of four Republicans, three Democrats, and two Independents, a committee of three persons is to be randomly selected. Let X_1 denote the number of Republicans and X_2 the number of Democrats on the committee.
(a) Find the joint probability distribution of X_1 and X_2.
(b) Find the marginal distributions of X_1 and X_2.
(c) Find the probability $P(X_1 = 1 | X_2 \geqslant 1)$.

5.44 For Exercise 5.41, find the conditional density of X_1 given $X_2 = x_2$. Are X_1 and X_2 independent?

5.45 For Exercise 5.42.
(a) Find the conditional density of X_1 given $X_2 = x_2$.
(b) Find the conditional density of X_2 given $X_1 = x_1$.
(c) Show that X_1 and X_2 are dependent.
(d) Find the probability $P(X_1 \leqslant 3/4 | X_2 = 1/2)$.

5.46 Let X_1 denote the amount of a certain bulk item stocked by a supplier at the beginning of a week and suppose that X_1 has a uniform distribution over the interval $0 \leqslant X_1 \leqslant 1$. Let X_2 denote the amount of this item sold by the supplier during the week and suppose that X_2 has a uniform distribution over the interval $0 \leqslant x_2 \leqslant x_1$, where x_1 is a specific value of X_1.

(a) Find the joint density function for X_1 and X_2.

(b) If the supplier stocks an amount of 1/2, what is the probability that he sells an amount greater than 1/4?

(c) If it is known that the supplier sold an amount equal to 1/4, what is the probability that he had stocked an amount greater than 1/2?

5.47 Let (X_1, X_2) denote the coordinates of a point at random inside a unit circle with center at the origin. That is, X_1 and X_2 have joint density function given by

$$f(x_1, x_2) = \begin{cases} 1/\pi & x_1^2 + x_2^2 \leqslant 1 \\ 0 & \text{elsewhere} \end{cases}$$

(a) Find the marginal density function of X_1.

(b) Find $P(X_1 \leqslant X_2)$.

5.48 Let X_1 and X_2 have the joint density function given by

$$f(x_1, x_2) = \begin{cases} x_1 + x_2 & 0 \leqslant x_1 \leqslant 1, 0 \leqslant x_2 \leqslant 1 \\ 0 & \text{elsewhere} \end{cases}$$

(a) Find the marginal density functions of X_1 and X_2.

(b) Are X_1 and X_2 independent?

(c) Find the conditional density of X_1 given $X_2 = x_2$.

5.49 Let X_1 and X_2 have the joint density function given by

$$f(x_1, x_2) = \begin{cases} K & 0 \leqslant x_1 \leqslant 2, 0 \leqslant x_2 \leqslant 1; 2x_2 \leqslant x_1 \\ 0 & \text{elsewhere} \end{cases}$$

(a) Find the value of K that makes the function a probability density.

(b) Find the marginal densities of X_1 and X_2.

(c) Find the conditional density of X_1 given $X_2 = x_2$.

(d) Find the conditional density of X_2 given $X_1 = x_1$.

(e) Find $P(X_1 \leqslant 1.5, X_2 \leqslant 0.5)$.

(f) Find $P(X_2 \leqslant 0.5 | X_1 \leqslant 1.5)$.

5.50 Let X_1 and X_2 have a joint distribution that is uniform over the region shaded in the diagram.

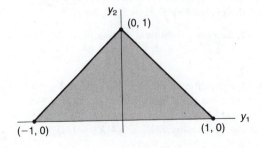

(a) Find the marginal density for X_2.
(b) Find the marginal density for X_1.
(c) Find $P[(X_1 - X_2) \geqslant 0]$.

5.51 Refer to Exercise 5.41.
(a) Find $E(X_1)$.
(b) Find $V(X_1)$.
(c) Find $Cov(X_1, X_2)$.

5.52 Refer to Exercise 5.42. Find $Cov(X_1, X_2)$.

5.53 Refer to Exercise 5.43.
(a) Find $Cov(X_1, X_2)$.
(b) Find $E(X_1 + X_2)$ and $V(X_1 + X_2)$ by finding the probability distribution of $X_1 + X_2$.
(c) Find $E(X_1 + X_2)$ and $V(X_1 + X_2)$ by using Theorem 5.2.

5.54 Refer to Exercise 5.48.
(a) Find $Cov(X_1, X_2)$.
(b) Find $E(3X_1 - 2X_2)$.
(c) Find $V(3X_1 - 2X_2)$.

5.55 Refer to Exercise 5.49.
(a) Find $E(X_1 + 2X_2)$.
(b) Find $V(X_1 + 2X_2)$.

5.56 A quality-control plan calls for randomly selecting three items from the daily production (assumed large) of a certain machine and observing the number of defectives. However, the proportion p of defectives produced by the machine varies from day to day and is assumed to have a uniform distribution on the interval $(0, 1)$. For a randomly chosen day find the unconditional probability that exactly two defectives are observed in the sample.

5.57 The number of defects per yard, denoted by X, for a certain fabric is known to have a Poisson distribution with parameter λ. However, λ is not known and is assumed to be random with its probability density function given by

$$f(\lambda) = \begin{cases} e^{-\lambda} & \lambda \geqslant 0 \\ 0 & \text{elsewhere} \end{cases}$$

Find the unconditional probability function for X.

5.58 The length of life X of a fuse has a probability density,

$$f(x) = \begin{cases} e^{-\lambda/\theta}/\theta & x > 0; \theta > 0 \\ 0 & \text{elsewhere} \end{cases}$$

Three such fuses operate independently. Find the joint density of their lengths of life, X_1, X_2, and X_3.

5.59 A retail grocery merchant figures that his daily gain from sales X is a normally distributed random variable with $\mu = 50$ and $\sigma^2 = 10$ (measurements in dollars). X could be negative if he is forced to dispose of perishable goods. Also he figures daily overhead costs Y to have a gamma distribution with $\alpha = 4$ and $\beta = 2$. If X and Y are independent, find the expected value and variance of his net daily gain. Would you expect his net gain for tomorrow to go above \$70?

5.60 Refer to Exercises 5.42 and 5.45.
  (a) Find $E(X_2|X_1 = x_1)$.
  (b) Use Theorem 5.3 to find $E(X_2)$.
  (c) Find $E(X_2)$ directly from the marginal density of X_2.

5.61 Refer to Exercise 5.57.
  (a) Find $E(X)$ by first finding the conditional expectation of X for given λ and then using Theorem 5.3.
  (b) Find $E(X)$ directly from the probability distribution of X.

5.62 Refer to Exercise 5.46. If the supplier stocks an amount equal to 3/4, what is the expected amount sold during the week?

5.63 Let X be a continuous random variable with distribution function $F(x)$ and density function $f(x)$. We can then write, for $x_1 \leqslant x_2$,

$$P(X \leqslant x_2 | X \geqslant x_1) = \frac{F(x_2) - F(x_1)}{1 - F(x_1)}$$

As a function of x_2 for fixed x_1, the right-hand side of this expression is called the conditional distribution function of X given that $X \geqslant x_1$. On taking the derivative with respect to x_2, we see that the corresponding conditional density function is given by

$$\frac{f(x_2)}{1 - F(x_1)} \qquad x_2 \geqslant x_1$$

Suppose that a certain type of electronic component has lifelength X with the density function (life length measured in hours)

$$f(x) = \begin{cases} (1/200)e^{-x/200} & x \geqslant 0 \\ 0 & \text{elsewhere} \end{cases}$$

Find the expected length of life for a component of this type that has already been in use for 100 hours.

*5.64 Let X_1, X_2, and X_3 be random variables, either continuous or discrete. The joint moment-generating function of X_1, X_2, and X_3 is defined by

$$M(t_1, t_2, t_3) = E(e^{t_1 X_1 + t_2 X_2 + t_3 X_3})$$

  (a) Show that $M(t, t, t)$ gives the moment-generating function of $X_1 + X_2 + X_3$.
  (b) Show that $M(t, t, 0)$ gives the moment-generating function of $X_1 + X_2$.
  (c) Show that

$$\frac{\partial^{k_1 + k_2 + k_3} M(t_1, t_2, t_3)}{\partial t_1^{k_1} \partial t_2^{k_2} \partial t_3^{k_3}}\bigg|_{t_1 = t_2 = t_3 = 0} = E(X_1^{k_1} X_2^{k_2} X_3^{k_3})$$

*5.65 Let X_1, X_2, and X_3 have multinomial distribution with probability function

$$p(x_1, x_2, x_3) = \frac{n!}{x_1! x_2! x_3!} p_1^{x_1} p_2^{x_2} p_3^{x_3} \qquad \sum_{i=1}^{n} x_i = n$$

Employ the results of Exercise 5.65 to answer the following.
  (a) Find the joint moment-generating function of X_1, X_2, and X_3.
  (b) Use the joint moment-generating function to find $Cov(X_1, X_2)$.

5.66 The negative binomial variable X was defined as the number of the trial on which the rth success occurs in a sequence of independent trials with constant probability p of success on each trial. Let X_i denote a geometric random variable, defined as the number of the trial on which the first success occurs. Then we can write

$$X = \sum_{i=1}^{n} X_i$$

for independently random variables $X_1, \ldots, X_r$. Use Theorem 5.2 to show that $E(X) = r/p$ and $V(X) = r(1 - p)/p^2$.

5.67 A box contains four balls, numbered 1 through 4. One ball is selected at random from this box. Let

$X_1 = 1$ if ball 1 or ball 2 is drawn

$X_2 = 1$ if ball 1 or ball 3 is drawn

$X_3 = 1$ if ball 1 or ball 4 is drawn

and the X_i's are zero otherwise. Show that any two of the random variables X_1, X_2, and X_3 are independent, but the three together are not.

6

Statistics and Sampling Distributions

ABOUT THIS CHAPTER

In the process of making an inference from a sample to a population we usually calculate one or more statistics, such as the mean or variance. Since samples are randomly selected, the values that such statistics assume may change from sample to sample. Thus sample statistics are, themselves, random variables, and their behavior can be modeled by probability distributions. The probability distribution of a sample statistic is called its *sampling distribution*. We will study the nature and properties of sampling distributions in this chapter.

CONTENTS

6.1 INTRODUCTION

Chapter 1 contained a discussion of ways to summarize data so that we can derive some useful, perhaps even important, information from them. We will now return to this idea of deriving information from data, but from this point on each data set under study will be viewed as a *sample* from a much larger set of data points that could be studied, called the *population*. A sample of 50 household incomes could be selected from all households in your community. A sample of 5 measurements could be taken on the diameter of a machined rod, these 5 being a small part of the nearly infinite number of diameter measurements that could have been taken if enough time were available. As you consider these two examples, notice that the first population (households in your community) is real (these households actually exist), while the second (measurements of rod diameters) is only conceptual (no list of rod diameter measurements really exists). In either case, however, we can visualize the nature of a population of values of which our sample data constitute only a small part. Our job is to describe populations as accurately as possible given only the data in samples from these populations. In other words we want to decide how closely our sample mirrors the characteristics of the population from which it came. As we have seen, numerical descriptive measures of a *population* are called *parameters*. The *sample* counterparts of these quantities, which form numerical descriptive measures of a sample, are called *statistics*.

DEFINITION 6.1	A **statistic** is a function of sample observations that contains no unknown parameters.

The households in your community do actually have an average, or mean, annual income (a parameter) even though it is unknown. A sample of households would have a sample mean income (a statistic) that can actually be calculated after the sampling is completed. Will the sample and population means be equal? Probably not, although we'll never be sure. We hope that the sample mean will be close to the population mean and, as we'll see, if the sampling is done well that hope is generally fulfilled. So we will now embark on a study of questions like "How great is the difference between a sample mean and its corresponding population mean likely to be?" The sample mean is only one of many possible statistics we might study, since the median, the quantiles, and the IQR (to name a few) are also important statistics. We begin with a careful study of the mean because of its central role in classical statistical inference procedures.

6.2 THE SAMPLE MEAN AND VARIANCE

To get a view of some essential features of sample means, we will generate samples from a population whose structure is known to us. That population is the set of random digits (0 through 9) in a large random-number table (real) or that could be generated by a computer (conceptual). If the digits really are randomly generated, an ideal sample of 100 digits, a sample that truly mirrors the population, would look like the one whose stem-and-leaf plot is given in Figure 6.1. (The leaves here are the tenths digit, all 0 in this case.) The mean of this ideal sample is 4.50, right in the middle of [0, 9]. This ideal sample mean can be called the *population mean*, since it reflects the average value of all digits in the population. But a true random sample of 100 digits will actually look more like Figure 6.2. A little irregularity slips into most random samples, and now the sample mean is 4.58, a little off from the ideal 4.50. This mean is based on 100 observations, a pretty large sample! What will happen to the discrepancies between the sample and population means for smaller samples? That's the question we'll investigate next.

FIGURE 6.1
Ideal Sample of Digits 0–9

10	0 | 0000000000
20	1 | 0000000000
30	2 | 0000000000
40	3 | 0000000000
50	4 | 0000000000
50	5 | 0000000000
40	6 | 0000000000
30	7 | 0000000000
20	8 | 0000000000
10	9 | 0000000000

6|0 denotes 6.0 or 6.

MEAN   =   4.5000

FIGURE 6.2
Real Sample of Digits 0–9

11	0 | 00000000000
20	1 | 000000000
27	2 | 0000000
35	3 | 00000000
49	4 | 00000000000000
(8)	5 | 00000000
43	6 | 0000000000000
30	7 | 00000000000
19	8 | 00000000000
8	9 | 00000000

MEAN   =   4.5800

Figure 6.3 displays the data for 100 *averages* of samples of size two from a random-digit generator. In other words two random digits were sampled and their mean (average) was recorded; then the process was repeated for a total of 100 sample averages. Notice first the shape of the distribution. The averages tend to pile up in the middle, with few values close to 0 or 9. But there is still a chance of getting a sample mean more than 4 units away from the ideal mean of 4.50.

The same idea is repeated for samples of size 3, 4, 5, and 10, and the resulting sample means are displayed in Figures 6.4, 6.5, 6.6, and 6.7. respectively.

FIGURE 6.3	5	0 | 55555
Sampling	8	1 | 000
Distribution for	21	2 | 0000000555555
Means of 2	37	3 | 0000000055555555
	(19)	4 | 00000000005555555555
	44	5 | 00000000000555555555
	25	6 | 000000555555
	13	7 | 000055
	7	8 | 0005
	3	9 | 000

MEAN = 4.4450

FIGURE 6.4	4	1 | 3366
Sampling	20	2 | 0000333366666666
Distribution for	37	3 | 00003333333366666
Means of 3	49	4 | 000003333666
	(26)	5 | 0003333333333333333336666666
	25	6 | 0000003333366666
	9	7 | 03366
	4	8 | 0033

MEAN = 4.6367

FIGURE 6.5	4	1 | 2557
Sampling	12	2 | 25555777
Distribution for	36	3 | 002222222255555555577777
Means of 4	(22)	4 | 0222225555557777777777
	42	5 | 000000022255555577777
	20	6 | 00000222257777
	6	7 | 0225
	2	8 | 05

MEAN = 4.6075

FIGURE 6.6	1	1 | 2
Sampling	10	2 | 044448888
Distribution for	27	3 | 00022444444666888
Means of 5	(37)	4 | 0000000222222224444444466666666666888
	36	5 | 0000002224444666688
	17	6 | 000222226888
	5	7 | 00022

MEAN = 4.5660

FIGURE 6.7
Sampling Distribution for Means of 10

3	2	558
25	3	2222223355667889999999
(41)	4	0011111111112233344555566666666667777888999
34	5	00000112444444555566688 9999
7	6	0123368

MEAN = 4.5980

Notice that the sample means pile up more and more in the middle as the sample size increases. Also the distribution of the sample means becomes quite symmetric around 4.50, the population mean. By the time the sample size gets to 5, there is virtually no chance of seeing a difference as large as 4 between a sample mean and its population mean. When the sample size reaches 10, there is virtually no chance of seeing a difference larger than 2.

To summarize, the *sampling distributions* of sample means (which is what Figures 6.2 through 6.7 are often called) tend to be symmetric about the true population mean and tend to pile up around the population mean, with small tail areas, as the sample size increases. This assumes that the samples are taken randomly and independently of one another. The box plots of Figure 6.8 are another way to show the shrinking variability and the concentration of values around the population mean with increasing sample size.

We have a principle of centrality for sampling distributions of means; the sample means tend to center around the population mean. Can we find a principle of variability more precise than "the variability among the sample means decreases as the sample size increases?" The study of this problem will be more easily presented if we introduce some notation.

We have used $\mu = E(X)$ to denote the expected, or mean, value of a random variable X and

$$\sigma^2 = V(X) = E(X - \mu)^2$$

to denote the variance. If the random variable X can take on any of N values with probability $1/N$, then the mean becomes

$$\mu = \frac{1}{N} \sum_{i=1}^{N} x_i$$

and the variance becomes

$$\sigma^2 = \frac{1}{N} \sum_{i=1}^{N} (x_i - \mu)^2$$

where x_i denotes the *i*th possible value for X. For a sample of size n the sample mean, which we will denote by $\bar{x}$, is

$$\bar{x} = \frac{1}{n} \sum_{i=1}^{n} x_i$$

where, in this case, x_i is the *i*th sample observation. So the sample copy of μ, namely $\bar{x}$, has a distribution in repeated sampling that centers about μ. We might say, then,

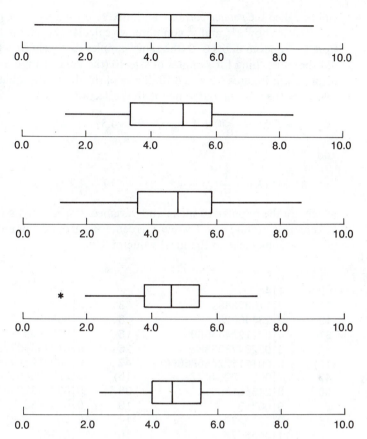

FIGURE 6.8
Box Plots of the Sampling Distributions of Averages (Sample sizes 2, 3, 4, 5, and 10 from top to bottom)

that $\bar{x}$ is a good estimator of μ. (The average annual household income calculated from a sample of 50 households should be close to the true average for all households in your community.)

To measure variability in a sample, we might try a sample copy of the variance, namely,

$$s^{*2} = \frac{1}{n} \sum_{i=1}^{n} (x_i - \bar{x})^2$$

As it turns out, this is not a very good estimator. Whereas random samples tend to center at their population mean, these same samples tend to have *less* variability than the population that they came from. This should be intuitively reasonable. A relatively small sample is not likely to capture the extremes of the population and, hence, will show less variability than is present in the population. To compensate for this fact, we change the divisor of s^{*2} from n to $n - 1$, thereby making the quantity larger. Thus

$$s^2 = \frac{1}{n-1} \sum_{i=1}^{n} (x_i - \bar{x})^2$$

will be called the *sample variance* and used to estimate σ^2, the population variance.

To see how s^{*2} and s^2 compare, we calculated each for 100 samples of size 5 from the random digits 0–9 (the same samples that went into Figure 6.6). Then we did the same thing for samples of size 10 (the data of Figure 6.7). The results are displayed in Figures 6.9 and 6.10. If any of the digits are equally likely on any one selection, then X, the outcome for that selection, has

$$\mu = E(X) = \frac{1}{10}(0 + 1 + \cdots + 9) = 4.5$$

and

$$\sigma^2 = V(X) = \frac{1}{10}[(0 - 4.5)^2 + \cdots + (9 - 4.5)^2] = 8.25$$

which are the population mean and variance, respectively. Note that s^{*2} seems to be centering around 6 or 7, while s^2 is centering around 8 or more. The mean values of s^2 are quite close to the ideal value of 8.25.

FIGURE 6.9
*Sampling
Distributions of s*²*

1	0|5			
2	1|4			
11	2|115699999	3	2|278	
19	3|23446678	8	3|56688	
31	4|001122445689	15	4|0012458	
42	5|02234788888	26	5|01244446778	
(16)	6|0011112245666668	47	6|000012344456666668889	
42	7|000222234477	(15)	7|222223466667888	
30	8|245555569	38	8|0000222344566888899	
21	9|00348	19	9|0445668	
16	10|2222228	12	10|68	
9	11|446677	10	11|04469	
3	12|49	5	12|044	
1	13|4	2	13|24	

MEAN   =   6.6248   MEAN   =   7.3702

Samples of size 5   Samples of size 10

We now know that s^2 is a good estimator of σ^2; for that reason s^2 (or s) will be used as the standard measure of variability in a sample. But what about the variability of the sample means? If we let X_i denote the random outcome of the ith sample observation to be selected, then

$$\bar{X} = \frac{1}{n} \sum_{i=1}^{n} X_i$$

and, from properties of expectations of linear functions,

$$E(\bar{X}) = \frac{1}{n} \sum_{i=1}^{n} E(X_i)$$

$$= \frac{1}{n} \sum_{i=1}^{n} \mu = \mu$$

FIGURE 6.10
Sampling
Distributions of
Sample Variances
(s²) (with box
plots)

1	0	7
2	1	8
4	2	77
11	3	2377777
19	4	02335578
29	5	0022335578
36	6	0235578
50	7	23333355777788
50	8	02333335888
39	9	000023377
30	10	35777778
22	11	23378
17	12	3888888
10	13	5
9	14	335577
3	15	5
2	16	28

MEAN = 8.28

Samples of size 5

1	2	4
4	3	029
13	4	012244679
18	5	04677
32	6	00012444677789
47	7	011223333345667
(19)	8	0000112444567779999
34	9	122234566788899
19	10	0446679
12	11	8
11	12	13779
6	13	2478
2	14	79

MEAN = 2.17

Samples of size 10

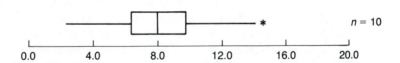

where $\mu = E(X_i)$, $i = 1, \ldots, n$. If, in addition, the sampled observations are selected independently of one another,

$$V(\bar{X}) = \frac{1}{n^2} V\left(\sum_{i=1}^{n} X_i\right)$$

$$= \frac{1}{n^2} \sum_{i=1}^{n} V(X_i)$$

$$= \frac{1}{n^2} \sum_{i=1}^{n} \sigma^2 = \frac{\sigma^2}{n}$$

Thus the standard deviation within any one sampling distribution of sample means should be $\sigma/\sqrt{n}$. To check this out for the data generated in Figures 6.3 through 6.7, we calculated $\sigma/\sqrt{n}$ for each case and compared it to the sample standard deviation calculated for the 100 sample means in that particular figure. The results are as follows (with $\sigma = \sqrt{8.25} = 2.87$):

Figure	Sample size	Calculated Standard Deviation	$\sigma/\sqrt{n}$
6.3	2	2.05	2.03
6.4	3	1.74	1.66
6.5	4	1.49	1.44
6.6	5	1.27	1.28
6.7	10	0.92	0.91

We see that in each case the observed variability for the 100 sample means is quite close to the theoretical value of $\sigma/\sqrt{n}$. In summary, then, the variability in the sample means decreases as the sample size increases, with the variance of a sample mean given by σ^2/n. The quantity $\sigma/\sqrt{n}$ is sometimes called the *standard error* of the mean. The σ^2 can be estimated by the sample variance s^2.

A sample that contains independent observations selected from the same population will be called a *random sample*. Suppose n resistors of the same type are placed into service, and their lifelengths measured. If the lifelengths are independent of one another, then the resulting measurements would constitute an observed random sample. The assumption of independence is difficult to justify in some cases, but careful thought should be given to the matter in every statistical problem. If the resistors are all in the same system and fail because of a power surge in the system, then their lifelengths are clearly *not* independent. If, however, they fail only from normal usage, then their lifelengths may be independent even if they are being used in the same system. *Most of the work in this book will assume that sample measurements come from random samples.*

The term *random sample* suggests the manner in which experimental items should be selected to ensure some degree of independence in the resulting measurements. The items should be selected *at random*. One method of random selection involves the use of a random-number table, as shown in Table 1 of the Appendix. Suppose 250 circuit boards constitute one day's production, and 10 are to be sampled for quality inspection. The boards are numbered from 1 to 250. Then 10 three-digit random numbers between 001 and 250 are selected from a random-number table (or computer). The sample consists of the 10 boards bearing the selected numbers. The 10 quality measurements (perhaps the number of properly functioning circuits per board) then make up a random sample of measurements.

EXAMPLE 6.1

_____

One hundred employees of a firm have annual salaries (in thousands of dollars) with a mean of 22, a standard deviation of 3, and a median of 20. Use Tchebysheff's Theorem, which holds for sample data, to find an interval that contains at least 75% of the salaries. What interpretation can you give to the median salary?

Solution   Tchebysheff's Theorem states that at least $(1 - 1/k^2)$ of the measurements must be within k standard deviations of their mean. For $k = 2$

$$1 - \frac{1}{k^2} = 1 - \frac{1}{4} = \frac{3}{4}$$

or 75%. Now $\bar{x} = 22$ and $s = 3$, and $\bar{x} \pm 2s$ yields

$$22 \pm 2(3)$$

or

$$22 \pm 6$$

At least 75% of the salaries must be between 16 and 28, inclusive.

The median of 20 tells us that 50% of the salaries must be below 20. These two pieces of information might suggest that there are many salaries (but not more than 50%) in the 16 to 20 interval, among the original 100. ∎

Suppose measurements $x_1, x_2, \ldots, x_n$ are transformed to new measurements $y_1, y_2, \ldots, y_n$ by

$$y_i = ax_i + b$$

for constants a and b. If $x_1, \ldots, x_n$ has mean $\bar{x}$ and variance s_x^2, then it is easy to show that $y_1, \ldots, y_n$ has mean $\bar{y}$ and variance s_y^2 given by

$$\bar{y} = a\bar{x} + b$$

and

$$s_y^2 = a^2 s_x^2$$

These properties, along with the calculation of $\bar{x}$ and s^2, are illustrated in Example 6.2.

EXAMPLE 6.2

For a simple scaffold structure made of steel, it is important to study the increase in length of tension members under load. For a load of 2000 kilograms ten similar tension members showed length increases as follows, with measurements in centimeters:

2.5, 2.2, 3.0, 2.1, 2.7, 2.5, 2.8, 1.9, 2.2, 2.6

(a) Find the mean and standard deviation of these measurements.

(b) Suppose the measurements had been taken in meters rather than centimeters. Find the mean and standard deviation of the corresponding measurements in meters.

Solution   (a) The mean is given by

$$\bar{x} = \frac{1}{n} \sum_{i=1}^{n} x_i = \frac{1}{10} (2.5 + 2.2 + \cdots + 2.6)$$

$$= \frac{1}{10} (24.5) = 2.45$$

and the variance is given by

$$s_x^2 = \frac{1}{n-1} \sum_{i=1}^{n} (x_i - \bar{x})^2$$

which is equivalent to

$$s_x^2 = \left(\frac{1}{n-1}\right)\left[\sum_{i=1}^{n} x_i^2 - \frac{1}{n}\left(\sum_{i=1}^{n} x_i\right)^2\right]$$

$$= \frac{1}{9}\left[61.09 - \frac{1}{10}(24.5)^2\right]$$

$$= \frac{1}{9}(61.09 - 60.025) = 0.1183$$

Then the standard deviation is

$$s_x = \sqrt{s_x^2} = \sqrt{0.1183} = 0.34$$

(b)   If y_i denotes the corresponding measurement in meters, then

$$y_i = (0.01)x_i$$

It follows that

$$\sum_{i=1}^{10} y_i = \sum_{i=1}^{10} (0.01)x_i = (0.01) \sum_{i=1}^{10} x_i$$

and

$$\bar{y} = (0.01)\bar{x} = (0.01)(2.45) = 0.0245$$

Also

$$s_y^2 = \frac{1}{9} \sum_{i=1}^{10} (0.01x_i - 0.01\bar{x})^2$$

$$= \frac{1}{9} \sum_{i=1}^{10} (0.01)^2(x_i - \bar{x})^2$$

$$= (0.01)^2 \left(\frac{1}{9}\right) \sum_{i=1}^{10} (x_i - \bar{x})^2$$

$$= (0.01)^2 s_x^2$$

Hence

$$s_y = \sqrt{s_y^2} = (0.01)s_x = (0.01)(0.34) = 0.0034$$ ■

EXAMPLE 6.3 _____

Calculate $\bar{x}$ and s^2 for the fifty lifelength observations in Table 4.1. Also approximate the variance of $\bar{X}$.

Solution   From the data in Table 4.1 we have

$$\bar{x} = \frac{1}{n} \sum_{i=1}^{n} x_i = \frac{1}{50} (113.296) = 2.266$$

Using

$$s^2 = \frac{1}{n-1} \left[\sum_{i=1}^{n} x_i^2 - \frac{1}{n} \left(\sum_{i=1}^{n} x_i \right)^2 \right]$$

we have

$$s^2 = \frac{1}{49} \left[440.2332 - \frac{1}{50} (113.296)^2 \right]$$

$$= 3.745$$

or

$$s = 1.935$$

Note that the exponential model suggested in Section 4.1,

$$f(x) = \frac{1}{2} e^{-x/2} \qquad x > 0$$

gives $E(X) = 2$, which is close to the observed value of $\bar{x}$. Also from the exponential model $V(X) = \sigma^2 = 4$, which is quite close to the observed s^2 value. $V(\bar{X}) = \sigma^2/n$ can be approximated by s^2/n, or $3.745/50 = 0.075$.   ∎

To illustrate further the behavior of $\bar{x}$ and s^2, we have selected one hundred samples each of size $n = 25$ from an exponential distribution with a mean of 10. That is, the probabilistic model for the population is given by

$$f(x) = \begin{cases} \frac{1}{10} e^{-x/10} & x > 0 \\ 0 & \text{elsewhere} \end{cases}$$

For this model $E(X) = \mu = 10$ and $V(X) = \sigma^2 = (10)^2$, or $\sigma = 10$. The values for $\bar{x}$ and s^2 were calculated for each of the 100 samples. The average of the 100 $\bar{x}$s turned out to be 9.88, and the average of the 100 values of s is 9.70. Note that both are reasonably close to 10. The standard deviation for the 100 values of $\bar{x}$ was calculated and found to be 2.17. Theoretically the standard deviation of $\bar{X}$ is

$$\sqrt{V(\bar{X})} = \frac{\sigma}{\sqrt{n}} = \frac{10}{\sqrt{25}} = 2.0$$

which is not far from the observed 2.17. A relative frequency histogram for the 100 values of $\bar{x}$ is shown in Figure 6.11. We pursue the notions connected with the shape of this distribution in the following section.

FIGURE 6.11
Relative Frequency
Histogram for $\bar{x}$
from 100 Samples
Each of Size 25

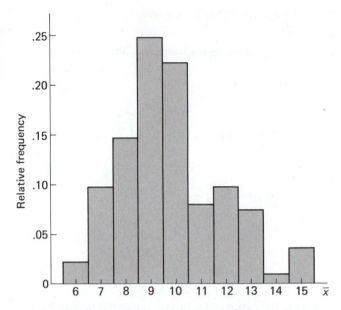

EXERCISES

6.1 Concentrations of uranium 238 were measured in twelve soil samples from a certain region, with the following results in pCi/g (picoCuries per gram):

0.76, 1.90, 1.84, 2.42, 2.01, 1.77,

1.89, 1.56, 0.98, 2.10, 1.41, 1.32

(a) Construct a stem-and-leaf display and box plot.
(b) Calculate $\bar{x}$ and s^2 for these data.
(c) If another soil sample is randomly selected from this region, find an interval in which the uranium concentration measurement should lie with probability at least 0.75. (Assume the sample mean and variance are good approximations to the population mean and variance. Use Tchebysheff's Theorem.)

6.2 The following data show the percent change in production of crude petroleum, 1976 to 1977, for selected countries in America, Western Europe, and the Middle East:

-1.4, $+0.3$, $+8.0$, -13.6, -4.1, -2.1, $+8.1$, -7.5, -2.0,

$+4.5$, 0.0, $+205.2$, -0.2, -6.5, -7.0, -7.6, $+7.7$, $+4.5$

(Source: *The World Almanac and Book of Facts*, 1979.)

(a) Construct a stem-and-leaf display and box plot.
(b) Calculate $\bar{x}$ and s^2, the sample mean and variance.
(c) Construct a relative frequency histogram for the data. Do the calculations in (b) look reasonable?
(d) If another country is chosen at random from America, Western Europe, or the Middle East, find an interval in which its change in petroleum production should lie with probability at least 0.89.

6.3   Refer to Exercise 6.2. The figure $+205.2$ is from the United Kingdom, and is exceptionally high because a major new source of petroleum was located in the North Sea. Eliminate this unusual figure and answer the questions of Exercise 6.2 with the reduced data set. Note the effect that this large value has on the mean and variance.

6.4   Five daily high temperature readings for a city in Florida were

22, 20, 24, 21, 26

in degrees Celsius.
   (a)   Find the mean and variance for these daily temperatures.
   (b)   Degrees Celsius (x) are transformed to degrees Fahrenheit (y) by the equation

$$y = 1.8x + 32$$

Find the mean and variance for the five temperature readings on the Fahrenheit scale.

6.5   The *Los Angeles Times* of September 20, 1981, contains the following statement: "According to the latest enrollment analysis by age-categories, half of the (Los Angeles Community College) district's 128,000 students are over the age of 24. The average student is 29." Can this statement be correct? Explain.

6.6   "In the region we are traveling west of Whitney, precipitation drops off and the average snow depth on April 1 for the southern Sierra is a modest 5 to 6 feet. And two winters out of three, the snow pack is below average." (E. Bowen, *The High Sierra*, Time-Life Books, 1972, p. 142.) Explain how this statement can be true.

6.7   Show that

$$s^2 = \frac{1}{n-1} \sum_{i=1}^{n} (x_i - \bar{x})^2$$

$$= \frac{1}{n-1} \left[\sum_{i=1}^{n} x_i^2 - \frac{1}{n} \left(\sum_{i=1}^{n} x_i \right)^2 \right]$$

6.8   If $X_1, \ldots, X_n$ is a random sample with $E(X_i) = \mu$ and $V(X_i) = \sigma^2$, show that $E(S^2) = \sigma^2$, where

$$S^2 = \frac{1}{n-1} \sum_{i=1}^{n} (X_i - \bar{X})^2$$

[*Hint:* Write $\sum_{i=1}^{n} (X_i - \bar{X})^2 = \sum_{i=1}^{n} [(X_i - \mu) - (\bar{X} - \mu)]^2$. Square the term in brackets, carry out the summation, and then find the expectations.]

6.3   THE SAMPLING DISTRIBUTION OF X̄

We have seen that the approximated sampling distributions of a sample mean (Figures 6.3 through 6.7) tend to have a particular shape, being somewhat symmetric with a single mound in the middle. This result is not coincidental or unique to those particular examples. That sampling distributions for sample means always tend to be approximately normal in shape is a consequence of the *Central Limit Theorem*.

706

THEOREM 6.1   *The Central Limit Theorem.*   If a random sample of size n is drawn from a population with mean μ and variance σ^2, then the sample mean $\bar{X}$ has approximately a normal distribution with mean μ and variance σ^2/n. That is, the distribution function of

$$\frac{\bar{X} - \mu}{\sigma/\sqrt{n}}$$

is approximately a standard normal. The approximation improves as the sample size increases.

Theorem 6.1 will not be proved in this book; the proof belongs in a more advanced course.

We sometimes abbreviate the central tenet of Theorem 6.1 to the phrase "$\bar{X}$ is asymptotically normal with mean μ and variance σ^2/n." The practical importance of this result is that, for large n, the sampling distribution of $\bar{X}$ can be closely approximated by a normal distribution. More precisely,

$$P(\bar{X} \leqslant b) = P\left(\frac{\bar{X} - \mu}{\sigma/\sqrt{n}} \leqslant \frac{b - \mu}{\sigma/\sqrt{n}}\right)$$

$$\approx P\left(Z \leqslant \frac{b - \mu}{\sigma/\sqrt{n}}\right)$$

where Z is a standard normal random variable.

We saw some approximate sampling distributions in Figures 6.3 through 6.7, but now we will look at more computer simulations of sampling distributions with larger sample sizes. Samples of size n were drawn from a population having the probability density function

$$f(x) = \begin{cases} \frac{1}{10}e^{-x/10} & x > 0 \\ 0 & \text{elsewhere} \end{cases}$$

The sample mean was computed for each sample. The relative frequency histogram of these mean values for 1000 samples of size $n = 5$ is shown in Figure 6.12. Figures 6.13 and 6.14 show similar results for 1000 samples of size $n = 25$ and $n = 100$, respectively. Although all the relative frequency histograms have a sort of bell shape, notice that the tendency toward a symmetric normal curve is better for larger n. A smooth curve drawn through the bar graph of Figure 6.14 would be nearly identical to a normal density function with mean 10 and variance $(10)^2/100 = 1$.

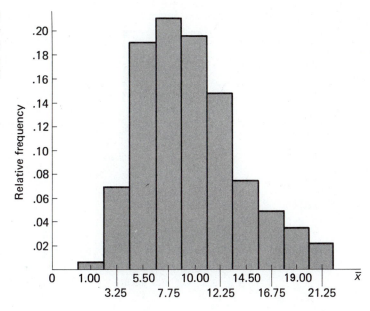

FIGURE 6.12
Relative Frequency Histogram for $\bar{x}$ *from 1000 Samples of Size* $n = 5$

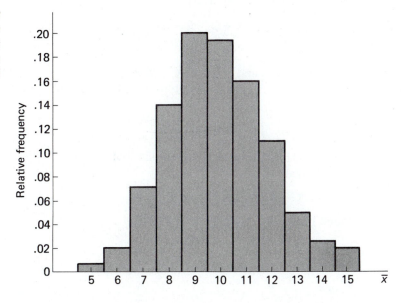

FIGURE 6.13
Relative Frequency Histogram for $\bar{x}$ *from 1000 Samples of Size* $n = 25$

The Central Limit Theorem provides a very useful result for statistical inference, for we now know not only that $\bar{X}$ has mean μ and variance σ^2/n if the population has mean μ and variance σ^2, but we know also that the probability distribution for $\bar{X}$ is approximately normal. For example, suppose we wish to find an interval (a, b) such that

$$P(a \leqslant \bar{X} \leqslant b) = 0.95$$

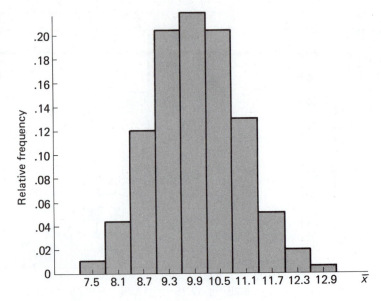

FIGURE 6.14
*Relative Frequency
Histogram for x̄
from 1000 Samples
of Size n = 100*

This probability is equivalent to

$$P\left(\frac{a - \mu}{\sigma/\sqrt{n}} \leqslant \frac{\bar{X} - \mu}{\sigma/\sqrt{n}} \leqslant \frac{b - \mu}{\sigma/\sqrt{n}}\right) = 0.95$$

for constants μ and σ. Since $(\bar{X} - \mu)/(\sigma/\sqrt{n})$ has approximately a standard normal distribution, the above equality can be approximated by

$$P\left(\frac{a - \mu}{\sigma/\sqrt{n}} \leqslant Z \leqslant \frac{b - u}{\sigma/\sqrt{n}}\right) = 0.95$$

where Z has a standard normal distribution. From Table 4 in the Appendix we know that

$$P(-1.96 \leqslant Z \leqslant 1.96) = 0.95$$

and hence

$$\frac{a - \mu}{\sigma/\sqrt{n}} = -1.96 \qquad \frac{b - \mu}{\sigma/\sqrt{n}} = 1.96$$

or

$$a = \mu - 1.96\sigma/\sqrt{n} \qquad b = \mu + 1.96\sigma/\sqrt{n}$$

EXAMPLE 6.4

The fracture strengths of a certain type of glass average 14 (thousands of pounds per square inch) and have a standard deviation of 2.

(a) What is the probability that the average fracture strength for 100 pieces of this glass exceeds 14.5?

(b) Find an interval that includes the average fracture strength for 100 pieces of this glass with probability 0.95.

Solution   (a) The average strength $\bar{X}$ has approximately a normal distribution with mean $\mu = 14$ and standard deviation

$$\frac{\sigma}{\sqrt{n}} = \frac{2}{\sqrt{100}} = 0.2$$

Thus

$$P(\bar{X} > 14.5) = P\left(\frac{\bar{X} - \mu}{\sigma/\sqrt{n}} > \frac{14.5 - \mu}{\sigma/\sqrt{n}}\right)$$

is approximately equal to

$$P\left(Z > \frac{14.5 - 14}{0.2}\right) = P\left(Z > \frac{0.5}{0.2}\right)$$

$$= P(Z > 2.5) = 0.5 - 0.4938 = 0.0062$$

from Table 4. The probability of seeing an average value (for $n = 100$) more than 0.5 unit above the population mean is, in this case, very small.

   (b) We have seen that

$$P\left(\mu - 1.96\,\frac{\sigma}{\sqrt{n}} \leqslant \bar{X} \leqslant \mu + 1.96\,\frac{\sigma}{\sqrt{n}}\right) = 0.95$$

for a normally distributed $\bar{X}$. In this problem

$$\mu - 1.96\,\frac{\sigma}{\sqrt{n}} = 14 - 1.96\,\frac{2}{\sqrt{100}} = 13.6$$

and

$$\mu + 1.96\,\frac{\sigma}{\sqrt{n}} = 14 + 1.96\,\frac{2}{\sqrt{100}} = 14.4$$

Approximately 95% of sample mean fracture strengths, for samples of size 100, should lie between 13.6 and 14.4. ∎

EXAMPLE 6.5

A certain machine used to fill bottles with liquid has been observed over a long period of time, and the variance in the amounts of fill is found to be approximately $\sigma^2 = 1$ oz. However, the mean ounces of fill μ depends on an adjustment that may change from day to day, or operator to operator. If $n = 25$ observations on ounces of fill dispensed are to be taken on a given day (all with the same machine setting), find the probability that the sample mean will be within 0.3 ounce of the true population mean for that setting.

Solution   We will assume $n = 25$ is large enough for the sample mean $\bar{X}$ to have approximately a normal distribution. Then

$$P(|\bar{X} - \mu| \leq 0.3) = P[-0.3 \leq (\bar{X} - \mu) \leq 0.3]$$

$$= P\left[-\frac{0.3}{\sigma/\sqrt{n}} \leq \frac{\bar{X} - \mu}{\sigma/\sqrt{n}} \leq \frac{0.3}{\sigma/\sqrt{n}}\right]$$

$$= P\left[-0.3\sqrt{25} \leq \frac{\bar{X} - \mu}{\sigma/\sqrt{n}} \leq 0.3\sqrt{25}\right]$$

$$= P\left[-1.5 \leq \frac{\bar{X} - \mu}{\sigma/\sqrt{n}} \leq 1.5\right]$$

Since $(\bar{X} - \mu)/(\sigma/\sqrt{n})$ has approximately a standard normal distribution, the above probability is approximately

$$P[-1.5 \leq Z \leq 1.5] = 0.8664$$

using Table 4 of the Appendix for the standard normal random variable Z. Thus chances are greater than 86% that the sample mean will fall within 0.3 ounce of the population mean. ∎

EXAMPLE 6.6

In the setting of Example 6.5 how many observations should be taken in the sample so that $\bar{X}$ would be within 0.3 oz of μ with probability 0.95?

Solution   Now we want

$$P[|\bar{X} - \mu| \leq 0.3] = P[-0.3 \leq (\bar{X} - \mu) \leq 0.3] = 0.95$$

We know that since $\sigma = 1$,

$$P\left[-0.3\sqrt{n} \leq \frac{\bar{X} - \mu}{\sigma/\sqrt{n}} \leq 0.3\sqrt{n}\right]$$

is approximately equal to

$$P[-0.3\sqrt{n} \leq Z \leq 0.3\sqrt{n}]$$

(on a standard normal random variable Z). But, using Table 4 of the Appendix,

$$P[-1.96 \leq Z \leq 1.96] = 0.95$$

and it must follow that

$$0.3\sqrt{n} = 1.96$$

or

$$n = \left(\frac{1.96}{0.3}\right)^2 = 42.68$$

Thus 43 observations will be needed for the sample mean to have a 95% chance of being within 0.3 ounce of the population mean. ∎

EXERCISES

Compute probability

6.9 Shear strength measurements for spot welds of a certain type have been found to have a standard deviation of approximately 10 psi. If 100 test welds are to be measured, find the approximate probability that the sample mean will be within 1 psi of the true population mean.

6.10 If shear strength measurements have a standard deviation of 10 psi, how many test welds should be used in the sample if the sample mean is to be within 1 psi of the population mean with probability approximately 0.95? *giving probability determine end*

6.11 The soil acidity is measured by a quantity called the pH, which may range from 0 to 14 for soils ranging from low to high acidity. Many soils have an average pH in the 5 to 8 range. A scientist wants to estimate the average pH for a large field from n randomly selected core samples by measuring the pH in each sample. If the scientist selects $n = 40$ samples, find the approximate probability that the sample mean of the 40 pH measurements will be within 0.2 unit of the true average pH for the field.

6.12 Suppose the scientist of Exercise 6.11 would like the sample mean to be within 0.1 of the true mean with probability 0.90. How many core samples should she take?

6.13 Resistors of a certain type have resistances that average 200 ohms with a standard deviation of 10 ohms. Twenty-five of these resistors are to be used in a circuit.
  (a) Find the probability that the average resistance of the 25 resistors is between 199 and 202 ohms.
  (b) Find the probability that the *total* resistance of the 25 resistors does not exceed 5,100 ohms.
    [*Hint*: Note that $P(\sum_{i=1}^{n} X_i > a) = P(n\bar{X} > a) = P(\bar{X} > a/n)$.]
  (c) What assumptions are necessary for the answers in (a) and (b) to be good approximations?

6.14 One-hour carbon monoxide concentrations in air samples from a large city average 12 ppm, with a standard deviation of 9 ppm. Find the probability that the average concentration in 100 samples selected randomly will exceed 14 ppm.

6.15 Unaltered bitumens, as commonly found in lead-zinc deposits, have atomic hydrogen/carbon ratios that average 1.4 with a standard deviation of 0.05. Find the probability that 25 samples of bitumen have an average H/C ratio below 1.3.

6.16 The downtime per day for a certain computing facility averages 4.0 hours with a standard deviation of 0.8 hour.
  (a) Find the probability that the average daily downtime for a period of 30 days is between 1 and 5 hours.
  (b) Find the probability that the *total* downtime for the 30 days is less than 115 hours.
  (c) What assumptions are necessary for the answers in (a) and (b) to be valid approximations?

6.17 The strength of a thread is a random variable with mean 0.5 lb and standard deviation of 0.2 lb. Assume the strength of a rope is the sum of the strengths of the threads in the rope.
  (a) Find the probability that a rope consisting of 100 threads will hold 45 pounds.
  (b) How many threads are needed for a rope that will hold 50 pounds with 99% assurance?

6.18   Many bulk products, such as iron ore, coal, and raw sugar, are sampled for quality by a method that requires many small samples to be taken periodically as the material is moving along a conveyor belt. The small samples are then aggregated and mixed to form one composite sample. Let Y_i denote the volume of the ith small sample from a particular lot, and suppose $Y_1, \ldots, Y_n$ constitutes a random sample with each Y_i having mean μ and variance σ^2. The average volume of the samples μ can be set by adjusting the size of the sampling device. Suppose the variance of sampling volumes σ^2 is known to be approximately 4 for a particular situation (measurements are to be in cubic inches). It is required that the total volume of the composite sample exceed 200 cubic inches with probability approximately 0.95 when $n = 50$ small samples are selected. Find a setting for μ that will allow the sampling requirements to be satisfied.

6.19   The service times for customers coming through a checkout counter in a retail store are independent random variables with a mean of 1.5 minutes and a variance of 1.0. Approximate the probability that 100 customers can be serviced in less than 2 hours of total service time by this one checkout counter.

6.20   Refer to Exercise 6.19. Find the number of customers n such that the probability of servicing all n customers in less than 2 hours is approximately 0.1.

6.21   Suppose that $X_1, \ldots, X_{n_1}$ and $Y_1, \ldots, Y_{n_2}$ constitute independent random samples from populations with means μ_1 and μ_2 and variances σ_1^2 and σ_2^2, respectively. Then the Central Limit Theorem can be extended to show that $\bar{X} - \bar{Y}$ is approximately normally distributed for large n_1 and n_2, with mean $\mu_1 - \mu_2$ and variance $(\sigma_1^2/n_1 + \sigma_2^2/n_2)$.

   Water flow through soils depends, among other things, on the porosity (volume proportion due to voids) of the soil. To compare two types of sandy soil, $n_1 = 50$ measurements are to be taken on the porosity of soil A, and $n_2 = 100$ measurements are to be taken on soil B. Assume that $\sigma_1^2 = 0.01$ and $\sigma_2^2 = 0.02$. Find the approximate probability that the difference between the sample means will be within 0.05 unit of the true difference between the population means, $\mu_1 - \mu_2$.

6.22   Refer to Exercise 6.21. Suppose samples are to be selected with $n_1 = n_2 = n$. Find the value of n that will allow the difference between the sample means to be within 0.04 unit of $\mu_1 - \mu_2$ with probability approximately 0.90.

6.23   An experiment is designed to test whether operator A or B gets the job of operating a new machine. Each operator is timed on 50 independent trials involving the performance of a certain task on the machine. If the sample means for the 50 trials differ by more than one second, the operator with the smaller mean gets the job. Otherwise the experiment is considered to end in a tie. If the standard deviations of times for both operators are assumed to be 2 seconds, what is the probability that operator A gets the job even though both operators have equal ability?

6.4   THE NORMAL APPROXIMATION TO THE BINOMIAL DISTRIBUTION

We have seen in Chapter 3 that a binomially distributed random variable Y can be written as a sum of independent Bernoulli random variables X_i. That is,

$$Y = \sum_{i=1}^{n} X_i$$

where $X_i = 1$ with probability p and $X_i = 0$ with probability $1 - p$, $i = 1, \ldots, n$. Y can represent the number of successes in a sample of n trials, or measurements, such as the number of thermistors conforming to standards in a sample of n thermistors.

Now the *fraction* of successes in the n trials is

$$\frac{Y}{n} = \frac{1}{n} \sum_{i=1}^{n} X_i = \bar{X}$$

so Y/n is a sample mean. In particular, for large n, Y/n has approximately a normal distribution with a mean of

$$E(X_i) = p$$

and a variance of

$$V(Y/n) = \frac{1}{n^2} \sum_{i=1}^{n} V(X_i)$$

$$= \frac{1}{n^2} \sum_{i=1}^{n} p(1 - p) = \frac{p(1 - p)}{n}$$

The normality follows from the Central Limit Theorem. Since $Y = n\bar{X}$, Y has approximately a normal distribution with mean np and variance $np(1 - p)$. Because of the fact that binomial probabilities are cumbersome to calculate for large n, we make extensive use of this normal approximation to the binomial distribution.

Figure 6.15 shows the histogram of a binomial distribution for $n = 20$ and $p = 0.6$. The heights of the bars represent the respective binomial probabilities. For this distribution the mean is $np = 20(0.6) = 12$, and the variance is $np(1 - p) = 20(0.6)(0.4) = 4.8$. Superimposed upon the binomial distribution is a normal distribution with mean $\mu = 12$ and variance $\sigma^2 = 4.8$. Notice how the normal curve closely approximates the binomial histogram.

For the situation displayed in Figure 6.15 suppose we wish to find $P(Y \leqslant 10)$. By the exact binomial probabilities found in Table 2 of the Appendix,

$$P(Y \leqslant 10) = 0.245$$

This value is the sum of the heights of the bars from $y = 0$ up to and including $y = 10$.

Looking at the normal curve in Figure 6.15, we can see that the areas in the bars at $y = 10$ and below are best approximated by the area under the curve to the left of 10.5. The extra 0.5 is added so that the total bar at $y = 10$ is included in the area under consideration. Thus if W represents a normally distributed random variable with $\mu = 12$ and $\sigma^2 = 4.8$ ($\sigma = 2.2$), then

$$P(Y \leqslant 10) \approx P(W \leqslant 10.5)$$

$$= P\left(\frac{W - \mu}{\sigma} \leqslant \frac{10.5 - 12}{2.2}\right) = P(Z \leqslant -0.68)$$

$$= 0.5 - 0.2517 = 0.2483$$

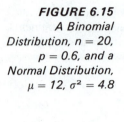

FIGURE 6.15
A Binomial Distribution, n = 20, p = 0.6, and a Normal Distribution, $\mu = 12$, $\sigma^2 = 4.8$

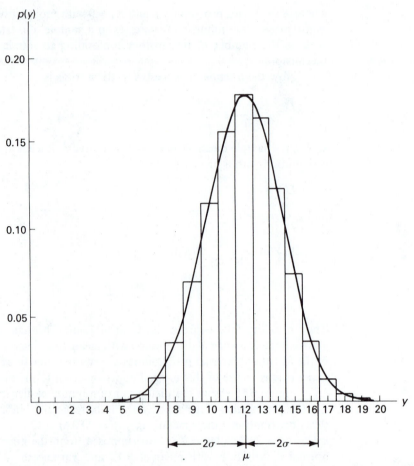

from Table 4 in the Appendix. We see that the normal approximation of 0.248 is close to the exact binomial probability of 0.245. The approximation would be even better if *n* were larger.

The normal approximation to the binomial distribution works well for even moderately large *n*, as long as *p* is not close to zero or one. A useful rule of thumb is to make sure *n* is large enough so that $p \pm 2\sqrt{p(1-p)/n}$ lies within the interval (0, 1) before the normal approximation is used. Otherwise the binomial distribution may be so asymmetric that the symmetric normal distribution cannot provide a good approximation.

EXAMPLE 6.7

Silicon wafers coming into a microchip plant are inspected for conformance to specifications. From a large lot of wafers $n = 100$ are inspected. If the number of nonconformances *Y* is no more than 12, the lot is accepted. Find the approximate probability of acceptance if the proportion of nonconformances in the lot is $p = 0.2$.

Solution The number of nonconformances Y has a binomial distribution if the lot is, indeed, large. Before using the normal approximation, we should check to see that

$$p \pm 2 \sqrt{\frac{p(1-p)}{n}} = 0.2 \pm 2 \sqrt{\frac{(0.2)(0.8)}{100}}$$

$$= 0.2 \pm 0.08$$

is entirely within the interval $(0, 1)$, which it is. Thus the normal approximation should work well.

Now the probability of accepting the lot is

$$p(Y \leqslant 12) \approx P(W \leqslant 12.5)$$

where W is a normally distributed random variable with $\mu = np = 20$ and $\sigma = \sqrt{np(1-p)} = 4$. It follows that

$$P(W \leqslant 12.5) = P\left(\frac{W-\mu}{\sigma} \leqslant \frac{12.5-20}{4}\right)$$

$$= P(Z \leqslant -1.88) = 0.5 - 0.4699 = 0.0301$$

There is only a small probability of accepting any lot that has 20% nonconforming wafers. ■

EXERCISES

6.24 The median age of residents of the U.S. is 31 years. If a survey of 100 randomly selected U.S. residents is taken, find the approximate probability that at least 60 of them will be under 31 years of age.

6.25 A lot of acceptance sampling plan for large lots, similar to that of Example 6.7, calls for sampling 50 items and accepting the lot if the number of nonconformances is no more than 5. Find the approximate probability of acceptance if the true proportion of nonconformances in the lot is
    (a) 10%   (b) 20%   (c) 30%

6.26 Of the customers entering a showroom for stereo equipment, only 30% make purchases. If 40 customers enter the showroom tomorrow, find the approximate probability that at least 15 make purchases.

6.27 The quality of computer disks is measured by the number of missing pulses. For a certain brand of disk 80% are generally found to contain no missing pulses. If 100 such disks are inspected, find the approximate probability that 15 or fewer contain missing pulses.

6.28 The capacitances of a certain type of capacitor are normally distributed with a mean of 53 μf and a standard deviation of 2 μf. If 64 such capacitors are to be used in an electronic system, approximate the probability that at least 12 of them will have capacitances below 50 μf.

6.29   The daily water demands for a city pumping station exceed 500,000 gallons with probability only 0.15. Over a thirty-day period find the approximate probability that demand for over 500,000 gallons per day occurs no more than twice.

6.30   At a specific intersection vehicles entering from the east are equally likely to turn left, turn right, or proceed straight ahead. If 500 vehicles enter this intersection from the east tomorrow, what is the approximate probability that
    (a)   150 or fewer turn right?
    (b)   at least 350 turn?

6.31   Waiting times at a service counter in a pharmacy are exponentially distributed with a mean of 10 minutes. If 100 customers come to the service counter in a day, approximate the probability that at least half of them must wait for more than 10 minutes.

6.32   A large construction firm has won 60% of the jobs for which it has bid. Suppose this firm bids on 25 jobs next month.
    (a)   Approximate the probability that it will win at least 20 of these.
    (b)   Find the exact binomial probability that it will win at least 20 of these. Compare to your answer in (a).
    (c)   What assumptions are necessary for your answers in (a) and (b) to be valid?

6.33   An auditor samples 100 of a firm's travel vouchers to check on how many of these vouchers are improperly documented. Find the approximate probability that more than 30% of the sampled vouchers will show up as being improperly documented if, in fact, only 20% of all the firm's vouchers are improperly documented.

6.5   THE SAMPLING DISTRIBUTION OF S^2

The beauty of the Central Limit Theorem lies in the fact that $\bar{X}$ will have approximately a normal sampling distribution no matter what the shape of the probabilistic model for the population, so long as n is large and σ^2 is finite. For many other statistics additional assumptions are needed before useful sampling distributions can be derived. A common assumption is that the probabilistic model for the population is itself normal. That is, we assume that if the population of measurements of interest could be viewed in histogram fashion, that histogram would have roughly the shape of a normal curve. This, incidentally, is not a bad assumption for many sets of measurements one is likely to come across in real-world experimentation. Examples were discussed in Section 4.6.

First note that if $X_1, \ldots, X_n$ are independent normally distributed random variables with common mean μ and variance σ^2, then $\bar{X}$ will be *precisely* normally distributed with mean μ and variance σ^2/n. No approximating distribution is needed in this case since linear functions of independent normal random variables are again normal.

Under this normality assumption for the population, a sampling distribution can be derived for S^2, but we do not present the derivation here. It turns out that $(n-1)S^2/\sigma^2$ has a sampling distribution that is a special case of the gamma density function. If we let $(n - 1)S^2/\sigma^2 = U$, then U will have the probability density function given by

$$f(u) = \begin{cases} \dfrac{1}{\Gamma\left(\dfrac{n-1}{2}\right) 2^{(n-1)/2}} u^{(n-1)/2 - 1} e^{-u/2} & u > 0 \\ 0 & \text{elsewhere} \end{cases}$$

The gamma density function with $\alpha = v/2$ and $\beta = 2$ is called a *chi-square density function* with parameter v. The parameter v is commonly known as the *degrees of freedom*. Thus when the sampled population is normal, $(n - 1)S^2/\sigma^2$ has a chi-square distribution with $n - 1$ degrees of freedom.

Specific values that cut off certain right-hand tail areas under the χ^2 density function are given in Table 6 of the Appendix. The value cutting off a tail area of α is denoted by $\chi_\alpha^2(v)$. A typical χ^2 density function is shown in Figure 6.16.

FIGURE 6.16
A χ^2 Distribution (Probability Density Function)

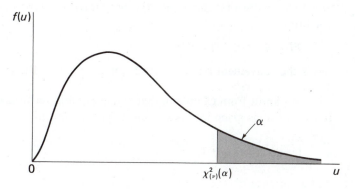

EXAMPLE 6.8 _____

For a machine dispensing liquid into bottles, the variance σ^2 in the ounces of fill was known to be approximately unity. For a sample of size $n = 10$ bottles find two positive numbers b_1 and b_2 so that the sample variance S^2 among the amounts of fill will satisfy

$$P[b_1 \leqslant S^2 \leqslant b_2] = 0.90$$

Assume that the population of amounts of liquid dispensed per bottle is approximately normally distributed.

Solution   Under the normality assumption, $(n - 1)S^2/\sigma^2$ has a $\chi^2(n - 1)$ distribution. Since

$$P[b_1 \leqslant S^2 \leqslant b_2] = P\left[\frac{(n-1)b_1}{\sigma^2} \leqslant \frac{(n-1)S^2}{\sigma^2} \leqslant \frac{(n-1)b_2}{\sigma^2}\right]$$

the desired values can be found by setting $(n - 1)b_2/\sigma^2$ equal to the value that cuts off an area of 0.05 in the upper tail of the χ^2 distribution, and $(n - 1)b_1/\sigma^2$ equal to the value that cuts off an area of 0.05 in the lower tail. Using Table 6 of the

Appendix yields, with $n - 1 = 9$ degrees of freedom,

$$\frac{(n - 1)b_2}{\sigma^2} = 16.919 = \chi^2_{0.05}(9)$$

and

$$\frac{(n - 1)b_1}{\sigma^2} = 3.325 = \chi^2_{0.95}(9)$$

Thus

$$b_2 = 16.919(1)/9 = 1.880$$

and

$$b_1 = 3.325(1)/9 = 0.369$$

Thus there is a 90% chance that the sample variance will fall between 0.369 and 1.880. Note that this is not the only interval that would satisfy the desired condition

$$P[b_1 \leqslant S^2 \leqslant b_2] = 0.90$$

but is the convenient one that has 5% in each tail of the χ^2 distribution.   ■

We know from Chapter 4 that a gamma random variable having the density function $f(u)$ as given in this section will have a mean of

$$\left(\frac{n - 1}{2}\right)2 = n - 1$$

and a variance of

$$\left(\frac{n - 1}{2}\right)2^2 = 2(n - 1)$$

Thus a χ^2 distribution has a mean equal to its degrees of freedom and a variance equal to twice its degrees of freedom. From this information we can find the mean and variance of S^2.

$$E\left[\frac{(n - 1)S^2}{\sigma^2}\right] = (n - 1)$$

$$\frac{(n - 1)}{\sigma^2} E(S^2) = (n - 1)$$

or

$$E(S^2) = (n - 1)\frac{\sigma^2}{(n - 1)} = \sigma^2$$

Actually S^2 is defined with a denominator of $(n - 1)$ so its expected value will equal σ^2. You saw this in Exercise 6.8.

Also

$$V\left[\frac{(n-1)}{\sigma^2} S^2\right] = 2(n-1)$$

$$\left(\frac{n-1}{\sigma^2}\right)^2 V(S^2) = 2(n-1)$$

or

$$V(S^2) = 2(n-1)\left(\frac{\sigma^2}{n-1}\right)^2$$

$$= \frac{2\sigma^4}{n-1}$$

The standard deviation of S^2 is $\sqrt{2}\sigma^2/\sqrt{n-1}$. Note that this value is valid for normal populations.

EXAMPLE 6.9

In designing mechanisms for hurling projectiles at targets, such as rockets, it is very important to study the variance of the distances by which the projectile misses the target center. (Obviously this variance should be as small as possible.) For a certain launching mechanism these distances are known to have a normal distribution with variance $\sigma^2 = 100\,\text{m}^2$. An experiment involving $n = 25$ launches is to be conducted. Let S^2 denote the sample variances of the distances between impact of the projectile and the target center.
    (a) Approximate $P(S^2 > 50)$.
    (b) Approximate $P(S^2 > 150)$.
    (c) Find $E(S^2)$ and $V(S^2)$.

Solution   Let $U = (n-1)S^2/\sigma^2$, which then has a $\chi^2(24)$ distribution for $n = 25$.

    (a) $P(S^2 > 50) = P\left(\dfrac{n-1}{\sigma^2}S^2 > \dfrac{24}{100}50\right)$

$$= P(U > 12)$$

Looking at the row for 24 degrees of freedom in Table 6 of the Appendix, we see

$$P(U > 12.4011) = 0.975$$

and

$$P(U > 10.8564) = 0.990$$

Thus $P(S^2 > 50)$ is a little larger than 0.975. We cannot find the exact probability since Table 6 provides only selected tail areas.

    (b) $P(S^2 > 150) = P\left(\dfrac{n-1}{\sigma^2}S^2 > \dfrac{24}{100}150\right)$

$$= P(U > 36)$$

From Table 6, with 24 degrees of freedom,

$$P(U > 36.4151) = 0.05$$

and

$$P(U > 33.1963) = 0.10$$

Thus $P(S^2 > 150)$ is a little larger than 0.05.

(c) We know that

$$E(S^2) = \sigma^2 = 100$$

and

$$V(S^2) = \frac{2\sigma^4}{n-1} = \frac{2(100)^2}{24} = \frac{(100)^2}{12}$$

Following up on this, the standard deviation of S^2 is $100/\sqrt{12} \approx 29$. Then by Tchebysheff's Theorem at least 75% of the values of S^2, in repeated sampling, should lie between

$$100 - 2(29) \quad \text{and} \quad 100 + 2(29)$$

or between

$$42 \quad \text{and} \quad 158$$

As a practical consideration this much range in distance variances might cause the engineers to reassess the design of the launching mechanism. ■

EXERCISES

6.34 The efficiency ratings (in lumens per watt) of light bulbs of a certain type have a population mean of 9.5 and a standard deviation of 0.5, according to production specifications. The specifications for a room in which eight of these bulbs are to be installed call for the average efficiency of the eight bulbs to exceed 10. Find the probability that this specification for the room will be met, assuming efficiency measurements are normally distributed.

6.35 In the setting of Exercise 6.34 what should the mean efficiency per bulb equal if the specification for the room is to be met with probability approximately 0.90? (Assume the standard deviation of efficiency measurements remains at 0.5.)

6.36 The Environmental Protection Agency is concerned with the problem of setting criteria for the amount of certain toxic chemicals to be allowed in freshwater lakes and rivers. A common measure of toxicity for any pollutant is the concentration of the pollutant that will kill half of the test species in a given amount of time (usually 96 hours for fish species). This measure is called the LC50 (lethal concentration killing 50% of the test species).

Studies of the effects of copper on a certain species of fish (say, species A) show the variance of LC50 measurements to be approximately 1.9, with concentration measured in milligrams per liter. If $n = 10$ studies on LC50 for copper are to be completed, find the probability that the sample mean LC50 will differ from the true population mean by no more than 0.5 unit. Assume that the LC50 measurements are approximately normal in their distribution.

6.37 If, in Exercise 6.36, it is desired that the sample mean differ from the population mean by no more than 0.5 with probability 0.95, how many tests should be run?

6.38 Suppose $n = 20$ observations are to be taken on normally distributed LC50 measurements, with $\sigma^2 = 1.9$. Find two numbers a and b such that $P(a \leqslant S^2 \leqslant b) = 0.90$. ($S^2$ is the sample variance of the 20 measurements.)

6.39 Ammeters produced by a certain company are marketed under the specification that the standard deviation of gauge readings is no larger than 0.2 amp. Ten independent readings on a test circuit of constant current, using one of these ammeters, gave a sample variance of 0.065. Does this suggest that the ammeter used does not meet the company's specification? [*Hint:* Find the approximate probability of a sample variance exceeding 0.065 if the true population variance is 0.04.]

6.40 A certain type of resistor is marketed with the specification that the variance of resistances produced is around 50 ohms. A sample of 15 of these resistors is to be tested for resistances produced. Find the approximate probability that the sample variance S^2
    (a) will exceed 80.
    (b) will be less than 20.
    (c) Find an interval in which at least 75% of such sample variances should lie.
    (d) What assumptions are necessary for your answers above to be valid?

6.41 Answer the questions posed in Exercise 6.40 if the sample size is 25 rather than 15.

6.42 In constructing an aptitude test for a job, it is important to plan for a fairly large variance in test scores so the best applicants can be easily identified. For a certain test, scores are assumed to be normally distributed with a mean of 80 and a standard deviation of 10. A dozen applicants are to take the aptitude test. Find the approximate probability that the sample standard deviation of the scores for these applicants will exceed 15.

6.43 For an aptitude test for quality-control technicians in an electronics firm, history shows scores to be normally distributed with a variance of 225. If 20 applicants are to take the test, find an interval in which the sample variance of test scores should lie with probability 0.90.

---

*6.6 A USEFUL METHOD OF APPROXIMATING DISTRIBUTIONS

We saw in the Central Limit Theorem of Section 6.3 that $\sqrt{n}(\bar{X} - \mu)/\sigma$ has, approximately, a standard normal distribution for large n, under certain general conditions. This notion of asymptotic normality actually extends to a large class of

*Optional section.

functions of $\bar{X}$ or, for that matter, to functions of *any* asymptotically normal random variable. The result holds because of properties of Taylor series expansions of functions, roughly illustrated as follows. Suppose $\bar{X}$ is a statistic based on a random sample of size n from a population with mean μ and finite variance σ^2. In addition, suppose we are interested in the behavior of a function of $\bar{X}$, say $g(\bar{X})$, where $g(x)$ is a real-valued function such that $g'(x)$ exists and is not zero in a neighborhood of μ. We can then write

$$g(\bar{X}) = g(\mu) + g'(\mu)(\bar{X} - \mu) + R$$

where R denotes the remainder term for the Taylor series. Rearranging terms and multiplying through by $\sqrt{n}$ yields

$$\sqrt{n}[g(\bar{X}) - g(\mu)] = \sqrt{n}g'(\mu)(\bar{X} - \mu) + \sqrt{n}R$$

In many cases $\sqrt{n}R$ approaches zero as n increases, and this term can be dropped to give the approximation

$$\sqrt{n}[g(\bar{X}) - g(\mu)] \approx \sqrt{n}(\bar{X} - \mu)g'(\mu)$$

$$= \frac{\sqrt{n}(\bar{X} - \mu)}{\sigma} \sigma g'(\mu)$$

Since $\sqrt{n}(\bar{X} - \mu)/\sigma$ is approximately standard normal, the right-hand side is approximately normal with mean 0 and variance $\sigma^2[g'(\mu)]^2$. Thus the term on the left side will also tend toward the same normal distribution. In summary,

$$\frac{\sqrt{n}[g(\bar{X}) - g(\mu)]}{|g'(\mu)|\sigma}$$

has a distribution function that will converge to the standard normal distribution function as $n \to \infty$. An example of the use of this property is given next.

EXAMPLE 6.10

The current I in an electrical circuit is related to the voltage E and the resistance R by Ohm's Law, $I = E/R$. Suppose that for circuits of a certain type, E is constant but the resistance R varies slightly from circuit to circuit. The resistance is to be measured independently in n circuits, yielding measurements $X_1, \ldots, X_n$. If $E(X_i) = \mu$ and $V(X_i) = \sigma^2$, $i = 1, \ldots, n$, approximate the distribution of $E/\bar{X}$, an approximation to the average current.

Solution   In this case $g(\bar{X}) = E/\bar{X}$ for a constant E. Thus

$$g'(\mu) = -\frac{E}{\mu^2}$$

and

$$\sqrt{n}\left[\frac{E}{\bar{X}} - \frac{E}{\mu}\right] \bigg/ \left(\frac{E\sigma}{\mu^2}\right)$$

has approximately a standard normal distribution for large n. If μ and σ were known, at least approximately, one could make probability statements about the behavior of $E/\bar{X}$ by using the table of normal curve areas. ∎

EXERCISES

6.44 Circuits of a certain type have resistances with a mean μ of 6 ohms and a standard deviation σ of 0.5 ohm. Twenty-five such circuits are to be used in a system, each with voltage E of 120. A specification calls for the average current in the twenty-five circuits to exceed 19 amps. Find the approximate probability that the specification will be met.

6.45 The distance d traveled in t seconds by a particle starting from rest is given by $d = \frac{1}{2}at^2$, where a is the acceleration. An engineer desires to study the acceleration of gravel particles rolling down an incline 10 meters in length. Observations on 100 particles showed the average time to travel the 10 meters was 20 seconds, and the standard deviation of these measurements was 1.6 seconds. Approximate the mean and variance of the average acceleration for these 100 particles.

6.46 A fuse used in an electric current has a lifelength X that is exponentially distributed with mean θ (measurements in hundreds of hours). The *reliability* of the fuse at time t is given by

$$R(t) = P(X > t) = \int_t^\infty \frac{1}{\theta} e^{-x/\theta} \, dx$$

$$= e^{-t/\theta}$$

If the lifelengths $X_1, \ldots, X_{100}$ of one hundred such fuses are to be measured, $e^{-t/\theta}$ can be approximated by $e^{-t/\bar{X}}$. Assuming θ is close to 4, approximate the probability that $e^{-t/\bar{X}}$ will be within 0.05 of $e^{-t/\theta}$ for $t = 5$.

6.7 CONCLUSION

It is important to remember that some of the sampling distributions we commonly deal with are exact and some are approximate. If the random sample under consideration comes from a population that can be modeled by a normal distribution, then $\bar{X}$ will have a normal distribution and $(n-1)S^2/\sigma^2$ a χ^2 distribution with *no* approximation involved. However, the Central Limit Theorem states that $\sqrt{n}(\bar{X} - \mu)/\sigma$ will have *approximately* a normal distribution for large n, under almost any probabilistic model for the population itself.

Sampling distributions for statistics will be used throughout the remainder of the text for the purpose of relating sample quantities (statistics) to parameters of the population.

SUPPLEMENTARY EXERCISES

6.47   The U.S. Census Bureau reports median ages of the population and median incomes for households. On the other hand the College Board reports average scores on the Scholastic Aptitude Test. Do you think each group is using an approximate measure? Why?

6.48   Twenty-five lamps are connected so that when one lamp fails, another takes over immediately. (Only one lamp is on at any one time.) The lamps operate independently, and each has a mean life of 50 hours and a standard deviation of four hours. If the system is not checked for 1300 hours after the first lamp is turned on, what is the probability that a lamp will be burning at the end of the 1300-hour period?

6.49   Suppose that $X_1, \ldots, X_{40}$ denotes a random sample of measurements on the proportion of impurities in samples of iron ore. Suppose that each X_t has probability density function

$$f(x) = \begin{cases} 3x^2 & 0 \leqslant x \leqslant 1 \\ 0 & \text{elsewhere} \end{cases}$$

The ore is to be rejected by a potential buyer if $\bar{X} > 0.7$. Find the approximate probability that the ore will be rejected, based on the 40 measurements.

6.50   Suppose that $X_1, \ldots, X_n$ and $Y_1, \ldots, Y_m$ are independent random samples, with the X_i's having mean μ_1 and variance σ_1^2 and the Y_i's having mean μ_2 and variance σ_2^2. The difference between the sample means $\bar{X} - \bar{Y}$ will again be approximately normally distributed since the Central Limit Theorem will apply to this difference.
    (a)   Find $E(\bar{X} - \bar{Y})$.   (b)   Find $V(\bar{X} - \bar{Y})$.

6.51   A study of the effects of copper on a certain species (say, species A) of fish show the variance of LC50 measurements (in milligrams per liter) to be 1.9. The effects of copper on a second species (say, species B) of fish produce a variance of LC50 measurements of 0.8. If the population means of LC50s for the two species are equal, find the probability that, with random samples of ten measurements from each species, the sample mean for species A exceeds the sample mean for species B by at least one unit.

6.52   If Y has an exponential distribution with mean θ, show that $U = 2Y/\theta$ has a χ^2 distribution with 2 degrees of freedom.

6.53   A plant supervisor is interested in budgeting for weekly repair costs for a certain type of machine. These repair costs over the past years tend to have an exponential distribution with a mean of 20 for each machine studied. Let $Y_1, \ldots, Y_5$ denote the repair costs for five of these machines for next week. Find a number c such that $P(\sum_{i=1}^{5} Y_i > c) = 0.05$, assuming the machines operate independently.

6.54   If Y has a $\chi^2(n)$ distribution, then Y can be represented as

$$Y = \sum_{i=1}^{n} X_i$$

where X_i has a $\chi^2(1)$ distribution and the X_i's are independent. It is not surprising, then, that Y will be approximately normally distributed for large n. Use this fact in the solution of the following.

A machine in a heavy-equipment factory produces steel rods of length Y, where Y is a normally distributed random variable with a mean of 6 inches and a variance of 0.2. The cost C of repairing a rod that is not exactly 6 inches in length is given by

$$C = 4(Y - \mu)^2$$

where $\mu = E(Y)$. If 50 rods with independent lengths are produced in a given day, approximate the probability that the total cost for repairs exceeds $48.

7

Estimation

ABOUT THIS CHAPTER

We now introduce one form of statistical inference by looking at the techniques of *estimation*. Sample statistics will be employed to estimate population parameters, such as means or proportions. By making use of the sampling distribution of the statistic employed, we construct intervals (confidence intervals) that include the unknown parameter value with high probability.

CONTENTS

7.1 INTRODUCTION

We have seen that probabilistic models for populations usually contain some unknown parameters. We have also seen in Chapter 5 that certain functions of sample observations, called statistics, have sampling distributions that contain some of these same unknown parameters. Thus it seems that we should be able to use statistics to obtain information on population parameters. For example, suppose a probabilistic model for a population of measurements is left unspecified except for the fact that it has mean μ and finite variance σ^2. A random sample X_1, ..., X_n, from this population gives rise to a sample mean $\bar{X}$ that, according to Section 5.3, will tend to have a normal distribution with mean μ and variance σ^2/n. How can we make use of this information to say something more specific about the unknown parameter μ? In this chapter we answer this question with a type of statistical inference called *estimation*. Chapter 8 will, in turn, consider *tests of hypotheses*.

DEFINITION 7.1	An **estimator** is a statistic that specifies how to use the sample data to estimate an unknown parameter of the population.

Note that an estimator is a random variable.

In the above example it seems obvious that the sample mean $\bar{X}$ could be used as an *estimator* of μ. But $\bar{X}$ will take on different numerical values from sample to sample, so some questions remain. What is the magnitude of the difference between $\bar{X}$ and μ likely to be? How does this difference behave as n gets larger and larger? In short, what properties does $\bar{X}$ have as an estimator of μ? This chapter provides answers for some of these basic questions regarding estimation.

7.2 PROPERTIES OF POINT ESTIMATORS

The statistics that will be used as estimators will, of course, have sampling distributions. Sometimes the sampling distribution is known, at least approximately, as in the case of the sample mean from large samples. In other cases the sampling distribution may not be completely specified, but we can still calculate the mean and variance of the estimator. We might, then, logically ask "What properties would we like this mean and variance to possess?"

Let the symbol θ denote an arbitrary population parameter, such as μ or σ^2, and let $\hat{\theta}$ denote an estimator of θ. We are then concerned with properties of $E(\hat{\theta})$ and $V(\hat{\theta})$. Different samples result in different numerical values for $\hat{\theta}$, but we would hope that some of these values underestimate θ and that some overestimate θ, so that the average value of $\hat{\theta}$ is close to θ. If $E(\hat{\theta}) = \theta$, then $\hat{\theta}$ is said to be an unbiased estimator of θ.

DEFINITION 7.2	An estimator $\hat{\theta}$ is **unbiased** for estimating θ if
	$$E(\hat{\theta}) = \theta$$

In the sampling distributions simulated in Section 6.2 we saw that the values of $\bar{X}$ tend to center at μ, the true population mean, when random samples are selected from the same population repeatedly. Similarly values of S^2 centered at σ^2, the true population variance. These are demonstrations of the fact that $\bar{X}$ is an unbiased estimator of μ and S^2 is an unbiased estimator of σ^2.

For an unbiased estimator $\hat{\theta}$ the sampling distribution of the estimator has mean value θ. How do we want the possible values for $\hat{\theta}$ to spread out to either side of θ for this unbiased estimator? Intuitively it would be desirable for all possible values of $\hat{\theta}$ to be very close to θ. That is, we want the variance of $\hat{\theta}$ to be as small as possible. At the level of this text it is not possible to prove that some of our estimators do indeed have the smallest variance among all unbiased estimators, but we will use this variance criterion for comparing estimators. That is, if $\hat{\theta}_1$ and $\hat{\theta}_2$ are both unbiased estimators of θ, then we would choose as the better estimator the one possessing the smaller variance. (See Figure 7.1.)

FIGURE 7.1
Distributions of Two Unbiased Estimators, $\hat{\theta}_1$ and $\hat{\theta}_2$, with $V(\hat{\theta}_2) < V(\hat{\theta}_1)$

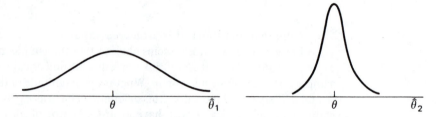

EXAMPLE 7.1

Suppose $X_1, \ldots, X_5$ denotes a random sample from some population with $E(X_i) = \mu$ and $V(X_i) = \sigma^2$, $i = 1, \ldots, 5$. The following are suggested as estimators for μ:

$$\hat{\theta}_1 = X_1 \qquad \hat{\theta}_2 = \frac{1}{2}(X_1 + X_5) \qquad \hat{\theta}_3 = \frac{1}{2}(X_1 + 2X_5)$$

$$\hat{\theta}_4 = \bar{X} = \frac{1}{5}(X_1 + X_2 + \cdots + X_5)$$

Which estimator would you use, and why?

Solution Looking at the means of these estimators, we have $E(\hat{\theta}_1) = \mu$, $E(\hat{\theta}_2) = \mu$, $E(\hat{\theta}_3) = \frac{3}{2}\mu$, and $E(\hat{\theta}_4) = \mu$. Thus $\hat{\theta}_1$, $\hat{\theta}_2$, and $\hat{\theta}_4$ are all unbiased, and we eliminate estimator $\hat{\theta}_3$ because it is biased (tends to overestimate μ).

Looking at variances of the unbiased estimators, we have

$$V(\hat{\theta}_1) = V(X_1) = \sigma^2$$

$$V(\hat{\theta}_2) = \frac{1}{4}[V(X_1) + V(X_5)] = \frac{\sigma^2}{2}$$

and

$$V(\hat{\theta}_4) = V(\bar{X}) = \frac{\sigma^2}{5}$$

Thus $\hat{\theta}_4$ would be chosen as the best estimator since it is unbiased and has smallest variance among the three unbiased estimators. ∎

We give a method for finding good point estimators in Section 7.6. In the meantime we discuss further details of estimation for intuitively reasonable point estimators.

EXERCISES

7.1 Suppose X_1, X_2, X_3 denotes a random sample from the exponential distribution with density function

$$f(x) = \begin{cases} \dfrac{1}{\theta} e^{-x/\theta} & x > 0 \\ 0 & \text{elsewhere} \end{cases}$$

Consider the following four estimators of θ:

$$\hat{\theta}_1 = X_1$$

$$\hat{\theta}_2 = \frac{X_1 + X_2}{2}$$

$$\hat{\theta}_3 = \frac{X_1 + 2X_2}{3}$$

$$\hat{\theta}_4 = \bar{X}$$

(a) Which of the above estimators are unbiased for θ?
(b) Among the unbiased estimators of θ, which has smallest variance?

7.2 The reading on a voltage meter connected to a test circuit is uniformly distributed over the interval $(\theta, \theta + 1)$, where θ is the true but unknown voltage of the circuit. Suppose that $X_1, \ldots, X_n$ denotes a random sample of readings from this voltage meter.
(a) Show that $\bar{X}$ is a biased estimator of θ.
(b) Find a function of $\bar{X}$ that is an unbiased estimator of θ.

7.3 The number of breakdowns per week for a certain minicomputer is a random variable X having a Poisson distribution with mean λ. A random sample $X_1, \ldots, X_n$ of observations on the number of breakdowns per week is available.
(a) Find an unbiased estimator of λ.
(b) The weekly cost of repairing these breakdowns is

$$C = 3Y + Y^2$$

Show that

$$E(C) = 4\lambda + \lambda^2$$

(c) Find an unbiased estimator of $E(C)$ that makes use of the entire sample, X_1, $\ldots, X_n$.

7.4 The *bias B* of an estimator $\hat{\theta}$ is given by

$$B = |\theta - E(\hat{\theta})|$$

The *mean squared error*, or MSE, of an estimator $\hat{\theta}$ is given by

$$\text{MSE} = E(\hat{\theta} - \theta)^2$$

Show that

$$\text{MSE}(\hat{\theta}) = V(\hat{\theta}) + B^2$$

[*Note*: $\text{MSE}(\hat{\theta}) = V(\hat{\theta})$ if $\hat{\theta}$ is an unbiased estimator of θ. Otherwise $\text{MSE}(\hat{\theta}) > V(\hat{\theta})$.]

7.5 Refer to Exercise 7.2. Find $\text{MSE}(\bar{X})$ when $\bar{X}$ is used to estimate θ.

7.6 Suppose $X_1, \ldots, X_n$ is a random sample from a normal distribution with mean μ and variance σ^2.
   (a) Show that $S = \sqrt{S^2}$ is a biased estimator of σ.
   (b) Adjust S to form an unbiased estimator of σ.

7.7 For a certain new model of microwave oven, it is desired to set a guarantee period so that only 5% of the ovens sold will have had a major failure in this length of time. Assuming length of time until the first major failure for an oven of this type is normally distributed with mean μ and variance σ^2, the guarantee period should end at $\mu - 1.645\sigma$. Use the results of Exercise 7.6 to find an unbiased estimator of $\mu - 1.645\sigma$ based upon a random sample $X_1, \ldots, X_n$ of measurements of time until the first major failure.

7.8 Suppose that $\hat{\theta}_1$ and $\hat{\theta}_2$ are each unbiased estimators of θ, with $V(\hat{\theta}_1) = \sigma_1^2$ and $V(\hat{\theta}_2) = \sigma_2^2$. A new unbiased estimator for θ can be formed by

$$\hat{\theta}_3 = a\hat{\theta}_1 + (1 - a)\hat{\theta}_2$$

$(0 \leqslant a \leqslant 1)$. If $\hat{\theta}_1$ and $\hat{\theta}_2$ are independent, how should a be chosen so as to minimize $V(\hat{\theta}_3)$?

7.3 CONFIDENCE INTERVALS: THE SINGLE-SAMPLE CASE

After knowing something about the mean and variance of $\hat{\theta}$ as an estimator of θ, it would be nice to know something about how small the distance between $\hat{\theta}$ and θ is likely to be. Answers to this kind of question require knowledge of the sampling distribution of $\hat{\theta}$ beyond the behavior of $E(\hat{\theta})$ and $V(\hat{\theta})$, and lead to *confidence intervals*.

The idea behind the construction of confidence intervals is as follows. If $\hat{\theta}$ is an estimator of θ (which has a known sampling distribution), and one can find two quantities that depend on $\hat{\theta}$, say $g_1(\hat{\theta})$ and $g_2(\hat{\theta})$, such that

$$P[g_1(\hat{\theta}) \leqslant \theta \leqslant g_2(\hat{\theta})] = 1 - \alpha$$

for some small positive number α, then we can say that $(g_1(\hat{\theta}),\, g_2(\hat{\theta}))$ forms an interval that has probability $1 - \alpha$ of capturing the true θ. This interval is referred to as a confidence interval with confidence coefficient $(1 - \alpha)$. The quantity $g_1(\hat{\theta})$ is called the *lower confidence limit* (LCL) and $g_2(\hat{\theta})$ the *upper confidence limit* (UCL). We generally want $(1 - \alpha)$ to be near unity and $(g_1(\hat{\theta}),\, g_2(\hat{\theta}))$ to be as short an interval as possible for a given sample size. We illustrate the construction of confidence intervals for some common parameters here, and other cases will be considered in Chapters 9 through 12.

7.3.1 General Distribution: Large-Sample Confidence Interval for μ

Suppose we are interested in estimating a mean μ for a population with variance σ^2, assumed, for the moment, to be known. We select a random sample $X_1, \ldots, X_n$ from this population and compute $\bar{X}$ as a point estimator of μ. If n is large (say, $n \geqslant 30$ as a rule of thumb), then $\bar{X}$ has approximately a normal distribution with mean μ and variance σ^2/n. From these facts we can state that the interval $\mu \pm 2\sigma/\sqrt{n}$ contains about 95% of the $\bar{x}$ values that could be generated in repeated random samplings from the population under study. For convenience let's call this middle 95% the "likely" values of $\bar{x}$. Now suppose we are to observe a single sample producing a single $\bar{x}$. A question of interest is "What possible values for μ would allow this $\bar{x}$ to lie in the likely range of possible sample means?" This set of possible values for μ is the confidence interval with confidence coefficient of approximately 0.95.

The main idea of confidence interval construction is shown in the diagram of Figure 7.2. For any value of μ the likely sample means are in the interval $\mu \pm 2\sigma/\sqrt{n}$, which shifts to the right as larger values of μ are considered. For a

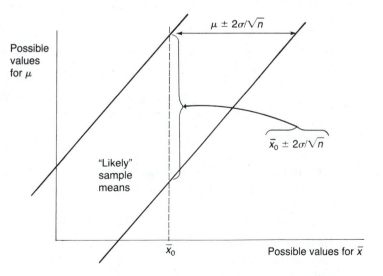

FIGURE 7.2
The Construction of a Confidence Interval

continuum of possible values for μ, the likely sample means fall between the two angled lines (which will be of 45° if the same numerical scale is used on both axes). Suppose we observe a single sample mean of $\bar{x}_0$. For what values of μ will $\bar{x}_0$ fall into the likely sample region? Precisely for those values of μ between the intersections of the dashed line and the angled lines. Considering both axes to have the same numerical scale, we can see that the intersection points are at $\bar{x}_0 - 2\sigma/\sqrt{n}$ and $\bar{x}_0 + 2\sigma/\sqrt{n}$. These two values form an approximate 95% confidence interval for μ (or a confidence interval with confidence coefficient equal to 0.95). Note that we are assuming σ to be a fixed value in this discussion.

More formally, under these conditions

$$Z = \frac{\bar{X} - \mu}{\sigma/\sqrt{n}}$$

has a standard normal distribution, approximately. Now for any prescribed α we can find from Table 4 in the Appendix a value $z_{\alpha/2}$ such that

$$P[-z_{\alpha/2} \leqslant Z \leqslant +z_{\alpha/2}] = 1 - \alpha$$

Rewriting this probability statement, we have

$$1 - \alpha = P\left[-z_{\alpha/2} \leqslant \frac{\bar{X} - \mu}{\sigma/\sqrt{n}} \leqslant +z_{\alpha/2}\right]$$

$$= P\left[-z_{\alpha/2} \frac{\sigma}{\sqrt{n}} \leqslant \bar{X} - \mu \leqslant +z_{\alpha/2} \frac{\sigma}{\sqrt{n}}\right]$$

$$= P\left[\bar{X} - z_{\alpha/2} \frac{\sigma}{\sqrt{n}} \leqslant \mu \leqslant \bar{X} + z_{\alpha/2} \frac{\sigma}{\sqrt{n}}\right]$$

The interval

$$\left(\bar{x} - z_{\alpha/2} \frac{\sigma}{\sqrt{n}}, \bar{x} + z_{\alpha/2} \frac{\sigma}{\sqrt{n}}\right)$$

forms a realization of a large-sample confidence interval for μ with confidence coefficient approximately $(1 - \alpha)$.

If σ is unknown, it can be replaced by s, the sample standard deviation, with no serious loss in accuracy for the large-sample case.

EXAMPLE 7.2

Consider once again the fifty lifelength observations of Table 4.1 on p. 125. Using these as the sample, find a confidence interval for the mean lifelength of batteries of this type, with confidence coefficient 0.95.

alpha $_{1-\alpha}^{\text{or}}$ $\frac{90}{2}$.

Solution   For these data $\bar{x} = 2.266$ and $s = 1.935$. Using the confidence interval

$$\bar{x} \pm z_{\alpha/2} \frac{\sigma}{\sqrt{n}}$$

$z_{1-\alpha/2} = z_{95\%} = \boxed{.4750} \gg 1.96$

with $(1 - \alpha) = 0.95$, we see that $z_{\alpha/2} = z_{0.025} = 1.96$ (from Table 4 of the Appendix). Substituting s for c, this interval yields

$\alpha = .95$

$$2.266 \pm 1.96 \frac{1.935}{\sqrt{50}}$$

$$2.266 \pm 0.536$$

or

$$(1.730, 2.802)$$

Thus we are 95% confident (similarly formed intervals will contain μ about 95% of the time in repeated sampling) that the true mean lies between 1.730 and 2.802. ∎

Take a careful look at the interpretation of confidence interval statements. The *interval* is random; the parameter is fixed. Before sampling, there is a probability of $(1 - \alpha)$ that the interval will include the true parameter value. After sampling, the resulting realization of the confidence interval either includes the parameter value or fails to include it, but we are quite confident that it will include the true parameter value if $(1 - \alpha)$ is large. The following computer simulation illustrates the behavior of a confidence interval for a mean in repeated sampling.

We started the simulation by selecting random samples of size $n = 100$ from a population having an exponential distribution with a mean value of $\mu = 10$. That is,

$$f(x) = \frac{1}{10} e^{-x/10} \qquad x > 0$$

$$= 0 \qquad\qquad \text{elsewhere}$$

For each sample a 95% confidence interval for μ was constructed by calculating $\bar{x} \pm 1.96 s/\sqrt{n}$. The sample mean, standard deviation, lower confidence limit, and upper confidence limit are given in Table 7.1 for each of 100 samples generated. There are seven intervals that do *not* include the true mean value. (These are marked with an asterisk.) Thus we found that 93 out of 100 intervals include the true mean in our simulation, whereas the theory says that 95 out of 100 should include μ. This is fairly good agreement between theory and application. Later in the chapter we will see other simulations that generate similar results.

We now move to a discussion of sample size selection. Since we know that a confidence interval for a mean μ will be of the form $\bar{X} \pm z_{\alpha/2}\sigma/\sqrt{n}$, we can choose the sample size to ensure a certain accuracy before carrying out an experiment. Suppose the goal is to achieve an interval of length $2B$, so the resulting confidence interval is of the form $\bar{X} \pm B$. We might, for example, want an estimate of the average assembly time for microcomputers to be within 2 minutes of the true

TABLE 7.1
Large-Sample Confidence Intervals (n = 100)

Sample	LCL	Mean	UCL	S
1	7.26064	8.6714	10.0822	7.1977
2	7.59523	9.5291	11.4630	9.8668
3	7.68938	9.3871	11.0849	8.6619
4	8.94686	11.1785	13.4100	11.3856
5	7.96793	10.1508	12.3336	11.1369
6	7.61223	10.0101	12.4080	12.2340
7	7.55475	9.2754	10.9961	8.7791
8	7.48558	9.1436	10.8016	8.4593
9	7.71919	9.5219	11.3245	9.1973
10	7.95952	9.7012	11.4428	8.8859
11	7.34959	8.7958	10.2420	7.3785
12	8.71580	10.8458	12.9759	10.8675
*13	6.41955	8.1029	9.7863	8.5885
14	7.27538	8.9598	10.6441	8.5938
15	8.30285	10.2902	12.2776	10.1398
16	9.48735	11.5540	13.6206	10.5441
17	8.35947	10.8521	13.3448	12.7176
18	7.63421	9.9196	12.2049	11.6599
19	7.97129	10.1196	12.2679	10.9608
20	8.86613	10.9223	12.9785	10.4906
21	7.16061	9.2142	11.2679	10.4778
22	6.99569	9.2782	11.5607	11.6454
23	8.13781	10.2701	12.4024	10.8791
24	7.70300	9.4910	11.2790	9.1226
25	9.61738	11.9111	14.2048	11.7026
26	7.23197	9.2257	11.2194	10.1720
27	7.85056	9.9053	11.9601	10.4836
28	8.82373	11.1849	13.5462	12.0470
29	8.36303	10.4008	12.4387	10.3970
30	8.74332	10.4288	12.1144	8.5996
31	7.80013	9.7064	11.6127	9.7259
32	7.88003	10.0706	12.2611	11.1763
33	7.01555	8.6994	10.3833	8.5913
34	8.31036	10.3074	12.3045	10.1891
35	7.91383	9.8344	11.7549	9.7986
36	8.80795	10.5208	12.2337	8.7390
37	7.27322	9.1728	11.0725	9.6919
38	7.47957	9.1333	10.7871	8.4375
39	7.77926	9.8034	11.8275	10.3272
40	7.09624	9.0955	11.0947	10.2001
41	8.32909	10.1639	11.9986	9.3610
42	8.23895	10.2593	12.2797	10.3080
43	8.44813	10.4284	12.4087	10.1034
44	7.11788	8.8708	10.6237	8.9436
45	8.14050	10.0992	12.0580	9.9936
46	7.53466	9.4198	11.3050	9.6183
47	8.93450	11.0165	13.0986	10.6227
48	7.43185	8.9012	10.3705	7.4965
49	7.83028	9.7367	11.6432	9.7268
50	7.61207	9.3698	11.1275	8.9680
51	7.97829	9.7442	11.5101	9.0098

TABLE 7.1 (*continued*)

Sample	LCL	Mean	UCL	S
52	8.97313	11.2890	13.6049	11.8158
*53	5.99726	7.7422	9.4871	8.9027
54	7.51538	9.4089	11.3025	9.6609
55	8.59019	10.4373	12.2843	9.4238
56	8.49077	10.3914	12.2920	9.6971
57	8.12970	10.4056	12.6814	11.6115
*58	6.85175	8.3456	9.8395	7.6217
59	8.87405	10.9482	13.0225	10.5827
60	8.70548	10.2597	11.8140	7.9298
61	6.37467	8.3475	10.3203	10.0653
62	8.64835	10.6703	12.6923	10.3161
63	8.37157	10.6196	12.8676	11.4696
64	7.76632	9.6919	11.6174	9.8241
65	9.19745	11.5565	13.9155	12.0359
66	7.54964	9.1818	10.8140	8.3275
67	8.57961	10.8918	13.2039	11.7967
68	8.23986	10.0088	11.7778	9.0254
69	8.08091	9.9399	11.7989	9.4846
70	7.52910	9.2224	10.9156	8.6391
71	7.52705	9.4949	11.4628	10.0401
72	8.61466	10.7728	12.9310	11.0110
73	9.13542	11.2654	13.3953	10.8670
*74	6.73171	8.1351	9.5386	7.1604
75	7.78546	10.1938	12.6021	12.2874
76	8.04925	9.6668	11.2843	8.2526
77	7.61919	9.6068	11.5943	10.1407
*78	5.91167	7.7383	9.5649	9.3194
79	8.36219	10.4003	12.4384	10.3984
80	8.78276	11.0946	13.4065	11.7953
81	7.75108	9.4840	11.2170	8.8415
82	8.41534	10.4168	12.4183	10.2117
83	7.56210	9.1461	10.7300	8.0815
*84	6.51658	8.0757	9.6349	7.9549
85	8.73332	10.7988	12.8643	10.5381
86	7.10357	8.9350	10.7664	9.3439
87	7.46386	9.5521	11.6403	10.6541
88	7.40553	9.0024	10.5993	8.1473
89	7.70464	9.6524	11.6002	9.9376
90	8.16395	10.3755	12.5870	11.2834
91	8.08567	9.8488	11.6120	8.9958
92	8.45424	10.2642	12.0741	9.2344
*93	6.78362	8.1950	9.6064	7.2010
94	8.56442	10.6626	12.7608	10.7050
95	7.49852	9.2474	10.9962	8.9226
96	8.01983	10.9016	13.7833	14.7028
97	7.40014	9.1321	10.8640	8.8363
98	7.80386	9.6444	11.4849	9.3903
99	8.47046	10.3403	12.2101	9.5399
100	8.08183	10.0133	11.9447	9.8542

average, with high probability. This tells us that B must equal 2 in the final result. If $\bar{X} \pm z_{\alpha/2}\sigma/\sqrt{n}$ must result in $\bar{X} \pm B$, then we must have

$$B = z_{\alpha/2} \frac{\sigma}{\sqrt{n}}$$

or

$$n = \left[\frac{z_{\alpha/2}\sigma}{B} \right]^2$$

Once the confidence coefficient $1 - \alpha$ is established, $z_{\alpha/2}$ is known. The only remaining problem is to find an approximation for σ. This can often be accomplished by prior experience, by taking a preliminary sample and using s to estimate σ, or by using any of a variety of other rough approximations to σ. Even if σ is not known exactly, working through the above equation to determine a sample size that will give desirable results is better than arbitrarily guessing what the sample size should be.

Sample size for establishing a confidence interval of the form $\bar{X} \pm B$ with confidence coefficient $1 - \alpha$.

$$n = \left[\frac{z_{\alpha/2}\sigma}{B} \right]^2$$

EXAMPLE 7.3

It is desired to estimate the average distance traveled to work for employees of a large manufacturing firm. Past studies of this type indicate that the standard deviation of these distances should be in the neighborhood of 2 miles. How many employees should be sampled if the estimate is to be within 0.1 mile of the true average, with confidence coefficient 0.95?

Solution The resulting interval is to be of the form $\bar{X} \pm 0.1$, with $1 - \alpha = 0.95$. Thus $B = 0.1$ and $z_{0.025} = 1.96$. It follows that

$$n = \left[\frac{z_{\alpha/2}\sigma}{B} \right]^2 = \left[\frac{(1.96)(2)}{0.1} \right]^2$$

$$= 1536.64 \quad \text{or} \quad 1537$$

Thus 1537 employees would have to be sampled to achieve the desired result. If a sample of this size is too costly, then either B must be increased or $1 - \alpha$ decreased. ∎

The confidence intervals discussed so far have been two-sided intervals. Sometimes we are interested in only a lower or upper limit for a parameter, but not both. A one-sided confidence interval can be formed in such cases.

Suppose in the study of mean lifelength of batteries, as in Example 7.2, we wanted only to establish a lower limit to this mean. That is, we want to find a statistic $g(\bar{X})$ so that

$$P[\mu > g(\bar{X})] = 1 - \alpha$$

Again we know that

$$P\left(\frac{\bar{X} - \mu}{\sigma/\sqrt{n}} \leqslant z_\alpha\right) = 1 - \alpha$$

and it follows that

$$P\left(\bar{X} - \mu \leqslant z_\alpha \frac{\sigma}{\sqrt{n}}\right) = 1 - \alpha$$

$$P\left(\bar{X} - z_\alpha \frac{\sigma}{\sqrt{n}} \leqslant \mu\right) = 1 - \alpha$$

or

$$\bar{X} - z_\alpha \frac{\sigma}{\sqrt{n}}$$

forms a lower confidence limit for μ with confidence coefficient $(1 - \alpha)$. Using the data from Example 7.2, a lower 95% confidence limit for μ is approximately

$$\bar{X} - z_{0.05} \frac{s}{\sqrt{n}}$$

or

$$2.266 - 1.645 \frac{1.935}{\sqrt{50}} = 2.266 - 0.450 = 1.816$$

Thus we are 95% confident that the mean battery lifelength exceeds 181.6 hours. Note that this lower confidence limit in the one-sided case is larger than the lower limit from the corresponding two-sided confidence interval.

7.3.2 Binomial Distribution: Large-Sample Confidence Interval for p

Estimating the parameter p for the binomial distribution is analogous to the estimation of a population mean. As seen in Section 3.3, this is because the random variable Y having a binomial distribution with n trials can be written as

$$Y = \sum_{i=1}^{n} X_i$$

where X_i, $i = 1, \ldots, n$, are independent Bernoulli random variables with common mean p. Thus $E(Y) = np$ or $E(Y/n) = p$, and Y/n is an unbiased estimator of p. Recall that $V(Y) = np(1 - p)$, and hence

$$V(Y/n) = \frac{p(1 - p)}{n}$$

Note that $Y/n = \sum_{i=1}^{n} X_i/n = \bar{X}$, with the X_i's having $E(X_i) = p$ and $V(X_i) = p(1 - p)$. The Central Limit Theorem then applies to Y/n, the sample fraction of successes, and we have that Y/n is approximately normally distributed with mean p and variance $p(1 - p)/n$. The large-sample confidence interval for p is then constructed by comparison with the corresponding result for μ. The observed fraction of successes, y/n, will be used as the estimate of p and will be noted by $\hat{p}$.

The interval

$$\hat{p} \pm z_{\alpha/2} \sqrt{\frac{\hat{p}(1 - \hat{p})}{n}}$$

forms a realization of a large-sample confidence interval for p with confidence coefficient approximately $1 - \alpha$, where $\hat{p} = y/n$.

EXAMPLE 7.4

In certain water-quality studies it is important to check for the presence or absence of various types of microorganisms. Suppose 20 out of 100 randomly selected samples of a fixed volume show the presence of a particular microorganism. Estimate the true probability p of finding this microorganism in a sample of this same volume, with a 90% confidence interval.

Solution   We assume that the number of samples showing the presence of the microorganism, out of n randomly selected samples, can be reasonably modeled by the binomial distribution. Using the confidence interval

$$\hat{p} \pm z_{\alpha/2} \sqrt{\frac{\hat{p}(1 - \hat{p})}{n}}$$

we have $1 - \alpha = 0.90$, $z_{\alpha/2} = z_{0.05} = 1.645$, $y = 20$, and $n = 100$. Thus the interval estimate of p is

$$0.20 \pm 1.645 \sqrt{\frac{0.2(0.8)}{100}}$$

or

$$0.20 \pm 0.066$$

We are about 90% confident that the true probability p is somewhere between 0.134 and 0.266.   ■

EXAMPLE 7.5

A firm wants to estimate the proportion of production line workers who favor a revised quality assurance program. The estimate is to be within 0.05 of the true proportion favoring the revised program, with a 90% confidence coefficient. How many workers should be sampled?

Solution Sample size is determined by

$$n = \left[\frac{z_{\alpha/2}\sigma}{B}\right]^2$$

but, in the binomial case, $\sigma = \sqrt{p(1 - p)}$. Thus

$$n = \left[\frac{z_{\alpha/2}\sqrt{p(1 - p)}}{B}\right]^2$$

In this problem $B = 0.05$ and $1 - \alpha = 0.90$, so $z_{0.05} = 1.645$. It remains to approximate $\sigma = \sqrt{p(1 - p)}$. The quantity $p(1 - p)$ is maximized at $p = 0.5$. Thus if no prior knowledge of p is available, we can use $p = 0.5$ in this calculation to determine the largest sample size necessary to ensure the desired results.

We have, under these conditions,

$$n = \left[\frac{1.645\sqrt{(0.5)(0.5)}}{0.05}\right]^2$$

$$= (16.45)^2 = 270.6 \quad \text{or} \quad 271$$

A sample of 271 workers is required to estimate the true proportion favoring the revised policy to within 0.05. ■

7.3.3 Normal Distribution: Confidence Interval for μ

If the samples we are dealing with are not large enough to ensure an approximately normal sampling distribution for $\bar{X}$, then the above results are not valid. What can we do in that case? One approach is to use techniques that do not depend upon distributional assumptions (so called distribution-free techniques). This will be considered in Chapter 12. Another approach, which we will pursue here, is to make an additional assumption concerning the nature of the probabilistic model for the population. In many real problems it is appropriate to assume that the random variable under study has a normal distribution. We can then develop an exact confidence interval for the mean μ.

The development of the confidence interval for μ in this section parallels the large-sample development in Section 7.3.1. We first must find a function of $\bar{X}$ and μ for which we know the sampling distribution. If $X_1, \ldots, X_n$ are independent normally distributed random variables, each with unknown mean μ and variance σ^2, then we know that

$$Z = \frac{\bar{X} - \mu}{\sigma/\sqrt{n}}$$

has a standard normal distribution. But σ is unknown, and a reasonable estimator is S. What, then, is the sampling distribution of

$$T = \frac{\bar{X} - \mu}{S/\sqrt{n}}$$

Since $E(\bar{X}) = \mu$, the sampling distribution of T should center at zero. Since σ, a constant, was replaced by S, a random variable, the sampling distribution of T should show more variability than the sampling distribution of Z. In fact, the sampling distribution of T is known and is called the t *distribution* (or Student's t distribution) with $n - 1$ degrees of freedom. A t probability density function is pictured in Figure 7.3. The t probability density function is symmetric about zero and has wider tails (is more spread out) than the standard normal distribution. Table 5 of the Appendix contains values t_α that cut-off an area of α in the right-hand tail of the distribution, as pictured in Figure 7.3. Only the upper tail values are tabled because of the symmetry of the density function. The tabled values depend upon the degrees of freedom since the distribution of T changes with changes in $(n - 1)$. As n gets large, the t distribution approaches the standard normal distribution. This t_α is approximately equal to z_α for large n.

FIGURE 7.3
*A t Distribution
(probability density
function)*

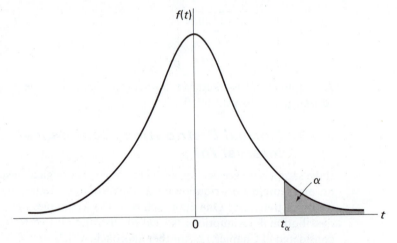

We now wish to construct a confidence interval for μ based on the t distribution. From Table 5 of the Appendix we can find a value $t_{\alpha/2}$, which cuts off an area of $\alpha/2$ in the right-hand tail of a t distribution with known degrees of freedom. Thus

$$P\left(-t_{\alpha/2} \leqslant \frac{\bar{X} - \mu}{S/\sqrt{n}} \leqslant t_{\alpha/2}\right) = 1 - \alpha$$

for some prescribed α. Reworking this inequality just as in the large-sample case of Section 7.3.1, we have

$$P\left(\bar{X} - t_{\alpha/2} \frac{S}{\sqrt{n}} \leqslant \mu \leqslant \bar{X} + t_{\alpha/2} \frac{S}{\sqrt{n}}\right) = 1 - \alpha$$

> If $X_1, \ldots, X_n$ denotes a random sample from a normal distribution, then
>
> $$\bar{x} \pm t_{\alpha/2} \frac{s}{\sqrt{n}}$$
>
> provides a realization of an exact confidence interval for μ with confidence coefficient $(1 - \alpha)$.

EXAMPLE 7.6

Prestressing wire for wrapping concrete pipe is manufactured in large rolls. A quality-control inspection required 5 specimens from a roll to be tested for ultimate tensile strength (UTS). The UTS measurements (in 1000 psi) turned out to be 253, 261, 258, 255, 256. Use these data to construct a 95% confidence interval estimate of the true mean UTS for the sampled roll.

Solution It is assumed that, if many wire specimens were tested, the relative frequency distribution of UTS measurements would be nearly normal. A confidence interval based on the t distribution can then be employed. With $1 - \alpha = 0.95$ and $n - 1 = 4$ degrees of freedom, $t_{0.025} = 2.776$.

From the observed data,

$$\bar{x} = 256.60 \qquad \text{and} \qquad s = 3.05$$

Thus

$$\bar{x} \pm t_{\alpha/2} \frac{s}{\sqrt{n}}$$

becomes

$$256.60 \pm (2.776) \frac{3.05}{\sqrt{5}}$$

or

$$256.60 \pm 3.79$$

We are 95% confident that the interval 252.81 to 260.39 includes the true mean UTS for the role.

If a computer with a statistical package is available to the reader, we would encourage its use to handle computations like those encountered above. The output for solving this problem on Minitab is shown below. The output for other software packages would be similar.

N	MEAN	STDEV	SE MEAN	95.0 PERCENT C.I.
5	256.60	3.05	1.36	(252.81, 260.39)

Note that if we are not justified in making the assumption of normality for the population of tensile strengths, then we would be uncertain of the confidence level for this interval. The Central Limit Theorem would not help in this small-sample case. We will show how to handle cases in which the normality assumption is *not* justified in Chapter 12.

7.3.4 Normal Distribution: Confidence Interval for σ^2

Still under the assumption that the random sample $X_1, \ldots, X_n$ comes from a normal distribution, we know from Section 5.5 that $(n-1)S^2/\sigma^2$ has a χ^2 distribution, with $\nu = n - 1$ degrees of freedom. We can employ this fact to establish a confidence interval for σ^2. Using Table 6 of the Appendix, we can find two values, $\chi^2_{\alpha/2}(\nu)$ and $\chi^2_{1-\alpha/2}(\nu)$, such that

$$1 - \alpha = P\left[\chi^2_{1-\alpha/2}(\nu) \leqslant \frac{(n-1)S^2}{\sigma^2} \leqslant \chi^2_{\alpha/2}(\nu)\right]$$

$$= P\left[\frac{(n-1)S^2}{\chi^2_{\alpha/2}(\nu)} \leqslant \sigma^2 \leqslant \frac{(n-1)S^2}{\chi^2_{1-\alpha/2}(\nu)}\right]$$

If $X_1, \ldots, X_n$ denotes a random sample from a normal distribution, then

$$\left(\frac{(n-1)s^2}{\chi^2_{\alpha/2}(\nu)}, \frac{(n-1)s^2}{\chi^2_{1-\alpha/2}(\nu)}\right)$$

provides a realization of an exact confidence interval for σ^2 with confidence coefficient $1 - \alpha$.

EXAMPLE 7.7 _____

In laboratory work it is desirable to run careful checks on the variability of readings produced on standard samples. In a study of the amount of calcium in drinking water undertaken as part of a water-quality assessment, the same standard was run through the laboratory six times at random intervals. The six readings, in parts per million, were 9.54, 9.61, 9.32, 9.48, 9.70, 9.26.

Estimate σ^2, the population variance for readings on this standard, using a 90% confidence interval.

Solution We must first calculate the sample variance s^2 as

$$s^2 = \left(\sum_{i=1}^{n} x_i^2 - n\bar{x}^2\right)\left(\frac{1}{n-1}\right) = [539.9341 - 6(9.485)^2]\left(\frac{1}{5}\right)$$

$$= (539.9341 - 539.7914)\left(\frac{1}{5}\right) = \frac{0.1427}{5} = 0.0285$$

Using Table 6 of the Appendix, we have that

$$\chi^2_{0.95}(5) = 1.1455 \qquad \chi^2_{0.05}(5) = 11.0705$$

Thus the confidence interval for σ^2 becomes

$$\left[\frac{(n-1)s^2}{\chi^2_{0.05}(v)}, \frac{(n-1)s^2}{\chi^2_{0.95}(v)} \right]$$

or

$$\left[\frac{0.1427}{11.0705}, \frac{0.1427}{1.1455} \right]$$

or

$$[0.0129, 0.1246]$$

That is, we are 90% confident that σ^2 is between 0.0129 and 0.1246. Note that this is a fairly wide interval, primarily because n is so small. ∎

CONFIDENCE INTERVALS FROM SINGLE SAMPLES

General Distribution: Large-sample confidence interval for μ:

$$\bar{x} \pm z_{\alpha/2}\sigma/\sqrt{n} \qquad (\sigma \text{ can be estimated by } s)$$

Binomial Distribution: Large-sample confidence interval for p:

$$\hat{p} \pm z_{\alpha/2}\sqrt{\frac{\hat{p}(1-\hat{p})}{n}} \qquad \text{where } \hat{p} = \frac{y}{n}$$

Normal Distribution: Confidence interval for μ:

$$\bar{x} \pm t_{\alpha/2}s/\sqrt{n}$$

Normal Distribution: Confidence interval for σ^2:

$$\left(\frac{(n-1)s^2}{\chi^2_{\alpha/2}(v)}, \frac{(n-1)s^2}{\chi^2_{1-\alpha/2}(v)} \right) \qquad (v = n-1)$$

EXERCISES

7.9 For a random sample of 50 measurements on the breaking strength of cotton threads, the mean breaking strength was found to be 210 grams and the standard deviation 18 grams. Obtain a confidence interval for the true mean breaking strength of cotton threads of this type, with confidence coefficient 0.90.

7.10 A random sample of 40 engineers was selected from among the large number employed by a corporation engaged in seeking new sources of petroleum. The hours worked in a particular week was determined for each engineer selected. These data had a mean of 46 hours and a standard deviation of 3 hours. For that particular week, estimate the mean hours worked for all engineers in the corporation, with a 95% confidence coefficient.

7.11 An important property of plastic clays is the percent of shrinkage on drying. For a certain type of plastic clay 45 test specimens showed an average shrinkage percentage of 18.4 and a standard deviation of 1.2. Estimate the true average percent of shrinkage for specimens of this type in a 98% confidence interval.

7.12 The breaking strength of threads has a standard deviation of 18 grams. How many measures on breaking strength should be used in the next experiment if the estimate of the mean breaking strength is to be within 4 grams of the true mean breaking strength, with confidence coefficient 0.90?

7.13 In the setting of Exercise 7.10 how many engineers should be sampled if it is desired to estimate the mean number of hours worked to within 0.5 hour with confidence coefficient 0.95?

7.14 Refer to Exercise 7.11. How many specimens should be tested if it is desired to estimate the percent of shrinkage to within 0.2 with confidence coefficient 0.98?

7.15 Upon testing 100 resistors manufactured by Company A, it is found that 12 fail to meet the tolerance specifications. Find a 95% confidence interval for the true fraction of resistors manufactured by Company A that fail to meet the tolerance specification. What assumptions are necessary for your answer to be valid?

7.16 Refer to Exercise 7.15. If it is desired to estimate the true proportion failing to meet tolerance specifications to within 0.05, with confidence coefficient 0.95, how many resistors should be tested?

7.17 Careful inspection of 70 precast concrete supports to be used in a construction project revealed 28 with hairline cracks. Estimate the true proportion of supports of this type with cracks in a 98% confidence interval.

7.18 Refer to Exercise 7.17. Suppose it is desired to estimate the true proportion of cracked supports to within 0.1, with confidence coefficient 0.98. How many supports should be sampled to achieve the desired accuracy?

7.19 In conducting an inventory and audit of parts in a certain stockroom, it was found that, for 60 items sampled, the audit value exceeded the book value on 45 items. Estimate, with confidence coefficient 0.90, the true fraction of items in the stockroom for which the audit value exceeds the book value.

7.20 The Environmental Protection Agency has collected data on the LC50 (concentration killing 50% of the test animals in a specified time interval) measurements for certain chemicals likely to be found in freshwater rivers and lakes. For a certain species of fish, the LC50 measurements for DDT in 12 experiments yielded the following:

16, 5, 21, 19, 10, 5, 8, 2, 7, 2, 4, 9

(Measurements are in parts per million.) Assuming such LC50 measurements to be approximately normally distributed, estimate the true mean LC50 for DDT with confidence coefficient 0.90.

7.21 The warpwise breaking strength measured on five specimens of a certain cloth gave a sample mean of 180 psi and a standard deviation of 5 psi. Estimate the true mean warpwise breaking strength for cloth of this type in a 95% confidence interval. What assumption is necessary for your answer to be valid?

7.22 Answer Exercise 7.21 if the same sample data had resulted from a sample of

(a) 10 specimens   (b) 100 specimens

7.23 Fifteen resistors were randomly selected from the output of a process supposedly producing 10-ohm resistors. The fifteen resistors actually showed a sample mean of 9.8 ohms and a sample standard deviation of 0.5 ohm. Find a 95% confidence interval for the true mean resistance of the resistors produced by this process. Assume resistance measurements are approximately normally distributed.

7.24 The variance of LC50 measurements is important because it may reflect an ability (or inability) to reproduce similar results in identical experiments. Find a 95% confidence interval for σ^2, the true variance of the LC50 measurements for DDT, using the data in Exercise 7.20.

7.25 In producing resistors, variability of the resistances is an important quantity to study, as it reflects upon the stability of the manufacturing process. Estimate σ^2, the true variance of the resistance measurements, in a 90% confidence interval, if a sample of 15 resistors showed resistances with a standard deviation of 0.5 ohm.

7.26 The increased use of wood for residential heating has caused some concern over the pollutants and irritants this process releases into the air. In one reported study in the northwestern U.S., the total suspended particulates (TSP) in 9 air samples had a mean of $72\,\mu g/m^3$ and a standard deviation of 23. Estimate the true mean TSP in a 95% confidence interval. What assumption is necessary for your answer to be valid? (See J. E. Core, J. A. Cooper, and R. M. Neulicht, *Journal of the Air Pollution Control Association*, 31, No. 2, 1984, pp. 138–143 for more details.)

7.27 The yield stress in steel bars (Grade FeB 400 HWL) reported by P. Booster (*Journal of Quality Technology*, 15, No. 4, 1983, p. 191) gave the following data for one test:

$$n = 150 \qquad \bar{x} = 477\ \text{N/mm}^2 \qquad s = 13$$

Estimate the true yield stress for bars of this grade and size in a 90% confidence interval. Are any assumptions necessary for your answer to be valid?

7.28 In the article quoted in Exercise 7.27 a sample of 12 measurements on the tensile strength of steel bars resulted in a mean of $585\,\text{N/mm}^2$ and a standard deviation of 38. Estimate the true mean tensile strength of bars of this size and type in a 95% confidence interval. List any assumptions you need to make.

7.29 Fatigue behavior of reinforced concrete beams in seawater was studied by T. Hodgkiess, et al. (*Materials Performance*, July 1984, pp. 27–29). The number of cycles to failure in seawater for beams subjected to certain bending and loading stress are as follows (in thousands):

774,   633,   477,   268,   407,   576,   659,   963,   193

Construct a 90% confidence interval estimate of the average number of cycles to failure for beams of this type.

7.30 Using the data given in Exercise 7.29, construct a 90% confidence interval for the variance of the number of cycles to failure for beams of this type. What assumptions are necessary for your answer to be valid?

7.31   The quantity

$$z_{\alpha/2}\frac{\sigma}{\sqrt{n}}\left(\text{or } z_{\alpha/2}\sqrt{\frac{p(1-p)}{n}}\right)$$

used in constructing confidence intervals for μ (or p) is sometimes called the *sampling error*.

Results of a public opinion poll reported in a news magazine (*Time*, January 2, 1984) show that 51% of the respondents completely agreed with the statement, "The Soviets are just as afraid of nuclear war as we are, and therefore it is in our mutual interest to find ways to negotiate." The article states that "The findings are based on a telephone survey of 1000 registered voters .... The potential sampling error is plus or minus 3%."

How was the 3% calculated, and what is its interpretation? Can we conclude that a majority of registered voters completely agree with the statement?

7.32   The results of a Louis Harris poll state that 36% of Americans list football as their favorite sport. A note then states: "In a sample of this size (1,091 adults) one can say with 95% certainty that the results are within plus or minus three percentage points of what they would be if the entire adult population had been polled." Do you agree? (Source: *Gainesville Sun*, May 7, 1961).

7.33   A. C. Nielsen Co. has electronic monitors hooked up to about 1200 of the 80 million American homes. The data from the monitors provide estimates of the proportion of homes tuned to a particular T.V. program. Nielsen offers the following defense of this sample size.

Mix together 70,000 white beans and 30,000 red beans and then scoop out a sample of 1,000. The mathematical odds are that the number of red beans will be between 270 and 330, or 27 to 33 percent of the sample, which translates to a "rating" of 30, plus or minus three, with a 20 to 1 assurance of statistical reliability. The basic statistical law wouldn't change even if the sampling came from 80 million beans rather than 100,000.

Interpret and justify this statement in terms of the results of this chapter. (Source: *Sky*, October 1982.)

7.4   CONFIDENCE INTERVALS: THE MULTIPLE-SAMPLE CASE

All methods for confidence intervals discussed in the preceding section considered only the case in which a single random sample was selected to estimate a single parameter. In many problems more than one population, and hence more than one sample, is involved. For example, we may desire to estimate the difference between mean daily yields for two industrial processes for producing a certain liquid fertilizer. Or we may want to compare the rates of defectives produced on two or more assembly lines within a factory.

7.4.1 General Distribution: Large-Sample Confidence Interval for Linear Function of Means

It is commonly of interest to compare two population means, and this is conveniently accomplished by estimating the *difference* between the two means. Suppose that the two populations of interest have, respectively, means and variances denoted by μ_1 and σ_1^2 and μ_2 and σ_2^2, with $\mu_1 - \mu_2$ the parameter to be estimated. A large sample of size n_1 from the first population has a mean of $\bar{X}_1$ and variance of S_1^2. An *independent* large sample from the second population has a mean of $\bar{X}_2$ and variance of S_2^2. If $\bar{X}_1 - \bar{X}_2$ is used to estimate μ_1 and μ_2, then

$$E(\bar{X}_1 - \bar{X}_2) = \mu_1 - \mu_2$$

(the estimator is unbiased) and

$$V(\bar{X}_1 - \bar{X}_2) = V(\bar{X}_1) + V(\bar{X}_2)$$

$$= \frac{\sigma_1^2}{n_1} + \frac{\sigma_2^2}{n_2}$$

Also $\bar{X}_1 - \bar{X}_2$ will have approximately a normal sampling distribution. Thus a confidence interval for $\mu_1 - \mu_2$ can be formed in the same manner as the one for μ in Section 7.3.1.

The interval

$$(\bar{x}_1 - \bar{x}_2) \pm z_{\alpha/2}\sqrt{\frac{\sigma_1^2}{n_1} + \frac{\sigma_2^2}{n_2}}$$

forms a realization of a large-sample confidence interval for $\mu_1 - \mu_2$ with confidence coefficient approximately $(1 - \alpha)$.

EXAMPLE 7.8

A farm equipment manufacturer wants to compare the average daily downtime for two sheet-metal stamping machines located in two different factories. Investigation of company records for 100 randomly selected days on each of the two machines gave the following results:

$$n_1 = 100 \qquad n_2 = 100$$

$$\bar{x}_1 = 12 \text{ minutes} \qquad \bar{x}_2 = 9 \text{ minutes}$$

$$s_1^2 = 6 \qquad s_2^2 = 4$$

Construct a 90% confidence interval estimate for $\mu_1 - \mu_2$.

Solution Using the form of the interval given above, with s_i^2 estimating σ_i^2, we have

$$(\bar{x}_1 - \bar{x}_2) \pm z_{0.05} \sqrt{\frac{s_1^2}{n_1} + \frac{s_2^2}{n_2}}$$

yielding

$$(12 - 9) \pm (1.645) \sqrt{\frac{6}{100} + \frac{4}{100}}$$

or

$$3 \pm 0.52$$

That is, we are about 90% confident that $\mu_1 - \mu_2$ is between $3 - 0.52 = 2.48$ and $3 + 0.52 = 3.52$. This evidence suggests that μ_1 must be larger than μ_2. ∎

Suppose we are interested in three populations, numbered 1, 2, and 3, with unknown means μ_1, μ_2, and μ_3, respectively, and variances σ_1^2, σ_2^2, and σ_3^2. If a random sample of size n_i is selected from the population with mean μ_i, $i = 1, 2, 3$, then any linear function of the form

$$\theta = a_1\mu_1 + a_2\mu_2 + a_3\mu_3$$

can be estimated unbiasedly by

$$\hat{\theta} = a_1\bar{X}_1 + a_2\bar{X}_2 + a_3\bar{X}_3$$

where $\bar{X}_i$ is the mean of the sample from population i. Also if the samples are independent of one another,

$$V(\hat{\theta}) = a_1^2 V(\bar{X}_1) + a_2^2 V(\bar{X}_2) + a_3^2 V(\bar{X}_3)$$

$$= a_1^2 \left(\frac{\sigma_1^2}{n_1}\right) + a_2^2 \left(\frac{\sigma_2^2}{n_2}\right) + a_3^2 \left(\frac{\sigma_3^2}{n_3}\right)$$

and so long as all of the n_i's are reasonably large, $\hat{\theta}$ will have approximately a normal distribution.

The interval

$$\hat{\theta} \pm z_{\alpha/2} \sqrt{V(\hat{\theta})}$$

will provide a large-sample confidence interval for θ with confidence coefficient approximately $(1 - \alpha)$.

EXAMPLE 7.9

A company has three machines for stamping sheet metal located in three different factories around the country. If μ_i denotes the average downtime (in minutes) per day for the ith machine, $i = 1, 2, 3$, then the expected daily cost for downtime on these three machines is

$$C = 3\mu_1 + 5\mu_2 + 2\mu_3$$

Investigation of company records for 100 randomly selected days on each of these machines showed the following:

$$n_1 = 100 \qquad n_2 = 100 \qquad n_3 = 100$$
$$\bar{x}_1 = 12 \qquad \bar{x}_2 = 9 \qquad \bar{x}_3 = 14$$
$$s_1^2 = 6 \qquad s_2^2 = 4 \qquad s_3^2 = 5$$

Estimate C in a 95% confidence interval.

Solution We have seen above that an unbiased estimator of C is

$$\hat{C} = 3\bar{X}_1 + 5\bar{X}_2 + 2\bar{X}_3$$

with variance

$$V(\hat{C}) = 9\left(\frac{\sigma_1^2}{100}\right) + 25\left(\frac{\sigma_2^2}{100}\right) + 4\left(\frac{\sigma_3^2}{100}\right)$$

We obtained an observed value of $\hat{C}$ equal to

$$3(12) + 5(9) + 2(14) = 109$$

Since the σ_i^2's are unknown, we can approximate them with s_i^2's and find an approximate variance of $\hat{C}$ to be

$$9\left(\frac{6}{100}\right) + 25\left(\frac{4}{100}\right) + 4\left(\frac{5}{100}\right) = 1.74$$

(This approximation works well only for large samples.) Thus a realization of the 95% confidence interval for C is

$$109 \pm 1.96\sqrt{1.74}$$

or

$$109 \pm 2.58$$

We are about 95% confident that the mean daily cost is between \$106.42 and \$111.58. ∎

7.4.2 Binomial Distribution: Large-Sample Confidence Interval for Linear Function of Proportions

Since we have seen that sample proportions behave like sample means, and that for large samples the sample proportions will tend to have normal distributions, we can adapt the above methodology to include estimation of linear functions of sample proportions. Again the most common practical problem concerns the estimation of the difference between two proportions. If Y_i has a binomial distribution with sample size n_i and probability of success p_i, then $p_1 - p_2$ can be estimated by $Y_1/n_1 - Y_2/n_2$. After the samples are taken, we estimate p_1 by $\hat{p}_1 = y_1/n_1$ and p_2 by $\hat{p}_2 = y_2/n_2$.

The interval

$$(\hat{p}_1 - \hat{p}_2) \pm z_{\alpha/2} \sqrt{\frac{\hat{p}_1(1 - \hat{p}_1)}{n_1} + \frac{\hat{p}_2(1 - \hat{p}_2)}{n_2}}$$

forms a realization of a large-sample confidence interval for $p_1 - p_2$ with confidence coefficient approximately $(1 - \alpha)$, where $\hat{p}_1 = y_1/n_1$ and $\hat{p}_2 = y_2/n_2$.

EXAMPLE 7.10

We want to compare the proportion of defective electric motors turned out by two shifts of workers. From the large number of motors produced in a given week, $n_1 = 50$ motors were selected from the output of shift I and $n_2 = 40$ motors were selected from the output of shift II. The sample from shift I revealed 4 to be defective, and the sample from shift II showed 6 faulty motors. Estimate the true difference between proportions of defective motors produced in a 95% confidence interval.

Solution   Starting with

$$(\hat{p}_1 - \hat{p}_2) \pm 1.96 \sqrt{\frac{\hat{p}_1(1 - \hat{p}_1)}{n_1} + \frac{\hat{p}_2(1 - \hat{p}_2)}{n_2}}$$

we have as an approximate 95% confidence interval

$$\left(\frac{4}{50} - \frac{6}{40}\right) \pm 1.96 \sqrt{\frac{0.08(0.92)}{50} + \frac{0.15(0.85)}{40}}$$

or

$$-0.07 \pm 0.13$$

Since the interval overlaps zero, there does not appear to be any significant difference between the rates of defectives for the two shifts. That is, zero cannot be ruled out as a plausible value of the true difference between proportions of defective motors, which implies that these proportions may be equal. ■

More general linear functions of sample proportions are sometimes of interest, as will be the case in the following example.

EXAMPLE 7.11

The personnel director of a large company wants to compare two different aptitude tests that are supposed to be equivalent in terms of which aptitudes they measure. He has reason to suspect that the degree of difference in test scores is not the same for women and men. Thus he sets up an experiment in which one hundred men and one hundred women, of nearly equal aptitude, are selected to take the tests. Fifty men take test I and 50 independently selected men take test II. Likewise, 50 women take test I and 50 independently selected women take test II. The proportions, out of 50, receiving passing scores are recorded for each of the four groups, and the data are as follows:

Test

	I	II
Male	0.7	0.9
Female	0.8	0.9

If p_{ij} denotes the true probability of passing in row i and column j, the director wants to estimate $(p_{12} - p_{11}) - (p_{22} - p_{21})$ to see if the change in probability for males is different from the corresponding change for females. Estimate this quantity in a 95% confidence interval.

Solution If $\hat{p}_{ij}$ denotes the observed proportion passing the test, out of 50, in row i and column j, then

$$(p_{12} - p_{11}) - (p_{22} - p_{21})$$

is estimated by

$$(\hat{p}_{12} - \hat{p}_{11}) - (\hat{p}_{22} - \hat{p}_{21})$$

which is observed to be

$$(0.9 - 0.7) - (0.9 - 0.8) = 0.1$$

Because of the independence of the samples, the estimated variance of $(\hat{p}_{12} - \hat{p}_{11}) - (\hat{p}_{22} - \hat{p}_{21})$ is

$$\frac{\hat{p}_{12}(1 - \hat{p}_{12})}{50} + \frac{\hat{p}_{11}(1 - \hat{p}_{11})}{50} + \frac{\hat{p}_{22}(1 - \hat{p}_{22})}{50} + \frac{\hat{p}_{21}(1 - \hat{p}_{21})}{50}$$

which is observed to be

$$\frac{1}{50}[(0.9)(0.1) + (0.7)(0.3) + (0.9)(0.1) + (0.9)(0.2)] = 0.011$$

The interval estimate is then

$$0.1 \pm 1.96\sqrt{0.011}$$

or

$$0.1 \pm 0.20$$

Since this interval overlaps zero, there is really no reason to suspect that the change for males is different from the change for females. ∎

7.4.3 Normal Distribution: Confidence Interval for Linear Function of Means

We must make some additional assumptions for the estimation of linear functions of means in small samples (where the Central Limit Theorem does not ensure approximate normality). If we have k populations and k independent random samples, we assume that all populations have approximately *normal distributions with a common unknown variance* σ^2. This assumption should be considered carefully, and if it does not seem reasonable, the methods outlined below should not be employed. The arguments for producing confidence intervals parallel those given for the interval based on the t distribution in Section 7.3.3.

As in the large-sample case of Section 7.4.1, the most commonly occurring function of means that we might want to estimate is a simple difference of the form $\mu_1 - \mu_2$. Suppose two populations of interest are normally distributed with means μ_1 and μ_2 and common variance σ^2. Independent random samples from these populations yield sample statistics of $\bar{X}_1$ and S_1^2 from the first population and $\bar{X}_2$ and S_2^2 from the second.

The obvious unbiased estimator of $\mu_1 - \mu_2$ is still $\bar{X}_1 - \bar{X}_2$. Now

$$V(\bar{X}_1 - \bar{X}_2) = V(\bar{X}_1) + V(\bar{X}_2)$$

$$= \frac{\sigma^2}{n_1} + \frac{\sigma^2}{n_2} = \sigma^2 \left(\frac{1}{n_1} + \frac{1}{n_2} \right)$$

because of the common variance assumption. Thus we must find a good estimator of the *common* variance σ^2. This estimator will be a function of S_1^2 and S_2^2, but the unbiased estimator with the smallest variance is produced by weighting each S_1^2 according to its degrees of freedom. That is, the pooled sample variance

$$\frac{(n_1 - 1)S_1^2 + (n_2 - 1)S_2^2}{n_1 + n_2 - 2} = S_p^2$$

is a good estimator of σ^2. The quantity

$$T = \frac{(\bar{X}_1 - \bar{X}_2) - (\mu_1 - \mu_2)}{S_p \sqrt{\dfrac{1}{n_1} + \dfrac{1}{n_2}}}$$

has a t distribution with $n_1 + n_2 - 2$ degrees of freedom. This quantity can then be turned into a confidence interval for $(\mu_1 - \mu_2)$ by finding $t_{\alpha/2}$ such that

$$P(-t_{\alpha/2} \leqslant T \leqslant t_{\alpha/2}) = 1 - \alpha$$

and reworking the inequality. We have

$$P\left(-t_{\alpha/2} \leqslant \frac{(\bar{X}_1 - \bar{X}_2) - (\mu_1 - \mu_2)}{S_p\sqrt{\dfrac{1}{n_1} + \dfrac{1}{n_2}}} \leqslant t_{\alpha/2}\right) = 1 - \alpha$$

$$P\left(-t_{\alpha/2}S_p\sqrt{\frac{1}{n_1} + \frac{1}{n_2}} \leqslant (\bar{X}_1 - \bar{X}_2) - (\mu_1 - \mu_2)\right.$$

$$\left. \leqslant t_{\alpha/2}S_p\sqrt{\frac{1}{n_1} + \frac{1}{n_2}}\right) = 1 - \alpha$$

or

$$P\left((\bar{X}_1 - \bar{X}_2) - t_{\alpha/2}S_p\sqrt{\frac{1}{n_1} + \frac{1}{n_2}} \leqslant \mu_1 - \mu_2 \leqslant (\bar{X}_1 - \bar{X}_2)\right.$$

$$\left. + t_{\alpha/2}S_p\sqrt{\frac{1}{n_1} + \frac{1}{n_2}}\right) = 1 - \alpha$$

which yields the confidence interval for $\mu_1 - \mu_2$.

> If both random samples come from independent normal distributions with common variance, then the interval
>
> $$(\bar{x}_1 - \bar{x}_2) \pm t_{\alpha/2}S_p\sqrt{\frac{1}{n_1} + \frac{1}{n_2}}$$
>
> provides a realization of an exact confidence interval for $\mu_1 - \mu_2$, with confidence coefficient $(1 - \alpha)$.

EXAMPLE 7.12

Copper produced by sintering (heating without melting) a powder under certain conditions is then measured for porosity (the volume fraction due to voids) in a certain laboratory. A sample of $n_1 = 4$ independent porosity measurements shows a mean of $\bar{x}_1 = 0.22$ and a variance of $s_1^2 = 0.0010$. A second laboratory repeats the same process on an identical powder and gets $n_2 = 5$ independent porosity measurements with $\bar{x}_2 = 0.17$ and $s_2^2 = 0.0020$. Estimate the true difference between the population means $(\mu_1 - \mu_2)$ for these two laboratories, with confidence coefficient 0.95.

Solution   First we assume that the population of porosity measurements in either laboratory could be modeled by a normal distribution, and that the population variances are approximately equal. Then a confidence interval based on the t distribution can be used, with

$$s_p^2 = \frac{(n_1 - 1)s_1^2 + (n_2 - 1)s_2^2}{n_1 + n_2 - 2}$$

$$= \frac{3(0.001) + 4(0.002)}{7} = 0.0016$$

or

$$s_p = 0.04$$

Using

$$(\bar{x}_1 - \bar{x}_2) + t_{\alpha/2}s_p\sqrt{\frac{1}{n_1} + \frac{1}{n_2}}$$

we have, with $\alpha/2 = 0.025$ and 7 degrees of freedom,

$$(0.22 - 0.17) \pm (2.365)(0.04)\sqrt{\frac{1}{4} + \frac{1}{5}}$$

or

$$0.05 \pm 0.06$$

Since the interval contains zero, we would say that there is not much evidence of any difference between the two population means. That is, zero is a plausible value for $\mu_1 - \mu_2$, and if $\mu_1 - \mu_2 = 0$, then $\mu_1 = \mu_2$. ■

To see how a statistical computer package (Minitab in this case) presents the output for a two-sample confidence interval, consider the following example.

EXAMPLE 7.13

To compare ultimate tensile strengths of two classes of wire (class II and class III), five specimens are randomly sampled from a roll of each type. The actual UTS measurements (in 1000 psi) are given as follows:

Class II   253, 261, 258, 255, 256

Class III   274, 275, 271, 277, 276

Estimate the difference between true mean UTS measurements for the two classes of wire, in a 95% confidence interval.

Solution   The computer output shows the summary statistics and the final computation of the confidence interval:

TWOSAMPLE T FOR classII VS classIII

	N	MEAN	STDEV	SE MEAN
classII	5	256.60	3.05	1.4
classIII	5	274.60	2.30	1.0

95 PCT CI FOR MU classII − MU classIII: (−21.9, −14.1)
POOLED STDEV =   2.70

Note that the true mean for III is apparently larger than the true mean for II, since the interval $(-21.9, -14.1)$ does not overlap zero. ∎

Suppose that we have three populations of interest, numbered 1, 2, and 3, all assumed approximately normal in distribution. The populations have means μ_1, μ_2, and μ_3, respectively, and common variance σ^2. We want to estimate a linear function of the form

$$\theta = a_1\mu_1 + a_2\mu_2 + a_3\mu_3$$

A good estimator is

$$\hat{\theta} = a_1\bar{X}_1 + a_2\bar{X}_2 + a_3\bar{X}_3$$

where $\bar{X}_i$ is the mean of the random sample from population i. Now

$$V(\hat{\theta}) = a_1^2 V(\bar{X}_1) + a_2^2 V(\bar{X}_2) + a_3^2 V(\bar{X}_3)$$

$$= \sigma^2 \left[\frac{a_1^2}{n_1} + \frac{a_2^2}{n_2} + \frac{a_3^2}{n_3} \right]$$

The problem now is to choose the best estimator of the common variance σ^2 from the three sample variances S_i^2, based on respective sample sizes n_1, n_2, and n_3. It can be shown that an unbiased and minimum variance estimator of σ^2 is

$$\frac{\sum\limits_{i=1}^{3} (n_i - 1)S_i^2}{\sum\limits_{i=1}^{3} (n_i - 1)} = S_p^2$$

(The subscript p denotes "pooled.") Since $\hat{\theta}$ will have a normal distribution and $\sum_{i=1}^{3} (n_i - 1)S_p^2/\sigma^2$ will have a χ^2 distribution with $\sum_{i=1}^{3} (n_i - 1)$ degrees of freedom,

$$T = \frac{\hat{\theta} - \theta}{S_p \sqrt{\dfrac{a_1^2}{n_1} + \dfrac{a_2^2}{n_2} + \dfrac{a_3^2}{n_3}}}$$

will have a t distribution with $\sum_{i=1}^{3} (n_i - 1)$ degrees of freedom. We can then use this statistic to derive a confidence interval for θ.

If all random samples come from normal distributions, then the interval

$$\hat{\theta} \pm t_{\alpha/2} S_p \sqrt{\frac{a_1^2}{n_1} + \frac{a_2^2}{n_2} + \frac{a_3^2}{n_3}}$$

provides an exact confidence interval for θ, with confidence coefficient $(1 - \alpha)$. Here $\theta = a_1\mu_1 + a_2\mu_2 + a_3\mu_3$ and $\hat{\theta} = a_1\bar{X}_1 + a_2\bar{X}_2 + a_3\bar{X}_3$.

EXAMPLE 7.14

Refer to Example 7.12. A third laboratory sinters a slightly different copper powder and then takes $n_3 = 10$ independent observations on the porosity. These yield a mean of $\bar{x}_3 = 0.12$ and a variance of $s_3^2 = 0.0018$. Compare the population mean for this powder with the mean for the type of powder used in Example 7.12. (Estimate the difference in a 95% confidence interval.)

Solution   Again we assume normality for the population of measurements that could be obtained for this third laboratory, and we assume the variances for these porosity measurements would be approximately equal for all three laboratories. Since the first two samples are conducted with the same type of powder, we assume that they have a common population mean μ. We estimate this parameter by using a weighted average of the first two sample means, namely

$$\frac{n_1 \bar{x}_1 + n_2 \bar{x}_2}{n_1 + n_2}$$

This weighted average is unbiased and will have smaller variance than the simple average $(\bar{x}_1 + \bar{x}_2)/2$. To estimate $\mu - \mu_3$, we take

$$\frac{n_1}{n_1 + n_2} \bar{x}_1 + \frac{n_2}{n_1 + n_2} \bar{x}_2 - \bar{x}_3 = \frac{4}{9}(0.22) + \frac{5}{9}(0.17) - 0.12 = 0.0722$$

The pooled estimate of σ^2 is

$$s_p^2 = \frac{\sum\limits_{i=1}^{3}(n_i - 1)s_i^2}{\sum\limits_{i=1}^{3}(n_i - 1)} = \frac{3(0.0010) + 4(0.0020) + 9(0.0018)}{16}$$

$$= 0.0017$$

and $s_p = 0.0412$.

Now the confidence interval estimate, with $\alpha/2 = 0.025$ and 16 degrees of freedom, is

$$\left(\frac{4}{9}\bar{x}_1 + \frac{5}{9}\bar{x}_2 - \bar{x}_3\right) \pm t_{\alpha/2} s_p \sqrt{\frac{(4/9)^2}{n_1} + \frac{(5/9)^2}{n_2} + \frac{(-1)^2}{n_3}}$$

yielding

$$0.0722 \pm (2.120)(0.0412)\sqrt{\frac{(4/9)^2}{4} + \frac{(5/9)^2}{5} + \frac{1}{10}}$$

or

$$0.0722 \pm 0.0401$$

This interval would suggest that the second type of powder seems to give a lower mean porosity than the first since the interval does *not* contain zero.   ■

7.4.4   Normal Distribution: Confidence Interval for σ_1^2/σ_2^2

When two samples from normally distributed populations are available in an experimental situation, it is sometimes of interest to compare the population variances. We generally do this by estimating the *ratio* of the population variances, σ_1^2/σ_2^2.

Suppose that two independent random samples from normal distributions with respective sizes n_1 and n_2 yield sample variances of S_1^2 and S_2^2. If the populations in question have variances σ_1^2 and σ_2^2, respectively, then

$$F = \frac{S_1^2/\sigma_1^2}{S_2^2/\sigma_2^2}$$

has a known sampling distribution, called an *F distribution*, with $v_1 = n_1 - 1$ and $v_2 = n_2 - 1$ degrees of freedom.

A probability density function of a random variable having an F distribution is shown in Figure 7.4. The shape of the distribution depends on both sets of degrees of freedom, v_1 and v_2. Values of an F variate that cut off right-hand tail areas of size α are given in Tables 7 and 8 of the Appendix for $\alpha = 0.05$ and $\alpha = 0.01$. Left-hand tail values of F can be found by the equation

$$F_{1-\alpha}(v_1, v_2) = \frac{1}{F_\alpha(v_2, v_1)}$$

where the subscripts still denote right-hand areas.

FIGURE 7.4
An F Distribution (probability density function)

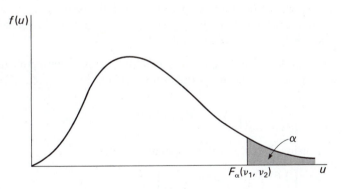

Knowledge of the F distribution allows us to construct a confidence interval for the ratio of two variances. We can find two values $F_{1-\alpha/2}(v_1, v_2)$ and $F_{\alpha/2}(v_1, v_2)$ so that

$$P(F_{1-\alpha/2}(v_1, v_2) \leqslant F \leqslant F_{\alpha/2}(v_1, v_2)) = 1 - \alpha$$

which yields

$$P\left(F_{1-\alpha/2}(v_1, v_2) \leqslant \frac{S_1^2}{S_2^2} \cdot \frac{\sigma_2^2}{\sigma_1^2} \leqslant F_{\alpha/2}(v_1, v_2)\right) = 1 - \alpha$$

$$P\left(\frac{S_2^2}{S_1^2} F_{1-\alpha/2}(v_1, v_2) \leqslant \frac{\sigma_2^2}{\sigma_1^2} \leqslant \frac{S_2^2}{S_1^2} F_{\alpha/2}(v_1, v_2)\right) = 1 - \alpha$$

Upon using the identity given above for lower-tailed F values, the confidence interval for σ_2^2/σ_1^2 is established.

If both random samples come from normal distributions, then the interval

$$\left[\frac{s_2^2}{s_1^2} \frac{1}{F_{\alpha/2}(v_2, v_1)}, \frac{s_2^2}{s_1^2} F_{\alpha/2}(v_1, v_2)\right]$$

provides a realization of an exact confidence interval for σ_2^2/σ_1^2, with confidence coefficient $(1 - \alpha)$.

EXAMPLE 7.15

A random sample of $n_1 = 10$ observations on breaking strengths of a type of glass gave $s_1^2 = 2.31$ (measurements were made in pounds per square inch). An independent random sample of $n_2 = 16$ measurements on a second machine, but with the same kind of glass, gave $s_2^2 = 3.68$. Estimate the true variance ratio, σ_2^2/σ_1^2, in a 90% confidence interval.

Solution   It is, of course, necessary to assume that the measurements in both populations can be modeled by normal distributions. Then the above confidence interval will yield

$$\left[\frac{s_2^2}{s_1^2} \frac{1}{F_{0.05}(15,9)}, \frac{s_2^2}{s_1^2} F_{0.05}(9,15)\right]$$

or, using Table 7 of the Appendix,

$$\left[\frac{3.68}{2.31}\left(\frac{1}{3.01}\right), \frac{3.68}{2.31}(2.59)\right]$$

or

(0.53, 4.13)

Thus we are about 90% confident that the ratio of σ_2^2 to σ_1^2 is between 0.53 and 4.13. Equality of σ_2^2 and σ_1^2 would imply a ratio of 1.00. Since our confidence interval includes 1.00, there is no real evidence that σ_2^2 differs from σ_1^2.   ∎

CONFIDENCE INTERVALS FROM TWO INDEPENDENT SAMPLES

General Distributions: Large-sample confidence interval for $\mu_1 - \mu_2$:

$$(\bar{x}_1 - \bar{x}_2) \pm z_{\alpha/2} \sqrt{\frac{\sigma_1^2}{n_1} + \frac{\sigma_2^2}{n_2}}$$

(σ_1 and σ_2 can be estimated by s_1 and s_2, respectively.)

Binomial Distributions: Large-sample confidence interval for $p_1 - p_2$:

$$(\hat{p}_1 - \hat{p}_2) \pm z_{\alpha/2} \sqrt{\frac{\hat{p}_1(1 - \hat{p}_1)}{n_1} + \frac{\hat{p}_2(1 - \hat{p}_2)}{n_2}}$$

where

$$\hat{p}_1 = y_1/n_2 \qquad \text{and} \qquad \hat{p}_2 = y_2/n_2$$

Normal Distributions with Common Variance: Confidence interval for $\mu_1 - \mu_2$:

$$(\bar{x}_1 - \bar{x}_2) \pm t_{\alpha/2} s_p \sqrt{\frac{1}{n_1} + \frac{1}{n_2}}$$

where

$$s_p^2 = \frac{(n_1 - 1)s_1^2 + (n_2 - 1)s_2^2}{n_1 + n_2 - 2}$$

and $t_{\alpha/2}$ depends on $n_1 + n_2 - 2$ degrees of freedom.

Normal Distributions: Confidence interval for σ_2^2/σ_1^2:

$$\left(\frac{s_2^2}{s_1^2} \frac{1}{F_{\alpha/2}(v_2, v_1)}, \frac{s_2^2}{s_1^2} F_{\alpha/2}(v_1, v_2) \right)$$

where

$$v_1 = n_1 - 1 \qquad \text{and} \qquad v_2 = n_2 - 1$$

EXERCISES

7.34 The abrasive resistance of rubber is increased by adding a silica filler and a coupling agent to chemically bond the filler to the rubber polymer chains. Fifty specimens of rubber made with a type I coupling agent gave a mean resistance measure of 92, the variance of the measurements being 20. Forty specimens of rubber made with a type II coupling agent gave a mean of 98 and a variance of 30 on resistance measurements. Estimate the true difference between mean resistances to abrasion in a 95% confidence interval.

7.35   Refer to Exercise 7.34. Suppose a similar experiment is to be run again with an equal number of specimens from each type of coupling agent. How many specimens should be used if we want to estimate the true difference between mean resistances to within 1 unit, with a confidence coefficient of 0.95?

7.36   Two different types of coating for pipes are to be compared with respect to their ability to aid in resistance to corrosion. The amount of corrosion on a pipe specimen is quantified by measuring the maximum pit depth. For coating A, 35 specimens showed an average maximum pit depth of 0.18 cm. The standard deviation of these maximum pit depths was 0.02 cm. For coating B the maximum pit depths in 30 specimens had a mean of 0.21 cm and a standard deviation of 0.03 cm. Estimate the true difference between mean depths in a 90% confidence interval. Do you think coating B does a better job of inhibiting corrosion?

7.37   Bacteria in water samples are sometimes difficult to count, but their presence can easily be detected by culturing. In fifty independently selected water samples from a certain lake, 43 contained certain harmful bacteria. After adding a chemical to the lake water, another fifty water samples showed only 22 with the harmful bacteria. Estimate the true difference between the proportions of samples containing the harmful bacteria with a 95% confidence coefficient. Does the chemical appear to be effective in reducing the amount of bacteria?

7.38   In studying the proportion of water samples containing harmful bacteria, how many samples should be selected before and after a chemical is added if we want to estimate the true difference between proportions to within 0.1 with a 95% confidence coefficient? (Assume the sample sizes are to be equal.)

7.39   A large firm made up of several companies has instituted a new quality-control inspection policy. Among 30 artisans sampled in Company A, only 5 objected to the new policy. Among 35 artisans sampled in Company B, 10 objected to the policy. Estimate the true difference between the proportions voicing *no* objection to the new policy for the two companies, with confidence coefficient 0.98.

7.40   A measurement of physiological activity important to runners is the rate of oxygen consumption. *Research Quarterly*, May 1979, reports on the differences between oxygen consumption rates for college males trained by two different methods, one involving continuous training for a period of time each day and the other involving intermittent training of about the same overall duration. The sample sizes, means, and standard deviations are as follows:

Continuous Training	Intermittent Training
$n_1 = 9$	$n_2 = 7$
$\bar{y}_1 = 43.71$	$\bar{y}_2 = 39.63$
$s_1 = 5.88$	$s_2 = 7.68$

(The measurements are in ml/kg·min.) If the measurements are assumed to come from normally distributed populations, estimate the difference between the population means with confidence coefficient 0.95.

Also find a 90% confidence interval for the ratio of the true variances for the two training methods. Does it appear that intermittent training gives more variable results?

7.41 For a certain species of fish the LC50 measurements for DDT in 12 experiments were as follows, according to the EPA:

16, 5, 21, 19, 10, 5, 8, 2, 7, 2, 4, 9

(in parts per million).

Another common insecticide, Diazinon, gave LC50 measurements of 7.8, 1.6, and 1.3 in three independent experiments. Estimate the difference between the mean LC50 for DDT and the mean LC50 for Diazinon in a 90% confidence interval. What assumptions are necessary for your answer to be valid? Also estimate the true variance ratio in a 90% confidence interval.

7.42 *Research Quarterly*, May 1979, reports on a study of impulses applied to the ball by tennis rackets of various construction. Three measurements on ball impulses were taken on each type of racket. For a Classic (wood) racket the mean was 2.41 and the standard deviation was 0.02. For a Yamaha (graphite) racket the mean was 2.22 and the standard deviation was 0.07. Estimate the difference between true mean impulses for the two rackets with confidence coefficient 0.95. What assumptions are necessary for the method used to be valid?

7.43 Seasonal ranges (in hectares) for alligators were monitored on a lake outside of Gainesville, Florida, by biologists from the Florida Game and Fish Commission. Six alligators monitored in the spring showed ranges of 8.0, 12.1, 8.1, 18.1, 18.2, and 31.7. Four different alligators monitored in the summer had ranges of 102.0, 81.7, 54.7, and 50.7. Estimate the difference between mean spring and summer ranges on a 95% confidence interval, assuming the data to come from normally distributed populations with common variance.

7.44 Unaltered, altered, and partly altered bitumens are found in carbonate-hosted lead-zinc deposits, and may aid in the production of sulfide necessary to precipitate ore bodies in carbonate rocks. (See T. G. Powell, *Science*, 6 April, 1984, p. 63). The atomic hydrogen/carbon ratios for 15 samples of altered bitumen had a mean of 1.02 and a standard deviation of 0.04. The ratios for 7 samples of partly altered bitumen had a mean of 1.16 and a standard deviation of 0.05. Estimate the difference between true mean H/C ratios for altered and partly altered bitumen, in a 98% confidence interval. Assume equal population standard deviations.

7.45 One-hour carbon monoxide concentrations in 45 air samples from a section of a city showed an average of 11.6 ppm and a variance of 82.4. After a traffic control strategy was put into place, 19 air samples showed an average carbon monoxide concentration of 6.2 ppm and a variance of 38.1. Estimate the true difference in average carbon monoxide concentrations in a 95% confidence interval. What assumptions are necessary for your answer to be valid? (See T. Zamurs, *Air Pollution Control Association Journal*, 34, No. 6, 1984, p. 637, for details of the measurements.)

7.46 The total amount of hydrogen chloride (HCl) in columns above an altitude of 12 kilometers was measured at selected sites over the El Chichon Volcano both before and after an eruption. (See W. G. Mankin and M. T. Coffey, *Science*, 12 October, 1984, p. 170.) Before-and-after data for one location were as follows (units are 10^{15} molecules per square centimeter):

Preeruption	Posteruption
$n_1 = 40$	$n_2 = 20$
$\bar{x}_1 = 1.26$	$\bar{x}_2 = 1.40$
$s_1 = 0.14$	$s_2 = 0.04$

Estimate the difference between mean HCl amounts with a confidence coefficient of 0.98.

7.47 Acid gases must be removed from other refinery gases in chemical production facilitities to minimize corrosion of the plants. Two methods for removing acid gases produced the corrosion rates (in mm/yr) listed below in experimental tests:

Method A: 0.3,   0.7,   0.5,   0.8,   0.9,   0.7,   0.8

Method B: 0.7,   0.8,   0.7,   0.6,   2.1,   0.6,   1.4,   2.3

Estimate the difference in mean corrosion rates for the two methods, using a confidence coefficient of 0.90. What assumptions must you make for your answer to be valid? (The data come from A. J. Kosseim, et al., *Chemical Engineering Progress*, 80, 1984, p. 64.)

7.48 Using the data presented in Exercise 7.47, estimate the ratio of variances for the two methods of removing acid gases in a 90% confidence interval.

7.49 The number of cycles to failure for reinforced concrete beams was measured in seawater and in air. The data (in thousands) are as follows:

Seawater: 774,   633,   477,   268,   407,   576,   659,   963,   193

Air: 734,   571,   520,   792,   773,   276,   411,   500,   672

Estimate the difference between mean cycles to failure for seawater and air, using a confidence coefficient of 0.95. Does seawater seem to lessen the number of cycles to failure? (See T. Hodgkiess, et al., *Materials Performance*, July 1984, pp. 27–29.)

7.50 The yield stresses were studied for two sizes of steel rods with results as follows:

10-mm-Diameter Rods	14-mm-Diameter Rods
$n_1 = 51$	$n_2 = 12$
$\bar{x}_1 = 485$ N/mm^2	$\bar{x}_2 = 499$ N/mm^2
$s_1 = 17.2$	$s_2 = 26.7$

Estimate the average increase in yield stress of the 14-mm rod over the 10-mm rod, with a confidence coefficient of 0.90.

7.51 Silicon wafers are scored and then broken into the many small microchips that will be mounted into circuits. Two breaking methods are being compared. Out of 400 microchips broken by method A, 32 are unusable because of faulty breaks. Out of 400 microchips broken by method B, only 28 are unusable. Estimate the difference between proportions of improperly broken microchips for the two breaking methods. Use a confidence coefficient of 0.95. Which method of breaking would you recommend?

7.52 Time-Yankelovich surveys, regularly seen in the news magazine *Time*, report on telephone surveys of approximately 1000 respondents. In December 1983, 60% of the respondents said that they worry about nuclear war. In a similar survey in June 1983, only 50% said that they worry about nuclear war. (See *Time*, January 2, 1984, p. 51.) The article reporting these figures says that when they are compared "the potential sampling error is plus or minus 4.5%." Explain how the 4.5% is obtained and what it means. Then estimate the true difference in these proportions in a 95% confidence interval.

7.53 An electric circuit contains three resistors, each of a different type. Tests on 10 type I resistors showed a sample mean resistance of 9.1 ohms with a sample standard deviation of 0.2 ohm. Tests on 8 type II resistors yielded a sample mean of 14.3 ohms and a sample standard deviation of 0.4 ohm, while tests on 12 type III resistors yielded a sample mean of 5.6 ohms and a sample standard deviation of 0.1 ohm. Find a 95% confidence interval for $\mu_I + \mu_{II} + \mu_{III}$, the expected resistance for the circuit. What assumptions are necessary for your answer to be valid?

7.5 PREDICTION INTERVALS

Previous sections of this chapter have considered the problem of *estimating* parameters, or population constants. A similar problem is that of *predicting* a value for a future observation of a random variable. Given a set of n lifelength measurements on components of a certain type, it may be of interest to form an interval (a prediction interval) in which we think the next observation on the lifelength of a similar component is quite likely to lie.

For independent random variables having a common normal distribution, a prediction interval is easily derived using the t distribution. Suppose we are to observe n independent, identically distributed normal random variables, and then use the information to predict where X_{n+1} might lie. We know from previous discussions that $\bar{X}_n - X_{n+1}$ is normally distributed with mean zero and variance $(\sigma^2/n) + \sigma^2 = \sigma^2[1 + (1/n)]$. If we use the variables $X_1, \ldots, X_n$ to calculate S^2 as an estimator of σ^2, then

$$T = \frac{\bar{X}_n - X_{n+1}}{S\sqrt{1 + \dfrac{1}{n}}}$$

will have a t distribution with $(n - 1)$ degrees of freedom. Thus

$$1 - \alpha = P\left[-t_{\alpha/2} \leq \frac{\bar{X}_n - X_{n+1}}{S\sqrt{1 + \dfrac{1}{n}}} \leq t_{\alpha/2}\right]$$

$$= P\left[\bar{x}_n - t_{\alpha/2}S\sqrt{1 + \dfrac{1}{n}} \leq X_{n+1} \leq \bar{X}_n + t_{\alpha/2}S\sqrt{1 + \dfrac{1}{n}}\right]$$

The interval

$$\bar{x}_n \pm t_{\alpha/2}S\sqrt{1 + \dfrac{1}{n}}$$

forms a realization of a prediction interval for X_{n+1}, with coverage probability $1 - \alpha$.

EXAMPLE 7.16

Ten independent observations are taken on bottles coming off a machine designed to fill them to 16 ounces. The $n = 10$ observations show a mean of $\bar{x} = 16.1$ oz and a standard deviation of $s = 0.01$ oz. Find a 95% prediction interval for the ounces of fill in the next bottle to be observed.

Solution   Assuming a normal probability model for ounces of fill, the interval

$$\bar{x} \pm t_{0.025} s \sqrt{1 + \frac{1}{n}}$$

will provide the answer. We obtain

$$16.1 \pm (2.262)(0.01) \sqrt{1 + \frac{1}{10}}$$

with $t_{0.025}$ based on 9 degrees of freedom, which yields

$$16.1 \pm 0.024$$

or

$$(16.076, 16.124)$$

We are about 95% confident that the next observation will lie between 16.076 and 16.124.   ■

Note that the prediction interval will *not* become arbitrarily small with increasing n. In fact, if n is very large, the term $1/n$ might be ignored, resulting in an interval of the form $\bar{x} \pm t_{\alpha/2} s$. There will be a certain amount of error in predicting the value of a random variable no matter how much information is available to estimate μ and σ^2.

This type of prediction interval is sometimes used as the basis for control charts when "sampling by variables." That is, if n observations on some important variable are made while a process is in control, then the next observation should lie in the interval $\bar{x} \pm t_{\alpha/2} s \sqrt{1 + 1/n}$. If it doesn't, then there may be some reason to suspect that the process has gone out of control. (Usually more than one observation would have to be observed outside of the appropriate interval before an investigation into causes of the abnormality would be launched.)

EXERCISES

7.54   In studying the properties of a certain type of resistor, the actual resistances produced were measured on a sample of 15 such resistors. These resistances had a mean of 9.8 ohms and a standard deviation of 0.5 ohm. One resistor of this type is to be used in a circuit. Find a 95% prediction interval for the resistance it will produce. What assumption is necessary for your answer to be valid?

7.55 For three tests with a certain species of freshwater fish, the LC50 measurements for Diazinon, a common insecticide, were 7.8, 1.6, and 1.3 (in parts per million). Assuming normality of the LC50 measurements, construct a 90% prediction interval for the LC50 of the next test with Diazinon.

7.56 It is extremely important for a business firm to be able to predict the amount of downtime for its computer system over the next month. A study of the past five months has shown the downtime to have a mean of 42 hours and a standard deviation of 3 hours. Predict the downtime for next month in a 95% prediction interval. What assumptions are necessary for your answer to be valid?

7.57 In designing concrete structures, the most important property of the concrete is its compressive strength. Six tested concrete beams showed compressive strengths (in thousand psi) of 3.9, 3.8, 4.4, 4.2, 3.8, and 5.4. Another beam of this type is to be used in a construction project. Predict its compressive strength in a 90% prediction interval. Assume compressive strengths to be approximately normally distributed.

7.58 A particular model of subcompact automobile has been tested for gas mileage 50 times. These mileage figures have a mean of 39.4 mpg and a standard deviation of 2.6 mpg. Predict the gas mileage to be obtained on the next test, with $1 - \alpha = 0.90$.

---

*7.6 THE METHOD OF MAXIMUM LIKELIHOOD

The estimators presented in previous sections of this chapter were justified merely by the fact that they seemed reasonable. For instance, our intuition says that a sample mean should be a good estimator of a population mean, or a sample proportion should somehow resemble the corresponding population proportion. There are, however, general methods for deriving estimators for unknown parameters of population models. One widely used method is called the *method of maximum likelihood*.

Suppose, for example, that a box contains four balls, of which an unknown number θ are white ($4 - \theta$ are nonwhite). We are to sample two balls at random and count X, the number of white balls in the sample. We know the probability distribution of X to be given by

$$P(X = x) = p(x) = \frac{\binom{\theta}{x}\binom{4 - \theta}{2 - x}}{\binom{4}{2}}$$

Now suppose we observe that $X = 1$. What value of θ will maximize the probability of this event? From the above distribution we have

$$p(1 \mid \theta = 0) = 0$$

*Optional section.

$$p(1|\theta = 1) = \frac{\binom{1}{1}\binom{3}{1}}{\binom{4}{2}} = \frac{3}{6} = \frac{1}{2}$$

$$p(1|\theta = 2) = \frac{2}{3}$$

$$p(1|\theta = 3) = \frac{1}{2}$$

$$p(1|\theta = 4) = 0$$

Hence $\theta = 2$ maximizes the probability of the observed sample, so we would choose this value, 2, as the maximum likelihood estimate for θ, given we observed that $X = 1$. You can show for yourself that if $X = 2$, then 4 will be the maximum likelihood estimate of θ.

In general suppose the random sample $X_1, X_2, \ldots, X_n$ results in observations $x_1, x_2, \ldots, x_n$. In the discrete case the *likelihood L* of the sample is the product of the marginal probabilities; that is,

$$L(\theta) = p(x_1; \theta)p(x_2; \theta) \cdots p(x_n; \theta)$$

where

$$P(X = x_i) = p(x_i; \theta)$$

θ being the parameter of interest. In the continuous case the likelihood of the sample is the product of the marginal density functions; that is,

$$L(\theta) = f(x_1; \theta)f(x_2; \theta) \cdots f(x_n; \theta)$$

In either case we maximize L as a function of θ to find the maximum likelihood estimator of θ.

If the form of the probability distribution is not too complicated, we can generally use methods of calculus to find the functional form of the maximum likelihood estimator, instead of working out specific numerical cases. We illustrate with the following examples.

EXAMPLE 7.17

Suppose that in a sequence of n independent, identical Bernoulli trials, Y successes are observed. Find the maximum likelihood estimator of p, the probability of successes on any single trial.

Solution   The Y is really the sum of n independent Bernoulli random variables $X_1, X_2, \ldots, X_n$. Thus the likelihood is given by

$$L(p) = p^{x_1}(1 - p)^{1 - x_1}p^{x_2}(1 - p)^{1 - x_2} \cdots p^{x_n}(1 - p)^{1 - x_n}$$

$$= p^{\sum_{i=1}^{n} x_i}(1 - p)^{n - \sum_{i=1}^{n} x_i}$$

$$= p^{y}(1 - p)^{n - y}$$

Since $\ln L(p)$ is a monotone function of $L(p)$, both $\ln L(p)$ and $L(p)$ will have their maxima at the same value of p. In many cases it is easier to maximize $\ln L(p)$, so that is the approach we follow here.

Now

$$\ln L(p) = y \ln p + (n - y) \ln(1 - p)$$

is a continuous function of $p(0 < p < 1)$, and thus the maximum value can be found by setting the first derivative equal to zero. We have

$$\frac{\partial \ln L(p)}{\partial p} = \frac{y}{p} - \frac{n - y}{1 - p}$$

Setting

$$\frac{y}{\hat{p}} - \frac{n - y}{1 - \hat{p}} = 0$$

we find

$$\hat{p} = \frac{y}{n}$$

and hence we take Y/n as the maximum likelihood estimator of p. In this case, as well as in many others, the maximum likelihood estimator agrees with our intuitive estimator given earlier in this chapter. ■

We now look at an example that is a bit more involved.

EXAMPLE 7.18
_____

Suppose we are to observe n independent lifelength measurements $X_1, \ldots, X_n$ from components known to have lifelengths exhibiting a Weibull model, given by

$$f(x) = \frac{\gamma x^{\gamma - 1}}{\theta} e^{-x^\gamma/\theta} \qquad x > 0$$

$$= 0 \qquad\qquad\qquad \text{elsewhere}$$

Assuming γ is known, find the maximum likelihood estimator of θ.

Solution In analogy with the discrete case given above, we write the likelihood $L(\theta)$ as the joint density function of $X_1, \ldots, X_n$, or

$$L(\theta) = f(x_1, x_2, \ldots, x_n)$$

$$= f(x_1) f(x_2) \cdots f(x_n)$$

$$= \frac{\gamma x_1^{\gamma - 1}}{\theta} e^{-x_1^\gamma/\theta} \cdots \frac{\gamma x_n^{\gamma - 1}}{\theta} e^{-x_n^\gamma/\theta}$$

$$= \left(\frac{\gamma}{\theta}\right)^n (x_1, x_2, \ldots, x_n)^{\gamma - 1} e^{-\Sigma_{i=1}^n x_i^\gamma/\theta}$$

Now

$$\ln L(\theta) = n \ln \gamma - n \ln \theta + (\gamma - 1) \ln(x_1 \cdots x_n) - \frac{1}{\theta} \sum_{i=1}^{n} x_i^{\gamma}$$

To find the θ that maximizes $L(\theta)$, we take

$$\frac{\partial \ln L(\theta)}{\partial \theta} = -\frac{n}{\theta} + \frac{1}{\theta^2} \sum_{i=1}^{n} x_i^{\gamma}$$

and setting this derivative equal to zero, we get

$$-\frac{n}{\theta} + \frac{1}{\theta^2} \sum_{i=1}^{n} x_i^{\gamma} = 0$$

or

$$\hat{\theta} = \frac{1}{n} \sum_{i=1}^{n} x_i^{\gamma}$$

Thus we take $(1/n)\sum_{i=1}^{n} X_i^{\gamma}$ as the maximum likelihood estimator of θ. The interested reader can check to see that this estimator is unbiased for θ. ∎

The maximum likelihood estimators shown in Examples 7.17 and 7.18 are both unbiased, but this is not always the case. For example, if X has a geometric distribution, as given in Chapter 3, then the maximum likelihood estimator of p, the probability of a success on any one trial, is $1/X$. In this case $E(1/X) \neq p$, so this estimator is not unbiased for p.

Many times the maximum likelihood estimator is simple enough that we can use its probability distribution to find a confidence interval, either exact or approximate, for the parameter in question. We illustrate the construction of an exact confidence interval for a nonnormal case with the exponential distribution. Suppose $X_1, \ldots, X_n$ represents a random sample from a population modeled by the density function

$$f(x) = \frac{1}{\theta} e^{-x/\theta} \qquad x > 0$$

$$= 0 \qquad \text{elsewhere}$$

Since this exponential density is a special case of the Weibull with $\gamma = 1$, we know from Example 7.18 that the maximum likelihood estimator of θ is $(1/n)\sum_{i=1}^{n} X_i$. We also know that $\sum_{i=1}^{n} X_i$ will have a gamma distribution, but we do not have percentage points of most gamma distributions in standard sets of tables, except for the special case of the χ^2 distribution. Thus we shall try to transform $\sum_{i=1}^{n} X_i$ into something that has a χ^2 distribution.

Let $U_i = 2X_i/\theta$. Then

$$F_U(u) = P(U_i \leqslant u) = P\left(\frac{2X_i}{\theta} \leqslant u\right)$$

$$= P\left(X_i \leqslant \frac{u\theta}{2}\right) = 1 - e^{-(1/\theta)(u\theta/2)}$$

$$= 1 - e^{-u/2}$$

Hence

$$f_U(u) = \frac{1}{2} e^{-u/2}$$

and U_i has an exponential distribution with mean 2, which is equivalent to the $\chi^2_{(2)}$ distribution. It then follows that

$$\sum_{i=1}^{n} U_i = \frac{2}{\theta} \sum_{i=1}^{n} X_i$$

has a χ^2 distribution with $2n$ degrees of freedom. Thus a confidence interval for θ with confidence coefficient $(1 - \alpha)$ can be formed by finding a $\chi^2_{\alpha/2}(2n)$ and a $\chi^2_{1-\alpha/2}(2n)$ value and writing

$$(1 - \alpha) = P\left[\chi^2_{1-\alpha/2}(2n) \leqslant \frac{2}{\theta} \sum_{i=1}^{n} X_i \leqslant \chi^2_{\alpha/2}(2n) \right]$$

$$= P\left[\frac{1}{\chi^2_{\alpha/2}(2n)} \leqslant \frac{\theta}{2 \sum_{i=1}^{n} X_i} \leqslant \frac{1}{\chi^2_{1-\alpha/2}(2n)} \right]$$

$$= P\left[\frac{2 \sum_{i=1}^{n} X_i}{\chi^2_{\alpha/2}(2n)} \leqslant \theta \leqslant \frac{2 \sum_{i=1}^{n} X_i}{\chi^2_{1-\alpha/2}(2n)} \right]$$

EXAMPLE 7.19

Consider the first ten observations from Table 4.1 on lifelengths of batteries. These are as follows: 0.406, 2.343, 0.538, 5.088, 5.587, 2.563, 0.023, 3.334, 3.491, and 1.267. Using these observations as a realization of a random sample of lifelength measurements from an exponential distribution with mean θ, find a 95% confidence interval for θ.

Solution   Using

$$\left[\frac{2 \sum_{i=1}^{n} x_i}{\chi^2_{\alpha/2}(2n)}, \frac{2 \sum_{i=1}^{n} x_i}{\chi^2_{1-\alpha/2}(2n)} \right]$$

as the confidence interval for θ, we have

$$\sum_{i=1}^{10} x_i = 24.640$$

$$\chi^2_{0.975}(20) = 9.591$$

and

$$\chi^2_{0.025}(20) = 34.170$$

Hence the realization of the confidence interval becomes

$$\left[\frac{2(24.640)}{34.170}, \frac{2(24.640)}{9.591} \right]$$

or (1.442, 5.138). We are about 95% confident that the true mean lifelength is between 144 and 514 hours. The rather wide interval is a reflection of both our relative lack of information in the sample of only $n = 10$ batteries and the large variability in the data.   ∎

It is informative to compare the exact intervals for the mean θ from the exponential distribution with what would have been generated if we had falsely assumed the random variables to have a normal distribution and used $\bar{x} \pm t_{\alpha/2} s/\sqrt{n}$ as a confidence interval. Table 7.2 shows 100 95% confidence intervals generated by exact methods (using the χ^2 table) for samples of size $n = 5$ from an exponential distribution with $\theta = 10$. Note that five of the intervals do not include the true θ, as expected. Four miss on the low side, and one misses on the high side.

Table 7.3 repeats the process of generating confidence intervals for θ, with the same samples of size $n = 5$, but now using $\bar{x} \pm t_{\alpha/2} s/\sqrt{n}$ with $\alpha/2 = 0.025$ and four degrees of freedom. Now note that ten of the intervals do not include the true θ, twice the number expected. Also the intervals that miss the true $\theta = 10$ all miss on the low side. That, coupled with the fact that the lower bounds are frequently negative even though $\theta > 0$, seems to indicate that the intervals based on normality are too far left, as compared to the true intervals based on the χ^2 distribution.

The lesson to be learned here is that *one should be careful of inflicting the normality assumption on data*, especially if a more reasonable and mathematically tractable model is available.

Before we leave the section on maximum likelihood estimation, we will show how we can carry out the estimation of certain functions of parameters. In general, if we have a maximum likelihood estimator, say Y, for a parameter θ, then any continuous function of θ, say $g(\theta)$, will have a maximum likelihood estimator $g(Y)$. Rather than going deeper into the theory here, we illustrate with a particular case.

Suppose $X_1, \ldots, X_n$ again denote independent lifelength measurements from a population modeled by the exponential distribution with mean θ. The *reliability* of each component is defined by

$$R(t) = P(X_i > t)$$

In the exponential case

$$R(t) = \int_t^\infty \frac{1}{\theta} e^{-x/\theta}\, dx = e^{-t/\theta}$$

Suppose we want to estimate $R(t)$ from $X_1, \ldots, X_n$. We know that the maximum likelihood estimator of θ is $\bar{X}$, and hence the maximum likelihood estimator of $R(t) = e^{-t/\theta}$ is $e^{-t/\bar{X}}$.

Now we can use the results from Section 5.6 to find an approximate confidence interval for $R(t)$ if n is large. Let

$$g(\theta) = e^{-t/\theta}$$

Then $g'(\theta) = (t/\theta^2) e^{-t/\theta}$ and, since $\bar{X}$ is approximately normally distributed with

TABLE 7.2
Exact Exponential Confidence Intervals (n = 5)

Sample	LCL	UCL	Sample	LCL	UCL
1	2.03600	12.8439	51	9.10805	57.4570
*2	1.49454	9.4281	52	4.20731	26.5413
3	4.15706	26.2243	53	5.08342	32.0681
4	3.07770	19.4153	54	5.04982	31.8561
5	5.89943	37.2158	55	3.60932	22.7689
6	6.49541	40.9755	56	5.84841	36.8940
7	2.19705	13.8598	57	5.19353	32.7627
8	2.37724	14.9965	58	4.18032	26.3710
9	5.90726	37.2652	59	4.66717	29.4422
10	7.67249	48.4009	60	5.39428	34.0291
11	6.64161	41.8977	61	6.06969	38.2899
12	4.78097	30.1601	62	3.45198	21.7764
13	3.17174	20.0085	63	5.19596	32.7781
*14	1.51473	9.5555	64	4.79768	30.2655
15	3.70163	23.3513	65	5.76487	36.3669
16	3.69998	23.3408	66	4.15674	26.2223
17	6.90343	43.5494	67	8.96348	56.5450
18	2.36005	14.8881	68	6.74431	42.5456
19	6.53294	41.2122	69	5.36919	33.8709
20	4.04756	25.5335	70	3.77090	23.7883
21	5.61737	35.4365	71	5.42996	34.2542
22	4.07102	25.6815	72	2.7016	17.0429
23	4.65284	29.3518	73	4.2962	27.1020
24	4.58284	28.9103	74	2.3106	14.5761
25	3.80908	24.0291	75	7.6502	48.2602
26	4.22984	26.6834	76	7.4693	47.1190
27	5.17128	32.6223	77	8.5458	53.9103
28	8.77035	55.3266	78	2.9840	18.8243
29	2.54689	16.0667	79	9.2678	58.4647
30	5.27866	33.2998	80	4.2078	26.5442
31	7.26216	45.8124	81	8.0267	50.6353
32	2.27762	14.3680	82	9.9686	62.8854
33	4.63549	29.2424	83	2.2581	14.2447
34	3.80281	23.9895	84	9.9971	63.0651
35	1.98777	12.5396	*85	10.2955	64.9478
36	4.63779	29.2569	86	6.0830	38.3740
37	7.36773	46.4784	87	3.0037	18.9483
38	3.55928	22.4533	88	2.3759	14.9881
39	2.99137	18.8707	89	4.0007	25.2376
40	5.79160	36.5356	*90	1.1737	7.4044
41	5.62684	35.4962	91	3.8375	24.2081
42	3.95994	24.9808	92	2.9560	18.6473
43	2.30394	14.5341	93	4.9714	31.3616
44	7.40844	46.7352	94	3.0063	18.9648
45	3.06875	19.3588	95	4.1684	26.2958
46	2.99999	18.9251	96	4.6269	29.1879
47	3.95912	24.9756	97	3.8026	23.9883
48	3.01418	19.0146	98	6.0056	37.8858
49	2.69137	16.9782	*99	1.5153	9.5592
50	4.28299	27.0187	100	7.0408	44.4161

TABLE 7.3
Confidence Intervals Based on the Normal Distributions (n = 5)

Sample	LCL	Mean	UCL	S
*1	1.975	4.1704	6.3656	1.7682
*2	−1.069	3.0613	7.1916	3.3269
3	−1.550	8.5150	18.5800	8.1074
4	−1.019	6.3041	13.6277	5.8991
5	4.183	12.0839	19.9850	6.3644
6	3.838	13.3046	22.7712	7.6253
7	−3.036	4.5002	12.0368	6.0707
*8	0.010	4.8693	9.7289	3.9144
9	1.004	12.0999	23.1954	8.9374
10	4.917	15.7156	26.5146	8.6986
11	8.340	13.6041	18.8683	4.2404
12	−0.397	9.7929	19.9832	8.2083
13	1.042	6.4967	11.9513	4.3937
*14	−0.274	3.1026	6.4790	2.7197
15	2.542	7.5821	12.6217	4.0594
16	3.398	7.5787	11.7595	3.3678
17	6.994	14.1404	21.2868	5.7564
*18	1.244	4.8341	8.4245	2.8920
19	−6.383	13.3815	33.1458	15.9202
20	1.122	8.2907	15.4591	5.7742
21	−3.593	11.5061	26.6049	12.1621
22	−0.433	8.3387	17.1102	7.0654
23	−5.925	9.5305	24.9856	12.4491
24	−11.200	9.3871	29.9740	16.5828
25	−4.166	7.8022	19.7701	9.6402
26	1.564	8.6640	15.7640	5.7191
27	3.818	10.5924	17.3667	5.4567
28	−3.492	17.9644	39.4208	17.2832
29	−0.656	5.2168	11.0900	4.7308
30	1.837	10.8123	19.7876	7.2296
31	2.581	14.8751	27.1690	9.9027
32	−1.941	4.6653	11.2716	5.3214
33	−10.734	9.4949	29.7238	16.2943
34	−2.725	7.7893	18.3035	8.4692
*35	−0.546	4.0716	8.6890	3.7193
36	0.841	9.4996	18.1584	6.9747
37	−1.467	15.0914	31.6496	13.3376
38	2.144	7.2905	12.4370	4.1455
39	−2.991	6.1272	15.2452	7.3446
40	−4.714	11.8630	28.4403	13.3530
41	4.356	11.5255	18.6946	5.7747
42	−4.436	8.1112	20.6585	10.1069
*43	1.221	4.7192	8.2178	2.8181
44	−0.877	15.1748	31.2261	12.9294
45	0.865	6.2858	11.7065	4.3665
46	0.751	6.1449	11.5392	4.3451
47	−1.325	8.1095	17.5442	7.5997
48	−1.116	6.1740	13.4641	5.8722
49	−2.837	5.5128	13.8623	6.7256
50	2.529	8.7729	15.0168	5.0294
51	−3.559	18.6561	40.8713	17.8944

TABLE 7.3 (continued)

Sample	LCL	Mean	UCL	S
52	−2.949	8.6179	20.1847	9.3171
53	1.648	10.4124	19.1769	7.0598
54	−2.852	10.3436	23.5391	10.6290
55	1.645	7.3930	13.1407	4.6298
56	2.344	11.9794	21.6150	7.7615
57	−8.525	10.6380	29.8011	15.4359
58	−0.781	8.5626	17.9059	7.5260
59	2.364	9.5598	16.7556	5.7962
60	−1.314	11.0491	23.4128	9.9589
61	1.465	12.4326	23.3998	8.8341
62	−2.978	7.0707	17.1193	8.0941
63	−6.065	10.6429	27.3514	13.4586
64	−4.378	9.8271	24.0327	11.4426
65	3.155	11.8082	20.4612	6.9700
66	0.298	8.5143	16.7307	6.6184
67	−4.655	18.3600	41.3751	18.5387
68	3.14	13.8144	24.4798	8.5910
69	−2.948	10.9978	24.9433	11.2331
70	0.332	7.7240	15.1161	5.9544
71	−3.383	11.1222	25.6275	11.8840
72	1.027	5.5338	10.0406	3.6302
73	−1.720	8.7999	19.3194	8.4735
74	−0.755	4.7328	10.2205	4.4203
75	−4.753	15.6699	36.0924	16.4503
76	−12.223	15.2994	42.8221	22.1695
77	−5.102	17.5045	40.1115	18.2099
78	0.504	6.1122	11.7204	4.5175
79	0.042	18.9833	37.9249	15.2575
80	−0.873	8.6188	18.1102	7.6453
81	7.232	16.4411	25.6501	7.4179
82	−0.289	20.4187	41.1267	16.6803
*83	−0.684	4.6252	9.9343	4.2765
84	1.162	20.4771	39.7917	15.5579
85	−13.471	21.0884	55.6473	27.8373
86	−0.086	12.4599	25.0060	10.1059
87	−3.659	6.1524	15.9641	7.9033
88	−4.660	4.8666	14.3930	7.6735
89	0.914	8.1946	15.4755	5.8648
*90	0.826	2.4042	3.9824	1.2712
91	−0.416	7.8603	16.1361	6.6662
92	−0.093	6.0547	12.2023	4.9519
93	2.451	10.1830	17.9150	6.2281
94	−6.613	6.1578	18.9282	10.2866
95	−0.470	8.5382	17.5461	7.2559
96	−0.036	9.4772	18.9902	7.6627
97	−2.201	7.7889	17.7785	8.0466
98	2.279	12.3014	22.3234	8.0727
*99	1.383	3.1038	4.8248	1.3862
100	2.804	14.4218	31.6476	13.8754

mean θ and standard deviation $\theta/\sqrt{n}$ (since $V(X_i) = \theta^2$), it follows that

$$\sqrt{n}\,\frac{[g(\bar{X}) - g(\theta)]}{|g'(\theta)|\theta} = \frac{\sqrt{n}[e^{-t/\bar{X}} - e^{-t/\theta}]}{\left(\dfrac{t}{\theta}\right)e^{-t/\theta}}$$

is approximately a standard normal random variable if n is large. Thus for large n, $e^{-t/\bar{X}}$ is approximately normally distributed with mean $e^{-t/\theta}$ and standard deviation $(1/\sqrt{n})(t/\theta)\,e^{-t/\theta}$. Note that this is only a large-sample approximation, as $E(e^{-t/\bar{X}}) \neq e^{-t/\theta}$. This result can be used to form an approximate confidence interval for $R(t) = e^{-t/\theta}$ of the form

$$e^{-t/\bar{x}} \pm z_{\alpha/2}\,\frac{1}{\sqrt{n}}\left(\frac{t}{\bar{x}}\right)e^{-t/\bar{x}}$$

using $\bar{x}$ as an approximation to θ in the standard deviation.

EXAMPLE 7.20

Using the fifty observations given in Table 4.1 as a realization of a random sample from an exponential distribution, estimate $R(5) = P(X_i > 5) = e^{-5/\theta}$ in an approximate 95% confidence interval.

Solution   For $(1 - \alpha) = 0.95$, $z_{\alpha/2} = 1.96$. Using the method indicated above, the interval becomes

$$e^{-t/\bar{x}} \pm (1.96)\,\frac{1}{\sqrt{n}}\left(\frac{t}{\bar{x}}\right)e^{-t/\bar{x}}$$

or

$$e^{-5/2.267} \pm (1.96)\,\frac{1}{\sqrt{50}}\left(\frac{5}{2.267}\right)e^{-5/2.267}$$

$$0.110 \pm 0.067$$

$$(0.043, 0.177)$$

That is, we are about 95% confident that the probability is between 0.043 and 0.177. For these data the true value of θ was 2, and $R(5) = e^{-5/2} = 0.082$ is well within this interval. The interval could be narrowed by selecting a larger sample.   ∎

EXERCISES

7.59   If $X_1, \ldots, X_n$ denotes a random sample from a Poisson distribution with mean λ, find the maximum likelihood estimator of λ.

7.60   Since $V(X_i) = \lambda$ in the Poisson case, it follows from the Central Limit Theorem that $\bar{X}$ will be approximately normally distributed with mean λ and variance λ/n, for large n.

(a) Use the above facts to construct a large-sample confidence interval for λ.

(b) Suppose that 100 reinforced concrete trusses were examined for cracks. The average number of cracks per truss was observed to be 4. Construct an approximate 95% confidence interval for the true mean number of cracks per truss for trusses of this type. What assumptions are necessary for your answer to be valid?

7.61 Suppose $X_1, \ldots, X_n$ denotes a random sample from the normal distribution with mean μ and variance σ^2. Find the maximum likelihood estimators of μ and σ^2.

7.62 Suppose $X_1, \ldots, X_n$ denotes a random sample from the gamma distribution with a known α but unknown β. Find the maximum likelihood estimator of β.

7.63 The stress resistances for specimens of a certain type of plastic tend to have a gamma distribution with $\alpha = 2$, but β may change with certain changes in the manufacturing process. For eight specimens independently selected from a certain process, the resistances (in psi) were

$$29.2, \quad 28.1, \quad 30.4, \quad 31.7, \quad 28.0, \quad 32.1, \quad 30.1, \quad 29.7$$

Find a 95% confidence interval for β. [*Hint:* Use the methodology outlined for Example 7.18.]

7.64 If X denotes the number of the trial on which the first defective is found in a series of independent quality-control tests, find the maximum likelihood estimator of p, the true probability of observing a defective.

7.65 The absolute errors in the measuring of the diameters of steel rods are uniformly distributed between 0 and θ. Three such measurements on a standard rod produced errors of 0.02, 0.06, and 0.05 centimeters. What is the maximum likelihood estimate of θ? [*Hint:* This problem cannot be solved by differentiation. Write down the likelihood, $L(\theta)$, for three independent measurements and think carefully about what value of θ produces a maximum. Remember, each measurement must be between 0 and θ.]

7.66 The number of improperly soldered connections per microchip in an electronics manufacturing operation follows a binomial distribution with $n = 20$ and p unknown. The cost of correcting these malfunctions, per microchip, is

$$C = 3X + X^2$$

Find the maximum likelihood estimate of $E(C)$ if $\hat{p}$ is available as an estimate of p.

---

*7.7 BAYES ESTIMATORS

In all previous sections on estimation, population parameters were treated as unknown constants. However, it is sometimes convenient and even necessary to treat an unknown parameter as a random variable in its own right. For example, the proportion of defectives p produced by an assembly line may change from day to day, or even hour to hour. It may be possible to model this changing behavior by

*Optional section.

allowing p to be a random variable with a probability density function, say $g(p)$. Even though we may not be able to choose the correct value of p for a given hour, we can find

$$P(a \leqslant p \leqslant b) = \int_a^b g(p)\, dp$$

Similarly the mean daily cost of production for the assembly line in question may vary from day to day, and could be assigned an appropriate probability density function to model the day-to-day variation.

Suppose we are to observe a random sample $Y_1, \ldots, Y_n$ from a probability density function with a single unknown parameter θ. If θ is a constant, then the density function is completely specified as $f(y|\theta)$. Note that $f(y|\theta)$ is now a *conditional* density function for y, *given* a fixed value of θ. The joint conditional density function for the random sample is given by

$$f(y_1, \ldots, y_n|\theta) = f(y_1|\theta) \cdots f(y_n|\theta)$$

Next suppose θ varies according to a probability density function $g(\theta)$. This density function $g(\theta)$ is referred to as the *prior* density for θ because it is assigned prior to the actual collection of data in a sample.

Knowledge of the above density functions allows us to compute

$$f(\theta|y_1, \ldots, y_n) = \frac{f(y_1, \ldots, y_n, \theta)}{f(y_1, \ldots, y_n)}$$

$$= \frac{f(y_1, \ldots, y_n|\theta)g(\theta)}{\int_{-\infty}^{\infty} f(y_1, \ldots, y_n|\theta)g(\theta)\, d\theta}$$

The density function $f(\theta|y_1, \ldots, y_n)$ is the conditional density of θ given the sample data, and is called the *posterior* density function for θ. We will call the *mean* of the posterior density function the *Bayes estimator* of θ. The computations are illustrated in the following example.

EXAMPLE 7.21

Let Y denote the number of defectives observed in a random sample of n items produced by a given machine in one day. The proportion p of defectives produced by this machine varies from day to day according to the probability density function

$$g(p) = \begin{cases} 1 & 0 \leqslant p \leqslant 1 \\ 0 & \text{elsewhere} \end{cases}$$

[In other words p is uniformly distributed over the interval $(0, 1)$.] Find the Bayes estimator of p.

Solution For a given p, Y will have a binomial distribution given by

$$f(y|p) = \binom{n}{y} p^y (1-p)^{n-y} \qquad y = 0, 1, \ldots, n$$

Then

$$f(y, p) = f(y|p)g(p)$$

$$= \binom{n}{y} p^y(1 - p)^{n - y}(1) \qquad \begin{array}{l} y = 0, 1, \ldots, n \\ 0 \leqslant p \leqslant 1 \end{array}$$

Now

$$f(y) = \int_{-\infty}^{\infty} f(y, p)\, dp$$

$$= \int_0^1 \binom{n}{y} p^y(1 - p)^{n - y}\, dp$$

$$= \binom{n}{y} \int_0^1 p^y(1 - p)^{n - y}\, dp$$

$$= \binom{n}{y} \frac{y!(n - y)!}{(n + 1)!}$$

$$= \frac{1}{n + 1} \qquad y = 0, 1, \ldots, n$$

The above integral is evaluated by recognizing it as a beta function of Chapter 4. Note that $f(y)$ says that each possible value of y has the same unconditional probability of occurring, since nothing is known about p other than the fact that it is in the interval $(0, 1)$.

It follows that

$$f(p|y) = \frac{f(y, p)}{f(y)}$$

$$= (n + 1)\binom{n}{y} p^y(1 - p)^{n - y}$$

$$= \frac{(n + 1)!}{y!(n - y)!} p^y(1 - p)^{n - y} \qquad 0 \leqslant p \leqslant 1$$

which is the beta density function.

The Bayes estimate of p is taken to be the mean of this posterior density function, namely

$$\int_0^1 pf(p|y)\, dp = \int_0^1 p \frac{(n + 1)!}{y!(n - y)!} p^y(1 - p)^{n - y}\, dp$$

$$= \frac{y + 1}{n + 2}$$

(Check Section 4.6 to find the mean of a beta density function.) Recall that the simple maximum likelihood estimator of p, when p is fixed but unknown, is Y/n.

∎

7.67 Refer to Example 7.21. Suppose that

$$g(p) = \begin{cases} 2 & 0 \leq p \leq \frac{1}{2} \\ 0 & \text{elsewhere} \end{cases}$$

If two items are produced on a given day and one is defective, find the Bayes estimate of p.

7.68 Let Y denote the number of defects per yard for a certain type of fabric. For a given mean λ, Y has a Poisson distribution. But λ varies from yard to yard according to the density function

$$g(\lambda) = \begin{cases} e^{-\lambda} & \lambda > 0 \\ 0 & \text{elsewhere} \end{cases}$$

Find the Bayes estimator of λ.

7.69 Suppose the lifelength Y of a certain component has probability density function

$$f(y|\theta) = \frac{\theta^\alpha}{\Gamma(\alpha)} y^{\alpha-1} e^{-y\theta} \qquad y > 0$$

$$= 0 \qquad\qquad\qquad \text{elsewhere}$$

Also suppose θ has a prior density function given by

$$g(\theta) = \begin{cases} e^{-\theta} & \theta > 0 \\ 0 & \text{elsewhere} \end{cases}$$

For a single observation Y, show that the Bayes estimator of θ is given by $(\alpha + 2)/(Y + 1)$.

7.8 CONCLUSION

We have developed confidence intervals for individual means and proportions, linear functions of means and proportions, individual variances, and ratios of variances. In a few cases we have shown how to calculate confidence intervals for other functions of parameters. We also touched on the notion of a prediction interval for a random variable. Most often these intervals were two-sided, in the sense that they provided both an upper and lower bound. Occasionally, a one-sided interval is in order.

Most of the estimators used in the construction of confidence intervals were selected on the basis of intuitive appeal. However, the method of maximum likelihood provides a technique for finding estimators with good properties. (Most of the intuitive estimators turned out to be maximum likelihood estimators.)

If it is appropriate to treat a population parameter as a random quantity, then Bayesian techniques can be used to find estimates.

Many of the estimators used in this chapter will be used in a slightly different setting in Chapter 8.

SUPPLEMENTARY EXERCISES

7.70 The diameter measurements of an armored electric cable, taken at 10 points along the cable, yield a sample mean of 2.1 centimeters and a sample standard deviation of 0.3 centimeter. Estimate the average diameter of the cable in a confidence interval with a confidence coefficient of 0.90. What assumptions are necessary for your answer to be valid?

7.71 Suppose the sample mean and standard deviation of Exercise 7.70 had come from a sample of 100 measurements. Construct a 90% confidence interval for the average diameter of the cable. What assumption must necessarily be made?

7.72 The Rockwell hardness measure of steel ingots is produced by pressing a diamond point into the steel and measuring the depth of penetration. A sample of 15 Rockwell hardness measurements on specimens of steel gave a sample mean of 65 and a sample variance of 90. Estimate the true mean hardness in a 95% confidence interval, assuming the measurements to come from a normal distribution.

7.73 Twenty specimens of a slightly different steel from that used in Exercise 7.72 were observed and yielded Rockwell hardness measurements with a mean of 72 and a variance of 94. Estimate the difference between the mean hardnesses for the two varieties of steel in a 95% confidence interval.

7.74 In the fabrication of integrated circuit chips, it is of great importance to form contact windows of precise width. (These contact windows facilitate the interconnections that make up the circuits.) Complementary metaloxide semiconductor (CMOS) circuits are fabricated by using a photolithography process to form windows. The key steps involve applying a photoresist to the silicon wafer, exposing the photoresist to ultraviolet radiation through a mask, and then dissolving away the photoresist from the exposed areas, revealing the oxide surface. The printed windows are then etched through the oxide layers down to the silicon. Key measurements that help determine how well the integrated circuits may function are the window widths before and after the etching process. (See M. S. Phadke, et al., *The Bell System Technical Journal*, 62, No. 5, 1983, pp. 1273–1309 for more details.)

Preetch window widths for a sample of ten test locations are as follows (in μm):

2.52,   2.50,   2.66,   2.73,   2.71,   2.67,   2.06,   1.66,   1.78,   2.56

Postetch window widths for ten independently selected test locations are

3.21,   2.49,   2.94,   4.38,   4.02,   3.82,   3.30,   2.85,   3.34,   3.91

One important problem is simply to estimate the true average window width for this process.
   (a) Estimate the average preetch window width, for the population from which this sample was drawn, in a 95% confidence interval.
   (b) Estimate the average postetch window width in a 95% confidence interval.

7.75 In the setting of Exercise 7.74 a second important problem is to compare the average window widths before and after etching. Using the data given there, estimate the true difference in average window widths in a 95% confidence interval.

7.76 Window sizes in integrated circuit chips must be of fairly uniform size in order for the circuits to function properly. Thus the variance of the window width measurements is an important quantity. Using the data of Exercise 7.74,

   (a) estimate the true variance of the postetch window widths in a 95% confidence interval.

   (b) estimate the ratio of the variance of postetch window widths to that of preetch window widths in a 95% confidence interval.

   (c) What assumptions are necessary for your answers in (a) and (b) to be valid?

7.77 The weights of aluminum grains follow a lognormal distribution, which means that the natural logarithms of the grain weights follow a normal distribution. A sample of 177 aluminum grains have logweights averaging 3.04 with a standard deviation of 0.25. (The original weight measurements were in 10^{-4} gram.) Estimate the true mean logweight of the grains from which this sample was selected, using a confidence coefficient of 0.90. (Source: Department of Materials Science, University of Florida.)

7.78 Two types of aluminum powders are blended before a sintering process is begun to form solid aluminum. The adequacy of the blending is gauged by taking numerous small samples from the blend and measuring the weight proportion of one type of powder. For adequate blending, the weight proportions from sample to sample should have small variability. Thus the variance becomes a key measure of blending adequacy.

   Ten samples from an aluminum blend gave the following weight proportions for one type of powder:

$$0.52, \quad 0.54, \quad 0.49, \quad 0.53, \quad 0.52, \quad 0.56, \quad 0.48, \quad 0.50, \quad 0.52, \quad 0.51$$

   (a) Estimate the population variance in a 95% confidence interval.

   (b) What assumptions are necessary for your answer in (a) to be valid? Are the assumptions reasonable for this case?

7.79 It is desired to estimate the proportion of defective items produced by a certain assembly line to within 0.1 with confidence coefficient 0.95. What is the smallest sample size that will guarantee this accuracy no matter where the true proportion of defectives might lie? [*Hint*: Find the value of p that maximizes the variance $p(1 - p)/n$, and choose the n that corresponds to it.]

7.80 Suppose that in a large-sample estimate of $\mu_1 - \mu_2$ for two populations with respective variances σ_1^2 and σ_2^2, a *total* of n observations is to be selected. How should these n observations be allocated to the two populations so that the length of the resulting confidence interval will be minimized?

7.81 Suppose the sample variances given in Exercise 7.34 are good estimates of the population variances. Using the allocation scheme of Exercise 7.80, find the number of measurements to be taken on each coupling agent to estimate the true difference in means to within 1 unit with confidence coefficient 0.95. Compare the answer to that of Exercise 7.35.

7.82 A factory operates with two machines of type A and one machine of type B. The weekly repair costs Y for type A machines are normally distributed with mean μ_1 and variance σ^2. The weekly repair costs X for machines of type B are also normally distributed but with mean μ_2 and variance $3\sigma^2$. The expected repair cost per week for the factory is then $2\mu_1 + \mu_2$. Suppose $Y_1, \ldots, Y_n$ denotes a random sample on costs for type A machines and $X_1, \ldots, X_m$ denotes an independent random sample on costs for type B machines. Use these data to construct a 95% confidence interval for $2\mu_1 + \mu_2$.

7.83 In polycrystalline aluminum the number of grain nucleation sites per unit volume is modeled as having a Poisson distribution with mean λ. Fifty unit-volume test specimens subjected to annealing under regime A showed an average of 20 sites per unit volume. Fifty independently selected unit-volume test specimens subjected to annealing regime B showed an average of 23 sites per volume. Find an approximate 95% confidence interval for the difference between the mean site frequencies for the two annealing regimes. Would you say that regime B tends to increase the number of nucleation sites?

7.84 A random sample of n items is selected from the large number of items produced by a certain production line in one day. The number of defectives X is observed. Find the maximum likelihood estimator of the ratio R of the number of defective to good items.

Hypothesis Testing

ABOUT THIS CHAPTER

We have learned how to estimate population parameters by using sample information. The second formal manner of making inferences from sample to population is through *tests of hypotheses*. The methodology allows the weight of sample evidence for or against the experimenter's hypotheses to be assessed probabilistically, and is basic to the scientific method. Again the sampling distributions of the statistics involved play a key role in the development.

CONTENTS

8.1 INTRODUCTION

One method of using sample data to formulate inferences about a population parameter, as seen in Chapter 7, is to produce a confidence interval estimate of the parameter in question. But it often happens that an experimenter is interested only in checking a claim, or hypothesis, concerning the value of a parameter, and is not really interested in the location or length of the confidence interval itself. For example, if an electronic component is guaranteed to possess a mean lifelength of at least 200 hours, then an investigator may be interested only in checking the hypothesis that the mean really is 200 hours or more against the alternative that the mean is less than 200 hours. A confidence interval on the mean is not, in itself, of great interest, although it can provide a mechanism for checking, or *testing*, the hypothesis of interest.

Hypothesis testing arises as a natural consequence of the scientific method. The scientist observes nature, formulates a theory, and then tests theory against observation. In the context of our problems, the experimenter theorizes that a population parameter takes on a certain value, or set of values. A sample is then selected from the population in question, and observation is compared with theory. If the observations disagree seriously with the theory, then the experimenter may reject the theory (hypothesis). If the observations are compatible with the theory, then the experimenter may proceed as if the hypothesis were true. Generally, the process does not stop in either case. New theories are posed, new experiments are completed, and new comparisons between theory and reality are made in a continuing cycle.

Note that hypothesis testing requires a decision when comparing the observed sample with theory. How do we decide whether the sample disagrees with the hypothesis? When should we reject the hypothesis, and when should we not reject? What is the probability that we will make a wrong decision? What function of the sample observations should we employ in our decision-making process? The answers to these questions lie in a study of statistical hypothesis testing.

8.2 HYPOTHESIS TESTING: THE SINGLE-SAMPLE CASE

In this section we present tests of hypotheses concerning the mean of an unspecified distribution, the probability of success in a binomial distribution, and the mean and variance of a normal distribution. Our first two tests require large samples, and are approximate in nature, while the second two are exact tests for any sample size.

8.2.1 General Distribution: Testing the Mean

One of the most common and, at the same time, easiest hypothesis-testing situations to consider is that of testing a population mean when a large random sample from the population in question is available for observation. Denoting the population mean by μ and the mean of the sample of size n by $\bar{X}$, we know from Chapter 7 that $\bar{X}$ is a good estimator of μ. Therefore it seems intuitively reasonable that our decision about μ should be based on $\bar{X}$, with extra information provided by the population variance σ^2 or its estimator S^2.

Suppose it is claimed that μ takes on the specific value μ_0, and we wish to test the validity of this claim. How can we tie this testing notion into the notion of a confidence interval established in Chapter 7? Upon reflection, it may seem reasonable that if the resulting confidence interval for μ does *not* contain μ_0, then we should *reject* the hypothesis that $\mu = \mu_0$. On the other hand, if μ_0 is within the confidence limits, then we cannot reject it as a plausible value for μ. Now if μ_0 is not within the interval $\bar{X} \pm z_{\alpha/2}\sigma/\sqrt{n}$, then either

$$\mu_0 < \bar{X} - z_{\alpha/2}\frac{\sigma}{\sqrt{n}}$$

or

$$\mu_0 > \bar{X} + z_{\alpha/2}\frac{\sigma}{\sqrt{n}}$$

The former equation can be rewritten as

$$\frac{\bar{X} - \mu_0}{\sigma/\sqrt{n}} > z_{\alpha/2}$$

and the latter as

$$\frac{\bar{X} - \mu_0}{\sigma/\sqrt{n}} < -z_{\alpha/2}$$

Assuming a large sample, so that $\bar{X}$ has approximately a normal distribution, we see that the hypothesized value μ_0 is rejected if

$$|Z| = \left|\frac{\bar{X} - \mu_0}{\sigma/\sqrt{n}}\right| > z_{\alpha/2}$$

for some prescribed α, where Z has approximately a standard normal distribution. In other words we reject μ_0 as a plausible value for μ if $\bar{X}$ is too many standard deviations (more than $z_{\alpha/2}$) from μ_0. Whereas $1 - \alpha$ was called the confidence coefficient in estimation problems, α is referred to as the *significance level* in hypothesis-testing problems.

EXAMPLE 8.1
_____

The depth setting on a certain drill press is two inches. One could then hypothesize that the average depth of all holes drilled by this machine is $\mu = 2$ inches. To check this hypothesis (and the accuracy of the depth gauge), a random sample of $n = 100$ holes drilled by this machine were measured and found to have a sample mean of $\bar{x} = 2.005$ inches with a standard deviation of $s = 0.03$ inch. With $\alpha = 0.05$, can the hypothesis be rejected based on this sample data?

Solution　With $n = 100$ it is assumed that $\bar{X}$ will be approximately normal in distribution. Thus we reject the hypothesis $\mu = 2$ if

$$|Z| = \left| \frac{\bar{X} - \mu_0}{\sigma/\sqrt{n}} \right| > z_{\alpha/2} = 1.96$$

For large n, S can be substituted for σ, and the approximate normality of Z still obtains. Thus the observed test statistic becomes

$$\frac{\bar{x} - \mu_0}{s/\sqrt{n}} = \frac{2.005 - 2.000}{0.03/\sqrt{100}} = 1.67$$

(This is actually a t statistic, but can be approximated by z for large n.) Since the observed value of $(\bar{x} - \mu_0)/s/\sqrt{n}$ is less than 1.96, we cannot reject the hypothesis that 2 is a plausible value of μ. (The 95% confidence interval for μ is 1.991 to 2.011, which includes the value 2.) Note that *not rejecting* the hypothesis that $\mu = 2$ is not the same as *accepting* the hypothesis $\mu = 2$. When we do not reject $H_0: \mu = 2$, we are saying that 2 is a possible value of μ, but there are other equally plausible values for μ. We cannot conclude that μ is equal to 2, and 2 alone. ∎

We now introduce the standard terminology of hypothesis-testing problems. The hypothesis that specifies a particular value for the parameter being studied is called the *null hypothesis* and is denoted by H_0. This hypothesis usually represents the standard operating procedure of a system or known specifications. In Example 8.1, $\mu = 2$ inches is the null hypothesis; it gives a specific value for μ and represents what should happen when the machine is operating according to specification.

The hypothesis that specifies those values of the parameter that represent an important change from standard operating procedure or known specifications is called the *alternative hypothesis* and is denoted by H_a. In Example 8.1 values of μ less than 2 and values of μ greater than 2 both represent important departures from the specification of $\mu = 2$. Thus the alternative hypothesis is $H_a: \mu \neq 2$.

Sample observations are collected and analyzed to determine if the evidence supports H_0 or H_a. The sample quantity on which the decision to support H_0 or H_a is based is called the test statistic. The set of values of the test statistic that leads to rejection of the null hypothesis in favor of the alternative hypothesis is called the *rejection region* (or *critical region*). In Example 8.1 the sample mean serves as an appropriate statistic to estimate μ, but we really want to know if μ is larger or smaller than 2. Thus the test statistic becomes Z, as shown above, since Z actually measures the standardized difference between $\bar{X}$ and 2. We reject H_0 if the absolute value of this difference is large, and the size of this rejection region is determined by α.

Table 8.1 summarizes the terminology of hypothesis testing and provides an example for a large-sample test of a population mean.

Sometimes a null hypothesis is enlarged to include an interval of possible values. For example, a can for 12 ounces of cola is designed to hold only slightly more than 12 ounces, so overfilling may not be serious. However, underfilling of the cans could have serious implications. Suppose μ represents the mean ounces of cola

TABLE 8.1
Components of a Hypothesis-Testing Problem

Terminology	Example
Null hypothesis, H_0	H_0: $\mu = \mu_0$
Alternative hypothesis, H_a	H_a: $\mu \neq \mu_0$
Test statistic	$Z = \dfrac{\bar{X} - \mu_0}{\sigma/\sqrt{n}}$
Rejection region	$\lvert Z \rvert > z_{\alpha/2}$

per can dispensed by a filling machine. Under a standard operating procedure that meets specifications, μ must be 12 or more, so we set H_0 as $\mu \geqslant 12$. An important change occurs if μ drops below 12, so H_a becomes $\mu < 12$. Under H_0 the vendor is in compliance with the law, and customers are satisfied. Under H_a the vendor is violating a law, and customers are unhappy. The vendor wants to sample the process regularly to be reasonably sure that a shift from H_0 to H_a has not taken place.

When H_0 is an interval like $\mu \geqslant 12$, the equality at $\mu = 12$ still plays a key role. Sample data will be collected, and a value for $\bar{x}$ will be calculated. If $\bar{x}$ is much less than 12, H_0 will be rejected. If the absolute difference between $\bar{x}$ and 12 is large enough to reject H_0, then the differences between $\bar{x}$ and 12.5 or 12.9 will be even larger, and again lead to rejection of H_0. Thus we need to consider only the boundary point between H_0 and H_a (12 in this case) when calculating an observed value of a test statistic.

For testing H_0: $\mu \geqslant \mu_0$ versus H_a: $\mu < \mu_0$ we consider only the analogous one-sided confidence interval for μ. Here it is important to establish an upper confidence limit for μ, since we want to know how large μ is likely to be. If μ_0 is larger than the upper confidence limit for μ, then we will reject H_0: $\mu \geqslant \mu_0$. The one-sided upper limit for μ is $\bar{X} + z_\alpha \sigma/\sqrt{n}$. (See Section 7.3.) We reject H_0 in favor of H_a when $\bar{X} + z_\alpha \sigma/\sqrt{n} < \mu_0$ or, equivalently, when

$$\frac{\bar{X} - \mu_0}{\sigma/\sqrt{n}} < -z_\alpha$$

Notice that we will have the same test statistic but that the rejection region has changed somewhat.

The corresponding rejection region for testing H_0: $\mu \leqslant \mu_0$ versus H_a: $\mu > \mu_0$ is given by

$$\frac{\bar{X} - \mu_0}{\sigma/\sqrt{n}} > z_\alpha$$

EXAMPLE 8.2

A vice-president for a large corporation claims that the number of service calls on equipment sold by that corporation is no more than 15 per week, on the average. To check his claim, service records were checked for $n = 36$ randomly selected weeks, with the result that $\bar{x} = 17$ and $s^2 = 9$ for the sample data. Does the sample evidence contradict the vice-president's claim at the 5% significance level?

Solution

The vice-president states that, according to his specifications, the mean must be 15 or less. Thus we are testing

$$H_0: \mu \leqslant 15 \quad \text{vs.} \quad H_a: \mu > 15 \qquad H_0$$

Using as test statistic

$$\frac{17 - 15}{3\sqrt{}}$$

$$Z = \frac{\bar{X} - \mu_0}{\sigma/\sqrt{n}}$$

we reject H_0 for $z > z_{0.05} = 1.645$. Substituting s for σ, we calculate the observed statistic to be

$$\frac{\bar{x} - \mu_0}{s/\sqrt{n}} = \frac{17 - 15}{3/\sqrt{36}} = 4$$

That is, the observed sample mean is 4 standard deviations greater than the hypothesized value of $\mu_0 = 15$. With $\alpha = 0.05$, $z_\alpha = 1.645$. Since the test statistic exceeds z_α, we have sufficient evidence to reject the null hypothesis. It does appear than the mean number of service calls exceeds 15. ■

In the preceding hypothesis-testing problems we see that there are two ways that errors can be made in the decision process. We can reject the null hypothesis when it is true, called a *Type I error*, or we can fail to reject the null hypothesis when it is false, called a *Type II error*. The possible decisions and errors are displayed in Table 8.2.

TABLE 8.2
Decisions and Errors in Hypothesis Testing

	Decision	
	Reject H_0	Do Not Reject H_0
H_0 True	Type I error	Correct decision
H_0 False	Correct decision	Type II error

For the one-sided test $H_0: \mu \geqslant \mu_0$ versus $H_a: \mu < \mu_0$, we reject H_0 if

$$\frac{\bar{X} - \mu_0}{\sigma/\sqrt{n}} < -z_\alpha$$

The probability that we reject H_0 when $\mu = \mu_0$ is given by

$$P[\text{Reject } H_0 \text{ when } \mu = \mu_0] = P\left[\frac{\bar{X} - \mu_0}{\sigma/\sqrt{n}} < -z_\alpha \text{ when } \mu = \mu_0\right] = \alpha$$

So we see that

$$\alpha = P[\text{Type I error when } \mu = \mu_0]$$

Thus in the case of a null hypothesis like $\mu \geqslant \mu_0$, the significance level α is the same as the probability of a Type I error at the point $\mu = \mu_0$.

We will denote the probability of a Type II error by β. That is,

$$\beta = P[\text{Type II error when } \mu = \mu_a]$$

$$= P[\text{Do not reject } H_0 \text{ when } \mu = \mu_a].$$

For the large-sample test of a mean we can show α and β as in Figure 8.1.

FIGURE 8.1
α and β for a
Statistical Test

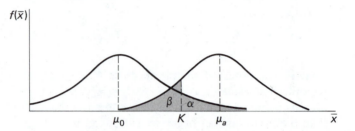

Referring to Figure 8.1, if μ_0 is the true value of μ under H_0 and μ_a is a specific alternative value of interest, then the area to the right of K under the normal curve centered at μ_0 is α. Note that α represents the chance that the null hypothesized value μ_0 will be rejected when, in fact, μ_0 is the true value of μ. The area to the left of K under the curve centered at μ_a is β for that particular alternative. Note that β represents the chance that the null hypothesis will be accepted when, in fact, μ_a is the true value of μ. How do we find the point denoted by K? Recall that for testing $H_0 : \mu = \mu_0$ versus $H_a : \mu > \mu_0$, we reject H_0 if

$$\frac{\bar{X} - \mu_0}{\sigma/\sqrt{n}} > z_\alpha$$

Thus

$$\alpha = P\left[\frac{\bar{X} - \mu_0}{\sigma/\sqrt{n}} > z_\alpha\right]$$

$$= P\left[\bar{X} > \mu_0 + z_\alpha \frac{\sigma}{\sqrt{n}}\right]$$

and therefore $K = \mu_0 + z_\alpha(\sigma/\sqrt{n})$.

We show how to calculate β in the following example.

EXAMPLE 8.3

In a power-generating plant, pressure in a certain line is supposed to maintain an average of 100 psi over any four-hour period. If the average pressure exceeds 103 psi for a four-hour period, serious complications can evolve. During a given four-hour period $n = 30$ measurements (assumed random) are to be taken. For testing $H_0: \mu \leqslant 100$ versus $H_a: \mu = 103$, α is to be 0.01. If $\sigma = 4$ psi for these measurements, calculate the probability of a Type II error.

Solution   For testing $H_0: \mu \leqslant 100$ vs. $H_a: \mu = 103$, we reject H_0 if

$$\frac{\bar{X} - \mu_0}{\sigma/\sqrt{n}} > z_{0.01} = 2.33$$

or if

$$\bar{X} > \mu_0 + z_{0.01}\frac{\sigma}{\sqrt{n}}$$

$$= 100 + 2.33\left(\frac{4}{\sqrt{30}}\right) = 101.7$$

Now if the true mean is really 103, then

$$\beta = P[\bar{X} < 101.7] = P\left[\frac{\bar{X} - \mu_a}{\sigma/\sqrt{n}} < \frac{101.7 - 103}{4/\sqrt{30}}\right]$$

$$= P[Z < -1.78] = 0.0375$$

Under these conditions the chance of observing a sample mean that is not in the rejection region when the true average pressure is 103 psi is quite small (less than 4%). Thus the operation can confidently continue if the sample mean of the tests supports H_0, that is, if $\bar{x}$ is less than 101.7.   ■

Notice that for a fixed sample size, increasing the size of the rejection region will increase α and decrease β. In many practical problems a specific value for an alternative will not be known, and consequently β cannot be calculated. In those cases we want to choose an appropriate significance level α and a test statistic that will make β as small as possible. Fortunately, most of the test statistics we use in this text have the property that, for fixed α and fixed sample size, β is nearly as small as possible for any alternative parameter value. In that sense we are using the best possible test statistics. Furthermore, we set up our hypotheses so that if the test statistic falls into the reject region, we reject H_0, knowing that the risk of a Type I error is fixed at α. However, if the test statistic is not in the rejection region for H_0, we hedge by stating that the evidence is insufficient to reject H_0. We *do not* affirmatively accept H_0.

For specified μ_0 and μ_a we can see from Figure 8.1 that if α is decreased, K will move to the right and β will increase. On the other hand, if α is increased, K will move to the left and β will decrease. However, if the sample size is increased, the sampling distributions will display less variability (become narrower and more sharply peaked), and *both* α and β can decrease. We will now show that it is possible to select a sample size to guarantee specified α and β, as long as μ_0 and μ_a are also specified.

For testing $H_0: \mu = \mu_0$ against $H_a: \mu = \mu_a$, where $\mu_a > \mu_0$, we have seen that H_0 is rejected when $\bar{X} > K$, where $K = \mu_0 + z_\alpha (\sigma/\sqrt{n})$. But from Figure 8.1 it is clear that we also have $K = \mu_a - z_\beta(\sigma/\sqrt{n})$. Thus

$$\mu_0 + z_\alpha(\sigma/\sqrt{n}) = \mu_a - z_\beta(\sigma/\sqrt{n})$$

$$(z_\alpha + z_\beta)(\sigma/\sqrt{n}) = \mu_a - \mu_0$$

or

$$\sqrt{n} = \frac{(z_\alpha + z_\beta)\sigma}{\mu_a - \mu_0}$$

This allows us to find an n that will produce a specified α and β for known μ_0 and μ_a.

Sample size for specified α and β when testing $H_0: \mu = \mu_0$ versus $H_a: \mu = \mu_a$:

$$n = \frac{(z_\alpha + z_\beta)^2 \sigma^2}{(\mu_a - \mu_0)^2}$$

EXAMPLE 8.4

For the power plant of Example 8.3 it is desired to test $H_0: \mu = 100$ vs. $H_a: \mu = 103$ with $\alpha = 0.01$ and $\beta = 0.01$. How many observations are needed to ensure this result?

Solution Since $\alpha = \beta = 0.01$, then $z_\alpha = z_\beta = 2.33$. Also $\sigma = 4$ for these measurements. We then have

$$n = \frac{(z_\alpha + z_\beta)^2 \sigma^2}{(\mu_a - \mu_0)^2}$$

$$= \frac{(2.33 + 2.33)^2 (4)^2}{(103 - 100)^2} = 38.6 \quad \text{or} \quad 39$$

By taking 39 measurements, we can reduce β to 0.01 while also holding α at 0.01. ∎

The hypothesis-testing procedure we have outlined above involves specifying a significance level α, finding $z_{\alpha/2}$ (or z_α for a one-sided test), calculating a value for the test statistic Z, and rejecting H_0 if $|z| > z_{\alpha/2}$ (or the appropriate one-sided adjustment). In this process α is somewhat arbitrarily determined. An alternative to specifying α is to find the *smallest* significance level at which the observed result would lead to rejection of H_0. This value is called the *P value* of the test, or the *attained significance level*. We would then reject H_0 for small *P* values.

In Example 8.1 the data produced $z = 1.67$ as the observed value of the test statistic for testing $H_0: \mu = 2$ vs. $H_a: \mu \neq 2$. The *smallest* significance level α that would lead to rejection of H_0 in this case must be such that $z_{\alpha/2} = 1.67$, since this is a two-sided test. Thus $\alpha/2$ must be the area under the standard normal curve to the right of 1.67, or $\alpha/2 = 0.5000 - 0.4525 = 0.0475$, which implies that $\alpha = 2(0.0475) = 0.095$. The *P* value for this test is 0.095, and we would reject H_0 for a significance level greater than or equal to this *P* value. Note that we did *not* reject H_0 at the 0.05 significance level in Example 8.1, and 0.05 is less than 0.095.

The *P* value can be thought of as the weight of evidence in the sample regarding H_0. If the *P* value is small (close to zero), there is not much evidence in favor of H_0, and it should be rejected. If the *P* value is large (say, close to 0.5), there is strong evidence in favor of H_0, and it should not be rejected.

8.2.2 Binomial Distribution: Testing the Probability of Success

Just as in the case of interval estimation, the large sample test for a mean can be easily transformed into a test for a binomial proportion. Recall that if Y has a binomial distribution with mean np, then Y/n is approximately normal for large n, with mean p and variance $p(1 - p)/n$. We illustrate with the following example.

EXAMPLE 8.5

A machine in a certain factory must be repaired if it produces more than 10% defectives among the large lot of items it produces in a day. A random sample of 100 items from the day's production contains 15 defectives, and the foreman says that the machine must be repaired. Does the sample evidence support his decision at the 0.01 significance level? Find the *P* value for this test.

Solution   We want to test $H_0: p \leq 0.10$ versus the alternative $H_a: p > 0.10$, where p denotes the proportion of defectives in the population. The test statistic will be based upon Y/n, where Y denotes the number of defectives observed. We will reject H_0 in favor of H_a if Y/n is suitably large. Following the procedure for testing a mean, we reject H_0 if

$$Y/n > K$$

or

$$Z = \frac{Y/n - p_0}{\sqrt{\dfrac{p_0(1 - p_0)}{n}}} > z_\alpha$$

Note that p_0 can be used in the variance of Y/n since we are assuming H_0 is true when we perform the test. Our observed value of the test statistic, with $\hat{p} = y/n = 0.15$, is

$$z = \frac{\hat{p} - p_0}{\sqrt{\dfrac{p_0(1 - p_0)}{n}}} = \frac{0.15 - 0.10}{\sqrt{\dfrac{(0.1)(0.9)}{100}}} = \frac{5}{3} = 1.67$$

Now $z_{0.01} = 2.33$, and since $1.67 < 2.33$, we will not reject H_0. It is quite likely that a sample fraction of 15% could occur even if p is in the neighborhood of 0.10. The evidence does not support the foreman's decision at the 0.01 significance level. The α value should be chosen to be quite small here since a Type I error (rejecting H_0 when it is true) has serious consequences, namely, the machine would be shut down unnecessarily for repairs.

Note that, even though the test statistic did not fall in the rejection region, we did not suggest that the null hypothesis be accepted (i.e., that the true proportion of defectives is 0.10). We simply stated that the sample does not refute the null hypothesis. To be more specific would be to risk a Type II error, and we have not computed a probability β of making such an error for any specified p_a.

The smallest significance level leading to rejection of H_0 in this case is the one producing $z_\alpha = 1.67$, since this is a one-sided test. From Table 4 of the Appendix the area to the right of 1.67 is 0.0475, which is the P value for this test. We would reject H_0 at any significance level at or above this value. ∎

8.2.3 Normal Distribution: Testing the Mean

When sample sizes are too small for the Central Limit Theorem to provide a good approximation to the distribution of $\bar{X}$, additional assumptions must be made on the nature of the probabilistic model for the population. Just as in the case of estimation, one common procedure is to assume, whenever it can be justified, that the population measurements fit the normal distribution reasonably well. Then $\bar{X}$, the sample mean for a random sample of size n, will have a normal distribution, and

$$\frac{\bar{X} - \mu}{S/\sqrt{n}}$$

will have a t distribution with $(n - 1)$ degrees of freedom. Thus the hypothesis-testing procedures for the population mean will have the same pattern as above, but the test statistic

$$T = \frac{\bar{X} - \mu_0}{S/\sqrt{n}}$$

will have a t distribution, rather than a normal distribution, under $H_0: \mu = \mu_0$. For testing $H_0: \mu = \mu_0$ versus $H_a: \mu \neq \mu_0$, the test statistic $T = (\bar{X} - \mu_0)/S/\sqrt{n}$ is calculated for the observed sample, and H_0 is rejected if $|T| \geqslant t_{\alpha/2}$, where $t_{\alpha/2}$ is the point cutting off an area of $\alpha/2$ in the upper tail of the t distribution with $(n - 1)$

degrees of freedom. (See Table 5 of the Appendix.) Corresponding one-sided tests can be conducted in a manner analogous to the large-sample case. The following examples indicate the methodology.

EXAMPLE 8.6

A corporation sets its annual budget for a new plant on the assumption that the average weekly cost for repairs is to be $\mu = \$1200$. To see if this claim is realistic, $n = 10$ weekly repair cost figures are obtained from similar plants. The sample is assumed to be random, and yields $\bar{x} = 1290$ and $s = 110$. Since a departure from the assumed average in either direction would be important to detect for budgeting purposes, it is desired to test $H_0: \mu = 1200$ versus $H_a: \mu \neq 1200$. Use $\alpha = 0.05$ and carry out the test.

Solution Assuming normality of weekly repair costs, the test statistic is

$$T = \frac{\bar{X} - \mu_0}{S/\sqrt{n}}$$

and the rejection region starts at $t_{\alpha/2} = t_{0.025} = 2.262$ for $n - 1 = 9$ degrees of freedom. The observed value of the test statistic is

$$t = \frac{\bar{x} - \mu_0}{s/\sqrt{n}} = \frac{1290 - 1200}{110/\sqrt{10}} = 2.587$$

which is greater than 2.262. Therefore we reject the null hypothesis. There is reason to suspect the 1200 is not a good assumed value of μ, and perhaps more investigation into these repair costs should be made before the budget is set. ■

The following illustrates a one-sided alternative-testing situation.

EXAMPLE 8.7

Muzzle velocities of eight shells tested with a new gunpowder yield a sample mean of $\bar{x} = 2959$ feet per second and a standard deviation $s = 39.4$. The manufacturer claims that the new gunpowder produces an average velocity of no less than 3000 feet per second. Does the sample provide enough evidence to contradict the manufacturer's claim? Use $= 0.05$.

Solution Here we are interested in testing $H_0: \mu \geq 3000$ versus $H_a: \mu < 3000$, since we want to see if there is evidence to refute the manufacturer's claim. Assuming that muzzle velocities can be reasonably modeled by a normal probability distribution, the test statistic is

$$T = \frac{\bar{X} - \mu_0}{S/\sqrt{n}}$$

observed to be

$$t = \frac{\bar{x} - \mu_0}{s/\sqrt{n}} = \frac{2959 - 3000}{39.4/\sqrt{8}} = -2.943$$

The rejection region starts at $-t_{0.05} = -1.895$ for 7 degrees of freedom. Since $t < -t_\alpha$, we reject the null hypothesis, and say that there appears to be good reason to doubt the manufacturer's claim. ∎

8.2.4 Normal Distribution: Testing the Variance

The variance of the underlying normal distributions in the above examples was assumed to be unknown, which is generally the case. If it is desired to test a hypothesis on the variance σ^2, it can be done by making use of the statistics used in constructing the confidence interval for σ^2. Recall that, for a random sample of size n from a normal distribution,

$$\frac{(n-1)S^2}{\sigma^2}$$

has a $\chi^2(n-1)$ distribution. Following the same principles as above, for testing $H_0: \sigma = \sigma_0^2$ versus $H_a: \sigma^2 \neq \sigma_0^2$, we calculate a value for the statistic $(n-1)S^2/\sigma_0^2$ and reject H_0, at level α, if this statistic is larger than $\chi^2_{\alpha/2}(n-1)$ or smaller than $\chi^2_{1-\alpha/2}(n-1)$. Suitable one-sided tests can also be employed.

EXAMPLE 8.8 _____

A machined engine part produced by a certain company is claimed to have a diameter variance no larger than 0.0002 inch. A random sample of 10 such parts gave a sample variance of $s^2 = 0.0003$. Assuming normality of diameter measurement, is there evidence here to refute the company's claim? Use $\alpha = 0.05$.

Solution   The test statistic is

$$\frac{(n-1)S^2}{\sigma_0^2}$$

which is observed to be

$$\frac{(n-1)s^2}{\sigma_0^2} = \frac{9(0.0003)}{0.0002} = 13.5$$

Here we are testing $H_0: \sigma^2 \leq 0.0002$ versus $H_a: \sigma^2 > 0.0002$. Thus we reject H_0 if $(n-1)S^2/\sigma_0^2$ is larger than $\chi^2_{0.05}(9) = 16.919$. Since this is not the case, we do not have sufficient evidence to refute the company's claim. ∎

Many other exact small-sample tests could be illustrated for other distributions, but these give the general idea for some commonly used cases. We will consider the small-sample tests for a binomial parameter in Chapter 13, in the context of lot acceptance sampling. The single-sample tests of this chapter are summarized in Table 8.3.

TABLE 8.3
Summary of Single-Sample Tests

	Null Hypothesis (H_0)	Alternative Hypothesis (H_a)	Test Statistic	Rejection Region
General Distribution (Large Sample)	$\mu = \mu_0$ $\mu \leqslant \mu_0$ $\mu \geqslant \mu_0$	$\mu \neq \mu_0$ $\mu > \mu_0$ $\mu < \mu_0$	$Z = \dfrac{\bar{X} - \mu_0}{\sigma/\sqrt{n}}$	$\lvert z \rvert > z_{\alpha/2}$ $z > z_\alpha$ $z < -z_\alpha$
Binomial Distribution (Large Sample)	$p = p_0$ $p \leqslant p_0$ $p \geqslant p_0$	$p \neq p_0$ $p > p_0$ $p < p_0$	$Z = \dfrac{Y/n - p_0}{\sqrt{\dfrac{p_0(1 - p_0)}{n}}}$	$\lvert z \rvert > z_{\alpha/2}$ $z > z_\alpha$ $z < -z_\alpha$
Normal Distribution	$\mu = \mu_0$ $\mu \leqslant \mu_0$ $\mu \geqslant \mu_0$	$\mu \neq \mu_0$ $\mu > \mu_0$ $\mu < \mu_0$	$T = \dfrac{\bar{X} - \mu_0}{S/\sqrt{n}}$ $(n - 1)$ df(degrees of freedom)	$\lvert t \rvert > t_{\alpha/2}$ $t > t_\alpha$ $t < -t_\alpha$
Normal Distribution	$\sigma^2 = \sigma_0^2$ $\sigma^2 \leqslant \sigma_0^2$ $\sigma^2 \geqslant \sigma_0^2$	$\sigma^2 \neq \sigma_0^2$ $\sigma^2 > \sigma_0^2$ $\sigma^2 < \sigma_0^2$	$\dfrac{(n - 1)S^2}{\sigma_0^2}$ $(n - 1)$ df	$\dfrac{(n - 1)s^2}{\sigma_0^2} < \chi^2_{1 - \alpha/2}$ $ > \chi^2_{\alpha/2}$ $\dfrac{(n - 1)s^2}{\sigma_0^2} > \chi^2_\alpha$ $\dfrac{(n - 1)s^2}{\sigma_0^2} < \chi^2_{1 - \alpha}$

EXERCISES

8.1 The output voltage for a certain electric circuit is specified to be 130. A sample of 40 independent readings on the voltage for this circuit gave a sample mean of 128.6 and a standard deviation of 2.1. Test the hypothesis that the average output voltage is 130 against the alternative that it is less than 130. Use a 5% significance level.

8.2 Refer to Exercise 8.1. If the voltage falls as low as 129, serious consequences may result. For testing $H_0: \mu \geqslant 130$ versus $H_a: \mu = 129$, find the probability of a Type II error, β, for the rejection region used in Exercise 8.1.

8.3 For testing $H_0: \mu = 130$ vs. $H_a: \mu = 129$ with $\sigma = 2.1$, as in Exercise 8.2, find the sample size that will yield $\alpha = 0.05$ and $\beta = 0.01$.

8.4 The Rockwell hardness index for steel is determined by pressing a diamond point into the steel and measuring the depth of penetration. For 50 specimens of a certain type of steel the Rockwell hardness index averaged 62 with a standard deviation of 8. The manufacturer claims that this steel has an average hardness index of at least 64. Test this claim at the 1% significance level. Find the P value for this test.

8.5 Steel is sufficiently hard for a certain use so long as the mean Rockwell hardness measure does not drop below 60. Using the rejection region found in Exercise 8.4, find β for the specific alternative $\mu = 60$.

8.6 For testing $H_0: \mu = 64$ vs. $H_a: \mu = 60$, with $\sigma = 8$, as in Exercises 8.4 and 8.5, find the sample size required for $\alpha = 0.01$ and $\beta = 0.05$.

8.7 The pH of water coming out of a certain filtration plant is specified to be 7.0. Thirty water samples independently selected from this plant show a mean pH of 6.8 and a standard deviation of 0.9. Is there any reason to doubt that the plant's specification is being maintained? Use $\alpha = 0.05$. Find the P value for this test.

8.8 A manufacturer of resistors claims that 10% fail to meet the established tolerance limits. A random sample of resistance measurements for 60 such resistors reveal 8 to lie outside the tolerance limits. Is there sufficient evidence to refute the manufacturer's claim, at the 5% significance level? Find the P value for this test.

8.9 For a certain type of electronic surveillance system the specifications state that the system will function for more than 1000 hours with probability at least 0.90. Checks on 40 such systems show that five failed prior to 1000 hours of operation. Does this sample provide sufficient information to conclude that the specification is not being met? Use $\alpha = 0.01$.

8.10 The hardness of a certain rubber (in degrees Shore) is claimed to be 65. Fourteen specimens are tested, resulting in an average hardness measure of 63.1 and a standard deviation of 1.4. Is there sufficient evidence to reject the claim, at the 5% level of significance? What assumption is necessary for your answer to be valid?

8.11 Certain rockets are manufactured with a range of 2500 meters. It is theorized that the range will be reduced after the rockets are in storage for some time. Six of these rockets are stored for a certain period of time and then tested. The ranges found in the tests are as follows: 2490, 2510, 2360, 2410, 2300, 2440. Does the range appear to be shorter after storage? Test at the 1% significance level.

8.12 For screened coke the porosity factor is measured by the difference in weight between dry and soaked coke. A certain supply of coke is claimed to have a porosity factor of 1.5 kilograms. Ten samples are tested, resulting in a mean porosity factor of 1.9 kilograms and a variance of 0.04. Is there sufficient evidence to indicate that the coke is more porous than is claimed? Use $\alpha = 0.05$ and assume the porosity measurements are approximately normally distributed.

8.13 The stress resistance of a certain plastic is specified to be 30 psi. The results from ten specimens of this plastic show a mean of 27.4 psi and a standard deviation of 1.1 psi. Is there sufficient evidence to doubt the specification at the 5% significance level? What assumption are you making?

8.14 Yield stress measurements on 51 steel rods with 10-mm diameters gave a mean of 485 N/mm^2 and a standard deviation of 17.2. (See P. Booster, *Journal of Quality Technology*, 1, No. 4, 1983, p. 191 for details.) Suppose the manufacturer claims that the yield stress for these bars is 490. Does the sample information suggest rejecting the manufacturer's claim at the 5% significance level?

8.15 The widths of contact windows in certain CMOS circuit chips have a design specification of 3.5 μm. (See M. S. Phadke, et al., *The Bell System Technical Journal*, 62,

No. 5, 1983, pp. 1273–1309 for details.) Postetch window widths on test specimens were as follows:

3.21, 2.49, 2.94, 4.38, 4.02, 3.82, 3.30, 2.85, 3.34, 3.91

(These data were also used in Exercise 7.74.) Can we reject the hypothesis that the design specification is being met at the 5% significance level? What assumptions are necessary for this test to be valid?

8.16 The variation in window widths for CMOS circuit chips must be controlled at a low level if the circuits are to function properly. Suppose specifications state that $\sigma = 0.30$ for window widths. Can we reject the claim that this specification is being met, using the data of Exercise 8.15? Use $\alpha = 0.05$.

8.17 The Florida Poll of February-March 1984 interviewed 871 adults from around the state. On one question 53% of the respondents favored strong support of Israel. Would you conclude that a majority of adults in Florida favor strong support of Israel? (Source: *Gainesville Sun*, April 1, 1984.)

8.18 A Yankelovich, Skelly, and White poll reported upon on November 30, 1984, showed that 54% of 2207 people surveyed thought the U.S. income tax system was too complicated. Can we conclude safely, at the 5% significance level, that the majority of Americans think the income tax system is too complicated?

8.19 Wire used for wrapping concrete pipe should have an ultimate tensile strength of 300,000 pounds, according to a design engineer. Forty tests of such wire used on a certain pipe showed a mean ultimate tensile strength of 295,000 pounds and a standard deviation of 10,000. Is there sufficient evidence to suggest that the wire used on the tested pipe does not meet the design specification, at the 10% level?

8.20 The break angle in a torsion test of wire used in wrapping concrete pipe is specified to be 40°. Fifty torsion tests of wire wrapping a certain malfunctioning pipe resulted in a mean break angle of 37° and a standard deviation of 6°. Can we say that the wire used on the malfunctioning pipe does not meet the specification at the 5% significance level?

8.21 Soil pH is an important variable when designing structures that will contact the soil. The pH at a potential construction site was said to average 6.5. Nine test samples of soil from the site gave readings of

7.3, 6.5, 6.4, 6.1, 6.0, 6.5, 6.2, 5.8, 6.7

Do these readings cast doubt upon the claimed average? (Test at the 5% significance level.) What assumptions are necessary for this test to be valid?

8.22 The resistances of a certain type of thermistor are specified to be normally distributed with a mean of 10,000 ohms and a standard deviation of 500 ohms, at a temperature of 25°C. Thus only about 2.3% of these termistors should produce resistances in excess of 11,000 ohms. The exact measurement of resistances produced is time consuming, but it is easy to determine if the resistance is larger than a specified value (say 11,000). For 100 thermistors tested, 5% showed resistances in excess of 11,000 ohms. Do you think the thermistors from which this sample was selected fail to meet the specifications? Why?

8.23 The dispersion, or variance, of haul times on a struction project are of great importance to the project foreman, since highly variable haul times cause problems in scheduling jobs. The foreman of the truck crews states that the range of haul times should not exceed 40 minutes. (The range is the difference between the longest and shortest times.) Assuming these haul times to be approximately normally distributed, the project foreman takes the statement on the range to mean that the standard deviation σ should be approximately 10 minutes. Fifteen haul times are actually measured, and show a mean of 142 minutes and a standard deviation of 12 minutes. Can the claim of $\sigma = 10$ be refuted at the 5% significance level?

8.24   Aptitude tests should produce scores with a large amount of variation so that an administrator can distinguish between persons with low aptitude and those with high aptitude. The standard test used by a certain industry has been producing scores with a standard deviation of 5 points. A new test is tried on 20 prospective employees and produces a sample standard deviation of 8 points. Are scores from the new test significantly more variable than scores from the standard? Use $\alpha = 0.05$.

8.25   For the rockets of Exercise 8.11 the variation in ranges is also of importance. New rockets have a standard deviation of range measurements equal to 20 kilometers. Does it appear that storage increases the variability of these ranges? Use $\alpha = 0.05$.

8.3   HYPOTHESIS TESTING: THE MULTIPLE-SAMPLE CASE

Just as we developed confidence interval estimates for linear functions of means when k samples from k different populations are available, we could develop the corresponding hypothesis tests. However, most tests of population means involve tests of simple differences of the form $\mu_1 - \mu_2$. Thus we will consider only this case. (The more general case of testing equality of k means will be covered in a discussion of analysis of variance in Chapter 11.)

8.3.1   General Distributions: Testing the Difference Between Two Means

We know from Chapter 7 that, in the large-sample case, the estimator $\bar{X}_1 - \bar{X}_2$ has an approximate normal distribution. Hence for testing $H_0: \mu_1 - \mu_2 = D_0$ versus $H_a: \mu_1 - \mu_2 \neq D_0$, we can use as a test statistic

$$Z = \frac{(\bar{X}_1 - \bar{X}_2) - D_0}{\sqrt{\dfrac{\sigma_1^2}{n_1} + \dfrac{\sigma_2^2}{n_2}}}$$

We reject H_0 for $|Z| \geq z_{\alpha/2}$ for a specified α. Similar one-sided tests can be constructed in the obvious way. If σ_1^2 and σ_2^2 are unknown, they can be estimated by the sample variances S_1^2 and S_2^2, respectively.

EXAMPLE 8.9 _____

A study was conducted to compare the length of time it took men and women to perform a certain assembly line task. Independent samples of 50 men and 50 women were employed in an experiment in which each person was timed on identical tasks. The results were as follows:

Men	Women
$n_1 = 50$	$n_2 = 50$
$\bar{x}_1 = 42$ sec	$\bar{x}_2 = 38$ sec
$s_1^2 = 18$	$s_2^2 = 14$

Do the data present sufficient evidence to suggest a difference between the true mean completion times for men and women at the 5% significance level?

Solution   Since we are interested in detecting a difference in either direction, we want to test $H_0: \mu_1 - \mu_2 = 0$ (no difference) versus $H_a: \mu_1 - \mu_2 \neq 0$. The test statistic is calculated to be (with s_i estimating σ_i)

$$z = \frac{(\bar{x}_1 - \bar{x}_2) - D_0}{\sqrt{\dfrac{s_1^2}{n_1} + \dfrac{s_2^2}{n_2}}} = \frac{42 - 38}{\sqrt{\dfrac{18}{50} + \dfrac{14}{50}}} = 5$$

That is, the difference between the sample means is 5 standard deviations from the hypothesized zero difference. Now $z_{0.025} = 1.96$, and since $|z| > 1.96$, we reject H_0. It does appear that the difference in times between men and women is real.   ■

8.3.2 Normal Distributions with Equal Variances: Testing the Difference Between Two Means

As before, the small-sample case necessitates an assumption on the nature of the probabilistic model for the random variables in question. If both populations seem to have normal distributions, and if the two variances are equal ($\sigma_1^2 = \sigma_2^2 = \sigma^2$), then t tests on hypotheses concerning $\mu_1 - \mu_2$ can be constructed. For testing $H_0: \mu_1 - \mu_2 = D_0$ versus $H_a: \mu_1 - \mu_2 \neq D_0$, the test statistic to be employed is

$$T = \frac{(\bar{X}_1 - \bar{X}_2) - D_0}{S_p \sqrt{\dfrac{1}{n_1} + \dfrac{1}{n_2}}}$$

where

$$S_p^2 = \frac{(n_1 - 1)S_1^2 + (n_2 - 1)S_2^2}{n_1 + n_2 - 2}$$

T has a t distribution with $n_1 + n_2 - 2$ degrees of freedom when H_0 is true. As in the large-sample case, we reject H_0 whenever $|T| > t_{\alpha/2}$. Corresponding one-sided tests are obtained easily.

EXAMPLE 8.10

The designer of a new sheet-metal stamping machine claims that his new machine can turn out a certain product faster than the machine now in use. Nine independent trials of stamping the same item on each machine gave the following results on times to completion:

Standard Machine	New Machine
$n_1 = 9$	$n_2 = 9$
$\bar{x}_1 = 35.22$ seconds	$\bar{x}_2 = 31.56$ seconds
$(n_1 - 1)s_1^2 = 195.50$	$(n_2 - 1)s_2^2 = 160.22$

At the 5% significance level, can the designer's claim be substantiated?

Solution   We are interested here in testing H_0: $\mu_1 - \mu_2 \leqslant 0$ versus H_a: $\mu_1 - \mu_2 > 0$ (or $\mu_2 < \mu_1$). The test statistic T is calculated as

$$t = \frac{(\bar{x}_1 - \bar{x}_2) - D_0}{s_p \sqrt{\dfrac{1}{n_1} + \dfrac{1}{n_2}}} = \frac{35.22 - 31.56}{4.71 \sqrt{\dfrac{1}{9} + \dfrac{1}{9}}} = 1.65$$

since

$$s_p^2 = \frac{(n - 1)s_1^2 + (n_2 - 1)s_2^2}{n_1 + n_2 - 2} = \frac{195.50 + 160.22}{16} = 22.24$$

Since $t_{0.05} = 2.130$ for 16 degrees of freedom, we cannot reject H_0. There is insufficient evidence here to substantiate the designer's claim.   ■

To see how a typical computer printout displays the computations from a standard two-sample test of equality of means, we can study the following example.

EXAMPLE 8.11

A quality-control inspector compares the ultimate tensile strength (UTS) measurements for class II and class III prestressing wire by taking a sample of 5 specimens from a roll of each class for laboratory testing. The sample data (in 1000 psi) are as follows:

Class II   253, 261, 258, 255, 256

Class III   274, 275, 271, 277, 276

Do the true mean UTS measurements for the two classes of wire appear to differ?

Solution   The computer output (Minitab) is as follows:

```
TWOSAMPLE T FOR classII VS classIII
            N      MEAN   STDEV    SE MEAN
classII     5     256.60   3.05      1.4
classIII    5     274.60   2.30      1.0
```

TTEST MU classII = MU classIII (VS NE): T = −10.53   P = 0.0000   DF = 8
POOLED STDEV =   2.70

For testing equality of means, the t statistic is -10.53 with a two-sided P value of essentially zero. Thus there is strong evidence to say that the mean UTS measurements differ for class II and class III wire.

8.3.3 Normal Distributions with Unequal Variances: Testing the Difference Between Two Means

If one suspects that the variances in the two normal populations under study are *not* equal, then the t test based on pooling the sample variances should not be used. What should be used in its place? Unfortunately the answer to that question is not easy since there is no exact way to solve this problem. There are, however, many approximate solutions, and one fairly simple one will be presented here.

   For this approximate procedure the first step is to calculate the T statistic similar to the Z statistic of Section 8.3.1, namely,

$$T = \frac{(\bar{X}_1 - \bar{X}_2) - D_0}{\sqrt{\dfrac{S_1^2}{n_1} + \dfrac{S_2^2}{n_2}}}$$

This statistic has approximately a t distribution under H_0: $\mu_1 - \mu_2 = D_0$, with degrees of freedom given by the integer part of

$$df = \frac{(S_1^2 + S_2^2)^2}{\left[\dfrac{S_1^2}{n_1 - 1}\right] + \left[\dfrac{S_2^2}{n_2 - 1}\right]}$$

where S_1^2 and S_2^2 are the sample variances. It is recommended that these computations be done by computer, and so we will show a Minitab calculation for an example of this type.

EXAMPLE 8.12 _____

The prestressing wire on each of two concrete pipes manufactured at different times was compared for torsion properties. Ten specimens randomly selected from each pipe were twisted in a laboratory apparatus until they broke, with the number of revolutions until complete failure being the measurement of interest. The results are as follows, with C1 and C2 denoting the two concrete pipes:

   C1:   6.38, 6.88, 7.00, 4.75, 1.50, 3.37, 7.63, 3.63, 4.00, 4.63

   C2:   3.38, 2.81, 3.00, 5.88, 5.25, 4.08, 4.25, 4.50, 4.13, 4.88

Is there evidence to suggest that the true mean revolutions to failure differ for the wire on the two pipes?

Solution   Careful study of the data sets shows that C1 results look to be more variable than the C2 results. This can be shown more dramatically on a stem-and-leaf plot of the two sets (here truncated to one decimal place):

```
      5 | 1 |
        | 2 | 8
     63 | 3 | 03
    760 | 4 | 01258
        | 5 | 28
     83 | 6 |
     60 | 7 |
     C1     C2
```

It doesn't seem reasonable, here, to assume that these two sets of measurements come from populations with the same variance. (We will see how to conduct a test for equal variances in Section 8.3.5.) Thus we will use the approximate procedure outlined above. The computer printout is shown below.

TWOSAMPLE T FOR C1 VS C2

	N	MEAN	STDEV	SE MEAN
C1	10	4.98	1.95	0.62
C2	10	4.216	0.974	0.31

TTEST MU C1 = MU C2 (VS NE): T = 1.10   P = 0.29   DF = 13

The observed t statistic is 1.10 with 13 degrees of freedom. The P value of 0.29 suggests that we not reject H_0, and state that there is no evidence to suggest the mean torsion values differ for the two pipes.

8.3.4   Normal Distributions: Testing the Difference Between Means for Paired Samples

The two-sample t test just given works only in the case of *independent* samples. On many occasions, however, two samples will arise in a dependent fashion. A commonly occurring example involves the situation in which repeated observations are taken on the *same* sampling unit, such as counting the number of accidents in various plants both before and after a safety awareness program is effected. The counts in one plant may be independent of the counts in another, but the two counts (before and after) within any one plant will be dependent. Thus we must develop a mechanism for analyzing measurements that occur in pairs.

Let $(X_1, Y_1), \ldots, (X_n, Y_n)$ denote a random sample of paired observations. That is, (X_i, Y_i) denotes two measurements taken in the same sampling unit, such as counts of accidents within a plant before and after a safety awareness program is put into effect, or lifelengths of two components within the same machine. Suppose it is of interest to test a hypothesis concerning the difference between $E(X_i)$ and $E(Y_i)$. The two-sample tests developed earlier in this chapter cannot be used for this purpose because of the dependence between X_i and Y_i. Observe, however, that $E(X_i) - E(Y_i) = E(X_i - Y_i)$. Thus the comparison of $E(X_i)$ with $E(Y_i)$ can be made through looking at the mean of the differences, $(X_i - Y_i)$. Letting $X_i - Y_i = D_i$, the hypothesis $E(X_i) - E(Y_i) = 0$ is equivalent to the hypothesis $E(D_i) = 0$.

To construct a test statistic with a known distribution, we assume $D_1, \ldots, D_n$ is a random sample of *differences*, each possessing a normal distribution with mean μ_D and variance σ_D^2. To test $H_0: \mu_D = 0$ versus the alternative $H_a: \mu_D \neq 0$, we employ the test statistics

$$T = \frac{\bar{D} - 0}{S_D / \sqrt{n}}$$

where

$$\bar{D} = \frac{1}{n} \sum_{i=1}^{n} D_i$$

and

$$S_D^2 = \frac{1}{n-1} \sum_{i=1}^{n} (D_i - \bar{D})^2$$

Since $D_1, \ldots, D_n$ have the same normal distribution, T will have a t distribution with $n - 1$ degrees of freedom when H_0 is true. Thus we have reduced the problem to that of a one-sample t test. Example 8.13 illustrates the use of this procedure.

EXAMPLE 8.13

Two methods of determining the percentage of iron in ore samples are to be compared by subjecting 12 ore samples to each method. The results of the experiment are as follows:

Ore Sample	Method A	Method B	d_i
1	38.25	38.27	-0.02
2	31.68	31.71	-0.03
3	26.24	26.22	$+0.02$
4	41.29	41.33	-0.04
5	44.81	44.80	$+0.01$
6	46.37	46.39	-0.02
7	35.42	35.46	-0.04
8	38.41	38.39	$+0.02$
9	42.68	42.72	-0.04
10	46.71	46.76	-0.05
11	29.20	29.18	$+0.02$
12	30.76	30.79	-0.03

Do the data provide evidence that method B has a higher average percentage than method A? Use $\alpha = 0.05$.

Solution We wish to test $H_0: \mu_D \geq 0$ versus $H_a: \mu_D < 0$, since μ_D will be negative if method B has the larger mean. For the given data

$$\bar{d} = \frac{1}{12}(-0.20) = -0.0167$$

and

$$s_D^2 = \frac{\sum_{i=1}^{n} d_i^2 - \frac{1}{n}\left(\sum_{i=1}^{n} d_i\right)^2}{n-1} = \frac{0.0112 - \frac{1}{12}(-0.20)^2}{11} = 0.0007$$

It follows that

$$t = \frac{\bar{d} - 0}{s_d/\sqrt{n}} = \frac{-0.0167}{\sqrt{0.0007}/\sqrt{12}} = -2.1865$$

Since $\alpha = 0.05$ and $n - 1 = 11$ degrees of freedom, the rejection region consists of those values of t smaller than $-t_{0.05} = -1.796$. Hence we reject the null hypothesis that $\mu_D \geq 0$ and conclude that the data support the alternative, $\mu_D < 0$. Note that the validity of this conclusion, especially the α level, depends upon the assumption that the probabilistic model underlying the differences is normal. ■

8.3.5 Normal Distributions: Testing the Ratio of Variances

One can compare the variances of two normal populations by looking at the ratio of sample variances. For testing $H_0: \sigma_1^2 = \sigma_2^2$ (or $\sigma_1^2/\sigma_2^2 = 1$) versus $H_a: \sigma_1^2 \neq \sigma_2^2$, we calculate a value for S_1^2/S_2^2 and reject if this statistic is either very large or very small. The precise rejection region can be found by observing that

$$F = \frac{S_1^2}{S_2^2}$$

has an F distribution when $H_0: \sigma_1^2 = \sigma_2^2$ is true. Thus we reject H_0 for $F > F_{\alpha/2}(v_1, v_2)$ or $F < F_{1-\alpha/2}(v_1, v_2)$. A one-sided test can always be constructed with the larger S_i^2 in the numerator, so that only the upper-tail critical F value needs to be found.

EXAMPLE 8.14

Suppose that the machined engine part of Example 8.8 is to be compared, with respect to diameter variance, with a similar part manufactured by a competitor. The former showed $s_1^2 = 0.0003$ for a sample of $n_1 = 10$ diameter measurements. A sample of $n_2 = 20$ of the competitor's showed $s_2^2 = 0.0001$. Is there evidence to conclude that $\sigma_2^2 < \sigma_1^2$ at the 5% significance level? Assume normality for both populations.

Solution   We are interested in testing $H_0: \sigma_1^2 \leqslant \sigma_2^2$ versus $H_a: \sigma_1^2 > \sigma_2^2$. The outcome of the statistic, $F = S_1^2/S_2^2$, which has an F distribution if $\sigma_1^2 = \sigma_2^2$, is calculated to be

$$\frac{s_1^2}{s_2^2} = \frac{0.0003}{0.0001} = 3$$

We reject H_0 if $F > F_\alpha(v_1, v_2) = F_{0.05}(9,19) = 2.42$. Since the observed ratio is 3, we reject H_0. It looks like the second manufacturer has less variability in the diameter measurements.   ■

The F test can be employed prior to the use of a two-sample t test to compare means so that we can tell which t test to use (the one for equal population variances or the one for unequal population variances). In Example 8.12 the F ratio for the sample variances is 4.0, which is in the rejection region for the 0.05 significance level. This is further evidence that we were correct in applying the approximate t test that does not require equality of population variances.

TABLE 8.4
Summary of Two-Sample Tests

	Null Hypothesis (H_0)	Alternative Hypothesis (H_a)	Test Statistic	Rejection Region
General Distribution (Large Sample)	$\mu_1 - \mu_2 = D_0$ $\mu_1 - \mu_2 \leqslant D_0$ $\mu_1 - \mu_2 \geqslant D_0$	$\mu_1 - \mu_2 \neq D_0$ $\mu_1 - \mu_2 > D_0$ $\mu_1 - \mu_2 < D_0$	$Z = \dfrac{(\bar{X}_1 - \bar{X}_2) - D_0}{\sqrt{\dfrac{\sigma_1^2}{n_1} + \dfrac{\sigma_2^2}{n_2}}}$	$|z| > z_{\alpha/2}$ $z > z_\alpha$ $z < -z_\alpha$
Normal Distributions (Equal Variances)	$\mu_1 - \mu_2 = D_0$ $\mu_1 - \mu_2 \leqslant D_0$ $\mu_1 - \mu_2 \geqslant D_0$	$\mu_1 - \mu_2 \neq D_0$ $\mu_1 - \mu_2 > D_0$ $\mu_1 - \mu_2 < D_0$	$T = \dfrac{(\bar{X}_1 - \bar{X}_2) - D_0}{S_p\sqrt{\dfrac{1}{n_1} + \dfrac{1}{n_2}}}$ $(n_1 + n_2 - 2)\,\mathrm{df}$	$|t| > t_{\alpha/2}$ $t > t_\alpha$ $t < -t_\alpha$
Normal Distributions (Paired Samples)	$\mu_D = \mu_{D0}$ $\mu_D \leqslant \mu_{D0}$ $\mu_D \geqslant \mu_{D0}$	$\mu_D \neq \mu_{D0}$ $\mu_D > \mu_{D0}$ $\mu_D < \mu_{D0}$	$T = \dfrac{\bar{D} - \mu_{D0}}{S_D/\sqrt{n}}$ $(n - 1)\,\mathrm{df}$	$|t| > t_{\alpha/2}$ $t > t_\alpha$ $t < -t_\alpha$
Normal Distributions	$\sigma_1^2 = \sigma_2^2$	$\sigma_1^2 \neq \sigma_2^2$	$F = \dfrac{S_1^2}{S_2^2}$	$\dfrac{S_1^2}{S_2^2} > F_{\alpha/2}(v_1, v_2)$ or $< F_{1-\alpha/2}(v_1, v_2)$
	$\sigma_1^2 \leqslant \sigma_2^2$	$\sigma_1^2 > \sigma_2^2$		$\dfrac{S_1^2}{S_2^2} > F_\alpha(v_1, v_2)$
	$\sigma_1^2 \geqslant \sigma_2^2$	$\sigma_1^2 < \sigma_2^2$	$v_1 = (n_1 - 1)\,\mathrm{df}$ $v_2 = (n_2 - 1)\,\mathrm{df}$	$\dfrac{S_2^2}{S_1^2} > F_\alpha(v_2, v_1)$

8.26 Two designs for a laboratory are to be compared with respect to the average amount of light produced on table surfaces. Forty independent measurements (in foot candles) are taken in each laboratory, with the following results:

Design I	Design II
$n_1 = 40$	$n_2 = 40$
$\bar{x}_1 = 28.9$	$\bar{x}_2 = 32.6$
$S_1^2 = 15.1$	$S_2^2 = 15.8$

Is there sufficient evidence to suggest that the designs differ with respect to the average amount of light produced? Use $\alpha = 0.05$.

8.27 Shear strength measurements derived from unconfined compression tests for two types of soils gave the following results (measurements in tons per square foot).

Soil Type I	Soil Type II
$n_1 = 30$	$n_2 = 35$
$\bar{x}_1 = 1.65$	$\bar{x}_2 = 1.43$
$S_1 = 0.26$	$S_2 = 0.22$

Do the soils appear to differ with respect to average shear strength, at the 1% significance level?

8.28 A study was conducted by the Florida Gama and Fish Commission to assess the amounts of chemical residues found in the brain tissue of brown pelicans. For DDT, random samples of $n_1 = 10$ juveniles and $n_2 = 13$ nestlings gave the following results (measurements in parts per million):

Juveniles	Nestlings
$n_1 = 10$	$n_2 = 13$
$\bar{y}_1 = 0.041$	$\bar{y}_2 = 0.026$
$s_1 = 0.017$	$s_2 = 0.016$

Test the hypothesis that there is no difference between mean amounts of DDT found in juveniles and nestlings versus the alternative that the juveniles have a larger mean. Use $\alpha = 0.05$. (This test has important implications regarding the build-up of DDT over time.)

8.29 The strength of concrete depends, to some extent, on the method used for drying. Two different drying methods showed the following results for independently tested specimens (measurements in psi):

Method I	Method II
$n_1 = 7$	$n_2 = 10$
$\bar{x}_1 = 3250$	$\bar{x}_2 = 3240$
$s_1 = 210$	$s_2 = 190$

Do the methods appear to produce concrete with different mean strengths? Use $\alpha = 0.05$. What assumptions are necessary in order for your answer to be valid?

8.30 The retention of nitrogen in the soil is an important consideration in cultivation practices, including the cultivation of forests. Two methods for preparing plots for planting pine trees after clear cutting were compared on the basis of the percentage of labeled nitrogen recovered. Method A leaves much of the forest floor intact, while method B removes most of the organic material. It is clear that method B will produce a much lower recovery of nitrogen from the forest floor. The question of interest is whether or not method B will cause more nitrogen to be retained in the microbial biomass, as a compensation for having less organic material available. Percentage of nitrogen recovered in the microbial biomass was measured on six test plots for each method. Method A plots showed a mean of 12 and a standard deviation of 1. Method B plots showed a mean of 15 and a standard deviation of 2. At the 10% significance level, should we say that the mean percentage of recovered nitrogen is larger for method B? (See P. M. Vitousek and P. A. Matson, *Science*, 225, 6 July 1984, p. 51 for more details.)

8.31 In the presence of reverberation, native as well as nonnative speakers of English have some trouble recognizing consonants. Random samples of 10 natives and 10 nonnatives were given the Modified Rhyme Test, and the percentages of correct responses were recorded. (See A. K. Nabelek and A. M. Donahue, *Journal of the Acoustical Society of America*, 75, 2, 1984, p. 633.) The data are as follows:

Natives: 93, 85, 89, 81, 88, 88, 89, 85, 85, 87

Nonnatives: 76, 84, 78, 73, 78, 76, 70, 82, 79, 77

Is there sufficient evidence to say that nonnative speakers of English have a smaller mean percentage of correct responses at the 5% significance level?

8.32 Biofeedback monitoring devices and techniques in control of physiologic functions help astronauts control stress. One experiment on this topic is discussed by G. Rotondo, et al., *Acta Astronautica*, 10, No. 8, 1983, pp. 591–598. Six subjects were placed in a stressful situation (using video games), followed by a period of biofeedback and adaptation. Another group of six subjects were placed under the same stress, and then simply told to relax. The first group had an average heartrate of 70.4 and a standard deviation of 15.3. The second group had an average heartrate of 74.9 and a standard deviation of 16.0. At the 10% significance level can we say that the average heartrate with biofeedback is lower than for those without biofeedback? What assumptions are necessary for your answer to be valid?

8.33 The t test for comparing means assumes that the population variances are equal. Is that a valid assumption based on the data of Exercise 8.32? (Test the equality of population variances at the 10% significance level.)

8.34 Data from the U.S. Department of Interior-Geological Survey gives flow rates for a small river in North Florida. It is of interest to compare flow rates for March and April, two relatively dry months. Thirty-one measurements for March showed a mean flow rate of 6.85 cubic feet per second and a standard deviation of 1.2. Thirty measurements for April showed a mean of 7.47 and a standard deviation of 2.3. Is there sufficient evidence to say that these two months have different average flow rates, at the 5% significance level?

8.35 An important measure of the performance of a machine is the mean time between failures (MTBF). A certain printer attached to a word processor was observed for a period of time during which ten failures were observed. The times between failures averaged 98 working hours with a standard deviation of 6 hours. A modified version of this printer was observed for 8 failures, the times between failures averaging 94 working hours with a standard deviation of 8 hours. Can we say, at the 1% significance level, that the modified version of the printer has a smaller MTBF? What assumptions are made in this test? Do you think the assumptions are reasonable for this situation?

8.36 It is claimed that tensile strength measurements for a 12-mm-diameter steel rod should, on the average, be at least 8 units (N/mm^2) higher than the tensile strength measurements for a 10-mm-diameter steel rod. Independent samples of 50 measurements each for the two sizes of rods gave the following results:

10-mm Rod	12-mm Rod
$\bar{x}_1 = 545$	$\bar{x}_2 = 555$
$s_1 = 24$	$s_2 = 18$

Is the claim justified at the 5% significance level?

8.37 Alloying is said to reduce the resistance in a standard type of electrical wire. Ten measurements on a standard wire yielded a mean resistance of 0.19 ohm and a standard deviation of 0.03. Ten independent measurements on an alloyed wire yielded a mean resistance of 0.11 ohm and a standard deviation of 0.02. Does alloying seem to reduce the mean resistance in the wire? Test at the 10% level of significance.

8.38 Two different machines, A and B, used for torsion tests of steel wire were tested on twelve pairs of different types of wire, with one member of each pair tested on each machine. The results (break angle measurements) were as follows:

Wire Type	1	2	3	4	5	6	7	8	9	10	11	12
Machine A	32	35	38	28	40	42	36	29	33	37	22	42
Machine B	30	34	39	26	37	42	35	30	30	32	20	41

(a) Is there evidence at the 5% significance level to suggest that machines A and B give different average readings?

(b) Is there evidence at the 5% significance level to suggest that machine B gives a lower average reading than machine A?

8.39 Six rockets, nominally with a range of 2500 meters, were stored for some time and then tested. The ranges found in the tests were 2490, 2510, 2360, 2410, 2300, and 2440. Another group of six rockets, of the same type, were stored for the same length of time but in a different manner. The ranges for these six were 2410, 2500, 2360, 2290, 2310, and 2340. Do the storage methods produce significantly different mean ranges? Use $\alpha = 0.05$, and assume range measurements to be approximately normally distributed with the same variance for each manner of storage.

8.40 The average depth of bedrock at two possible construction sites is to be compared by driving five piles at random locations within each site. The results, with depths in feet, are as follows:

Site A	Site B
$n_1 = 5$	$n_2 = 5$
$\bar{x}_1 = 142$	$\bar{x}_2 = 134$
$s_1 = 14$	$s_2 = 12$

Do the average depths of bedrock differ for the two sites at the 10% significance level? What assumptions are you making?

8.41 Gasoline mileage is to be compared for two automobiles, A and B, by testing each automobile on five brands of gasoline. Each car used one tank of each brand, with the following results (in miles per gallon):

Brand	Auto A	Auto B
1	28.3	29.2
2	27.4	28.4
3	29.1	28.2
4	28.7	28.0
5	29.4	29.6

Is there evidence to suggest a difference between true average mileage figures for the two automobiles? Use a 5% significance level.

8.42 The two drying methods for concrete introduced in Exercise 8.29 were used on seven different mixes, with each mix of concrete subjected to each drying method. The resulting strength test measurements (in psi) are given below. Is there evidence of a difference between average strengths for the two drying methods at the 10% significance level?

Mix	Method I	Method II
A	3160	3170
B	3240	3220
C	3190	3160
D	3520	3530
E	3480	3440
F	3220	3210
G	3120	3120

8.43 Two procedures for sintering copper are to be compared by testing each procedure on six different types of powder. The measurement of interest is the porosity (volume percentage due to voids) of each test specimen. The results of the tests were as follows:

Powder	Procedure I	Procedure II
1	21	23
2	27	26
3	18	21
4	22	24
5	26	25
6	19	16

Is there evidence of a difference between true average porosity measurements for the two procedures? Use $\alpha = 0.05$.

8.44 Refer to Exercise 8.29. Do the variances for strength measurements differ for the two drying methods? Test at the 10% significance level.

8.45 Refer to Exercise 8.39. Is there sufficient evidence to say that the variances among range measurements differ for the two storage methods? Use $\alpha = 0.10$.

8.46 Refer to Exercise 8.40. Does site A have significantly more variation among depth measurements than site B? Use $\alpha = 0.05$.

8.4 χ^2 TESTS ON FREQUENCY DATA

In Section 8.2 we saw how to construct tests of hypotheses concerning a binomial parameter p in the large-sample case. In that case the test statistic was based on Y, the *number* (or *frequency*) of successes in n trials of an experiment. The observation of interest there was a frequency count, rather than a continuous measurement such as a lifelength, reaction time, velocity, or weight. Now we want to take a detailed look at three types of situations in which hypothesis-testing problems arise with frequency, or count, data. All of the results in this section are approximations that work well only when samples are reasonably large. More about what we mean by "large" will be stated later.

8.4.1 Testing Parameters of the Multinomial Distribution

The first of the three situations involves testing a hypothesis concerning the parameters of a multinomial distribution. (See Section 3.8 for a description of this distribution and some of its properties.) Suppose we have n observations (trials) from a multinomial distribution with k possible outcomes per trial. Let X_i, $i = 1$, ..., k, denote the number of trials resulting in outcome i, with p_i denoting the probability that any one trial will result in outcome i. Recall that $E(X_i) = np_i$. Suppose we want to test the hypothesis that the p_i's have specified values, that is, $H_0: p_1 = p_{10}, p_2 = p_{20}, \ldots, p_k = p_{k0}$. The alternative will be the most general one that simply states "at least one equality fails to hold." To test the validity of this hypothesis, we can compare the observed count in cell i, X_i, with what we would expect that count to be if H_0 were true, namely $E(X_i) = np_{i0}$. So we will base our test statistic upon $X_i - E(X_i)$, $i = 1, \ldots, k$. We don't know if these differences are large or small unless we can standardize them in some way. It turns out that a good statistic to use squares these differences and divides them by $E(X_i)$, resulting in the test statistic

$$X^2 = \sum_{i=1}^{k} \frac{[X_i - E(X_i)]^2}{E(X_i)}$$

In repeated sampling from a multinomial distribution for which H_0 is true, X^2 will have approximately a $\chi^2(k-1)$ distribution. We would reject the null hypothesis for large values of $X^2(X^2 > \chi_\alpha^2(k-1))$ since X^2 will be large if there are large discrepancies between X_i and $E(X_i)$.

EXAMPLE 8.15

The ratio of number of items produced in a factory by three shifts, first, second, and third, is $4:2:1$, due primarily to the decreased number of employees on the later shifts. This means that 4/7 of the items produced come from the first shift, 2/7 from the second, and 1/7 from the third. It is hypothesized that the number of defectives produced should follow this same ratio. A sample of 50 defective items was tracked back to the shift that produced them, with the following results:

	Shift		
	1	2	3
Number of defectives	20	16	14

Test the hypothesis indicated above with $\alpha = 0.05$.

Solution   Using the X^2 statistic with X_i replaced by its observed value, we have, for H_0: $p_1 = 4/7$, $p_2 = 2/7$, $p_3 = 1/7$,

$$\sum_{i=1}^{3} \frac{[X_i - E(X_i)]^2}{E(X_i)} = \frac{[20 - 50(4/7)]^2}{50(4/7)} + \frac{[16 - 50(2/7)]^2}{50(2/7)}$$

$$+ \frac{[14 - 50(1/7)]^2}{50(1/7)} = 9.367$$

Now $\chi_{0.05}^2(2) = 5.991$, and since our observed X^2 is larger than this critical χ^2 value, we reject H_0.

This chi-square test is always two-tailed, and although the observed 14 is approximately twice the expected 50(1/7), the observed 20 is less than the expected 50(4/7), and both observed cells are out of line. This test does not decide which discrepancy is more serious. ∎

8.4.2 Testing Equality Among Binomial Parameters

The second situation in which a χ^2 statistic can be used on frequency data is in the testing of equality among binomial parameters for k separate populations. In this case k independent random samples are selected that result in k binomially distributed random variables $Y_1, \ldots, Y_k$, where Y_i is based on n_i trials with success probability p_i on each trial. The problem is to test the null hypothesis $H_0: p_1 = p_2 = \cdots = p_k$ against the alternative of at least one inequality. We now really have $2k$ cells to consider, as outlined in Figure 8.2.

FIGURE 8.2
Cells for k Binomial Observations

	Observation					
	1	2	3	$\cdots$	k	Total
Successes	y_1	y_2	y_3		y_k	y
Failures	$n_1 - y_1$	$n_2 - y_2$	$n_3 - y_3$		$n_k - y_k$	$n - y$
Total	n_1	n_2	n_3		n_k	n

$$n = \sum_{i=1}^{k} n_i \qquad y = \sum_{i=1}^{k} y_i$$

As in the case of testing multinomial parameters, the test statistic should be constructed by comparing the observed cell frequencies with the expected cell frequencies. However, the null hypothesis does not specify values for $p_1, p_2, \ldots, p_k$, and hence the expected cell frequencies must be estimated. The expected cell frequencies and their estimators will now be discussed.

Since each Y_i has a binomial distribution, we know from Chapter 3 that

$$E(Y_i) = n_i p_i$$

and

$$E(n_i - Y_i) = n_i - n_i p_i = n_i(1 - p_i)$$

Under the null hypothesis that $p_1 = p_2 = \cdots = p_k = p$, the common value of p is estimated by pooling the data from all k samples. The minimum variance unbiased estimator of p is

$$\frac{1}{n} \sum_{i=1}^{k} Y_i = \frac{Y}{n} = \frac{\text{total number of successes}}{\text{total sample size}}$$

The estimators of the expected cell frequencies are then taken to be

$$\hat{E}(Y_i) = n_i \left(\frac{Y}{n} \right)$$

and

$$\hat{E}(n_i - Y_i) = n_i \left(1 - \frac{Y}{n} \right) = n_i \left(\frac{n - Y}{n} \right)$$

In each case note that the estimate of an expected cell frequency is found by the following rule:

$$\text{estimated expected cell frequency} = \frac{(\text{column total})(\text{row total})}{\text{overall total}}$$

To test $H_0\colon p_1 = p_2 = \cdots = p_k = p$, we make use of the statistic

$$X^2 = \sum_{i=1}^{k} \left\{ \frac{[Y_i - \hat{E}(Y_i)]^2}{\hat{E}(Y_i)} + \frac{[(n_i - Y_i) - \hat{E}(n_i - Y_i)]^2}{\hat{E}(n_i - Y_i)} \right\}$$

This statistic has approximately a $\chi^2(k-1)$ distribution, so long as the sample sizes are reasonably large.

We illustrate with an example.

EXAMPLE 8.16
_____

A chemical company is experimenting with four different mixtures of a chemical designed to kill a certain species of insect. Independent random samples of 200 insects are subjected to one of the four chemicals, and the number of insects dead after one hour of exposure is counted. The results are as follows:

	1	2	3	4	Total
Dead	124 (141)	147 (141)	141 (141)	152 (141)	564
Not Dead	76 (59)	53 (59)	59 (59)	48 (59)	236
Total	200	200	200	200	

Estimated expected cell frequencies are shown in parentheses.

We want to test the hypothesis that the rate of kill is the same for all four mixtures. Test this claim at the 5% level of significance.

Solution
We are testing $H_0\colon p_1 = p_2 = p_3 = p_4 = p$, where p_i denotes the probability that an insect subjected to chemical i dies in the indicated length of time. Now the estimated cell frequencies under H_0 are found by

$$\frac{(n_1)(y)}{n} = \frac{1}{n}(\text{column 1 total})(\text{row 1 total})$$

$$= \frac{1}{800}(200)(564) = \frac{1}{4}(564) = 141$$

$$\frac{n_1(n-y)}{n} = \frac{1}{800}(200)(236) = \frac{1}{4}(236) = 59$$

and so on. Note that in this case all estimated frequencies in any one row are equal. Calculating an observed value for X^2, we get

$$\frac{[124 - 141]^2}{141} + \cdots + \frac{[48 - 59]^2}{59} = 10.72$$

Since $\chi^2_{0.05}(3) = 7.815$, we reject the hypothesis of equal kill rates for the four chemicals.   ■

8.4.3   Contingency Tables

The third type of experimental situation that makes use of a χ^2 test again makes use of the multinomial distribution. However, this time the cells arise from a double classification scheme. For example, employees could be classified according to sex and according to marital status, which results in four cells as follows:

	Male	Female
Married		
Unmarried		

A random sample of n employees would then give values of random frequencies to the four cells. In these two-way classifications a common question to ask is, "Does the row criterion depend upon the column criterion?" In other words "Is the row criterion *contingent* upon the column criterion?" In this context the two-way tables, like the one just illustrated, are referred to as *contingency tables*. We might want to know whether or not the marital status depends upon the sex of the employee, or if these two classification criteria seem to be independent of one another.

The null hypothesis in these two-way tables is that the row criterion and the column criterion are independent of each other. The alternative is simply that they are *not* independent. In looking at a 2×2 table, we find that an array such as

5	15
10	30

would support the hypothesis of independence since, within each column, the chance of being in row 1 is about 1 out of 3. That is, the chance of being in row 1 does *not* depend upon which column you happen to be in. On the other hand, an array like

5	30
10	15

would support the alternative hypothesis of dependence.

In general the cell frequencies and cell probabilities can be tabled as follows:

Row	Column 1	$\cdots$	j	c	
1	$X_{11}; p_{11}$			$X_{1c}; \; p_{1c}$	$X_{1.}$
$\vdots$	$\vdots$				
i		$\cdots$	$X_{ij}; p_{ij}$		$X_{i.}$
r	$X_{r1}; p_{r1}$			$X_{rc}; p_{rc}$	
	$X_{.1}$		$X_{.j}$		n

We will let $X_{i.}$ denote the frequency total for row i and $X_{.j}$ the total for column j. Also $p_{i.}$ will denote the probability of being in row i and $p_{.j}$ the probability of being in column j.

Under the null hypothesis of independence between rows and columns, $p_{ij} = p_{i.}p_{.j}$. Also $E(X_{ij}) = np_{ij} = np_{i.}p_{.j}$. Now $p_{i.}$ and $p_{.j}$ must be estimated, and the best estimators are

$$\hat{p}_{i.} = \frac{X_{i.}}{n}, \qquad \hat{p}_{.j} = \frac{X_{.j}}{n}$$

Thus the best estimator of an expected cell frequency is

$$\hat{E}(X_{ij}) = \left(n \frac{X_{i.}}{n}\right)\left(\frac{X_{.j}}{n}\right) = \frac{X_{i.}X_{.j}}{n}$$

Once again we see that the estimate of an expected cell frequency is given by

$$\frac{X_{i.}X_{.j}}{n} = \frac{1}{n} \text{(row } i \text{ total)(column } j \text{ total)}$$

The test statistic for testing the independence hypothesis is

$$X^2 = \sum_{ij} \frac{[X_{ij} - \hat{E}(X_{ij})]^2}{\hat{E}(X_{ij})}$$

which has approximately a $\chi^2(r-1)(c-1)$ distribution, where r is the number of rows and c the number of columns in the table.

EXAMPLE 8.17

A sample of 200 machined parts is selected from the one-week output of a machine shop that employs three machinists. The parts are inspected to determine whether or not they are defective, and are categorized according to which machinist did the work. The results are as follows:

	Machinist A	B	C	
Defective	10 (9.92)	8 (10.88)	14 (11.2)	32
Nondefective	52 (52.08)	60 (57.12)	56 (58.8)	168
	62	68	70	200

Is the defective/nondefective classification independent of machinist classification? Conduct a test at the 1% significance level.

Solution First we must find the estimated expected cell frequencies as follows:

$$\frac{x_1. x_{.1}}{n} = \frac{(32)(62)}{200} = 9.92$$

$$\frac{x_1. x_{.2}}{n} = \frac{(32)(68)}{200} = 10.88$$

$$\vdots$$

$$\frac{x_2. x_{.3}}{n} = \frac{(168)70}{200} = 58.8$$

(These numbers are shown in parentheses on the table.) Now the test statistic is observed to be

$$\frac{(10 - 9.22)^2}{9.22} + \cdots + \frac{(56 - 58.8)^2}{58.8} = 1.74$$

In this case the degrees of freedom are given by

$$(r - 1)(c - 1) = (2 - 1)(3 - 1) = 2$$

and

$$\chi^2_{0.01}(2) = 9.21$$

Thus we cannot reject the null hypothesis of independence between the machinists and defective/nondefective classifications. There is not sufficient evidence to say that the rate of defectives produced differs among machinists. ■

Recall that all of the above test statistics have only an approximate χ^2 distribution for large samples. A rough guideline for "large" is that each cell should contain an expected frequency count of at least two.

EXERCISES

8.47 Two types of defects, A and B, are frequently seen in the output of a certain manufacturing process. Each item can be classified into one of the four classes AB, $A\bar{B}$, $\bar{A}B$, $\bar{A}\bar{B}$, where $\bar{A}$ denotes the absence of the A-type defect. For 100 inspected items the following frequencies were observed:

AB 48

$A\bar{B}$ 18

$\bar{A}B$ 21

$\bar{A}\bar{B}$ 13

Test the hypothesis that the four categories, in the order listed, occur in the ratio $5:2:2:1$. (Use $\alpha = 0.05$.)

8.48   Vehicles can turn right, turn left, or continue straight ahead at a certain intersection. It is hypothesized that half of the vehicles entering this intersection will continue straight ahead. Of the other half, equal proportions will turn right and left. Fifty vehicles were observed to have the following behavior:

	Straight	Left Turn	Right Turn
Frequency	28	12	10

Test the stated hypothesis at the 10% level of significance.

8.49   A manufacturer of stereo amplifiers has three assembly lines. We want to test the hypothesis that the three lines do not differ with respect to the number of defectives produced. Independent samples of 30 amplifiers each are selected from the output of the lines, and the number of defectives is observed. The data are as follows:

Line	I	II	III
No. of Defectives	6	5	9
Sample Size	30	30	30

Conduct a test of the hypothesis given above at the 5% significance level.

8.50   Two inspectors are asked to rate independent samples of textiles from the same loom. Inspector A reports that 18 out of 25 samples fall in the top category while inspector B reports that 20 out of 25 samples merit the top category. Do the inspectors appear to differ in their assessments? Use $\alpha = 0.05$.

8.51   Two chemicals, A and B, are designed to protect pine trees from a certain disease. One hundred trees are sprayed with chemical A and one hundred are sprayed with chemical B. All trees are subjected to the disease, with the following results:

	A	B
Infected	20	16
Sample Size	100	100

Do the chemicals appear to differ in their ability to protect the trees? Test at the 1% significance level.

8.52   Refer to Exercise 8.47. Tese the hypothesis that the type A defects occur independently of the type B defects. Use $\alpha = 0.05$.

8.53   The *Sociological Quarterly* for Spring 1978 reports on a study of the relationship between athletic involvement and academic achievement for college students. The 852 students sampled were categorized according to amount of athletic involvement and grade-point averages at graduation, with results as follows:

		Athletic Involvement	
		None	1–3 Semesters	4 or More Semesters
GPA	Below Mean	290	94	42
	Above Mean	238	125	63
		528	219	105	852

(a) Do final grade-point averages appear to be independent of athletic involvement? (Use $\alpha = 0.05$.)

(b) For students with 4 or more semesters of athletic involvement, is the proportion with GPAs above the mean significantly different, at the 0.05 level, from the proportion with GPAs below the mean?

8.54 A new sick-leave policy is being introduced into a firm. A sample of employee opinions showed the following breakdowns by sex and opinion:

	Favor	Oppose	Undecided
Male	31	44	6
Female	42	36	8

Does the reaction to the new policy appear to be related to sex? Test at the 5% significance level.

8.55 A sample of 150 people are observed using one of four entrances to a commercial building. The data are as follows:

Entrance	1	2	3	4
No. of People	42	36	31	41

(a) Test the hypothesis that all four entrances are used equally often. Use $\alpha = 0.05$.

(b) Entrances 1 and 2 are on a subway level while 3 and 4 are on ground level. Test the hypothesis that subway and ground-level entrances are used equally often, again with $\alpha = 0.05$.

8.56 The National Maximum Speed Limit (NMSL) of 55 miles per hour was put in force in early 1974. (See D. B. Kamerud, *Transportation Research*, 17A, No. 1, p. 61 for details.) In experimental observation of trucks on rural secondary roads, 71% traveled under 55 mph in 1973 and 80% traveled under 55 mph in 1974. If 100 trucks were observed in each year, can we say that the proportion of trucks traveling under 55 mph was significantly different for the two years? Use $\alpha = 0.01$.

8.57 The use of aerosol dispensers has been somewhat controversial in recent years because of their possible effects on the ozone. According to T. Walsh (*Manufacturing Chemist*, 55, No. 12, 1984, pp. 43–45), a sample of 300 women in 1983 found 35% fearing the aerosol effects on ozone, whereas only 29% of a sample of 300 stated this fear in 1984. Is there a significant difference between the true proportions for the two years, at the 1% level of significance?

8.58   The same article referenced in Exercise 8.57 shows results of a four-year study, with 300 respondents per year. The percent of respondents stating that the aerosol dispensers did not spray properly were as follows:

1981	1982	1983	1984
73%	62%	65%	69%

Are there significant differences among these percentages at the 1% level of significance?

8.59   Industrial workers who are required to wear respirators were checked periodically to see that the equipment fit properly. From past records it was expected that 1.2% of those checked would fail (i.e., their equipment would not fit properly). The data from one series of checks were as follows:

Time After First Measurement (months)	Number of Workers Measured	Observed Failures	Expected Failures (1.2% of col. 2)
0– 6	28	1	0.34
6–12	92	3	1.12
12–18	270	1	3.29
18–24	702	9	8.56
24–30	649	10	7.92
30–36	134	0	1.63
>36	93	0	1.13
Total	1968	24	24

Can the assumption of the constant 1.2% failure rate be rejected at the 5% significance level?

8.60   In a study of 1000 major U.S. firms in 1976, 54% stated that quality of life was a major factor in locating corporate headquarters. A similar study of 1000 firms in 1981 showed 55% making this statement. (See B. E. Sullivan, *Transportation Research*, 18A, No. 2, 1984, p. 119.) Is this a significant change at the 5% significance level?

8.61   A Yankelovich, Skelly, and White poll carried out in December 1983 showed 60% of 1000 respondents saying that they "worry a lot" about the possibility of nuclear war. A similar poll conducted six months earlier showed only 50% responding this way. (See *Time*, January 2, 1984.) Is this a significant difference at the 1% level of significance?

8.62   A precedent-setting case in the use of statistics in courts of law was the U.S. Supreme Court case of Castaneda v. Partida [430 U.S. 482, 51 L. Ed. 2d 498 (1977)]. A county in Texas was accused of discrimination against Mexican-Americans in the selection of grand juries. Although 79% of the population of the county was Mexican-American, only 339 Mexican-Americans were selected among the 870 summoned grand jurors, over an 11-year period. Assuming a binomial model for the number of Mexican-Americans selected, one would expect to see $870(0.79) = 687.3$, or approximately 687, of them on the grand juries over the time in question. Do you think the Supreme Court was correct in stating that the small number of Mexican-Americans selected was probably not due to chance? (Conduct a statistical test at the 5% level of significance.)

8.5   GOODNESS-OF-FIT TESTS

Thus far our main emphasis has been that of choosing a probabilistic model for a measurement, or a set of measurements, taken on some natural phenomenon. We have talked about the basic probabilistic manipulations one can make, and discussed estimation of certain unknown parameters, as well as testing hypotheses concerning these unknown parameters. Up to this point, we have not discussed any rigorous way of checking to see whether the data we observe do, in fact, agree with the underlying probabilistic model we assumed for these data. That subject, called testing goodness of fit, will be discussed in this section. Goodness-of-fit testing is a broad subject, with entire textbooks devoted to it. We will merely skim the surface by presenting two techniques that have wide applicability and are fairly easy to calculate: the χ^2 goodness-of-fit test for discrete distributions and the Kolmogorov-Smirnov (K-S) test for continuous distributions.

Let Y denote a discrete random variable that can take on values $y_1, y_2, \ldots$. Under the null hypothesis H_0 we assume a probabilistic model for Y. That is, we assume $P(Y = y_i) = p_i$ where p_i may be completely specified or may be a function of other unknown parameters. For example, we could assume a Poisson model for Y, and thus

$$P(Y = y_i) = p_i = \lambda^{y_i} e^{-\lambda}/y_i!$$

In addition, we could assume λ is specified at λ_0, or we could assume nothing is known about λ, in which case λ would have to be estimated from the sample data. The alternative hypothesis is the general one that the model in H_0 does not fit.

The sample will consist of n independently observed values of the random variable Y. For this sample let F_i denote the number of times $Y = y_i$ is observed. Thus F_i is the frequency count for the value y_i. Under the null hypothesis that $P(Y = y_i) = p_i$, the expected frequency can be calculated as $E(F_i) = np_i$.

After the sample is taken, we will have an observed value of F_i in hand. At that point we can compare the observed value of F_i with what is expected under the null hypothesis. The test statistic will be based upon the differences $[F_i - E(F_i)]$, as in previously discussed χ^2 tests. In many cases one or more parameters will have to be estimated from sample data. In those cases $E(F_i)$ is also estimated and referred to as $\hat{E}(F_i)$, the estimated expected frequency.

For certain samples many of the frequency counts F_i may be quite small. (In fact many may be zero.) When this occurs, consecutive values of y_i are grouped together and the corresponding probabilities added, since

$$P(Y = y_i \text{ or } Y = y_j) = p_i + p_j$$

It follows that the frequencies F_i and F_j can be added, since $E(F_i + F_j) = n(p_i + p_j)$. As a handy rule of thumb, values of Y should be grouped so that the expected frequency count is at least two in every cell. After this grouping is completed, let k denote the number of cells obtained.

The test statistic for the null hypothesis that $P(Y = y_i) = p_i$, $i = 1, 2, \ldots$, is given by

$$X^2 = \sum_{i=1}^{k} \frac{[F_i - \hat{E}(F_i)]^2}{\hat{E}(F_i)}$$

For large n this statistic has approximately a χ^2 distribution with degrees of freedom given by

$$(k - 1) - (\text{the number of parameters estimated})$$

Thus if one parameter is estimated, the statistic would have $(k - 2)$ degrees of freedom.

We illustrate this procedure with a Poisson example.

EXAMPLE 8.18 _____

The number of accidents per week Y in a certain factory was checked for a random sample of $n = 50$ weeks with the following results:

y	Frequency
0	32
1	12
2	6
3 or more	0

Test the hypothesis that Y follows a Poisson distribution, with $\alpha = 0.05$.

Solution H_0 states that Y has a probability distribution given by

$$P(y_i) = \frac{\lambda^{y_i} e^{-\lambda}}{y_i!} \qquad y_i = 0, 1, 2, \ldots$$

Since λ is unknown, it must be estimated, and the best estimate is $\hat{\lambda} = \bar{Y}$. For the given data, $\bar{y} = [0(32) + 1(12) + 2(6)]/50 = 24/50 = 0.48$.

Using the guideline that each cell in a χ^2 test should have two or more observations, we must group the last two categories into one cell, giving us a total of three cells, indexed by the values $Y = 0$, $Y = 1$, and $Y \geqslant 2$. The corresponding probabilities for observing a value in these cells are

$$p_0 = P[Y = 0] = e^{-\lambda}$$

$$p_1 = P[Y = 1] = \lambda e^{-\lambda}$$

and

$$p_2 = P[Y \geqslant 2] = 1 - e^{-\lambda} - \lambda e^{-\lambda}$$

If F_0, F_1, F_2 denote the observed frequencies in the respective cells, then

$$E(F_i) = np_i$$

and

$$\hat{E}(F_i) = n\hat{p}_i$$

where $\hat{p}_i$ denotes p_i with λ replaced by $\hat{\lambda}$. The observed values of the estimated expected frequencies then become

$$50\,e^{-0.48} = 30.94$$

$$50(0.48)\,e^{-0.48} = 14.85$$

and

$$50[1 - e^{-0.48} - 0.48\,e^{-0.48}] = 4.21$$

The observed value of X^2 is then

$$\frac{[32 - 30.94]^2}{30.94} + \frac{[12 - 14.85]^2}{14.85} + \frac{[6 - 4.21]^2}{4.21} = 1.34$$

The test statistic has one degree of freedom since $k = 3$ and one parameter is estimated. Now $\chi^2_{0.05}(1) = 3.841$, and hence we do not have sufficient evidence to reject H_0. The data do appear to fit the Poisson model reasonably well. ∎

It is possible to construct χ^2 tests for goodness of fit to continuous distributions, but the procedure is a little more subjective since the continuous random variable does not provide natural cells into which the data can be grouped. For the continuous case we choose to use the Kolmogorov-Smirnov (K-S) statistic, which compares the empirical distribution function of a random sample with a hypothesized theoretical distribution function.

Before we define the K-S statistic, let us define the *empirical distribution function*. Suppose Y is a continuous random variable having distribution function $F(y)$. A random sample of n realizations of Y yields the observations $y_1, \ldots, y_n$. It is convenient to reorder these observed values from smallest to largest, and we denote the ordered y_i's by $y_{(1)} \leqslant y_{(2)} \leqslant \cdots \leqslant y_{(n)}$. That is, if $y_1 = 7$, $y_2 = 9$, and $y_3 = 3$, then $y_{(1)} = 3$, $y_{(2)} = 7$, and $y_{(3)} = 9$. Now the empirical distribution function is given by

$$F_n(y) = \text{fraction of the sample less than or equal to } y$$

$$= \begin{cases} \dfrac{(i-1)}{n} & \text{if } y_{(i-1)} \leqslant y < y_{(i)} \qquad i = 1, \ldots, n \\ 1 & \text{if } y \geqslant y_{(n)} \end{cases}$$

where we let $y_0 = -\infty$.

Suppose a continuous random variable Y is assumed, under the null hypothesis, to have a distribution function given by $F(y)$. The alternate hypothesis is that $F(y)$ is *not* the true distribution function for Y. After a random sample of n values of Y is observed, $F(y)$ should be "close" to $F_n(y)$ provided the null hypothesis is true. Our statistic, then, must measure the closeness of $F(y)$ to $F_n(y)$ over the whole range of y values. [See Figure 8.3 for a typical plot of $F(y)$ and $F_n(y)$.]

FIGURE 8.3
A Plot of $F_n(y)$ and $F(y)$

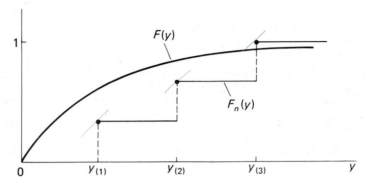

The K-S statistic D is based upon the *maximum* distance between $F(y)$ and $F_n(y)$. That is,

$$D = \max_y |F(y) - F_n(y)|$$

The null hypothesis is rejected if D is "too large."

Because $F(y)$ and $F_n(y)$ are nondecreasing and $F_n(y)$ is constant between sample observations, the maximum deviation between $F_n(y)$ and $F(y)$ will occur either at one of the observation points $y_1 \ldots y_n$ or immediately to the left of one of these points. To find the observed value of D, then, it is necessary to check only

$$D^+ = \max_{1 \le i \le n} \left[\frac{i}{n} - F(y_i) \right]$$

and

$$D^- = \max_{1 \le i \le n} \left[F(y_i) - \frac{i-1}{n} \right]$$

since

$$D = \max(D^+, D^-)$$

If H_0 hypothesizes the *form* of $F(y)$ but leaves some parameters unspecified, then these unknown parameters must be estimated from the sample data before the test can be carried out.

Values cutting off upper-tail areas of 0.15, 0.10, 0.05, 0.025, and 0.01 for a modified form of D are given by Stephens (1974) and are reproduced in Table 8.5 for three cases. These cases are for the null hypotheses of a fully specified $F(y)$, a normal $F(y)$ with unknown mean and variance, and an exponential $F(y)$ with unknown mean.

We illustrate the use of all three cases.

EXAMPLE 8.19 ——————————————————————————————

Consider the first ten observations of Table 4.1 on p. 125 to be a random sample from a continuous distribution. Test the hypothesis that these data are from an exponential distribution with mean 2, at the 0.05 significance level.

TABLE 8.5
Upper-Tail Percentage Points of Modified D

	Modified Form of D	Tail Area 0.15	0.10	0.05	0.025	0.01
Specified $F(y)$	$(D)(\sqrt{n} + 0.12 + 0.11/\sqrt{n})$	1.138	1.224	1.358	1.480	1.626
Normal $F(y)$ Unknown μ, σ^2	$(D)(\sqrt{n} - 0.01 + 0.85/\sqrt{n})$	0.775	0.819	0.895	0.955	1.035
Exponential $F(y)$ Unknown θ	$(D - 0.2/n)(\sqrt{n} + 0.26 + 0.5/\sqrt{n})$	0.926	0.990	1.094	1.190	1.308

Solution   We must order the ten observations and then find, for each $y_{(i)}$, the value of $F(y_i)$, where H_0 states that $F(y)$ is exponential with $\theta = 2$. Thus

$$F(y_i) = 1 - e^{-y_i/2}$$

The data and pertinent calculations are given in Table 8.6.

D^+ is the maximum value in column 6 and D^- the maximum in column 7. Thus $D^+ = 0.0886$ and $D^- = 0.2901$, giving $D = 0.2901$. To find the critical value from Table 8.5, we need to calculate

$$(D)\left(\sqrt{n} + 0.12 + \frac{0.11}{\sqrt{n}}\right) = (0.2901)(3.317) = 0.9623$$

At the $0.05 = \alpha$ significance level, the rejection region starts at 1.358. Thus we do not reject the null hypothesis. Notice we are saying that we cannot reject the exponential model ($\theta = 2$) as a plausible model for these data. This does not imply that the exponential is the *best* model for these data, as numerous other possible null hypotheses would not be rejected either.

TABLE 8.6
Data and Calculations for Example 8.19

i	$y_{(i)}$	$F(y_i)$	i/n	$(i-1)/n$	$i/n - F(y_i)$	$F(y_i) - (i-1)/n$
1	0.023	0.0114	0.1	0	0.0886	0.0114
2	0.406	0.1838	0.2	0.1	0.0162	0.0838
3	0.538	0.2359	0.3	0.2	0.0641	0.0359
4	1.267	0.4693	0.4	0.3	-0.0693	0.1693
5	2.343	0.6901	0.5	0.4	-0.1901	0.2901
6	2.563	0.7224	0.6	0.5	-0.1224	0.2224
7	3.334	0.8112	0.7	0.6	-0.1112	0.2112
8	3.491	0.8254	0.8	0.7	-0.0254	0.1254
9	5.088	0.9214	0.9	0.8	-0.0214	0.1214
10	5.587	0.9388	1.0	0.9	0.0612	0.0388

EXAMPLE 8.20

The following data are an ordered sample of observations on the amount of pressure (in pounds per square inch) needed to fracture a certain type of glass. Test the hypothesis that these data fit a Weibull distribution with $\gamma = 2$ but unknown θ.

04.90	08.60	11.42	15.46	19.19	20.69
40.29	41.19	43.55	44.62	53.56	77.61

Solution We see the form of the Weibull probability density function in Section 4.8. We also see there that when $\gamma = 2$, Y^2 will have an exponential distribution under the null hypothesis that Y has a Weibull distribution. Hence we want to test the hypothesis that Y^2 has an exponential distribution with unknown θ. The best estimator of θ is the average of the Y^2 values, observed to be 1446.93. $F(y)$ is then estimated to be

$$\hat{F}(y) = 1 - e^{-y/1446.93}$$

The information needed to complete the test is given in Table 8.7.

TABLE 8.7
Data and Calculations for Example 8.20

i	$u_i = y_{(i)}^2$	$\hat{F}(u_i)$	i/n	$i/n - \hat{F}(u_i)$	$\hat{F}(u_i) - (i-1)/n$
1	24.01	0.0165	0.0833	0.0668	0.0165
2	73.96	0.0498	0.1667	0.1169	-0.0335
3	130.42	0.0862	0.2500	0.1638	-0.0805
4	239.01	0.1523	0.3333	0.1810	-0.0977
5	538.26	0.2247	0.4167	0.1920	-0.1086
6	428.08	0.2561	0.5000	0.2439	-0.1606
7	1623.28	0.6743	0.5833	-0.0910	0.1743
8	1696.62	0.6904	0.6667	-0.0237	0.1071
9	1896.60	0.7304	0.7500	0.0196	0.0637
10	1990.94	0.7474	0.8333	0.0859	-0.0026
11	2868.67	0.8623	0.9167	0.0544	0.0290
12	6023.31	0.9844	1.0000	0.0156	0.0677

From the last two columns we see that $D = 0.2439$. Thus the modified D is

$$\left(D - \frac{0.2}{n}\right)\left(\sqrt{n} + 0.26 + \frac{0.5}{\sqrt{n}}\right) = (0.2439 - 0.0167)(3.8684)$$

$$= 0.8789$$

If we select $\alpha = 0.05$, we have from Table 8.5 that the critical value is 1.094. Thus we cannot reject the null hypothesis that the data fit a Weibull distribution with $\gamma = 2$. ∎

EXAMPLE 8.21

Soil-water flux measurements (in centimeters/day) were taken at 20 experimental plots in a field. The soil-water flux was measured in a draining soil profile in which steady-state soil-water flow conditions had been established. Theory and empirical evidence have suggested that the measurements should fit a lognormal distribution. The data recorded as y_i in Table 8.8 are the logarithms of the actual measurements. Test the hypothesis that the y_i come from a normal distribution. Use $\alpha = 0.05$.

TABLE 8.8
Data and Calculations for Example 8.21

i	$y_{(i)}$	$u_i = \dfrac{y_{(i)} - \bar{y}}{s}$	i/n	$F(u_i)$	$i/n - F(u_i)$	$F(u_i) - (i-1)/n$
1	0.3780	-2.0224	0.0500	0.0216	0.0284	0.0216
2	0.5090	-1.5215	0.1000	0.0641	0.0359	0.0141
3	0.6230	-1.0856	0.1500	0.1388	0.0112	0.0388
4	0.6860	-0.8448	0.2000	0.1991	0.0009	0.0491
5	0.7350	-0.6574	0.2500	0.2555	-0.0055	0.0555
6	0.7520	-0.5924	0.3000	0.2768	0.0232	0.0268
7	0.7580	-0.5695	0.3500	0.2845	0.0655	-0.0155
8	0.8690	-0.1451	0.4000	0.4423	-0.0423	0.0923
9	0.8890	-0.0686	0.4500	0.4726	-0.0226	0.0726
10	0.8890	-0.0686	0.5000	0.4726	0.0274	0.0226
11	0.8990	-0.0304	0.5500	0.4879	0.0621	-0.0121
12	0.9370	0.1149	0.6000	0.5457	0.0543	-0.0043
13	0.9820	0.2869	0.6500	0.6129	0.0371	0.0129
14	1.0220	0.4399	0.7000	0.6700	0.0300	0.0200
15	1.0370	0.4972	0.7500	0.6905	0.0595	-0.0095
16	1.0880	0.6922	0.8000	0.7556	0.0444	0.0056
17	1.1230	0.8260	0.8500	0.7956	0.0544	-0.0044
18	1.2060	1.1434	0.9000	0.8736	0.0264	0.0236
19	1.3340	1.6328	0.9500	0.9487	0.0013	0.0487
20	1.4230	1.9730	1.0000	0.9758	0.0242	0.0258

Solution

The basic idea of the K-S test in this case is to transform the observed data to new observations that should look like standard normal variates, if the null hypothesis is true. If random variable Y has a normal distribution with mean μ and variance σ^2, then $(Y - \mu)/\sigma$ will have a standard normal distribution. Since μ and σ are unknown, we will estimate them by $\bar{y}$ and s, respectively. Then we transform each y_i value to a u_i, where

$$u_i = \frac{y_i - \bar{y}}{s}$$

and test to see if the u_i's fit a standard normal distribution. The $F(u_i)$ values in Table 8.8 come from cumulative probabilities under a standard normal curve, and can be obtained from a table such as Table 4 of the Appendix, or a computer subroutine.

From Table 8.8 we see that $D = 0.0923$. Using Table 8.5, the modified D is

$$(D)\left(\sqrt{n} - 0.10 + \frac{0.85}{\sqrt{n}}\right) = (0.0923)(4.6522) = 0.4294$$

From Table 8.5 the critical value at the 5% significance level is 0.895, and thus we do not reject the null hypothesis. The data appear to fit the normal distribution. Figure 8.4 shows graphically how $F_n(u)$ and $F(u)$ compare. The figure is a plot of the empirical distribution function for $n = 20$ soil-water flux measurements and a normal distribution function. The statistics D^- and D^+ are illustrated in the inset.

FIGURE 8.4
F(u) and $F_n(u)$ for Example 8.21

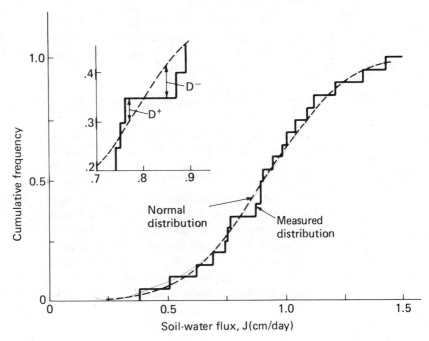

Example 8.21 has a more appropriate sample size for a goodness-of-fit test. Samples of size less than 20 do not allow for such discrimination among distributions; that is, many different distributions may all appear to fit equally well.

There are numerous goodness-of-fit statistics that could be considered in addition to the Kolmogorov-Smirnov. Some of these are the Cramér-von Mises, the Anderson-Darling, and the Watson statistics. The interested reader can look into these procedures by referring to the references given in the bibliography, particularly the 1974 paper by Stephens. We will, however, present one graphical technique that provides a rough check on goodness-of-fit to normality.

A graphical check for normality is derived from the fact that, if X is normally distributed with mean μ and standard deviation σ, then

$$\frac{X - \mu}{\sigma} = Z$$

where Z has a standard normal distribution, or

$$X = \mu + \sigma Z$$

Now suppose $X_{(1)} \leqslant X_{(2)} \leqslant \cdots \leqslant X_{(n)}$ represent an ordered random sample from a distribution function $F(x)$. If $F(x)$ is a normal distribution function with mean μ and variance σ^2, then

$$X_{(i)} = \mu + \sigma Z_{(i)}, \ i = 1, \ldots, n$$

and

$$E(X_{(i)}) = \mu + \sigma E(Z_{(i)})$$

If $x_{(1)} \leqslant x_{(2)} \leqslant \cdots \leqslant x_{(n)}$ represents an actual set of ordered sample observations from the distribution with distribution function $F(x)$, then the observed $x_{(i)}$ can be used to estimate $E(X_{(i)})$, for $i = 1, \ldots, n$. From the standard normal distribution $E(Z_{(i)})$ can be approximated by

$$E(Z_{(i)}) \approx \left[\left(\frac{i - \frac{3}{8}}{n + \frac{1}{4}} \right) \text{ percentage point of the standard normal distribution} \right]$$

The percentage points on the right side of the above expression are called the *normal scores*. Thus if $F(x)$ is a normal distribution,

$$x_{(i)} \approx \mu + \sigma(i\text{th normal score})$$

and the points (ith normal score, $x_{(i)}$), $i = 1, \ldots, n$, should fall nearly on a straight line with intercept μ and slope σ. If $F(x)$ is not normal, the points should deviate substantially from a straight line. Many computer packages compute the normal scores directly, so that the plot of $x_{(i)}$ versus normal scores is easily made.

Figure 8.5 shows the plot (from Minitab) of the sample data versus normal scores for the soil-flux measurements of Table 8.8. To see how normal scores are calculated, let's look at the second ordered pair ($i = 2$). Here the normal score is the

FIGURE 8.5
*Normal Scores Plot
of Data from
Table 8.8*

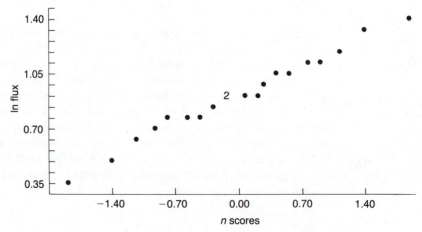

$$\frac{i - \frac{3}{8}}{n + \frac{1}{4}} = \frac{2 - \frac{3}{8}}{20 + \frac{1}{4}} = 0.08$$

percentage point of the standard normal distribution, or the z value that cuts off an area of 0.08 in the lower tail. From Table 4 of the Appendix (or your computer) we can find this value to be around -1.40. Thus the second smallest point on Figure 8.5 is $(-1.40, 0.509)$. The other 19 points are determined in a similar way.

The points in Figure 8.5 do fall rather close to a straight line, which substantiates the claim that the data come from a normal distribution. The intercept (ln flux measurement at normal score of 0) is approximately 0.90, which is close to the sample mean for the 20 data points of 0.907. The slope of the line can be approximated by looking at two points such as $(1.40, 1.33)$ and $(-1.40, 0.50)$, which gives a slope of

$$\frac{1.33 - 0.50}{1.40 - (-1.40)} = 0.296$$

This is close to the calculated sample standard deviation of 0.262.

To see that normality is required in order to get the straight-line fit, look at Figure 8.6. This shows the data from Table 8.6 plotted against normal scores. The plot shows large deviations from a straight line, especially at the lower end, and these data do not appear to come from a normal distribution. (We have already shown that these data fit an exponential model.) Such departures from a straight line often appear in plots of data versus normal scores, readily signaling that the data do *not* come from a normal distribution.

FIGURE 8.6
Normal Scores Plot of Data from Table 8.6

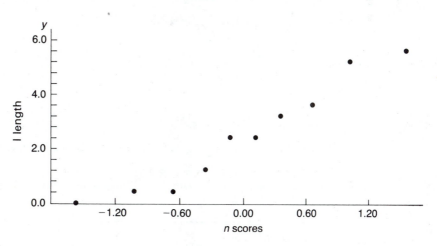

EXERCISES

8.63 For fabric coming off a certain loom, the number of defects per square yard is counted on 50 sample specimens, each one square yard in size. The results are as follows:

Number of Defects	Frequency of Observation
0	0
1	3
2	5
3	10
4	14
5	8
6 or more	10

Test the hypothesis that the data come from a Poisson distribution. Use $\alpha = 0.05$.

8.64   The following data show the frequency counts for 400 observations on the number of bacterial colonies within the field of a microscope, using samples of milkfilm. (Source: Bliss and Owens, *Biometrics*, 9, 1953.)

Number of Colonies per Field	Frequency of Observations
0	56
1	104
2	80
3	62
4	42
5	27
6	9
7	9
8	5
9	3
10	2
11	0
19	1
	400

Test the hypothesis that the data fit the Poisson distribution. Use $\alpha = 0.05$.

8.65   The number of accidents experienced by machinists in a certain industry were observed for a certain period of time with the following results. (Source: Bliss and Fisher, *Biometrics*, 9, 1953.)

Accidents per Machinist	0	1	2	3	4	5	6	7	8
Frequency of Observation (Number of Machinists)	296	74	26	8	4	4	1	0	1

At the 5% level of significance, test the hypothesis that the data came from a Poisson distribution.

8.66 Counts on the number of items per cluster (or colony or group) must necessarily be greater than or equal to one. Thus the Poisson distribution does not generally fit these kinds of counts. For modeling counts on phenomena such as number of bacteria per colony, number of people per household, and number of animals per litter, the *logarithmic series* distribution often proves useful. This discrete distribution has probability function given by

$$p(y) = -\frac{1}{\ln(1-\alpha)} \frac{\alpha^y}{y} \qquad y = 1, 2, 3, \ldots \qquad 0 < \alpha < 1$$

where α is an unknown parameter.

   (a) Show that the maximum likelihood estimator, $\hat{\alpha}$, of α satisfies the equation

$$\bar{y} = \frac{\hat{\alpha}}{-(1-\hat{\alpha})\ln(1-\hat{\alpha})}$$

   where $\bar{y}$ is the mean of the sampled observations $y_1, \ldots, y_n$.

   (b) The following data give frequencies of observation for counts on the number of bacteria per colony, for a certain type of soil bacteria. (Source: Bliss and Fisher, *Biometrics*, 9, 1953.)

Bacteria per Colony	1	2	3	4	5	6	7 or more
Number of Colonies Observed	359	146	57	41	26	17	29

Test the hypothesis that these data fit a logarithmic series distribution. Use $\alpha = 0.05$. (Note that $\bar{y}$ must be approximated because we do not have exact information on counts greater than six. Treat the "7 or more" as "exactly 7.")

8.67 The following data are observed LC50 values on copper in a certain species of fish (measurements in parts per million):

0.075   0.10   0.23   0.46   0.10   0.15   1.30   0.29

0.31   0.32   0.33   0.54   0.85   1.90   9.00

It has been hypothesized that the natural logarithms of these data fit the normal distribution. Check this claim at the 5% significance level. Also, make a normal scores plot if a computer is available.

8.68 The time (in seconds) between vehicle arrivals at a certain intersection was measured for a certain time period with the following results:

9.0   10.1   10.2   9.3   9.5   9.8   14.2   16.1

8.9   10.5   10.0   18.1   10.6   16.8   13.6   11.1

   (a) Test the hypothesis that these data come from an exponential distribution. Use $\alpha = 0.05$.

   (b) Test the hypothesis that these data come from an exponential distribution with a mean of 12 seconds. Use $\alpha = 0.05$.

8.69 The weights of copper grains are hypothesized to follow a lognormal distribution, which implies that the logarithms of the weight measurements should follow a normal distribution. Twenty weight measurements, in 10^{-4} gram, are as follows:

$$2.0, \quad 3.0, \quad 3.1, \quad 4.3 \quad 4.4, \quad 4.8, \quad 4.9, \quad 5.1, \quad 5.4, \quad 5.7$$

$$6.1, \quad 6.6, \quad 7.3, \quad 7.6, \quad 8.3, \quad 9.1, \quad 11.2, \quad 14.4, \quad 16.7, \quad 19.8$$

Test the hypothesis stated above at the 5% significance level.

8.70   Fatigue life, in hours, for ten bearings of a certain type were as follows:

$$152.7, \quad 172.0, \quad 172.5, \quad 173.3, \quad 193.0$$

$$204.7, \quad 216.5, \quad 234.9, \quad 262.6, \quad 422.6$$

(Source: A. C. Cohen, et al., *Journal of Quality Technology*, 16, No. 3, 1984, p. 165.) Test the hypothesis, at the 5% level, that these data follow a Weibull distribution with $\gamma = 2$.

8.71   The times to failure of eight turbine blades in jet engines, in 10^3 hours, were as follows:

$$3.2, \quad 4.7, \quad 1.8, \quad 2.4, \quad 3.9, \quad 2.8, \quad 4.4, \quad 3.6$$

(Source: Pratt & Whitney Aircraft, PWA 3001, 1967.) Do the data appear to fit a Weibull distribution with $\gamma = 3$? Use $\alpha = 0.025$.

8.72   Records were kept on the time of successive failures of the air conditioning system of a Boeing 720 jet airplane. If the air conditioning system has a constant failure rate, then the intervals between successive failures must have an exponential distribution. The observed intervals, in hours, between successive failures are as follows:

$$23, \quad 261, \quad 87, \quad 7, \quad 120, \quad 14, \quad 62, \quad 47, \quad 225, \quad 71$$

$$246, \quad 21, \quad 42, \quad 20, \quad 5, \quad 12, \quad 120, \quad 11, \quad 3, \quad 14$$

$$71, \quad 11, \quad 14, \quad 11, \quad 16, \quad 90, \quad 1, \quad 16, \quad 52, \quad 95$$

(Source: F. Proschan, *Technometrics*, 5, No. 3, 1963, p. 376.) Do these data seem to follow an exponential distribution at the 5% significance level? If possible, construct a normal scores plot for these data. Does the plot follow a straight line?

8.6   CONCLUSION

For the simple cases that we have considered, hypothesis-testing problems parallel estimation problems in that similar statistics are used. Some exceptions are noted, such as the χ^2 tests on frequency data and goodness-of-fit tests. These latter tests are not motivated by confidence intervals.

This might prompt the question of how one chooses appropriate test statistics. We did not discuss a general method of choosing a test statistic as we did for choosing an estimator (maximum likelihood principle). General methods of choosing text statistics with good properties do, however, exist. The reader interested in the theory of hypothesis testing should consult a text on mathematical statistics.

The tests given in this chapter have some intuitive appeal, as well as good theoretical properties.

SUPPLEMENTARY EXERCISES

8.73 An important quality characteristic of automobile batteries is their weight, since that characteristic is sensitive to the amount of lead in the battery plates. A certain brand of heavy-duty battery has a weight specification of 69 pounds. Forty batteries selected from recent production have an average weight of 64.3 pounds and a standard deviation of 1.2 pounds. Would you suspect that something has gone wrong with the production process? Why?

8.74 Quality of automobile batteries can also be measured by cutting them apart and measuring plate thicknesses. A certain brand of batteries is specified to have positive plate thicknesses of 120 thousandths of an inch and negative plate thicknesses of 100 thousandths of an inch. A sample of batteries from recent production was cut apart and plate thicknesses were measured. Sixteen positive plate thickness measurements averaged 111.6 with a standard deviation of 2.5. Nine negative plate thickness measurements averaged 99.44 with a standard deviation of 3.7. Do the positive plates appear to meet the thickness specification? Do the negative plates appear to meet the thickness specification?

8.75 The mean breaking strength of cotton threads must be at least 215 grams for the thread to be used in a certain garment. A random sample of 50 measurements on a certain thread gave a mean breaking strength of 210 grams and a standard deviation of 18 grams. Should this thread be used on the garments?

8.76 Prospective employees of an engineering firm are told that engineers in the firm work at least 45 hours per week, on the average. A random sample of 40 engineers in the firm showed that, for a particular week, they averaged 44 hours of work with a standard deviation of 3 hours. Are the prospective employees being told the truth?

8.77 Company A claims that no more than 8% of the resistors it produces fail to meet the tolerance specifications. Tests on 100 randomly selected resistors yielded 12 that failed to meet the specifications. Is the company's claim valid? Test at the 10% significance level.

8.78 An approved standard states that the average LC50 for DDT should be 10 parts per million for a certain species of fish. Twelve experiments produced LC50 measurements of

$$16, \quad 5, \quad 21, \quad 19, \quad 10, \quad 5, \quad 8, \quad 2, \quad 7, \quad 2, \quad 4, \quad 9$$

Do the data cast doubt on the standard at the 5% significance level?

8.79 A production process is supposed to be producing 10-ohm resistors. Fifteen randomly selected resistors showed a sample mean of 9.8 ohms and a sample standard deviation of 0.5 ohm. Are the specifications of the process being met? Should one use a two-tailed test here?

8.80 For the resistors of Exercise 8.79 the claim is made that the standard deviation of the resistors produced will not exceed 0.4 ohm. Can the claim be refuted at the 10% significance level?

8.81 The abrasive resistance of rubber is increased by adding a silica filler and a coupling agent to chemically bond the filler to the rubber polymer chains. Fifty specimens of rubber made with a type I coupling agent gave a mean resistance measure of 92, the variance of the measurements being 20. Forty specimens of rubber made with a type II coupling agent gave a mean of 98 and a variance of 30 on resistance measurements.

Is there sufficient evidence to say that the mean resistance to abrasion differs for the two coupling agents at the 5% level of significance?

8.82 Two different types of coating for pipes are to be compared with respect to their ability to aid in resistance to corrosion. The amount of corrosion on a pipe specimen is quantified by measuring the maximum pit depth. For coating A, 35 specimens showed an average maximum pit depth of 0.18 cm. The standard deviation of these maximum pit depths was 0.02 cm. For coating B the maximum pit depths in 30 specimens had a mean of 0.21 cm and a standard deviation of 0.03 cm. Do the mean pit depths appear to differ for the two types of coating? Use $\alpha = 0.05$.

8.83 Bacteria in water samples are sometimes difficult to count, but their presence can easily be detected by culturing. In fifty independently selected water samples from a certain lake, 43 contained certain harmful bacteria. After adding a chemical to the lake water, another fifty water samples showed only 22 with the harmful bacteria. Does the addition of the chemical significantly reduce the proportion of samples containing the harmful bacteria? Use $\alpha = 0.025$.

8.84 A large firm made up of several companies has instituted a new quality-control inspection policy. Among 30 artisans sampled from Company A, only 5 objected to the new policy. Among 35 artisans sampled in Company B, 10 objected to the policy. Is there a significant difference, at the 5% level, between the proportions voicing *no* objection to the new policy?

8.85 *Research Quarterly*, May 1979, reports on a study of impulses applied to the ball by tennis rackets of various construction. Three measurements on ball impulses were taken on each type of racket. For a Classic (wood) racket the mean was 2.41 and the standard deviation was 0.02. For a Yamaha (graphite) racket the mean was 2.22 and the standard deviation was 0.07. Is there evidence of a significant difference between the mean impulses for the two rackets at the 5% significance level?

8.86 For the impulse measurements of Exercise 8.85 is there sufficient evidence to say that the graphite racket gives more viable results than the wood racket? Use $\alpha = 0.05$.

8.87 An interesting and practical use of the χ^2 test comes about in the testing for segregation of species of plants or animals. Suppose that two species of plants, say A and B, are growing on a test plot. To assess whether or not the species tend to segregate, n plants are randomly sampled from the plot, and the species of each sampled plant *and* the species of its *nearest* neighbor are recorded. The data are then arranged on a table as follows:

		Nearest Neighbor	
		A	B
	A	a	b
Sampled Plant			
	B	c	d
			n

If a and d are large relative to b and c, we would be inclined to say that the species tend to segregate. (Most of A's neighbors are of type A, and most of B's neighbors are of type B.) If b and c are large compared to a and d, we would say that the species tend to be overly mixed. In either of these cases (segregation or overmixing) a χ^2 test should yield a large value, and the hypothesis of random mixing would be rejected. For each of the following cases, test the hypothesis of random mixing (or, equivalently, the hypothesis that the species of a sampled plant is independent of the species of its nearest neighbor). Use $\alpha = 0.05$ in each case.

(a)　$a = 20, b = 4, c = 8, d = 18$　　(b)　$a = 4, b = 20, c = 18, d = 8$

(c)　$a = 20, b = 4, c = 18, d = 8$

9

Simple Regression

ABOUT THIS CHAPTER

In previous chapters we have estimated and tested hypotheses concerning a population mean by making use of a set of measurements on a single variable from that population. Frequently, however, the mean of one variable is dependent upon one or more related variables. For example, the average amount of energy required to heat houses of a certain size depends upon the air temperature during the days of the study. In this chapter we begin to build models in which the mean of one variable can be written as a linear function of another variable. This is sometimes referred to as the *regression* of one variable upon another.

CONTENTS

9.1 INTRODUCTION

Many engineering applications involve *modeling* the relationships among sets of variables. For example, a chemical engineer may want to model the yield of a chemical process as a function of the temperature and pressure at which the reactions take place. An electrical engineer may be interested in modeling the daily peak load of a power plant as a function of the time of year and number of customers. An environmental engineer may want to model the dissolved oxygen of samples from a large lake as a function of the algal and nitrogen content of the sample.

One method of modeling the relationship between variables is called *regression analysis*. In this chapter we discuss this important topic.

9.2 PROBABILISTIC MODELS

A problem facing every power plant is the estimation of the daily peak power load. Suppose we wanted to model the peak power load as a function of the maximum temperature for the day. The first question to be answered is this: "Does an exact relationship exist between these variables?" That is, is it possible to predict the exact peak load if the maximum temperature is known? We might see that this is not possible for several reasons. Peak load depends on other factors beside the maximum temperature. For example, the number of customers the plant services, the geographical location of the plant, the capacity of the plant, and the average temperature over the day, all probably affect the daily peak power load. However, even if all these factors were included in a model, it is still unlikely that we would be able to predict the peak load *exactly*. There will almost certainly be some variation in peak load due strictly to *random phenomena* that cannot be modeled or explained.

If we were to construct a model that hypothesized an exact relationship between variables, it would be called a *deterministic model*. For example, if we believe that y, the peak load (in megawatts), will be exactly five times x, the maximum temperature (degrees Fahrenheit), we write

$$y = 5x$$

This represents a deterministic relationship between the variables y and x. It implies that y can always be determined exactly when the value of x is known. There is no allowance for error in this prediction.

The primary usefulness of deterministic models is in the description of physical laws. For example Ohm's law describes the relationship between current and resistance in a deterministic manner. Newton's laws of motion are other examples of deterministic models. However, it should be noted that in all examples, the laws hold precisely only under ideal conditions. Laboratory experiments rarely reproduce these laws exactly. There will usually be random error introduced by the experiment that will cause the laws to provide only approximations to reality.

Returning to the peak power load example, if we believe there will be unexplained variation in peak power load, we will discard the deterministic model and use a model that accounts for this *random error*. This *probabilistic model* includes both a deterministic component and a random error component. For example, if we hypothesize that the peak load Y, now a random variable, is related to the maximum temperature x by

$$Y = 5x + \text{random error}$$

we are hypothesizing a *probabilistic relationship* between Y and x. Note that the deterministic component of this probabilistic model is $5x$.

Figure 9.1(a) shows possible peak loads for five different values of x when the model is deterministic. All the peak loads must fall exactly on the line because the deterministic model leaves no room for error.

FIGURE 9.1
Deterministic and Probabilistic Models

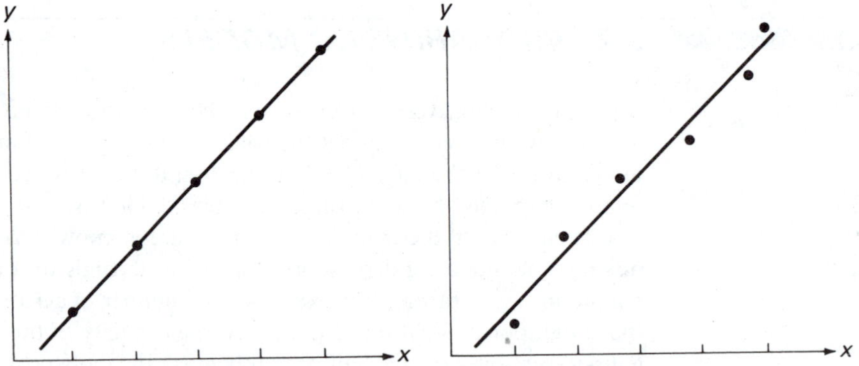

EXTRA RELATIONSHIP Between VARS

(a) Deterministic model
$y = 5x$

(b) Probabilistic model
$y = 5x + \text{random error}$

Figure 9.1(b) shows a possible set of responses for the same values of x when we are using a probabilistic model. Note that the deterministic part of the model (the straight line itself) is the same. Now, however, the inclusion of a random error component allows the peak loads to vary from this line. Since we believe that the peak load will vary randomly for a given value of x, the probabilistic model provides a more realistic model for Y than does the deterministic model.

GENERAL FORM OF PROBABILISTIC MODELS

$Y = $ deterministic component + random error

where Y is the random variable to be predicted. We will always assume that the mean value of the random error equals zero. This is equivalent to assuming that the mean value of Y, $E(Y)$, equals the deterministic component of the model:

$E(Y) = $ deterministic component

In the previous two chapters we discussed the simplest form of a probabilistic model. We showed how to make inferences about the mean of Y, $E(Y)$, when

$$E(Y) = \mu$$

where μ is a constant. However, we realized that this did not imply that Y would equal μ exactly, but instead would be equal to μ plus or minus a random error. In particular if we assume that Y is normally distributed with mean μ and variance σ^2, then we may write the probabilistic model

$$Y = \mu + \varepsilon$$

where the random component ε (epsilon) is normally distributed with mean 0 and variance σ^2. Some observations generated by this model might appear as shown in Figure 9.2.

FIGURE 9.2
The Probabilistic Model $Y = \mu + \varepsilon$

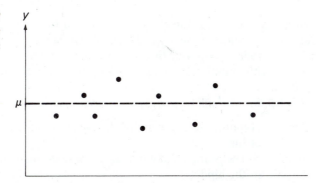

The purpose of this chapter is to generalize this model to allow $E(Y)$ to be a function of other variables. For example, if we want to model daily peak power load Y as a function of the maximum temperature x for the day, we might hypothesize that the mean of Y is a straight-line function of x, as shown in the following box.

THE STRAIGHT-LINE PROBABILISTIC MODEL

$$Y = \beta_0 + \beta_1 x + \varepsilon$$

where $Y =$ dependent variable (variable to be modeled)
            $x =$ independent[1] variable (variable used as a predictor of Y)
            $\varepsilon =$ random error component
            $\beta_0 = y$ intercept of the line, that is, point at which the line
                     intercepts or cuts through the y axis (see Figure 9.3)
            $\beta_1 =$ slope of the line, that is, amount of increase (or decrease)
                     in the mean of Y for every 1 unit increase in x
                     (see Figure 9.3)

[1]The word *independent* should not be interpreted in a probabilistic sense, as defined in Chapter 2. The phrase *independent variable* is used in regression analysis to refer to a predictor variable for the response Y.

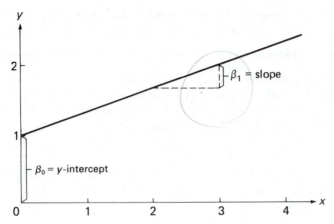

FIGURE 9.3
Straight-Line Probabilistic Model

Note that we use the Greek symbols β_0 amd β_1 to represent the y intercept and slope of the model, as we used the Greek symbol μ to represent the constant mean in the model $Y = \mu + \varepsilon$. In each case these symbols represent population parameters with numerical values that will need to be estimated using sample data.

It is helpful to think of regression analysis as a five-step procedure:

Step 1. Hypothesize the form for the mean $E(Y)$ (deterministic component of the model).

Step 2. Collect sample data and use them to estimate unknown parameters in the model.

Step 3. Specify the probability distribution of ε, the random error component, and estimate any unknown parameters of this distribution.

Step 4. Statistically check the adequacy of the model.

Step 5. When satisfied with the model's adequacy, use it for prediction, estimation, and so on.

In this chapter we simplify the more difficult step 1 and introduce the concepts of regression analysis via the straight-line model. In Chapter 10 we discuss how to build more complex models.

9.3 FITTING THE MODEL: THE LEAST-SQUARES APPROACH

Suppose we want to estimate the mean daily peak load for a power plant given the sample of peak loads for ten days in Table 9.1. We hypothesize the model

$$Y = \mu + \varepsilon$$

and wish to use the sample data to estimate μ. One method, the *least-squares approach*, chooses the estimator that minimizes the sum of squared errors (SSE).

TABLE 9.1
Sample of 10 Days Peak Power
Load

Day	Peak Load (y_i)
1	214
2	152
3	156
4	129
5	254
6	266
7	210
8	204
9	213
10	150

That is, we choose the estimator $\hat{\mu}$ so that

$$\text{SSE} = \sum_{i=1}^{n} (y_i - \hat{\mu})^2$$

is minimized. The form of this estimator can be obtained by differentiating SSE with respect to $\hat{\mu}$, setting it equal to zero, and solving for $\hat{\mu}$. Thus

$$\frac{d(\text{SSE})}{d\hat{\mu}} = -2 \sum_{i=1}^{n} (y_i - \hat{\mu}) = 0$$

Simplifying,

$$-2 \sum_{i=1}^{n} y_i + 2n\hat{\mu} = 0$$

Solving for $\hat{\mu}$,

$$\hat{\mu} = \frac{\sum_{i=1}^{n} y_i}{n} = \bar{y}$$

Thus the sample mean $\bar{Y}$ is the estimator that minimizes the sum of squared errors, and is called the *least-squares estimator* of μ.

For the peak power load data in Table 9.1 we calculate

$$\bar{y} = \frac{1948}{10} = 194.8$$

and

$$\text{SSE} = \sum_{i=1}^{n} (y_i - \bar{y})^2 = 19{,}263.6$$

We know that no other estimate of μ will yield as small an SSE as this because $\bar{Y}$ is the *least-squares* estimator.

Now suppose we decide that the mean peak power load can be modeled as a function of the maximum temperature x for the day. Specifically we model the mean peak power load $E(Y)$ as a straight-line function of x. Thus we record the maximum temperature for the days in Table 9.1, and obtain the data in Table 9.2. A plot of this data, called a *scattergram*, is shown in Figure 9.4.

TABLE 9.2
Data for Peak Power Load y and Maximum Temperature x for Ten Days

Day	Maximum Temperature x	Peak Power Load y
1	95	214
2	82	152
3	90	156
4	81	129
5	99	254
6	100	266
7	93	210
8	95	204
9	93	213
10	87	150

FIGURE 9.4
Scattergram for
Data in Table 9.2

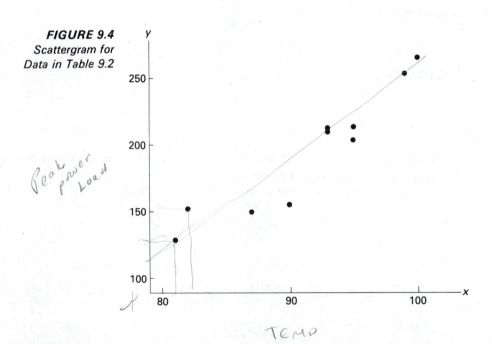

We hypothesize the straight-line probabilistic model

$$Y = \beta_0 + \beta_1 x + \varepsilon$$

and want to use the sample data to estimate the y intercept β_0 and the slope β_1. We use the same principle to estimate β_0 and β_1 in the straight-line model that we used to estimate μ in the constant mean model: the least-squares approach. Thus we choose the estimate

$$\hat{y} = \hat{\beta}_0 + \hat{\beta}_1 x$$

so that

$$\text{SSE} = \sum_{i=1}^{n} (y_i - \hat{y})^2 = \sum_{i=1}^{n} (y_i - \hat{\beta}_0 - \hat{\beta}_1 x_i)^2$$

is minimized. We differentiate SSE with respect to $\hat{\beta}_0$ and $\hat{\beta}_1$, set the results equal to zero, and then solve for $\hat{\beta}_0$ and $\hat{\beta}_1$.

$$\frac{\partial(\text{SSE})}{\partial \hat{\beta}_0} = -2 \sum_{i=1}^{n} (y_i - \hat{\beta}_0 - \hat{\beta}_1 x_i) = 0$$

$$\frac{\partial(\text{SSE})}{\partial \hat{\beta}_1} = -2 \sum_{i=1}^{n} x_i(y_i - \hat{\beta}_0 - \hat{\beta}_1 x_i) = 0$$

and the solution is

$$\hat{\beta}_1 = \frac{\sum_{i=1}^{n} (x_i - \bar{x})(y_i - \bar{y})}{\sum_{i=j}^{n} (x_i - \bar{x})^2} = \frac{\text{SS}_{xy}}{\text{SS}_{xx}}$$

$$\hat{\beta}_0 = \bar{y} - \hat{\beta}_1 \bar{x}$$

For the data in Table 9.3 we find

$$\text{SS}_{xy} = \sum_{i=1}^{n} (x_i - \bar{x})(y_i - \bar{y}) = \sum_{i=1}^{n} x_i y_i - \frac{\left(\sum_{i=1}^{n} x_i\right)\left(\sum_{i=1}^{n} y_i\right)}{n}$$

$$= 180{,}798 - \frac{(915)(1948)}{10} = 2556$$

$$\text{SS}_{xx} = \sum_{i=1}^{n} (x_i - \bar{x})^2 = \sum_{i=1}^{n} x_i^2 - \frac{\left(\sum_{i=1}^{n} x_i\right)^2}{n}$$

$$= 84{,}103 - \frac{(915)^2}{10} = 380.5$$

and

$$\bar{x} = \frac{915}{10} = 91.5$$

$$\bar{y} = 194.8$$

TABLE 9.3
Data, Predicted Values, and Errors for the Peak Power Load Data

x	y	$\hat{y} = \hat{\beta}_0 + \hat{\beta}_1 x$	$(y - \hat{y})$	$(y - \hat{y})^2$
95	214	218.31	−4.31	18.58
82	152	130.98	21.02	441.84
90	156	184.72	−28.72	824.84
81	129	124.27	4.73	22.37
99	254	245.18	8.82	77.79
100	266	251.90	14.10	198.81
93	210	204.88	5.12	26.21
95	204	218.31	−14.31	204.78
93	213	204.88	8.12	65.93
87	150	164.57	−14.57	212.28
915	1948			SSE = 2093.43

Then the least-squares estimates are

$$\hat{\beta}_1 = \frac{SS_{xy}}{SS_{xx}} = \frac{2556}{380.5} = 6.7175$$

$$\hat{\beta}_0 = \bar{y} - \hat{\beta}_1 \bar{x} = 194.8 - (6.7175)(91.5)$$

$$= -419.85$$

The least-squares line is therefore

$$\hat{y} = -419.85 + 6.7175x$$

as shown in Figure 9.5. Note that the errors are the vertical distances between the observed points and the prediction line, $(y_i - \hat{y}_i)$. The predicted values $\hat{y}_i$, the error

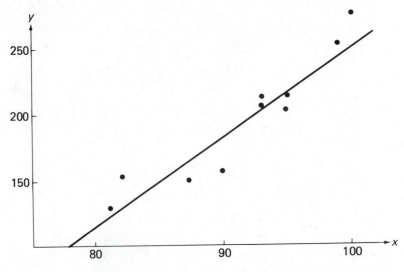

FIGURE 9.5
Scattergram and Least-Squares Line for Data in Table 9.2

$(y_i - \hat{y}_i)$, and the squared error $(y_i - \hat{y}_i)^2$ are shown in Table 9.3. You can see that the sum of squared errors, SSE, is 2093.43. We know that no other straight line will yield as small an SSE as this one.

To summarize, we have defined the best-fitting straight line to be the one that satisfies the least-squares criterion; that is, the sum of squared errors will be smaller than for any other straight-line model. This line is called the *least-squares line*, and its equation is called the *least-squares prediction equation*. The prediction equation can be determined by substituting the sample values of x and y into the formulas given in the box. We will discuss the adequacy of the model later.

FORMULAS FOR THE LEAST-SQUARES LINE

Slope: $\hat{\beta}_1 = \dfrac{SS_{xy}}{SS_{xx}}$

y intercept: $\hat{\beta}_0 = \bar{y} - \hat{\beta}_1\bar{x}$

where

$$SS_{xy} = \sum_{i=1}^{n} x_i y_i - \frac{\left(\sum_{i=1}^{n} x_i\right)\left(\sum_{i=1}^{n} y_i\right)}{n}$$

$$SS_{xx} = \sum_{i=1}^{n} x_i^2 - \frac{\left(\sum_{i=1}^{n} x_i\right)^2}{n}$$

n = number of pairs of observations (sample size)

EXERCISES

9.1 Use the method of least squares to fit a straight line to the following six data points:

x	1	2	3	4	5	6
y	1	2	2	3	5	5

(a) What are the least-squares estimates of β_0 and β_1?
(b) Plot the data points and graph the least-squares line. Does the line pass through the data points?

9.2 Use the method of least squares to fit a straight line to the following data points:

x	-2	-1	0	1	2
y	4	3	3	1	-1

(a) What are the least-squares estimates of β_0 and β_1?
(b) Plot the data points and graph the least-squares line. Does the line pass through the data points?

9.3   The elongation of a steel cable is assumed to be linearly related to the amount of force applied. Five identical specimens of cable gave the following results when varying forces were applied:

Force (x)	1.0	1.5	2.0	2.5	3.0
Elongation (y)	3.0	3.8	5.4	6.9	8.4

Use the method of least squares to fit the line

$$Y = \beta_0 + \beta_1 x + \varepsilon$$

9.4   A company wants to model the relationship between its sales and the sales for the industry as a whole. For the following data fit a straight line by the method of least squares.

Year	Company Sales y (millions of $)	Industry Sales x (millions of $)
1982	0.5	10
1983	1.0	12
1984	1.0	13
1985	1.4	15
1986	1.3	14
1987	1.6	15

9.5   It is thought that abrasion loss in certain steel specimens should be a linear function of the Rockwell hardness measure. A sample of eight specimens gave the following results:

Rockwell Hardness (x)	60	62	63	67	70	74	79	81
Abrasion Loss (y)	251	245	246	233	221	202	188	170

Fit a straight line to these measurements.

9.6   Due primarily to the price controls of the Organization of Petroleum Exporting Countries (OPEC), a cartel of crude oil suppliers, the price of crude oil has risen dramatically since the early 1970s. As a result motorists have been confronted with a similar upward spiral of gasoline prices. The data in the table are typical prices for a gallon of regular leaded gasoline and a barrel of crude oil (refiner acquisition cost) for the indicated years.

Year	Gasoline (cents/gal) y	Crude Oil (dollars/bbl) x
1973	38.8	4.15
1975	56.7	10.38
1976	59.0	10.89
1977	62.2	11.96
1978	62.6	12.46
1979	85.7	17.72
1980	119.1	28.07
1981	133.3	36.11

Source: *Statistical Abstract of U.S.*: 1981, p. 476.

(a) Use these data to calculate the least-squares line that describes the relationship between the price of a gallon of gas and the price of a barrel of crude oil.

(b) Plot your least-squares line on a scattergram of the data. Does your least-squares line appear to be an appropriate characterization of the relationship between y and x? Explain.

(c) If the price of crude oil fell to $20 per barrel, to what level (approximately) would the price of regular gasoline fall? Justify your response.

9.7   A study is made of the number of parts assembled as a function of the time spent on the job. Twelve employees are divided into three groups, assigned to three time intervals, with the following results:

Time (x)	Number of parts assembled (y)
10 minutes	27, 32, 26, 34
15 minutes	35, 30, 42, 47
20 minutes	45, 50, 52, 49

4 data pts. in each line

y entry
4 entry —

Fit the model

$$Y = \beta_0 + \beta_1 x + \varepsilon$$

by the method of least squares.

9.8   Laboratory experiments designed to measure LC50 values for the effect of certain toxicants on fish are run by basically two different methods. One method has water continuously flowing through laboratory tanks, and the other has static water conditions. For purposes of establishing criteria for toxicants, the Environmental Protection Agency (EPA) wants to adjust all results to the flow-through condition. Thus a model is needed to relate the two types of observations. Observations on certain toxicants examined under both static and flow-through conditions yielded the following measurements (in parts per million):

Toxicant	LC50 flow-through (y)	LC50 static (x)
1	23.00	39.00
2	22.30	37.50
3	9.40	22.20
4	9.70	17.50
5	0.15	0.64
6	0.28	0.45
7	0.75	2.62
8	0.51	2.36
9	28.00	32.00
10	0.39	0.77

Fit the model $Y = \beta_0 + \beta_1 x + \varepsilon$ by the method of least squares.

9.4 THE PROBABILITY DISTRIBUTION OF THE RANDOM ERROR COMPONENT

We have now completed the first two steps of regression modeling: we have hypothesized the form of $E(Y)$, and used the sample data to estimate unknown parameters in the model. The hypothesized model relating peak power load Y to daily maximum temperature x is

$$Y = \beta_0 + \beta_1 x + \varepsilon$$

and the least-squares estimate of $E(Y) = \beta_0 + \beta_1 x$ is

$$\hat{y} = -419.85 + 6.7175x$$

Step 3. Specify the probability distribution of the random error term, and estimate any unknown parameters of this distribution.

Recall that when we wanted to make inferences about a population mean and had only a small sample with which to work, we used a t statistic and assumed that the data were normally distributed. Similarly when we want to make inferences about the parameters of a regression model, we assume that the error component ε is normally distributed, and the assumption is most important when the sample size is small. The mean of ε is zero, since the deterministic component of the model describes $E(Y)$. Finally we assume that the variance of ε is σ^2, a constant for all values of x, and that the errors associated with different observations are independent.

The assumptions about the probability distribution of ε are summarized in the box, and are shown in Figure 9.6.

FIGURE 9.6
The Probability Distribution of ε

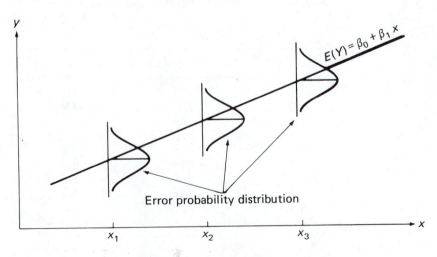

Error probability distribution

PROBABILITY DISTRIBUTION OF THE RANDOM
ERROR COMPONENT ε

> The error component is normally distributed with mean zero and constant variance σ^2. The errors associated with different observations are independent.

Various techniques exist for testing the validity of the assumptions, and there are some alternative techniques to be used when they appear to be invalid. These techniques often require us to return to Step 2 and to use a method other than least squares to estimate the model parameters. Much current statistical research is devoted to alternative methodologies. Most are beyond the scope of this text, but we will mention a few in later chapters. In actual practice the assumptions need not hold exactly for the least-squares techniques to be useful.

Note that the probability distribution of ε would be completely specified if the variance σ^2 were known. (We already know that ε is normally distributed with a mean of zero.) To estimate σ^2, we make use of the SSE for the least-squares model. The estimate s^2 of σ^2 is calculated by dividing SSE by the number of degrees of freedom associated with the error component. We use 2 df to estimate the y intercept and slope in the straight-line model, leaving $(n - 2)$ df for the error variance estimation. Thus

$$s^2 = \frac{\text{SSE}}{n - 2}$$

where

$$\text{SSE} = \sum_{i=1}^{n} (y_i - \hat{y}_i)^2 = \text{SS}_{yy} - \hat{\beta}_1 \text{SS}_{xy}$$

and

$$\text{SS}_{yy} = \sum_{i=1}^{n} (y_i - \bar{y})^2 = \sum_{i=1}^{n} y_i^2 - \frac{\left(\sum_{i=1}^{n} y_i\right)^2}{n}$$

In our peak power load example we calculated SSE = 2093.43 for the least-squares line. Recalling that there were $n = 10$ data points, we have $n - 2 = 8$ df for estimating σ^2. Thus

$$s^2 = \frac{\text{SSE}}{n - 2} = \frac{2093.43}{8} = 261.68$$

is the estimated variance, and

$$s = \sqrt{261.68} = 16.18$$

is the estimated standard deviation of ε.

You may be able to obtain an intuitive feeling for s by recalling that about 95% of the observations from a normal distribution lie within two standard deviations of the mean. Thus we expect approximately 95% of the observations to fall within $2s$ of the estimated mean $\hat{y} = \hat{\beta}_0 + \hat{\beta}_1 x$. For our peak power load example note that all ten observations fall within $2s = 2(16.18) = 32.36$ of the least-squares line. In Section 9.8 we show how to use the estimated standard deviation to evaluate the prediction error when $\hat{y}$ is used to predict a value of y to be observed for a given value of x.

EXERCISES

9.9 Calculate SSE and s^2 for the data of Exercises 9.1–9.8.

9.10 Matis and Wehrly (*Biometrics*, 35, No. 1, March 1979) report the following data on the proportion of green sunfish that survive a fixed level of thermal pollution for varying lengths of time:

Proportion of survivors (y)	1.00	0.95	0.95	0.90	0.85	0.70	0.65	0.60	0.55	0.40
Scaled time (x)	0.10	0.15	0.20	0.25	0.30	0.35	0.40	0.45	0.50	0.55

(a) Fit the linear model $Y = \beta_0 + \beta_1 x + \varepsilon$ by the method of least squares.
(b) Plot the data points and graph the line found in (a). Does the line fit through the points?
(c) Compute SSE and s^2.

9.5 ASSESSING THE ADEQUACY OF THE MODEL: MAKING INFERENCES ABOUT SLOPE β_1

Step 4. Statistically check the adequacy of the model.

Now that we have specified the probability distribution of ε and estimated the variance σ^2, we are prepared to make statistical inferences about the model's adequacy for describing the mean $E(Y)$ and for predicting the values of Y for given values of x.

Refer again to the peak power load example and suppose that the daily peak load is *completely unrelated* to the maximum temperature that day. This implies that the mean $E(Y) = \beta_0 + \beta_1 x$ does not change as x changes. In this straight-line model this means that the true slope β_1 is equal to zero (see Figure 9.7). Note that if β_1 is equal to zero, the model is just $Y = \beta_0 + \varepsilon$, the constant mean model used in previous chapters. (Note that the symbol β_0 is the same as the symbol for the mean μ when $\beta_1 = 0$.) Therefore to test the null hypothesis that x contributes no information for the prediction of Y, against the alternative hypothesis that these

FIGURE 9.7
Graph of the Model
$Y = \beta_0 + \varepsilon$

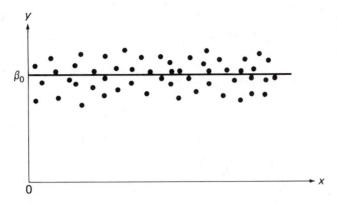

variables are linearly related with a slope differing from zero, we test

$$H_0: \beta_1 = 0 \quad \text{vs.} \quad H_a: \beta_1 \neq 0$$

If the data support the alternative hypothesis, we conclude that x does contribute information for the prediction of Y using the straight-line model (although the true relationship between $E(Y)$ and x could be more complex than a straight line). Thus to some extent this is a test of the adequacy of the hypothesized model.

The appropriate test statistic is found by considering the sampling distribution of $\hat{\beta}_1$, the least-squares estimator of the slope β_1.

SAMPLING DISTRIBUTION OF $\hat{\beta}_1$

If we assume that the error components are independent normal random variables with mean zero and constant variance σ^2, the sampling distribution of the least-squares estimator $\hat{\beta}_1$ of the slope will be normal, with mean β_1 (the true slope) and standard deviation

$$\sigma_{\hat{\beta}_1} = \frac{\sigma}{\sqrt{SS_{xx}}} \quad \text{(see Figure 9.8)}$$

[*Note*: Proof of the unbiasedness of $\hat{\beta}_1$ and a derivation of its standard deviation are given next.]

FIGURE 9.8
Sampling
Distribution of $\hat{\beta}_1$

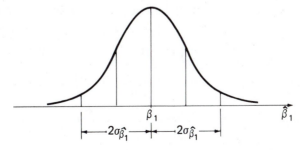

PROOF OF UNBIASEDNESS OF $\hat{\beta}_1$

Recall that

$$\hat{\beta}_1 = \frac{SS_{xy}}{SS_{xx}} = \frac{\sum_{i=1}^{n} (x_i - \bar{x})(\bar{Y}_i - \bar{Y})}{\sum_{i=1}^{n} (x_i - \bar{x})^2}$$

Using the facts:
1. $Y_i = \beta_0 + \beta_1 x + \varepsilon_i$
2. $\bar{Y} = \beta_0 + \beta_1 \bar{x} + \bar{\varepsilon}$
3. x_i's are fixed[2]

we find

$$E(\hat{\beta}_1) = E\left\{ \frac{\sum_{i=1}^{n} (x_i - \bar{x})(Y_i - \bar{Y})}{\sum_{i=1}^{n} (x_i - \bar{x})^2} \right\}$$

$$= \frac{1}{\sum_{i=1}^{n} (x_i - \bar{x})^2} \sum_{i=1}^{n} (x_i - \bar{x}) E(Y_i - \bar{Y})$$

$$= \frac{1}{SS_{xx}} \sum_{i=1}^{n} (x_i - \bar{x}) E[(\beta_0 + \beta_1 x_i + \varepsilon_i) - (\beta_0 + \beta_1 \bar{x} + \bar{\varepsilon})]$$

$$= \frac{1}{SS_{xx}} \sum_{i=1}^{n} (x_i - \bar{x})[\beta_1 (x_i - \bar{x})]$$

$$= \frac{SS_{xx}}{SS_{xx}} \beta_1$$

$$= \beta_1$$

showing that $\hat{\beta}_1$ is an unbiased estimator of β_1.

DERIVATION OF THE VARIANCE OF $\hat{\beta}_1$

Using the formula derived for the variance of a linear function,

$$\sigma_{\hat{\beta}_1}^2 = \mathrm{Var}(\hat{\beta}_1) = \mathrm{Var}\left\{ \frac{\sum_{i=1}^{n} (x_i - \bar{x})(Y_i - \bar{Y})}{\sum_{i=1}^{n} (x_i - \bar{x})^2} \right\}$$

[2]If the independent variable X is random, we use expectations *conditional* on the observed values of X.

$$= \frac{1}{\left[\sum_{i=1}^{n} (x_i - \bar{x})^2\right]^2} \operatorname{Var}\left\{\sum_{i=1}^{n} (x_i - \bar{x})(Y_i - \bar{Y})\right\}$$

$$= \frac{1}{SS_{xx}^2} \left[\sum_{i=1}^{n} (x_i - \bar{x})^2 \operatorname{Var}(Y_i - \bar{Y})\right.$$

$$\left. + \sum_{\substack{i=1 \\ i \neq j}}^{n} \sum_{j=1}^{n} (x_i - \bar{x})(x_j - \bar{x}) \operatorname{Cov}\{(Y_i - \bar{Y}), (Y_j - \bar{Y})\}\right]$$

Now

$$\operatorname{Var}(Y_i - \bar{Y}) = \operatorname{Var}\{(\beta_0 + \beta_1 x_i + \varepsilon_i) - (\beta_0 + \beta_1 \bar{x} + \bar{\varepsilon})\}$$

$$= \operatorname{Var}(\varepsilon_i - \bar{\varepsilon}) = \sigma^2 + \frac{\sigma^2}{n} - 2\frac{\sigma^2}{n}$$

$$= \sigma^2 - \frac{\sigma^2}{n}$$

and

$$\operatorname{Cov}\{(Y_i - \bar{Y}), (Y_j - \bar{Y})\} = \operatorname{Cov}\{(\varepsilon_i - \bar{\varepsilon}), (\varepsilon_j - \bar{\varepsilon})\}$$

$$= -\operatorname{Cov}(\varepsilon_i, \bar{\varepsilon}) - \operatorname{Cov}(\varepsilon_j, \bar{\varepsilon}) + \operatorname{Var}(\bar{\varepsilon})$$

$$= -\frac{\sigma^2}{n} - \frac{\sigma^2}{n} + \frac{\sigma^2}{n}$$

$$= -\frac{\sigma^2}{n}$$

where we have used the following facts:

1. $\operatorname{Var}(\varepsilon_i) = \sigma^2$

2. $\operatorname{Cov}(\varepsilon_i, \varepsilon_j) = 0, \ i \neq j$

3. $\operatorname{Var}(\bar{\varepsilon}) = \operatorname{Var}\left(\frac{\sum_{i=1}^{n} \varepsilon_i}{n}\right) = \frac{1}{n^2} \sum_{i=1}^{n} \operatorname{Var}(\varepsilon_i)$

$$= \frac{n\sigma^2}{n^2} = \frac{\sigma^2}{n}$$

4. $\operatorname{Cov}(\varepsilon_i, \bar{\varepsilon}) = \operatorname{Cov}\left(\varepsilon_i, \frac{\sum_{j=1}^{n} \varepsilon_j}{n}\right) = \frac{\operatorname{Var}(\varepsilon_i)}{n} = \frac{\sigma^2}{n}$

Thus

$$\sigma_{\hat{\beta}_1}^2 = \frac{1}{SS_{xx}^2} \left[\sum_{i=1}^{n} (x_i - \bar{x})^2 \left(\sigma^2 - \frac{\sigma^2}{n}\right) - \sum_{\substack{i=1 \\ i \neq j}}^{n} \sum_{j=1}^{n} (x_i - \bar{x})(x_j - \bar{x}) \right.$$

$$\left. \times \left(-\frac{\sigma^2}{n}\right)\right]$$

$$= \frac{1}{SS_{xx}^2} \left\{ SS_{xx}\sigma^2 - \frac{\sigma^2}{n} \left[\sum_{i=1}^{n} (x_i - \bar{x}) \right]^2 \right\}$$

$$= \frac{\sigma^2}{SS_{xx}} \quad \text{using } \sum_{i=1}^{n} (x_i - \bar{x}) = 0$$

The standard error of $\hat{\beta}_1$ is therefore $\sigma_{\hat{\beta}_1} = \sigma/\sqrt{SS_{xx}}$.

Since the standard deviation of ε, σ, will usually be unknown, the appropriate test statistic will generally be a Student's t statistic formed as follows:

$$t = \frac{\hat{\beta}_1 - (\text{hypothesized value of } \beta_1)}{s_{\hat{\beta}_1}}$$

where $s_{\hat{\beta}_1} = s/\sqrt{SS_{xx}}$, the estimated standard deviation of the sampling distribution of $\hat{\beta}_1$. We will usually be testing the null hypothesis $H_0: \beta_1 = 0$, so that the t statistic becomes

$$t = \frac{\hat{\beta}_1}{s/\sqrt{SS_{xx}}}$$

For the peak power load example with $H_a: \beta_1 \neq 0$, we choose $\alpha = 0.05$, and since $n = 10$, our rejection region is

$$t < -t_{0.025}(10 - 2) = -2.306$$

$$t > t_{0.025}(8) = 2.306$$

We previously calculated $\hat{\beta}_1 = 6.7175$, $s = 16.18$, and $SS_{xx} = 380.5$. Thus

$$t = \frac{\hat{\beta}_1}{s/\sqrt{SS_{xx}}} = \frac{6.7175}{16.18/\sqrt{380.5}} = 8.10$$

Since this calculated t value falls in the upper-tail rejection region, we reject the null hypothesis and conclude that the slope β_1 is not zero. The sample evidence indicates that the peak power load tends to increase as a day's maximum temperature increases.

A TEST OF MODEL UTILITY

One-tailed test	Two-tailed test
$H_0: \beta_1 = 0$	$H_0: \beta_1 = 0$
$H_a: \beta_1 < 0 \quad (\text{or } H_a: \beta_1 > 0)$	$H_a: \beta_1 \neq 0$
Test statistic:	Test statistic:
$t = \dfrac{\hat{\beta}_1}{s_{\hat{\beta}_1}} = \dfrac{\hat{\beta}_1}{s/\sqrt{SS_{xx}}}$	$t = \dfrac{\hat{\beta}_1}{s_{\hat{\beta}_1}} = \dfrac{\hat{\beta}_1}{s/\sqrt{SS_{xx}}}$

Rejection region: Rejection region:

$$t < -t_\alpha(n-2)$$ $$t < -t_{\alpha/2}(n-2) \quad \text{or}$$

(or $t > t_\alpha(n-2)$ when H_α: $\beta_1 > 0$) $t > t_{\alpha/2}(n-2)$

Assumptions: The four assumptions about ε listed in Section 9.4.

What conclusion can be drawn if the calculated t value does not fall in the rejection region? We know from previous discussions of the philosophy of hypothesis testing that such a t value does not lead us to accept the null hypothesis. That is, we do not conclude that $\beta_1 = 0$. Additional data might indicate that β_1 differs from zero, or a more complex relationship may exist between x and Y, requiring the fitting of a model other than the simple linear model. We discuss several such models in Chapter 10.

Another way to make inferences about the slope β_1 is to estimate it by using a confidence interval. This interval is formed as shown next.

A 100(1−α) PERCENT CONFIDENCE INTERVAL FOR THE SLOPE $\hat{\beta}_1$

$$\hat{\beta}_1 \pm t_{\alpha/2}(n-2)s_{\hat{\beta}_1}$$

where

$$s_{\hat{\beta}_1} = \frac{s}{\sqrt{SS_{xx}}}$$

Assumptions: The four assumptions about ε listed in Section 9.4.

For the peak power load example a 95% confidence interval for the slope β_1 is

$$\hat{\beta}_1 \pm t_{0.025}(8)s_{\hat{\beta}_1} = 6.7175 \pm 2.306\left(\frac{16.18}{\sqrt{380.5}}\right)$$

$$= 6.7175 \pm 1.913 = (4.80, 8.63)$$

This confidence interval confirms the conclusion of the test of the null hypothesis H_0: $\beta_1 = 0$, since it implies that the true slope is between 4.80 and 8.63.

EXERCISES

9.11 Refer to Exercise 9.1. Is there sufficient evidence to say that the slope of the line is significantly different from zero? Use $\alpha = 0.05$.

9.12 Refer to Exercise 9.2. Estimate β_1 in a confidence interval with confidence coefficient 0.90.

9.13   Refer to Exercise 9.3. Estimate the amount of elongation per unit increase in force. Use a confidence coefficient of 0.95.

9.14   Refer to Exercise 9.4. Does industry sales appear to contribute any information to the prediction of company sales? Test at the 10% level of significance.

9.15   Refer to Exercise 9.5. Does Rockwell hardness appear to be linearly related to abrasion loss? Use $\alpha = 0.05$.

9.16   In Exercise 9.6 the price of a gallon of gasoline y was modeled as a function of the price of a barrel of crude oil x using the following equation: $Y = \beta_0 + \beta_1 x + \varepsilon$. Estimates of β_0 and β_1 were found to be 25.883 and 3.115, respectively.
   (a)   Estimate the variance and standard deviation of ε.
   (b)   Suppose the price of a barrel of crude oil is \$15. Estimate the mean and standard deviation of the price of a gallon of gas under these circumstances.
   (c)   Repeat part (b) for a crude oil price of \$30.
   (d)   What assumptions about ε did you make in answering parts (b) and (c)?

9.17   Refer to Exercise 9.7. Estimate the increase in expected number of parts assembled per 5-minute increase in time spent on the job. Use a 95% confidence coefficient.

9.18   Refer to Exercise 9.8. Does it appear that flow-through LC50s are linearly related to static LC50s? Use $\alpha = 0.05$.

9.19   It is well known that large bodies of water have a mitigating effect on the temperature of the surrounding land masses. On a cold night in central Florida, temperatures were recorded at equal distances along a transect running in the downwind direction from a large lake. The data are as follows:

Site (x)	1	2	3	4	5	6	7	8	9	10
Temperature y	37.00	36.25	35.41	34.92	34.52	34.45	34.40	34.00	33.62	33.90

   (a)   Fit the linear model $Y = \beta_0 + \beta_1 x + \varepsilon$ by the method of least squares.
   (b)   Find SSE and s^2.
   (c)   Does it appear that temperatures decrease significantly as distance from the lake increases? Use $\alpha = 0.05$.

9.6   THE COEFFICIENT OF CORRELATION

A common joint probability density function used to model the joint behavior of continuous random variables X and Y is the *bivariate normal*. This density function is given by

$$f(x, y) = \frac{1}{2\pi\sigma_1\sigma_1\sqrt{1 - \rho^2}}\, e^{-Q/2} \qquad \begin{array}{l} -\infty < x < \infty \\ -\infty < y < \infty \end{array}$$

where

$$Q = \frac{1}{1 - \rho^2}\left[\frac{(x - \mu_1)^2}{\sigma_1^2} - 2\rho\,\frac{(x - \mu_1)(y - \mu_2)}{\sigma_1\sigma_2} + \frac{(y - \mu_2)^2}{\sigma_2^2} \right]$$

For this distribution the marginal distributions will both be normal, with X having mean μ_1 and variance σ_1^2, while Y has mean μ_2 and variance σ_2^2.

One parameter of this distribution is of particular interest in the context of simple regression modeling: the correlation coefficient ρ. The reason for this importance is most easily understood by considering the conditional expectation (Chapter 2) of Y for given $X = x$,

$$E(Y|X = x) = \mu_y + \rho\sigma_y \left(\frac{x - \mu_x}{\sigma_x} \right)$$

where μ_y and μ_x are the unconditional means of Y and X, respectively, and σ_y and σ_x are the corresponding standard deviations. This may be rewritten in the more familiar form of a simple regression model,

$$E(Y|X = x) = \beta_0 + \beta_1 x$$

where

$$\beta_0 = \mu_y - \rho \frac{\mu_x \sigma_y}{\sigma_x}$$

$$\beta_1 = \rho \frac{\sigma_y}{\sigma_x}$$

Since σ_y/σ_x is a positive quantity, the sign of the slope β_1 and the correlation coefficient ρ will always agree, and β_1 will be zero if and only if ρ is zero. It should not be surprising then that inferences concerning the slope in a straight-line model relating bivariate normal random variables are equivalent to inferences about the correlation coefficient ρ. In addition, ρ will always have a numerical value between -1 and $+1$.

The sample correlation coefficient r can be computed from the least-squares slope $\hat{\beta}_1$ by inverting the above formula and substituting sample estimates:

$$r = \hat{\beta}_1 \frac{s_x}{s_y}$$

where s_y and s_x are the sample standard deviations of y and x. The calculation formula for r is

$$r = \frac{SS_{xy}}{\sqrt{SS_{xx} SS_{yy}}}$$

Note that r is computed by using the same quantities used in fitting the least-squares line. Since both r and $\hat{\beta}_1$ provide information about the utility of the model, it is not surprising that there is a similarity in their computational formulas. Particularly note that SS_{xy} appears in the numerators of both expressions and, since both denominators are always positive, r and $\hat{\beta}_1$ will always be of the same sign (either both positive or both negative). A value of r near or equal to zero implies little or no linear relationship between y and x. In contrast, the closer r is to 1 or -1, the stronger the linear relationship between y and x. And if $r = 1$ or $r = -1$, all the points fall exactly on the least-squares line. Positive values of r

FIGURE 9.9
Values of r and
Their Implications

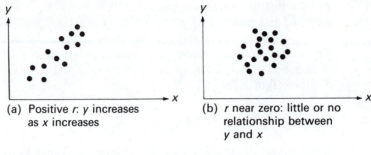

(a) Positive *r*: *y* increases
as *x* increases

(b) *r* near zero: little or no
relationship between
y and *x*

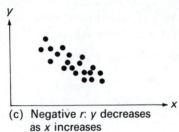

(c) Negative *r*: *y* decreases
as *x* increases

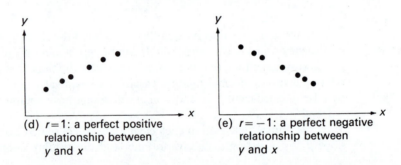

(d) *r*=1: a perfect positive
relationship between
y and *x*

(e) *r*=−1: a perfect negative
relationship between
y and *x*

imply that *y* increases as *x* increases; negative values imply that *y* decreases as *x* increases. Each of these situations is portrayed in Figure 9.9.

In the peak power load example we related peak daily power usage *y* to the day's maximum temperature *x*. For these data (Table 9.2) we found

$$SS_{xy} = 2556$$

and

$$SS_{xx} = 380.5$$

where

$$SS_{yy} = \sum_{i=1}^{n} (y_i - \bar{y})^2 = 19{,}263.6$$

and therefore

$$r = \frac{2556}{\sqrt{(380.5)(19{,}263.6)}} = 0.94$$

The fact that the value of r is positive and near 1 indicates that the peak power load tends to increase as the daily maximum temperature increases *for this sample of ten days*. This is the same conclusion we reached upon finding that the least-squares slope $\hat{\beta}_1$ was positive.

If we want to use the sample information to test the null hypothesis that the normal random variables X and Y are uncorrelated, that is, $H_0: \rho = 0$, we can use the results of the test that the slope of the straight-line model is zero, that is, $H_0: \beta_1 = 0$. The test given in Section 9.2 for the latter hypothesis is identical to the one used for the former hypothesis. When we tested the null hypothesis $H_0: \beta_1 = 0$ in connection with the peak power load example, the data led to a rejection of the null hypothesis at the $\alpha = 0.05$ level. This implies that the null hypothesis of a zero linear correlation between the two variables (maximum temperature and peak power load) can also be rejected at the $\alpha = 0.05$ level. The only real difference between the least-squares slope $\hat{\beta}_1$ and the coefficient of correlation r is the measurement scale. Therefore the information they provide about the utility of the least-squares model is to some extent redundant. However, the test for correlation requires the additional assumption that the variables X and Y have a bivariate normal distribution, while the test of slope does not require that x be a random variable. For example, we could use the straight-line model to describe the relationship between the yield of a chemical process Y and the temperature at which the reaction takes place x, even if the temperature were controlled and therefore nonrandom. However, the correlation coefficient would have no meaning in this case, since the bivariate normal distribution requires that both variables be marginally normal.

We close the section with a caution: do not infer a causal relationship on the basis of high sample correlation. Although it is probably true that a high maximum daily temperature *causes* an increase in peak power demand, the same would not necessarily be true of the relationship between availability of oil and peak power load, even though they are probably positively correlated. That is, we would not be willing to state that low availability of oil *causes* a lower peak power load. Rather, the scarcity of oil tends to cause an increase in conservation awareness, which in turn tends to cause a decrease in peak power load. Thus a large sample correlation indicates only that the two variables tend to change together, but causality must be determined by considering other factors in addition to the size of the correlation.

9.7 THE COEFFICIENT OF DETERMINATION

Another way to measure the contribution of x in predicting Y is to consider how much the errors of prediction of Y were reduced by using the information provided by x. If you do not use x, the best prediction for any value of Y would be $\bar{y}$, and the sum of squares of the deviations of the observed y values about $\bar{y}$ is the familiar

$$\mathrm{SS}_{yy} = \sum (y_i - \bar{y})^2$$

On the other hand if you use x to predict Y, the sum of squares of the deviations of the y values about the least-squares line is

$$SSE = \sum (y_i - \hat{y}_i)^2$$

To illustrate, suppose a sample of data has the scattergram shown in Figure 9.10(a). If we assume that x contributes no information for the prediction of y, the best prediction for a value of y is the sample mean $\bar{y}$, which is shown as the horizontal line in Figure 9.10(b). The vertical line segments in Figure 9.10(b) are the deviations of the points about the mean $\bar{y}$. Note that the sum of squares of deviations for the model $\hat{y} = \bar{y}$ is

$$SS_{yy} = \sum (y_i - \bar{y})^2$$

Now suppose you fit a least-squares line to the same set of data and locate the deviations of the points about the line as shown in Figure 9.10(c). Compare the deviations about the prediction lines in parts (b) and (c) of Figure 9.10. You can see that

1. If x contributes little or no information for the prediction of y, the sums of squares of deviations for the two lines,

$$SS_{yy} = \sum (y_i - \bar{y})^2 \quad \text{and} \quad SSE = \sum (y_i - \hat{y}_i)^2$$

will be nearly equal.

2. If x does contribute information for the prediction of y, the SSE will be smaller than SS_{yy} as shown in Figure 9.10. In fact, if all the points fall on the least-squares line, then SSE $= 0$.

It can be shown that

$$\sum (y_i - \bar{y})^2 = \sum (y_i - \hat{y}_i)^2 + \sum (\hat{y}_i - \bar{y})^2$$

We label the third sum $\sum (\hat{y}_i - \bar{y})^2$ the Sum of Squares for Regression and denote it by SSR. Thus

$$SS_{yy} = SSE + SSR$$

and it follows that

$$SS_{yy} - SSE \geqslant 0$$

The SSR assesses differences between $\hat{y}_i$ and $\bar{y}$. If the regression line has slope near zero, then $\hat{y}_i$ and $\bar{y}$ will be close together for any i, and SSR will be small. If, on the other hand, the slope is far from zero, SSR will be large.

Then the reduction in the sum of squares of deviations that can be attributed to x, expressed as a proportion of SS_{yy}, is

$$\frac{SS_{yy} - SSE}{SS_{yy}} = \frac{SSR}{SS_{yy}}$$

It can be shown that this quantity is equal to the square of the simple linear coefficient of correlation.

FIGURE 9.10
A Comparison of
the Deviations for
Two Models

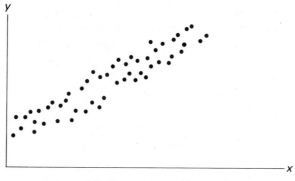

(a) Scattergram of data

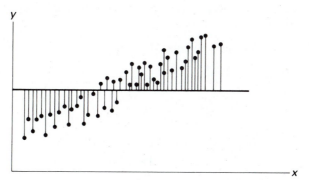

(b) Assumption: x contributes no information for predicting y,
$\hat{y} = \bar{y}$

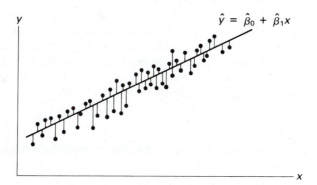

(c) Assumption: x contributes information for predicting y,
$\hat{y} = \hat{\beta}_0 + \hat{\beta}_1 x$

DEFINITION 9.1	The square of the coefficient of correlation is called the **coefficient of determination**. It represents the proportion of the sum of squares of deviations of the y values about their mean that can be attributed to a linear relation between y and x.
	$$r^2 = \frac{SS_{yy} - SSE}{SS_{yy}} = 1 - \frac{SSE}{SS_{yy}} = \frac{SSR}{SS_{yy}}$$

Note that r^2 is always between 0 and 1. Thus $r^2 = 0.60$ means that 60% of the sum of squares of deviations of the observed y values about their mean is attributable to the linear relation between y and x.

EXAMPLE 9.1 _____

Calculate and interpret the coefficient of determination for the peak power load example.

Solution We have previously calculated

$$SS_{yy} = 19,263.6$$

$$SSE = 2093.4$$

so that

$$r^2 = \frac{SS_{yy} - SSE}{SS_{yy}} = \frac{19,263.6 - 2093.4}{19,263.6}$$

$$= 0.89$$

Note that we could have obtained $r^2 = (0.94)^2 = 0.89$ more simply by using the correlation coefficient calculated in the previous section. However, use of the above formula is worthwhile because it reminds us that r^2 represents the fraction reduction in variability from SS_{yy} to SSE. In the peak power load example this fraction is 0.89, meaning that the sample variability of the peak loads about their mean is reduced by 89% when the mean peak load is modeled as a linear function of daily high temperature. Remember, though, that this statement holds true only for the sample at hand. We have no way of knowing how much of the population variability can be explained by the linear model. ■

EXERCISES ▨▨

9.20 Compute r and r^2 for the data of Exercises 9.1–9.8.

9.21 In the summer of 1981 the Minnesota Department of Transportation installed a state-of-the-art weigh-in-motion scale into the concrete surface of the eastbound lanes of Interstate 494 in Bloomington, Minnesota. The system is computerized and monitors traffic continuously. It is capable of distinguishing among 13 different types of vehicles (car, 5-axle semi, 5-axle twin trailer, etc.). The primary purpose of the system is to provide traffic counts and weights for use in the planning and design of future roadways. Following installation, a study was undertaken to determine whether the scale's readings corresponded with the static weights of the vehicles being monitored. Studies of this type are known as *calibration studies*. After some preliminary comparisons using a two-axle, six-tire truck carrying different loads (see table), calibration adjustments were made in the software of the weigh-in-motion system, and the weigh-in-motion scales were reevaluated (Wright, Owen, and Pena, "Status of MN/DOT's Weigh-In-Motion Program," Minnesota DOT, 1983).

(All Weights in Thousands of Pounds)

Trial No.	Static Weight of Truck, x	Weigh-In-Motion Reading y_1 (prior to calibration adjustment)	Weigh-In-Motion y_2 (after calibration adjustment)
1	27.9	26.0	27.8
2	29.1	29.9	29.1
3	38.0	39.5	37.8
4	27.0	25.1	27.1
5	30.3	31.6	30.6
6	34.5	36.2	34.3
7	27.8	25.1	26.9
8	29.6	31.0	29.6
9	33.1	35.6	33.0
10	35.5	40.2	35.0

(a) Construct two scattergrams, one of y_1 vs. x, and the other of y_2 vs. x.

(b) Use the scattergrams of part (a) to evaluate the performance of the weigh-in-motion scale both before and after the calibration adjustment.

(c) Calculate the correlation coefficient for both sets of data and interpret their values. Explain how these correlation coefficients can be used to evaluate the weigh-in-motion scale.

(d) Suppose the sample correlation coefficient for y_2 and x were 1. Could this happen if the static weights and the weigh-in-motion readings disagreed? Explain.

9.22 A problem of both economic and social concern in the United States is the importation and sale of illicit drugs. The data shown below are a part of a larger body of data collected by the Florida Attorney General's Office in an attempt to relate the incidence of drug seizures and drug arrests to the characteristics of the Florida counties. They show the number y of drug arrests per county in 1982, the density x_1 of the county (population per square mile), and the number x_2 of law enforcement employees. To simplify the calculations, we show data for only 10 counties.

(a) Fit a least-squares line to relate the number y of drug arrests per county in 1982 to the county population density x_1.

(b) We might expect the mean number of arrests to increase as the population density increases. Do the data support this theory? Test by using $\alpha = 0.05$.

(c) Calculate the coefficient of determination for this regression analysis and interpret its value.

County	1	2	3	4	5	6	7	8	9	10
Population Density, x_1	169	68	278	842	18	42	112	529	276	613
Number of Law Enforcement Employees, x_2	498	35	772	5788	18	57	300	1762	416	520
Number of Arrests in 1982, y	370	44	716	7416	25	50	189	1097	256	432

9.23 Repeat the instructions of Exercise 9.22 except let the independent variable be the number x_2 of county law enforcement employees.

9.24 Refer to Exercise 9.22.
   (a) Calculate the correlation coefficient r between the county population density x_1 and the number of law enforcement employees x_2.
   (b) Does the correlation between x_1 and x_2 differ significantly from 0? Test by using $\alpha = 0.05$.
   (c) Note that the regression lines of Exercises 9.22 and 9.23 both possess positive slopes. How might the value for the correlation coefficient, calculated in part (a), be related to this phenomenon?

9.8 USING THE MODEL FOR ESTIMATION AND PREDICTION

If we are satisfied that a useful model has been found to describe the relationship between peak power load and maximum daily temperature, we are ready to accomplish the original objective in constructing the model.

Step 5. When satisfied that the model is adequate, use it for prediction, estimation, and so on.

The most common uses of a probabilistic model for making inferences can be divided into two categories. The first is the use of the model for estimating the mean value of Y, $E(Y)$, for a specific value of x. For our peak power load example we may want to estimate the mean peak power load for days during which the maximum temperature is 90°F. The second use of the model entails predicting a particular value for a given x. That is, we may want to predict the peak power load for a particular day during which the maximum temperature will be 90°F.

In the first case we are attempting to estimate the mean value of Y over a very large number of days. In the second case we are trying to predict the Y value for a single day. Which of these model uses, estimating the mean value of Y or predicting an individual value of Y (for the same value of x), can be accomplished with the greater accuracy?

Before answering this question, first we consider the problem of choosing an estimator (or predictor) of the mean (or individual) Y value. We use the least-squares model

$$\hat{y} = \hat{\beta}_0 + \hat{\beta}_1 x$$

both to estimate the mean value of Y and to predict a particular value of Y for a given value of x. For our example we found

$$\hat{y} = -419.85 + 6.7175x$$

so the estimated mean peak power load for all days when $x = 90$ (maximum temperature is 90°F) is

$$\hat{y} = -419.85 + 6.7175(90) = 184.72$$

The identical value is used to predict the Y value when $x = 90$. That is, both the estimated mean and the predicted value of Y are $\hat{y} = 184.72$ when $x = 90$, as shown in Figure 9.11.

FIGURE 9.11
Estimated Mean Value and Predicted Individual Value of Peak Power Load Y for x = 90

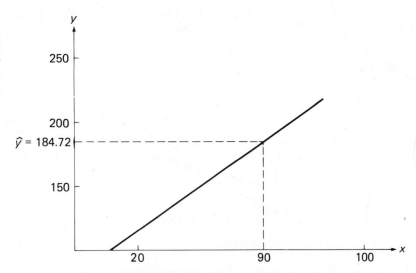

The difference between the uses of these two models lies in the relative precision of the estimate and prediction. The precisions are measured by the expected squared distances between the estimator/predictor $\hat{y}$ and the quantity being estimated or predicted. In the case of estimation we are trying to estimate

$$E(Y) = \beta_0 + \beta_1 x_p$$

where x_p is the particular value of x at which the estimate is being made. The estimator is

$$\hat{Y} = \hat{\beta}_0 + \hat{\beta}_1 x_p$$

so the sources of error in estimating $E(Y)$ are the estimators $\hat{\beta}_0$ and $\hat{\beta}_1$. The variance can be shown to be

$$\sigma_{\hat{Y}}^2 = E\{[\hat{Y} - E(\hat{Y})]^2\} = \sigma^2 \left[\frac{1}{n} + \frac{(x_p - \bar{x})^2}{SS_{xx}} \right]$$

In the case of predicting for a particular y value we are trying to predict

$$Y_p = \beta_0 + \beta_1 x_p + \varepsilon_p$$

where ε_p is the error associated with the particular Y value Y_p. Thus the sources of prediction error are the estimators $\hat{\beta}_0$ and $\hat{\beta}_1$ *and* the error ε_p. The variance associated with ε_p is σ^2, and it can be shown that this is the additional term in the expected squared prediction error:

$$\sigma^2(Y_p - \hat{Y}) = E[(\hat{Y} - Y_p)^2] = \sigma^2 + \sigma_{\hat{Y}}^2$$

$$= \sigma^2 \left[1 + \frac{1}{n} + \frac{(x_p - \bar{x})^2}{\mathrm{SS}_{xx}} \right]$$

The sampling errors associated with the estimator and predictor are summarized next. Figures 9.12 and 9.13 graphically depict the difference between the error of estimating a mean value of Y and the error of predicting a future value of Y for $x = x_p$. Note that the error of estimation is just the vertical difference between the least-squares line and the true line of means, $E(Y) = \beta_0 + \beta_1 x$. However, the error

FIGURE 9.12
Error of Estimating the Mean Value of Y for a Given Value of x

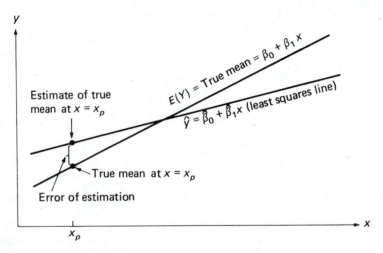

FIGURE 9.13
Error of Predicting a Future Value of Y for a Given Value of x

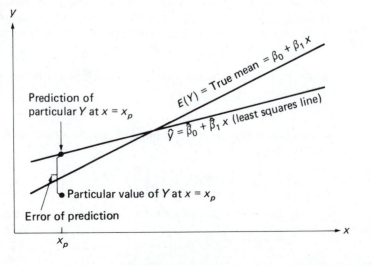

in predicting some future value of Y is the sum of two errors: the error of estimating the mean of Y, $E(Y)$, shown in Figure 9.12, plus the random error ε_p that is a component of the value Y_p to be predicted. Note that both errors will be smallest when $x_p = \bar{x}$. The farther x_p lies from the mean, the larger the errors of estimation and prediction.

SAMPLING ERRORS FOR THE ESTIMATOR OF THE MEAN OF Y AND THE PREDICTOR OF AN INDIVIDUAL Y

1. The standard deviation of the sampling distribution of the estimator $\hat{Y}$ of the mean value of Y at a fixed x is

$$\sigma_{\hat{Y}} = \sigma \sqrt{\frac{1}{n} + \frac{(x - \bar{x})^2}{SS_{xx}}}$$

where σ is the standard deviation of the random error ε.

2. The standard deviation of the prediction error for the predictor $\hat{Y}$ of an individual Y value at a fixed x is

$$\sigma_{(Y - \hat{Y})} = \sigma \sqrt{1 + \frac{1}{n} + \frac{(x - \bar{x})^2}{SS_{xx}}}$$

where σ is the standard deviation of the random error ε.

The true value of σ will rarely be known. Thus we estimate σ by s and calculate the estimation and prediction intervals as shown next.

A 100(1 − α) PERCENT CONFIDENCE INTERVAL FOR THE MEAN VALUE OF Y AT A FIXED x

$$\hat{y} \pm t_{\alpha/2}(n - 2) \text{ (estimated standard deviation of } \hat{Y})$$

or

$$\hat{y} \pm t_{\alpha/2}(n - 2)s \sqrt{\frac{1}{n} + \frac{(x - \bar{x})^2}{SS_{xx}}}$$

EXAMPLE 9.2

Find a 95% confidence interval for the mean peak power load when the maximum daily temperature is 90°F.

Solution A 95% confidence interval for estimating the mean peak load at a high temperature of 90° is given by

$$\hat{y} \pm t_{\alpha/2}(n - 2)s \sqrt{\frac{1}{n} + \frac{(x - \bar{x})^2}{SS_{xx}}}$$

We have previously calculated $\hat{y} = 184.72$, and $s = 16.18$, $\bar{x} = 91.5$, and $SS_{xx} = 380.5$. With $n = 10$ and $\alpha = 0.05$, $t_{0.025}(8) = 2.306$ and thus the 95% confidence interval is

$$\hat{y} \pm t_{\alpha/2}(n - 2)s \sqrt{\frac{1}{n} + \frac{(x - \bar{x})^2}{SS_{xx}}}$$

$$184.72 \pm (2.306)(16.18) \sqrt{\frac{1}{10} + \frac{(90 - 91.5)^2}{380.5}}$$

or

$$184.72 \pm 12.14$$

Thus we can be 95% confident that the *mean* peak power load is between 172.58 and 196.86 megawatts for days with a maximum temperature of 90°F. ■

A 100(1 − α) PERCENT PREDICTION INTERVAL[3] FOR AN INDIVIDUAL Y AT A FIXED x

$$\hat{y} \pm t_{\alpha/2}(n - 2)[\text{estimated standard deviation of } (Y - \hat{Y})]$$

or

$$\hat{y} \pm t_{\alpha/2}(n - 2)s \sqrt{1 + \frac{1}{n} + \frac{(x - \bar{x})^2}{SS_{xx}}}$$

EXAMPLE 9.3

Predict the peak power load for a day during which the maximum temperature is 90°F.

Solution   Using the same values as in the previous example, we find

$$\hat{y} \pm t_{\alpha/2}(n - 2)s \sqrt{1 + \frac{1}{n} + \frac{(x - \bar{x})^2}{SS_{xx}}}$$

$$184.72 \pm (2.306)(16.18) \sqrt{1 + \frac{1}{10} + \frac{(90 - 91.5)^2}{380.5}}$$

or

$$184.72 \pm 39.24$$

Thus we are 95% confident that the peak power load will be between 145.48 and 223.96 on a particular day when the maximum temperature is 90°F. ■

A comparison of the confidence interval for the mean peak power load and the prediction interval for a single day's peak power load for a maximum

---

[3]The term *prediction interval* is used when the interval is intended to enclose the value of a random variable. The term *confidence interval* is reserved for estimation of population parameters (such as the mean).

temperature of 90°F ($x = 90$) is illustrated in Figure 9.14. It is important to note that the prediction interval for an individual value of Y will always be wider than the corresponding confidence interval for the mean value of Y.

FIGURE 9.14
A 95% Confidence Interval for Mean Peak Power Load and a Prediction Interval for Peak Power Load When x = 90

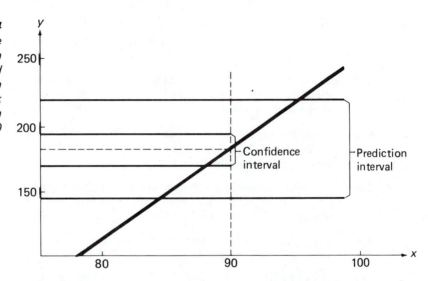

Be careful not to use the least-squares prediction equation to estimate the mean value of Y or to predict a particular value of Y for values of x that fall outside the range of the values of x contained in your sample data. The model might provide a good fit to the data over the range of x values contained in the sample but a very poor fit for values of x outside this region (see Figure 9.15). Failure to heed this warning may lead to errors of estimation and prediction that are much larger than expected.

FIGURE 9.15
The Danger of Using a Model to Predict Outside the Range of the Sample Values of x

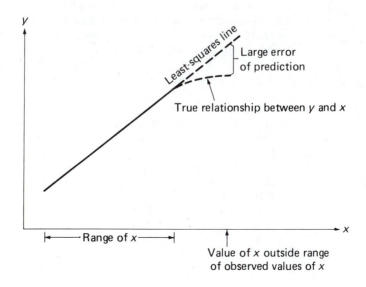

EXERCISES

9.25 Refer to Exercise 9.1.
   (a) Estimate the mean value of Y for $x = 2$, using a 95% confidence interval.
   (b) Find a 95% prediction interval for a response Y at $x = 2$.

9.26 Refer to Exercise 9.3. Estimate the expected elongation for a force of 1.8, using a 90% confidence interval.

9.27 Refer to Exercise 9.4. If in 1978 the industry sales figure is 16, find a 95% prediction interval for the company sales figure. Do you see any possible difficulties with the solution to this problem?

9.28 Refer to Exercise 9.5. Estimate, in a 95% confidence interval, the mean abrasion loss for steel specimens with a Rockwell hardness measure of 75. What assumptions are necessary for your answer to be valid?

9.29 Refer again to Exercise 9.5. Predict, in a 95% prediction interval, the abrasion loss for a particular steel specimen with a Rockwell hardness measure of 75. What assumptions are necessary for your answer to be valid?

9.30 Refer to Exercise 9.7. Estimate, in a 90% confidence interval, the expected number of parts assembled by employees working for 12 minutes.

9.31 Refer to Exercise 9.8. Find a 95% prediction interval for a flow-through measurement corresponding to a static measurement of 21.0.

9.32 Refer to Exercise 9.3. If no force is applied, the expected elongation should be zero (that is, $\beta_0 = 0$). Construct a test of $H_0: \beta_0 = 0$ versus $H_a: \beta_0 \neq 0$, using the data of Exercise 9.3. Use $\alpha = 0,05$.

9.33 In planning for an initial orientation meeting with new electrical engineering majors, the chairman of the Electrical Engineering Department wants to emphasize the importance of doing well in the major courses to get better-paying jobs after graduation. To support this point, the chairman plans to show that there is a strong positive correlation between starting salaries for recent electrical engineering graduates and their grade-point averages in the major courses. Records for seven of last year's electrical engineering graduates are selected at random and given in the table:

Grade-Point Average in Major Courses x	Starting Salary y ($ thousands)
2.58	16.5
3.27	18.8
3.85	19.5
3.50	19.2
3.33	18.5
2.89	16.6
2.23	15.6

(a) Find the least-squares prediction equation.

(b) Plot the data and graph the line as a check on your calculations. Do the data provide sufficient evidence that grade-point average provides information for predicting starting salary?

(c) Find a 95% prediction interval for the starting salary of a graduate whose grade-point average is 3.2.

(d) What is the mean starting salary for graduates with grade-point averages equal to 3.0? Use a 95% confidence interval.

9.34 Managers are an important part of any organization's resource base. Accordingly the organization should be just as concerned about forecasting its future managerial needs as it is with forecasting its needs for, say, the natural resources used in its production processes. According to William F. Glueck (*Management*, Dryden Press, 1977), one commonly used procedure for forecasting the demand for managers is to model the relationship between sales and the number of managers needed, the theory being that "... the demand for managers is the result of the increases and decreases in the demand for products and services that an enterprise offers its customers and clients" (p. 274). To develop this relationship, data such as those shown in the table can be collected from a firm's records.

Date	Monthly Sales (Units) x	Number of Managers on 15th of the Month y
3/80	5	10
6/80	4	11
9/80	8	10
12/80	7	10
3/81	9	9
6/81	15	10
9/81	20	11
12/81	21	17
3/82	25	19
6/82	24	21
9/82	30	22
12/82	31	25
3/83	36	30
6/83	38	30
9/83	40	31
12/83	41	31
3/84	51	32
6/84	40	30
9/84	48	32
12/84	47	32

(a) Use simple linear regression to model the relationship between the number of managers and the number of units sold for the data in the table.

(b) Plot your least-squares line on a scattergram of the data. Does it appear that the relationship between x and y is linear? If not, does it appear that your least-squares model will provide a useful approximation to the relationship? Explain.

(c)   Test the usefulness of your model. Use $\alpha = 0.05$. State your conclusion in the context of the problem.

(d)   The company projects that next May they will sell 39 units. Use your least-squares model to construct (1) a 90% confidence interval for the mean number of managers needed when sales is at a level of 39, and (2) a 90% prediction interval for the number of managers needed next May.

(e)   Compare the sizes and the interpretations of the two intervals you constructed in part (d).

9.9 COMPUTATION OF MORE DETAILED ANALYSES

As you can see, the computations for a simple regression analysis are quite complicated and cumbersome. This load is lightened by any of a number of readily available statistical computer packages. We illustrate how computer printouts can produce the analyses explained in the earlier sections of this chapter, and extend them a bit, by the use of Minitab.

The standard computer printout for a regression analysis of the peak power load data looks as follows:

The regression equation is
$y = -420 + 6.72 \times$

Predictor	Coef	Stdev	t-ratio	p
Constant	−419.85	76.06	−5.52	0.000
x	6.7175	0.8294	8.10	0.000

$s = 16.18$ R-sq $= 89.1\%$ $R - sq(adj) = 87.8\%$

Analysis of Variance

SOURCE	DF	SS	MS	F	P
Regression	1	17170	17170	65.60	0.000
Error	8	2094	262		
Total	9	19264			

We identify the components of the printout with the terminology used in this chapter in the following list: (SS can be read "sum of squares" and MS can be read "mean square.")

$$\hat{\beta}_0 = \text{Predictor constant coef.} = -419.85$$

$$\hat{\beta}_1 = \text{Predictor } x \text{ coef.} = 6.7175$$

$$SS_{yy} = \text{Total SS (under the "Analysis of Variance")} = 19{,}264$$

$$SSE = \text{Error SS} = 2094$$

$$SS_{yy} - SSE = SSR = 17{,}170$$

$$s^2 = \text{Error MS} = 262$$

$$s = \sqrt{s^2} = 16.18$$

$$s_{\hat{\beta}_1} = \text{Predictor } x \text{ st. dev.} = 0.8294$$

$$s_{\hat{\beta}_0} = \text{Predictor constant st. dev.} = 76.06$$

$$t = \frac{\hat{\beta}_i}{s_{\hat{\beta}_1}} = \text{Predictor } x \text{ } t \text{ ratio} = 8.10$$

(and the p column gives the observed significance level, 0.000 in this case)

$$r^2 = R - sq = 0.891 = \frac{SSR}{\text{Total SS}}$$

The R-sq (adj.) will not be used in this book, and the F column will be explained in Chapter 10. The regression equation is printed at the top, and all pertinent computations are displayed concisely.

If we ask for an estimate of $E(Y)$ at a particular x, say $x = 90$, and to predict a specific Y at $x = 90$, the printout is as follows:

Fit	Stdev. Fit	95% C.I.	95% P.I.
184.72	5.26	(172.58, 196.87)	(145.48, 223.97)

The "fit" value is $\hat{y}$ at $x = 90$ and

$$\text{St. dev. fit} = s \sqrt{\frac{1}{n} + \frac{(x - \bar{x})^2}{SS_{xx}}}$$

The confidence interval and prediction interval are based on the t statistic, as developed earlier in this chapter.

A computer allows for quick and detailed analyses of the "residuals," the deviations $y - \hat{y}$ for each value of x (y denoting the observed value of the dependent variable). Regression analysis as developed earlier makes certain assumptions about these residuals:

1. They should be "random" quantities.
2. They should have variation that does not depend on x.
3. For estimation and hypothesis testing they should be approximately normally distributed.

The first two assumptions sometimes can be checked by looking at a plot of the residuals graphed against the values of x. Such a plot for the peak power load data is given in Figure 9.16. These residuals look fairly random, but there is a hint of a pattern in that those for the middle x values are mostly negative, while those at either end are positive. We will see a possible reason for this in Chapter 10. The variation in the residuals seems to be about the same for small x values as for large ones, but this is difficult to check here because of so few repeated x values.

FIGURE 9.16
Residuals versus Values of x

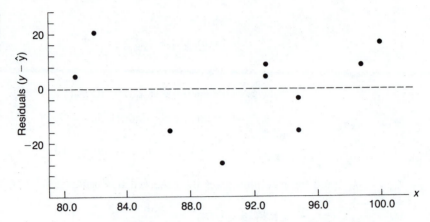

For the normality assumption we could look at a stem-and-leaf plot of the residuals (Figure 9.17), which does exhibit some mound-shapedness. A more discriminating way to check for normality is to use the normal scores plot introduced in Chapter 8. For these residuals the normal scores plot (Figure 9.18) shows quite a good fit to a straight line, which supports the assumption of normality of residuals.

What if the assumptions are not met or the simple regression model does not fit? Then we must work harder to find an acceptable probabilistic model. To illustrate one way to improve the model, we will look at a data set for which a transformation will solve the problem.

FIGURE 9.17
Stem-and-Leaf Plot of Residuals

1	-2	8
3	-1	44
4	-0	4
(4)	0	4588
2	1	4
1	2	1

FIGURE 9.18
Plot of Residuals versus Normal Scores (Peak Power Load Data)

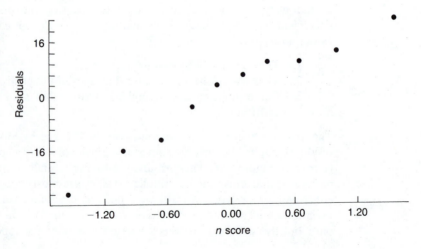

TABLE 9.4
Average Population/Sq. Mile

Row	Year	Apop	ln Apop
1	1790	4.5	1.50408
2	1800	6.1	1.80829
3	1810	4.3	1.45862
4	1820	5.5	1.70475
5	1830	7.4	2.00148
6	1840	9.8	2.28238
7	1850	7.9	2.06686
8	1860	10.6	2.36085
9	1870	10.9	2.38876
10	1880	14.2	2.65324
11	1890	17.8	2.87920
12	1900	21.5	3.06805
13	1910	26.0	3.25810
14	1920	29.9	3.39786
15	1930	34.7	3.54674
16	1940	37.2	3.61631
17	1950	42.6	3.75185
18	1960	50.6	3.92395
19	1970	57.5	4.05178
20	1980	64.0	4.15888

(Source: *World Almanac and Book of Facts*, 1988.)

Table 9.4 shows the average population per square mile in the United States for each census year since 1790. (The table also gives the natural logs of these figures, which will be used later.) Since these average population figures increase with year, we may start building a model of the relationship between population per square mile and year by fitting a simple linear regression model. The Minitab output for such a model is shown below. (In this analysis the row value—1, 2, ..., 20—was used as a coded year.)

The regression equation is
apop = −8.42 + 3.01 x

Predictor	Coef	Stdev	t-ratio	p
Constant	−8.424	2.968	−2.84	0.011
x	3.0071	0.2478	12.14	0.000

s = 6.390   R-sq = 89.1%   R-sq(adj) = 88.5%

Analysis of Variance

SOURCE	DF	SS	MS	F	p
Regression	1	6013.2	6013.2	147.28	0.000
Error	18	734.9	40.8		
Total	19	6748.2			

To the casual observer the linear model may seem to fit well, with a highly significant positive slope and a large r^2. But a graph of the data and a plot of the residuals will show that the "good" fit is not nearly good enough. The actual data (Figure 9.19) show a definite curvature; the rate of increase seems larger in later years. This curvature is magnified in the plot of residuals from the straight regression line, as seen in Figure 9.20. Here the residuals have a definite pattern; they cannot be regarded as random fluctuations.

FIGURE 9.19
Average Population per Square Mile versus Year

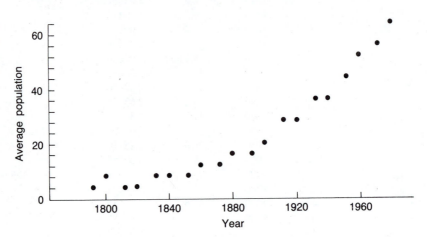

FIGURE 9.20
Residuals versus Values of x for the average Population Data

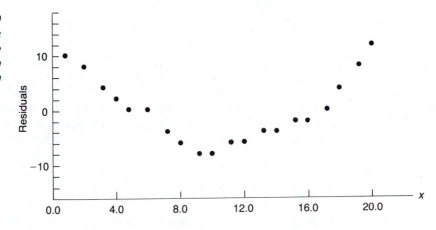

Since the rate of increase in the population seems to be growing larger with time, an exponential relationship between population size and year may be appropriate. If

$$y = ae^{bx}$$

then

$$\ln y = \ln a + bx$$

FIGURE 9.21
Logarithm of
Average Population
per Square Mile
versus Year

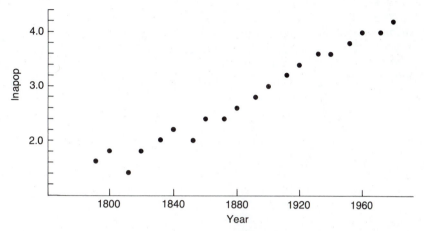

FIGURE 9.21
Logarithm of
Average Population
per Square Mile
versus Year

and there will be a linear relationship between ln y and x. The graph of ln y versus year for the population data is shown in Figure 9.21. This transformation does appear to "linearize" the data.

A simple linear model fit to the transformed data produced the following results:

The regression equation is
lnapop $= 1.24 + 0.148\,x$

Predictor	Coef	Stdev	t-ratio	p
Constant	1.23582	0.05930	20.84	0.000
x	0.148408	0.004950	29.98	0.000

$s = 0.1277$ R-sq $= 98.0\%$ R-sq(adj) $= 97.9\%$

Analysis of Variance

SOURCE	DF	SS	MS	F	p
Regression	1	14.647	14.647	898.84	0.000
Error	18	0.293	0.016		
Total	19	14.940			

This is a much better-fitting straight line as is shown by the higher r^2 and the fact (see Figure 9.22) that the pattern in the residuals is removed.

The stem-and-leaf plot of the residuals (Figure 9.23) does show some mound-shapedness and symmetry, so that the normality assumption may be justified.

So the exponential model fits well and the assumptions for regression analysis seem to be satisfied. However, the fit seems to be better for larger values of x than for the smaller ones (looking at Figures 9.21 and 9.22). Can you suggest a historical reason for this?

We were fortunate in that a rather simple transformation solved our "linearization" problem in the population growth example. Sometimes, however, such a transformation on y or x, or both, is much more difficult to find (and may

FIGURE 9.22
Residuals versus Values of x for the ln Average Population Data

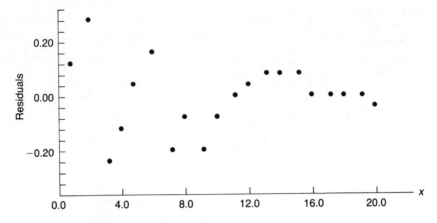

FIGURE 9.23
Stem-and-Leaf Plot of Residuals for ln Average Population Data

2	−2	20
3	−1	8
4	−1	2
6	−0	66
9	−0	400
(4)	0	0112
7	0	5889
3	1	1
2	1	5
1	2	
1	2	7

TABLE 9.5
Tree Basal Areas and Volumes

Basal Area (x) (square feet)	Volume (y) (cubic feet)
0.3	6
0.5	9
0.4	7
0.9	19
0.7	15
0.2	5
0.6	12
0.5	9
0.8	20
0.4	9
0.8	18
0.6	13

not produce such dramatic improvement) or may not even exist. We will give one more example of a case in which a reasonable transformation makes a small improvement. One technique for handling problems in which a simple linearizing transformation is not sufficient will be discussed in Chapter 10.

Table 9.5 shows the basal area (in square feet) for certain experimental trees along with their trunk volume (in cubic feet). A plot of the volume against basal area (Figure 9.24) shows a linear trend and suggests a simple linear regression model to predict volume from basal area. The Minitab output for such a model is given below.

The regression equation is
$y = -1.23 + 23.4x$

Predictor	Coef	Stdev	t-ratio	p
Constant	-1.234	1.080	-1.14	0.280
x	23.404	1.815	12.90	0.000

$s = 1.295$ R-sq = 94.3%   R-sq(adj) = 93.8%

Analysis of Variance

SOURCE	DF	SS	MS	F	p
Regression	1	278.90	278.90	166.35	0.000
Error	10	16.77	1.68		
Total	11	295.67			

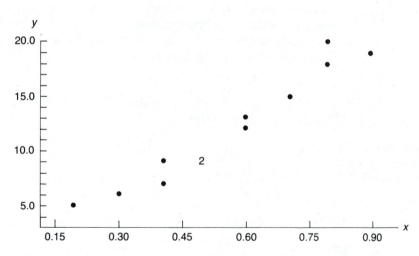

FIGURE 9.24
Volume versus Basal Area of Trees

The straight line fits the data rather well, with a significantly positive slope and a large r^2. However, the residual plot (Figure 9.25) suggests a hint of curvature as we move from lower to higher x values. Also the residuals do not have a symmetric distribution, as shown in Figure 9.26.

How might we reduce this curvature? Here there is no reason to suggest an exponential model of the type that worked for the population data. On thinking

FIGURE 9.25
Residuals versus
Basal Areas of Trees

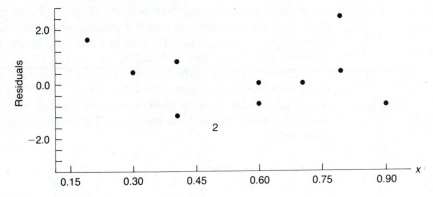

FIGURE 9.26
Stem-and-Leaf Plot
of Residuals for
Tree Volume Data

3	−1 441
5	−0 88
6	−0 1
6	0 12
4	0 58
2	1
2	1 5
1	2
1	2 5

carefully, though, we observe that x is measured in square units and y in cubic units. Couldn't we do better, perhaps, if the units were "matched up," so that we were predicting volume from a cubic measure? Thus we might try raising x to the 3/2 power, and modeling $E(Y)$ as

$$E(Y) = \beta_0 + \beta_1 x^{1.5}$$

The display of y plotted against $x^{1.5}$ is shown in Figure 9.27, and a strong linear trend is still evident.

FIGURE 9.27
Volume versus (Tree
Basal Area)$^{1.5}$

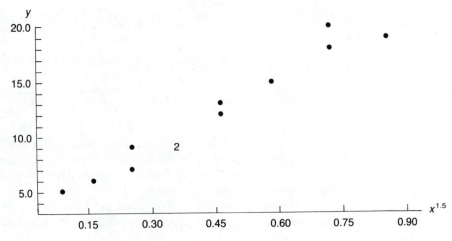

The regression analysis below shows, again, a strong positive slope and an r^2 slightly higher than that for the untransformed data. The value of the transformation, however, is more clearly seen in the residuals. Figure 9.28 shows that the transformation on x removed most of the curvature from the pattern of residuals (they now look more random), and Figure 9.29 shows that the distribution of residuals is now more symmetric about zero. The gain here is not dramatic, but the linear model does fit the transformed data a little better, and the assumptions underlying regression modeling are better fulfilled after transforming the x scale.

The regression equation is
$$y = 2.53 + 21.2 x^{**}1.5$$

Predictor	Coef	Stdev	t-ratio	p
Constant	2.5317	0.7123	3.55	0.005
x**1.5	21.193	1.439	14.73	0.000

$s = 1.141$ R-sq = 95.6%   R-sq(adj) = 95.2%

Analysis of Variance

SOURCE	DF	SS	MS	F	p
Regression	1	282.64	282.64	216.95	0.000
Error	10	13.03	1.30		
Total	11	295.67			

FIGURE 9.28
Residuals versus
(Tree Basal Area)[1.5]

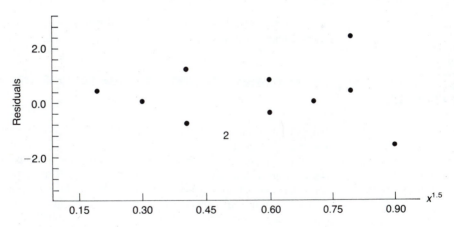

FIGURE 9.29
Stem-and-Leaf Plot
of Residuals for the
Transformed Tree
Volume Data

1	−1	6
3	−1	00
4	−0	8
6	−0	30
6	0	03
4	0	56
2	1	1
1	1	
1	2	3

As you see, there may not be a "best" model for all situations. The statistical goal is to find a good model that will serve the purposes of the investigator for estimation and prediction, but, at the same time, will satisfy the assumptions underlying regression theory. In the exercises you may try your hand at coming up with good models.

9.10 CONCLUSION

Many engineering problems involve modeling the relationship between a dependent variable Y and one or more independent variables. In this chapter we have discussed the case of a single independent variable x. The regression analysis presented here considers only the situation in which $E(Y)$ is a linear function of x, but the general steps to follow are these:

Step 1. Hypothesize the form of $E(Y)$.
Step 2. Use the sample data to estimate unknown parameters in the model.
Step 3. Specify the probability distribution of the random component of the model.
Step 4. Statistically check the adequacy of the model.
Step 5. When satisfied with the model's adequacy, use it for prediction and estimation.

In Chapter 10 we consider the situation in which we can use more than one independent variable to model $E(Y)$.

SUPPLEMENTARY EXERCISES

9.35 Two different elements, nickel and iron, can be used in bonding zircaloy components. An experiment was conducted to determine whether there is a correlation between the strengths of the bonds for the two elements. Two pairs of components were randomly selected from each of seven different batches of the zircaloy components. One pair in each batch was bonded by nickel and the other pair by iron, and the strength of the bond was determined for each pair by measuring the amount of pressure (in thousands of pounds per square inch) required to separate the bonded pair. The data from the experiment are shown below:

Batch	Nickel	Iron
1	71.3	76.2
2	71.8	73.1
3	80.3	86.9
4	77.0	82.4
5	77.4	78.5
6	79.9	88.2
7	70.1	75.1

(a) Find the correlation coefficient for the data.
(b) Find the coefficient of determination and interpret it.
(c) Is there sufficient evidence to indicate that the correlation between the strengths of the bonds of the two elements differs from 0? Use $\alpha = 0.05$.

9.36 Labor and material costs are two basic components in analyzing the cost of construction. Changes in the component costs, of course, will lead to changes in total construction costs.

Month	Construction Cost[a] y	Index of All Construction Materials[b] x
January	193.2	180.0
February	193.1	181.7
March	193.6	184.1
April	195.1	185.3
May	195.6	185.7
June	198.1	185.9
July	200.9	187.7
August	202.7	189.6

[a]Source: United States Department of Commerce, Bureau of the Census.
[b]Source: United States Department of Labor, Bureau of Labor Statistics.
Tables were given in Tables E-1 (p. 43) and E-2 (p. 44), respectively, in *Construction Review*. United States Department of Commerce, Oct. 1976, 22 (8).

(a) Use the data in the table to find a measure of the importance of the materials component. Do this by determining the fraction of reduction in the variability of the construction cost index that can be explained by a linear relationship between the construction cost index and the material cost index.
(b) Do the data provide sufficient evidence to indicate a nonzero correlation between Y and X?

9.37 Use the method of least squares and the sample data in the table to model the relationship between the number of items produced by a particular manufacturing process and the total variable cost involved in production. Find the coefficient of determination and explain its significance in the context of this problem.

Total Output y	Total Variable Cost x dollars
10	10
15	12
20	20
20	21
25	22
30	20
30	19

9.38   Does a linear relationship exist between the Consumer Price Index (CPI) and the Dow Jones Industrial Average (DJA)? A random sample of 10 months selected from several years in the 1970s produced the following corresponding DJA and CPI data:

DJA y	CPI x
660	13.0
638	14.2
639	13.7
597	15.1
702	12.6
650	13.8
579	15.7
570	16.0
725	11.3
738	10.4

(a) Find the least-squares line relating the DJA, y, to the CPI, x.

(b) Do the data provide sufficient evidence to indicate that x contributes information for the prediction of Y? Test by using $\alpha = 0.05$.

(c) Find a 95% confidence interval for β_1 and interpret your result.

(d) Suppose you want to estimate the value of the DJA when the CPI is at 15.0. Should you calculate a 95% prediction interval for a particular value of the DJA, or a 95% confidence interval for the mean value of the DJA? Explain the difference.

(e) Calculate both intervals considered in part (d) when the CPI is 15.0.

9.39   Spiraling energy costs have generated interest in energy conservation in businesses of all sizes. Consequently firms planning to build new plants or make additions to existing facilities have become very conscious of the energy efficiency of proposed new structures. Such firms are interested in knowing the relationship between a building's yearly energy consumption and the factors that influence heat loss. Some of these factors are the number of building stories above and below ground, the materials used in the construction of the building shell, the number of square feet of building shell, and climatic conditions (American Society of Heating, Refrigerating, and Air Conditioning Engineers [ASHRAE] *Guide and Data Book*, 1968). The table below lists the energy consumption in British thermal units (a BTU is the amount of heat required to raise 1 pound of water 1°F) for 1980 for twenty-two buildings that were all subjected to the same climatic conditions.

(a) Find the least-squares line for BTUs as a function of shell area. (You may round the data to the thousands digit.)

(b) Find a 95% confidence interval for the slope of the line estimated in (a) and interpret your result.

BTU/Year (thousands)	Shell Area (square feet)	BTU/Year (thousands)	Shell Area (square feet)
1,371,000	13,530	337,500	5,650
2,422,000	26,060	567,500	8,001
672,200	6,355	555,300	6,147
233,100	4,576	239,400	2,660
218,900	24,680	2,629,000	19,240
354,000	2,621	1,102,000	10,700
12,220,000	59,660	2,680,000	23,680
3,135,000	23,350	423,500	9,125
1,470,000	18,770	423,500	6,510
1,408,000	12,220	1,691,000	13,530
2,201,000	25,490	1,870,000	18,860

9.40 Does the national 55-mile-per-hour highway speed limit provide a substantial savings in fuel? To investigate the relationship between automobile gasoline consumption and driving speed, a small economy car was driven twice over the same stretch of an interstate freeway at each of six different speeds. The numbers of miles per gallon measured for each of the twelve trips are shown below:

Miles per hour	50	55	60	65	70	75
Miles per gallon	34.8, 33.6	34.6, 34.1	32.8, 31.9	32.6, 30.0	31.6, 31.8	30.9, 31.7

(a) Fit a least-squares line to the data.
(g) Is there sufficient evidence to conclude there is a linear relationship between speed and gasoline consumption?
(c) Construct a 90% confidence interval for β_1.
(d) Construct a 95% confidence interval for miles per gallon when the speed is 72 miles per hour.
(e) Construct a 95% prediction interval for miles per gallon when the speed is 58 miles per hour.

9.41 At temperatures approaching absolute zero (273 degrees below zero Celsius), helium exhibits traits that defy many laws of conventional physics. An experiment has been conducted with helium in solid form at various temperatures near absolute zero. The solid helium is placed in a dilution refrigerator along with a solid impure substance, and the fraction (in weight) of the impurity passing through the solid helium is recorded. (The phenomenon of solids passing directly through solids is known as *quantum tunneling*.) The data are given in the table.

(a) Fit a least-squares line to the data.
(b) Test the null hypothesis $H_0: \beta_1 = 0$ against the alternative hypothesis $H_a: \beta_1 < 0$ at the $\alpha = 0.01$ level of significance.

(c)   Compute r^2 and interpret your results.

(d)   Find a 95% prediction interval for the percentage of the solid impurity passing through solid helium at $-273°C$. (Note that this value of x is outside the experimental region, where use of the model for prediction may be dangerous.)

Temperature $x°C$	Proportion of Impurity Passing Through Helium y
-262.0	0.315
-265.0	0.202
-256.0	0.204
-267.0	0.620
-270.0	0.715
-272.0	0.935
-272.4	0.957
-272.7	0.906
-272.8	0.985
-272.9	0.987

9.42   The data in the table were collected to calibrate a new instrument for measuring interocular pressure. The interocular pressure for each of ten glaucoma patients was measured by the new instrument and by a standard, reliable, but more time-consuming method.

Patient	Reliable Method x	New Instrument y
1	20.2	20.0
2	16.7	17.1
3	17.1	17.2
4	26.3	25.1
5	22.2	22.0
6	21.8	22.1
7	19.1	18.9
8	22.9	22.2
9	23.5	24.0
10	17.0	18.1

(a)   Fit a least-squares line to the data.

(b)   Calculate r and r^2. Interpret each of these qualities.

(c)   Predict the pressure measured by the new instrument when the reliable method gives a reading of 20.0. Use a 90% prediction interval.

9.43   During June, July, and early August of 1981 a total of ten bids were made by DuPont, Seagram, and Mobil to take over Conoco. Finally, on August 5, DuPont announced that it had succeeded. The total value of the offer accepted by Conoco was $7.54 billion, making it the largest takeover in the history of American business at that time. As part of an analysis of the Conoco takeover, Richard S. Ruback ("The Conoco Takeover and Stockholder Returns," *Sloan Management Review* 23, 1982) used regression analysis to examine whether movements in the rate of return of each of the above-mentioned companies' common stock could be explained by movements in the rate of return of the stock market as a whole. He used the following model: $Y = \beta_0 + \beta_1 X + \varepsilon$, where Y is the daily rate of return of a stock, X is the daily rate of return of the stock market as a whole as measured by the daily rate of return of Standard & Poor's 500 Composite Index, and ε is assumed to possess the characteristics described in the assumptions of Section 9.4. This model is known in the finance literature as the *market model*. Notice that the parameter β_1 reflects the sensitivity of the stock's rate of return to movements in the stock market as a whole. Using daily data from the beginning of 1979 through the end of 1980 ($n = 504$), Ruback obtained the following least-squares line for the four firms in question:

Firm	Estimated Market Model	
Conoco	$\hat{Y} = 0.0010 + 1.40X$	($t = 21.93$)
DuPont	$\hat{Y} = -0.0005 + 1.21X$	($t = 18.76$)
Mobil	$\hat{Y} = 0.0010 + 1.62X$	($t = 16.21$)
Seagram	$\hat{Y} = 0.0013 + 0.76X$	($t = 6.05$)

The t statistics associated with values of $\hat{\beta}_1$ are shown to the right of each least-squares prediction equation.

(a) For each of the above models test $H_0: \beta_1 = 0$ vs. $H_a: \beta_1 \neq 0$. Use $\alpha = 0.01$. Draw the appropriate conclusions regarding the usefulness of the market model in each case.

(b) If the rate of return of Standard & Poor's 500 Composite Index increased by 0.10, how much change would occur in the mean rate of return of Conoco's common stock? How much change would occur in the mean rate of return of Seagram's common stock?

(c) Which of the two stocks, Conoco or Seagram, appears to be more responsive to changes in the market as a whole? Explain.

9.44 A study was conducted to determine whether there is a linear relationship between the breaking strength y of wooden beams and the specific gravity x of the wood. Ten randomly selected beams of the same cross-sectional dimensions were stressed until they broke. The breaking strengths and the density of the wood are shown below for each of the ten beams:

Beam	Specific Gravity x	Strength y
1	0.499	11.14
2	0.558	12.74
3	0.604	13.13
4	0.441	11.51
5	0.550	12.38
6	0.528	12.60
7	0.418	11.13
8	0.480	11.70
9	0.406	11.02
10	0.467	11.41

(a) Fit the model $Y = \beta_0 + \beta_1 x + \varepsilon$.

(b) Test $H_0: \beta_1 = 0$ against the alternative hypothesis $H_a: \beta_1 \neq 0$.

(c) Estimate the mean strength for beams with specific gravity 0.590, using a 90% confidence interval.

9.45 A firm's *demand curve* describes the quantity of its product that the firm can sell at different possible prices, other things being equal (Leftwich, *The Price System and Resource Allocation*, Dryden Press, 1973). Over a period of a year a tire company varied the price of one of its radial tires to estimate the firm's demand curve for the tire. They observed that when the price was set very low or very high, they sold few tires. The latter result they understood; the former they determined was due to consumers' misperception that the tire's low price must be linked to poor quality. The data in the table describe the tire's sales over the experimental period.

Tire Price x (dollars)	Number Sold y (hundreds)
20	13
35	57
45	85
60	43
70	17

(a) Calculate a least-squares line to approximate the firm's demand function.

(b) Construct a scattergram and plot your least-squares line as a check on your calculations.

(c) Test $H_0: \beta_1 = 0$, using a two-tailed test and $\alpha = 0.05$. Draw the appropriate conclusions in the context of the problem.

(d) Does your nonrejection of H_0 in part (c) imply that no relationship exists between tire price and sales volume? Explain.

(e) Calculate the coefficient of determination for the least-squares line of part (a) and interpret its value in the context of the problem.

9.46 The octane number y of refined petroleum is related to the temperature x of the refining process, but is also related to the particle size of the catalyst. An experiment with a small particle catalyst gave a fitted least-squares line of

$$\hat{y} = 9.360 + 0.115x$$

with $n = 31$ and $s_{\beta_1} = 0.0225$. An independent experiment with a large particle catalyst gave

$$\hat{y} = 4.265 + 0.190x$$

with $n = 11$ and $s_{\beta_1} = 0.0202$. (*Source*: Gweyson and Cheasley, *Petroleum Refiner*, August, 1959, p. 135.)

(a) Test the hypotheses that the slopes are significantly different from zero, with each test at the 5% significance level.

(b) Test, at the 5% level, that the two types of catalyst produce the same slope in the relationship between octane number and temperature. Assume that the true variances of the slope estimates are equal.

9.47 The data in the table give the mileages per gallon obtained by a test automobile when using gasolines of varying levels of octane.

Mileage, y (miles per gallon)	Octane, x
13.0	89
13.2	93
13.0	87
13.6	90
13.3	89
13.8	95
14.1	100
14.0	98

(a) Calculate r and r^2.
(b) Do the data provide sufficient evidence to indicate a correlation between octane level and miles per gallon for the test automobile?

For the following data sets find a good simple linear regression model to answer the problems under study. You should always test the slope of a line for significance, and use the r^2 values to judge whether or not one model fits the data better than another. Then study the residuals to assess goodness of fit and to check the underlying assumptions. Some data sets will require transforming the dependent variable, the independent variable, or both. The answers in the back of this book will provide some suggested solutions, but remember, there often is no single "best" answer to any one problem.

9.48 The following table gives average height and weight measurements for children. Find regression models for each of the following:
(a) Predicting a boy's height from his age.
(b) Predicting a boy's age from his height.
(c) Predicting a girl's height from her age.
(d) Predicting a boy's weight from his height.
(e) Predicting a girl's weight from her age.
(f) Predicting a girl's weight from the weight of a boy.

Average Height and Weight of Children

Age (yrs)	Boys Height (cm)	Boys Weight (kg)	Girls Height (cm)	Girls Weight (kg)
1	73.6	9.5	73.6	9.1
2	83.8	11.8	83.8	11.3
3	91.4	14.0	91.4	13.6
4	99.0	15.4	99.0	15.0
5	106.6	17.7	104.1	17.2
6	114.2	20.9	111.7	20.4
7	119.3	23.1	119.3	22.2

Average Height and Weight of Children (cont)

	Boys		Girls	
Age (yrs)	Height (cm)	Weight (kg)	Height (cm)	Weight (kg)
8	127.0	25.9	127.0	25.4
9	132.0	28.6	132.0	28.1
10	137.1	31.3	137.1	31.3
11	142.2	34.9	142.2	34.9
12	147.3	37.7	147.3	39.0
13	152.4	41.7	152.4	45.5
14	157.5	48.5	157.5	48.5

(Source: *World Almanac and Book of Facts*, 1988, p. 825)

9.49 The U.S. Federal Highway Administration publishes data on reaction distance for drivers and braking distance for a typical automobile traveling on dry pavement. Some of the data are as follows:

Speed (mph)	Reaction Distance (feet)	Braking Distance (feet)
20	22	20
30	33	40
40	44	72
50	55	118
60	66	182
70	77	266

Total stopping distance is the sum of the reaction distance and braking distance. Find simple linear regression models for estimating each average:

(a) reaction distance from speed.
(b) braking distance from speed.
(c) total stopping distance from speed.

Use these answers to fill in a line of the table for an automobile traveling 55 mph.

9.50 The number of years required for a planet to complete a revolution around the sun (called its *sidereal year*) is related to the planet's distance from the sun. The following table shows average distance from the sun (in millions of miles) and sidereal year measurements for the planets. Use the data to find a linear regression model relating these two measurements, with sidereal year as the dependent variable. [*Hint*: Knowledge of Kepler's Third Law may help you suggest an appropriate transformation.]

Planet	Average Distance from Sun	Sidereal Year
Mercury	36.0	0.241
Venus	67.0	0.615
Earth	93.0	1.000
Mars	141.5	1.880
Jupiter	483.0	11.900
Saturn	886.0	29.500
Uranus	1782.0	84.000
Neptune	2793.0	165.000
Pluto	3670.0	248.000

9.51 The data below show the amount of suspended particulate matter (in millions of metric tons) emitted into the air for various years (in the United States).

1940	22.8	1976	9.7
1950	24.5	1977	9.1
1960	21.1	1978	9.2
1970	18.1	1979	9.0
1971	16.7	1980	8.5
1972	15.2	1981	7.9
1973	14.1	1982	7.0
1974	12.4	1983	6.7
1975	10.4	1984	7.0
		1985	7.3

(Source: *USA by Numbers*, Zero Population Growth, Inc., 1988)

Find a regression model for predicting amount of particulate matter as a function of the year. Use your model to predict the amount of suspended particulate matter for 1986. Do you see any difficulty with this approach to prediction?

9.52 The Materials Science Department of the University of Florida carried out a research project on the properties of self-lubricating bronze bearings manufactured by sintering copper and tin powders. Four variables of importance are the sintering time, change in weight of the bearings, Rockwell hardness of the bearings, and porosity as measured by the weight of liquid wax taken up by the bearing. Data on these four variables are given below. (Note that not all measurements were taken on each sample.) Using linear regression techniques, model the following:
    (a) change in weight as a function of sintering time.
    (b) Rockwell hardness as a function of sintering time.
    (c) weight of wax absorbed as a function of sintering time.
Do you see any sample values that are questionable in this study?

Sample Bearing	Sintering Time (min)	Change in Weight (g)	Rockwell Hardness	Weight of Wax (g)
1	1.0	−0.224		
2	1.0	−0.197		
3	2.0	−0.259		
4	2.0	−0.257		
5	2.0	−0.258		
6	2.0	−0.260		
7	4.0	−0.262		
8	4.0	−0.264		
9	4.0	−0.266		
10	4.0	−0.267		
11	5.5	−0.263		
12	5.5	−0.252		
13	7.0	−0.276	48.18	
14	7.0	−0.276	48.18	0.615
15	7.0	−0.280	48.18	0.606
16	7.0	−0.276	48.18	0.611
17	9.0	−0.281	45.01	
18	9.0	−0.279	45.01	0.586
19	11.0	−0.282	45.45	
20	11.0	−0.282	45.45	0.511
21	11.0	−0.284	45.45	0.454
22	11.0	−0.285	45.45	0.440
23	13.0	−0.289	47.80	
24	13.0	−0.282	47.80	0.393
25	15.0	−0.288	46.06	
26	15.0	−0.292	46.06	0.322
27	15.0	−0.289	46.06	0.343
28	15.0	−0.286	46.06	0.341

10

Multiple Regression Analysis

ABOUT THIS CHAPTER

Following up on the work of Chapter 9, we now proceed to build models in which the mean of one variable can be written as a function of two or more independent variables. For example, the average amount of energy required to heat a house depends not only on the air temperature but also on the size of the house, the amount of insulation, and the type of heating unit, among other things. We will use *multiple regression* models for estimating means and for predicting future values of key variables under study.

CONTENTS

10.1 INTRODUCTION

Most practical applications of regression analysis utilize models that are more complex than the simple straight-line model. For example, a realistic model for a power plant's peak power load would include more than just the daily high temperature. Factors such as humidity, day of the week, and season are a few of the many variables that might be related to peak load. Thus we would want to incorporate these and other potentially important independent variables into the model to make more accurate predictions.

Probabilistic models that include terms involving x^2, x^3 (or higher-order terms), or more than one independent variable are called *multiple regression models*. The general form of these models is

$$Y = \beta_0 + \beta_1 x_1 + \beta_2 x_2 + \cdots + \beta_k x_k + \varepsilon$$

The dependent variable Y is now written as a function of k independent variables $x_1, x_2, \ldots, x_k$. The random error term is added to allow for deviation between the deterministic part of the model, $\beta_0 + \beta_1 x_1 + \cdots + \beta_k x_k$, and the value of the dependent variable Y. The random component makes the model probabilistic rather than deterministic. The value of the coefficient β_i determines the contribution of the independent variable x_i, and β_0 is the Y intercept. The coefficients $\beta_0, \beta_1, \ldots, \beta_k$ will usually be unknown, since they represent population parameters.

At first glance it might appear that the regression model shown above would not allow for anything other than straight-line relationships between Y and the independent variables, but this is not true. Actually $x_1, x_2, \ldots, x_k$ can be functions of variables as long as the functions do not contain unknown parameters. For example the yield Y of a chemical process could be a function of the independent variables

x_1 = temperature at which the reaction takes place

x_2 = (temperature)^2 = x_1^2

x_3 = pressure at which the reaction takes place

We might hypothesize the model

$$E(Y) = \beta_0 + \beta_1 x_1 + \beta_2 x_1^2 + \beta_3 x_3$$

Although this model contains a second-order (or "quadratic") term x_1^2 and therefore is curvilinear, the model is still *linear in the unknown parameters* $\beta_0, \beta_1, \beta_2$, and β_3. Multiple regression models that are linear in the unknown parameters (the βs) are called *linear models*, even though they may contain nonlinear independent variables.

The same steps we followed in developing a straight-line model are applicable to the multiple regression model.

Step 1. First, hypothesize the form of the model. This involves the choice of the independent variables to be included in the model.

Step 2. Next, estimate the unknown parameters $\beta_0, \beta_1, \ldots, \beta_k$.

Step 3. Then, specify the probability distribution of the random error component ε and estimate its variance σ^2.

Step 4. Fourth, check the adequacy of the model.

Step 5. Finally, use the fitted model to estimate the mean value of Y or to predict a particular value of Y for given values of the independent variables.

First we consider steps 2–5, leaving the more difficult problem of model construction until last.

10.2 FITTING THE MODEL: THE LEAST-SQUARES APPROACH

The method of fitting multiple regression models is identical to that of fitting the simple straight-line model: the method of least squares. That is, we choose the estimated model

$$\hat{y} = \hat{\beta}_0 + \hat{\beta}_1 x_1 + \cdots + \hat{\beta}_k x_k$$

that minimizes

$$\text{SSE} = \sum_{i=1}^{n} (y_i - \hat{y}_i)^2$$

As in the case of the simple linear model, the sample estimates $\hat{\beta}_0, \hat{\beta}_1, \ldots, \hat{\beta}_k$ are obtained as a solution of a set of simultaneous linear equations.

The primary difference between fitting the simple and multiple regression models is computational difficulty. The $(k + 1)$ simultaneous linear equations that must be solved to find the $(k + 1)$ estimated coefficients $\hat{\beta}_0, \hat{\beta}_1, \ldots, \hat{\beta}_k$ are difficult (sometimes nearly impossible) to solve with a pocket or desk calculator. Consequently, we resort to the use of computers. Many computer packages have been developed to fit a multiple regression model using the method of least squares. We will present output from several of the more popular computer packages instead of presenting the tedious hand calculations required to fit the models. Most package regression programs have similar output, so you should have little trouble interpreting regression output from other packages. As in previous chapters, we will make most of our computations by using Minitab.

Recall the peak power load data from Chapter 9, repeated in Table 10.1. We previously used a straight-line model to describe the peak power load/daily high temperature relationship. Now suppose we want to hypothesize a curvilinear relationship:

$$Y = \beta_0 + \beta_1 x + \beta_2 x^2 + \varepsilon$$

Note that the scattergram in Figure 10.1 provides some support for the inclusion of the term $\beta_2 x^2$ in the model, since there appears to be some curvature present in the relationship.

TABLE 10.1
Peak Power Load Data

Daily High Temperature (x)	Peak Power Load (y)
95	214
82	152
90	156
81	129
99	254
100	266
93	210
95	204
93	213
87	150

FIGURE 10.1
Scattergram of
Power Load Data,
Table 10.1

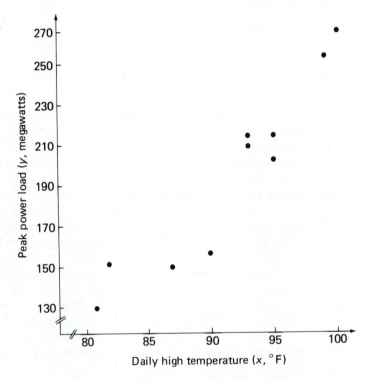

Part of the output from the Minitab multiple regression routine for the peak power load data is reproduced in Figure 10.2. The least-squares estimates of the β parameters appear in the column labeled "Coef." You can see that $\hat{\beta}_0 = 1784.2$, $\hat{\beta}_1 = -42.39$, and $\hat{\beta}_2 = 0.27$. Therefore the equation that minimizes the SSE for the

FIGURE 10.2
Computer Output for the Peak Power Load Example

The regression equation is
y = 1784 − 42.4 x + 0.272 x-sq

Predictor	Coef	Stdev	t-ratio	p
Constant	1784.2	944.1	1.89	0.101
x	−42.39	21.00	−2.02	0.083
x-sq	0.2722	0.1163	2.34	0.052

s = 12.96   R-sq = 93.9%   R-sq(adj) = 92.2%

Analysis of Variance

SOURCE	DF	SS	MS	F	p
Regression	2	18088.5	9044.3	53.88	0.000
Error	7	1175.1	167.9		
Total	9	19263.6			

data is

$$\hat{y} = 1784.2 - 42.39x + 0.27x^2$$

The value of the SSE, 1175.1, also appears in the printout. We will discuss the rest of the printout as the chapter progresses.

Note that the graph of the multiple regression model (Figure 10.3) provides a good fit to the data of Table 10.1. However before we can more formally measure the utility of the model, we need to estimate the variance of the error component ε.

FIGURE 10.3
Plot of Curvilinear Model

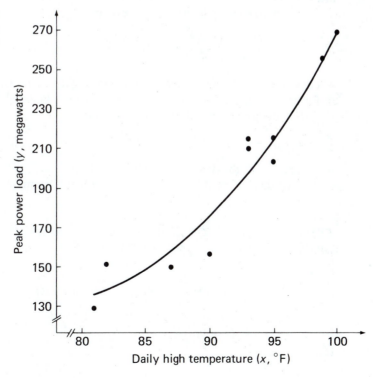

10.3   ESTIMATION OF σ^2, THE VARIANCE OF ε

The specification of the probability distribution of the random error component ε of the multiple regression model follows the same general outline as for the straight-line model. We assume that ε is normally distributed with mean zero and constant variance σ^2 for any set of values for the independent variables $x_1, x_2, \ldots,$ x_k. Furthermore the errors are assumed to be independent. Given these assumptions the remaining task in specifying the probability distribution of ε is to estimate σ^2.

For example, in the quadratic model describing peak power load as a function of daily high temperature, we found a minimum SSE = 1175.1. Now we want to use this quantity to estimate the variance of ε. Recall that the estimator for the straight-line model was $s^2 = \text{SSE}/(n-2)$ and note that the denominator is $n -$ (number of estimated β parameters), which is $n - (2)$ in the straight-line model. Since we must estimate one more parameter β_2 for the quadratic model $Y = \beta_0 + \beta_1 x + \beta_2 x^2 + \varepsilon$, the estimate of σ^2 is

$$s^2 = \frac{\text{SSE}}{n-3}$$

That is, the denominator becomes $(n-3)$ because there are now three β parameters in the model.

The numerical estimate for this example is

$$s^2 = \frac{\text{SSE}}{10-3} = \frac{1175.1}{7} = 167.87$$

where s^2 is called the mean square for error, or MSE. This estimate of σ^2 is shown in Figure 10.2 in the column titled "MS" and the row titled "Error."

For the general multiple regression model

$$Y = \beta_0 + \beta_1 x_1 + \beta_2 x_2 + \cdots + \beta_k x_k + \varepsilon$$

we must estimate the $(k+1)$ parameters $\beta_0, \beta_1, \beta_2, \ldots, \beta_k$. Thus the estimator of σ^2 is the SSE divided by the quantity $n -$ (number of estimated β parameters).

ESTIMATOR OF σ^2 FOR MULTIPLE REGRESSION MODEL WITH k INDEPENDENT VARIABLES

$$\text{MSE} = \frac{\text{SSE}}{n - (\text{number of estimated }\beta\text{ parameters})}$$

$$= \frac{\text{SSE}}{n - (k+1)}$$

We use the estimator of σ^2 both to check the adequacy of the model (Sections 10.4 and 10.5) and to provide a measure of the reliability of predictors and estimates when the model is used for those purposes (Section 10.6). Thus you can see that the estimation of σ^2 plays an important part in the development of a regression model.

10.4 A TEST OF MODEL ADEQUACY: THE COEFFICIENT OF DETERMINATION

To find a statistic that measures how well a multiple regression model fits a set of data, we use the multiple regression equivalent of r^2, the coefficient of determination for the straight-line model (Chapter 9). Thus we define the *multiple coefficient of determination R^2* as

$$R^2 = 1 - \frac{\sum_{i=1}^{n} (y_i - \hat{y}_i)^2}{\sum_{i=1}^{n} (y_i - \bar{y})^2} = 1 - \frac{\text{SSE}}{\text{SS}_{yy}}$$

where $\hat{y}_i$ is the predicted value of Y_i for the model. Just as for the simple linear model, R^2 represents the fraction of the sample variation of the y values (measured by SS_{yy}) that is explained by the least-squares prediction equation. Thus $R^2 = 0$ implies a complete lack of fit of the model to the data, and $R^2 = 1$ implies a perfect fit, with the model passing through every data point. In general the larger the value of R^2, the better the model fits the data.

To illustrate, the value of $R^2 = 0.939$ for the peak power load data is indicated on Figure 10.2. This value of R^2 implies that using the independent variable daily high temperature in a quadratic model results in a 93.9% reduction in the total *sample* variation (measured by SS_{yy}) of peak power load Y. Thus R^2 is a sample statistic that tells us how well the model fits the data, and thereby represents a measure of the adequacy of the model.

The fact that R^2 is a sample statistic implies that it can be used to make inferences about the utility of the entire model for predicting the population of y values at each setting of the independent variables. In particular for the peak power load data, the test

H_0: $\beta_1 = \beta_2 = 0$

H_a: At least one of the coefficients is nonzero

would formally test the global utility of the model. The test statistic used to test this null hypothesis is

Test statistic: $F = \dfrac{R^2/k}{(1 - R^2)/[n - (k + 1)]}$

where n is the number of data points and k is the number of parameters in the model, not including β_0. The test statistic F will have the F probability distribution with k degrees of freedom in the numerator and $[n - (k + 1)]$ degrees of freedom in the denominator. The tail values of the F distribution are given in Tables 7 and 8 in the Appendix.

The F-test statistic becomes large as the coefficient of determination R^2 becomes large. To determine how large F must be before we can conclude at a given significance level that the model is useful for predicting Y, we set up the rejection region as follows:

Rejection region: $F > F_\alpha(k, n - (k + 1))$

For the electrical usage example ($n = 10$, $k = 2$, $n - (k + 1) = 7$, and $\alpha = 0.05$), we reject $H_0: \beta_1 = \beta_2 = 0$ if

$F > F_{0.05}(2, 7)$

or

$F > 4.74$

From the computer printout (Figure 10.2) we find that the computed F is 53.88. Since this value greatly exceeds the tabulated value of 4.74, we conclude that at least one of the model coefficients β_1 and β_2 is nonzero. Therefore this global F test indicates that the quadratic model $Y = \beta_0 + \beta_1 x + \beta_2 x^2 + \varepsilon$ is useful for predicting peak load.

TESTING THE UTILITY OF A MULTIPLE REGRESSION MODEL: THE GLOBAL F TEST

$H_0: \beta_1 = \beta_2 = \cdots = \beta_k = 0$

H_a: At least one of the β parameters does not equal zero

Test statistic: $F = \dfrac{R^2/k}{(1 - R^2)/[n - (k + 1)]}$

Assumptions: See Section 9.4 for the four assumptions about the random component ε.

Rejection region: $F > F_\alpha(k, n - (k + 1))$

where

n = number of data points

k = number of β parameters in the model, excluding β_0

EXAMPLE 10.1

A study is conducted to determine the effects of company size and presence or absence of a safety program on the number of work-hours lost due to work-related accidents. A total of 40 companies are selected for the study, 20 randomly chosen from companies having no active safety programs, and the other 20 from companies that have enacted active safety programs. Each company is monitored for a 1-year period, and the following model is proposed:

$$E(Y) = \beta_0 + \beta_1 x_1 + \beta_2 x_2$$

where

Y = lost work-hours over the 1-year study period

x_1 = number of employees

$$x_2 = \begin{cases} 1 & \text{if an active safety program is used} \\ 0 & \text{if no active safety program is used} \end{cases}$$

The variable x_2 is called a *dummy* or *indicator variable*. Dummy variables are used to represent categorical or qualitative independent variables, like "presence or absence of a safety program." Note that the coefficient β_2 of the dummy variable x_2 represents the expected difference in lost work-hours between companies with safety programs and those without safety programs, assuming the companies have the same number of employees. For example the expected number of lost work-hours for a company with 500 employees and no safety program is, according to the model,

$$E(Y) = \beta_0 + \beta_1(500) + \beta_2(0)$$
$$= \beta_0 + 500\beta_1$$

while the expected lost work-hours for a company with 500 employees and an active safety program is

$$E(Y) = \beta_0 + \beta_1(500) + \beta_2(1)$$
$$= \beta_0 + 500\beta_1 + \beta_2$$

You can see that the difference in the means is β_2. Thus the use of the dummy variable allows us to assess the effects of a qualitative variable on the mean response.

The data are shown in Table 10.2.
(a) Test the utility of the model by testing $H_0: \beta_1 = \beta_2 = 0$.
(b) Give and interpret the least-squares coefficients $\hat{\beta}_1$ and $\hat{\beta}_2$.

Solution   (a)   The computer printout corresponding to the least-squares fit of the model

$$Y = \beta_0 + \beta_1 x_1 + \beta_2 x_2 + \varepsilon$$

is shown in Figure 10.4. We want to use this information to test

$$H_0: \beta_1 = \beta_2 = 0$$

H_a: At least one β is nonzero, that is,
the model is useful for predicting Y

TABLE 10.2

Number of employees (x_1)	Safety Program (x_2) ($x_2 = 0$, no program $x_2 = 1$, active program)	Lost Hours Due to Accidents (y) (annually, in thousands of hours)
6490	0	121
7244	0	169
7943	0	172
6478	0	116
3138	0	53
8747	0	177
2020	0	31
4090	0	94
3230	0	72
8786	0	171
1986	0	23
9653	0	177
9429	0	178
2782	0	65
8444	0	146
6316	0	129
2363	0	40
7915	0	167
6928	0	115
5526	0	123
3077	1	44
6600	1	73
2732	1	8
7014	1	90
8321	1	71
2422	1	37
9581	1	111
9326	1	89
6818	1	72
4831	1	35
9630	1	86
2905	1	40
6308	1	44
1908	1	36
8542	1	78
4750	1	47
6056	1	56
7052	1	75
7794	1	46
1701	1	6

FIGURE 10.4
Computer Printout of Least-Squares Fit for the Safety Program Example

The regression equation is
y = 31.7 + 0.143 x1 − 58.2 x2

Predictor	Coef	Stdev	t-ratio	p
Constant	31.670	8.560	3.70	0.001
x1	0.014272	0.001222	11.68	0.000
x2	−58.223	6.316	−9.22	0.000

s = 19.97   R-sq = 85.9%   R-sq(adj) = 85.2%

Analysis of Variance

SOURCE	DF	SS	MS	F	p
Regression	2	90058	45029	112.94	0.000
Error	37	14752	399		
Total	39	104811			

Test statistic:

$$F = \frac{R^2/k}{(1 - R^2)/[n - (k + 1)]} = \frac{R^2/2}{(1 - R^2)/37}$$

Rejection region: For $\alpha = 0.05$,

$$F > F_{0.05}(2, 37)$$

The value of R^2 is 0.86. Then

$$F = \frac{0.86/2}{(1 - 0.86)/37} = 113.6$$

Thus we can be very confident that this model contributes information for the prediction of lost work-hours, since $F = 113.6$ greatly exceeds the tabled value, 3.25. The F value is also given on the printout, so we do not have to perform the calculation. Using the printout value also helps to avoid rounding errors; note the difference between our calculated F value and the printout value of 112.94.

We should be careful not to be too excited about this very large F value, since we have only concluded that the model contributes *some* information about Y. The widths of the prediction intervals and confidence intervals are better measures of the amount of information the model provides about Y. Note that the estimated standard deviation of this model is 19.97. Since most (usually 90% or more) of the values of Y will fall within two standard deviations of the predicted value, we have the notion that the model predictions will usually be accurate to within about 40 thousand hours. We will make these notions more precise in Section 10.5, but the standard deviation helps us obtain a preliminary idea of the amount of information the model contains about Y.

(b) The estimated values of β_1 and β_2 are shown on the printout, Figure 10.4. The least-squares model is

$$\hat{y} = 31.67 + 0.0143x_1 - 58.22x_2$$

The coefficient $\hat{\beta}_1 = 0.0143$ represents the estimated slope of the number of lost hours (in thousands) versus number of employees. In other words we estimate that, on average, each employee loses 14.3 work-hours (0.0143 thousand hours) annually due to accidents. Note that this slope applies both to plants with and plants without safety programs. We will show how to construct models that allow the slopes to differ later in this chapter.

The coefficient $\hat{\beta}_2 = -58.2$ is *not* a slope, because it is the coefficient of the dummy variable x_2. Instead $\hat{\beta}_2$ represents the estimated mean change in lost hours between plants with no safety programs ($x_2 = 0$) and plants with safety programs ($x_2 = 1$), assuming both plants have the same number of employees. Our estimate indicates that an average of 58.2 thousand fewer work-hours are lost by plants with safety programs than by those without them. ■

10.5 ESTIMATING AND TESTING HYPOTHESES ABOUT INDIVIDUAL β PARAMETERS

After determining that the model is useful by testing and rejecting $H_0: \beta_1 = \beta_2 = \cdots \beta_k = 0$, we may be interested in making inferences about particular β parameters that have practical significance. In the peak power load example we fit the model

$$E(Y) = \beta_0 + \beta_1 x + \beta_2 x^2$$

where Y is the peak load and x is the daily maximum temperature. A test of particular interest would be

$H_0: \beta_2 = 0$ (no quadratic relationship exists)

$H_a: \beta_2 > 0$ (the peak power load increases at an increasing rate as the daily maximum temperature increases)

A test of this hypothesis can be performed using a Student's t test. The t test utilizes a test statistic analogous to that used to make inferences about the slope of the simple straight-line model (Section 9.5). The t statistic is formed by dividing the sample estimate $\hat{\beta}_2$ of the population coefficient β_2 by the estimated standard deviation of the repeated sampling distribution of $\hat{\beta}_2$:

Test statistic: $\quad t = \dfrac{\hat{\beta}_2}{s_{\hat{\beta}_2}}$

We use the symbol $s_{\hat{\beta}_2}$ to represent the estimated standard deviation of $\hat{\beta}_2$. Most computer packages list the estimated standard deviation $s_{\hat{\beta}_1}$ for each of the estimated model coefficients β_i. In addition, they usually give the calculated t values for each coefficient in the model.

The rejection region for the test is found in exactly the same way as the rejection regions for the t tests in previous chapters. That is, we consult the t table in the Appendix to obtain an upper-tail value of t. This is a value t_α such that $P(t > t_\alpha) = \alpha$. Then we can use this value to construct rejection regions for either one- or two-tailed tests. To illustrate with the power load example, the error degrees of freedom is $(n - 3) = 7$, the denominator of the estimate of σ^2. Then the rejection region (shown in Figure 10.5) for a one-tailed test with $\alpha = 0.05$ is

Rejection region: $t > t_\alpha(n - 3)$

$$t > 1.895$$

FIGURE 10.5
Rejection Region
for Test of H_0:
$\beta_2 = 0$

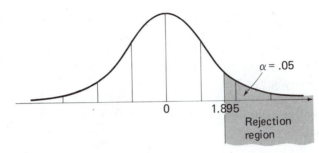

In Figure 10.6 we again show a portion of the computer printout for the peak power load example.

The estimated standard deviations for the model coefficient appear under the column "Stdev." The t statistic for testing the null hypothesis that the true coefficients are equal to zero appear under the column headed "t-ratio." The t value corresponding to the test of the null hypothesis $H_0: \beta_2 = 0$ is the last one in the column, that is, $t = 2.34$. Since this value is greater than 1.895, we conclude that the quadratic term $\beta_2 x^2$ makes a contribution to the predicted value of peak power load, and that there is evidence that the peak power load increases at an increasing rate as the daily maximum temperature increases.

The Minitab printout shown in Figure 10.6 also lists the observed two-tailed significance levels[1] for each t value. These values appear under the column headed "p." The significance level 0.052 corresponds to the quadratic term, and this implies

FIGURE 10.6
Computer Printout
for the Peak Power
Load Example

Predictor	Coef	Stdev	t-ratio	p
Constant	1784.2	944.1	1.89	0.101
x	−42.39	21.00	−2.02	0.083
x-sq	0.2722	0.1163	2.34	0.052

$s = 12.96$ R-sq = 93.9%   R-sq(adj) = 92.2%

Analysis of Variance

SOURCE	DF	SS	MS	F	p
Regression	2	18088.5	9044.3	53.88	0.000
Error	7	1175.1	167.9		
Total	9	19263.6			

[1]See Chapter 8 for a discussion of significance levels in tests of hypotheses.

that we would reject $H_0: \beta_2 = 0$ in favor of $H_a: \beta_2 \neq 0$ at any α level larger than 0.052. Since our alternative was one-sided, $H_a: \beta_2 > 0$, the significance level is half that given in the printout, that is, $\frac{1}{2}(0.052) = 0.026$. Thus there is evidence at the $\alpha = 0.05$ level that the peak power load increases more quickly per unit increase in daily maximum temperature for high temperatures than for low ones.

We can also form a confidence interval for the parameter β_2 as follows:

$$\hat{\beta}_2 \pm t_{\alpha/2}(n-3)s_{\hat{\beta}_2} = 0.272 \pm (1.895)(0.116)$$

or (0.052, 0.492). Note that the t value 1.895 corresponds to $\alpha/2 = 0.05$ and $(n-3) = 7\,df$. This interval constitutes a 90% confidence interval for β_2 and represents an estimate of the rate of curvature in mean peak power load as the daily maximum temperature increases. Note that all values in the interval are positive, reconfirming the conclusion of our test.

Testing a hypothesis about a single β parameter that appears in any multiple regression model is accomplished in exactly the same manner as described for the quadratic electrical usage model. The form of that test is shown here.

TEST OF AN INDIVIDUAL PARAMETER COEFFICIENT IN THE MULTIPLE REGRESSION MODEL

One-Tailed Test	Two-Tailed Test
$H_0: \beta_i = 0$	$H: \beta_i = 0$
$H_a: \beta_i < 0$ (or $H_a: \beta_i > 0$)	$H_a: \beta_i \neq 0$

Test statistic:

$$t = \frac{\hat{\beta}_i}{s_{\hat{\beta}_1}}$$

Test statistic:

$$t = \frac{\hat{\beta}_i}{s_{\hat{\beta}_1}}$$

Rejection region:

$$t < -t_{\alpha}[n - (k+1)]$$
$$(\text{or } t > t_{\alpha}[n - (k+1)]$$
$$\text{when } H_a: \beta_i > 0)$$

Rejection region:

$$t < -t_{\alpha/2}[n - (k+1)]$$
or
$$t > t_{\alpha/2}[n - (k+1)]$$

where

$n =$ number of observations

$k =$ number of β parameters in the model, excluding β_0

Assumptions: See Section 9.4 for the assumptions about the probability distribution for the random error component ε.

If all the tests of model utility indicate that the model is useful for predicting Y, can we conclude that the best prediction model has been found? Unfortunately we cannot. There is no way of knowing (without further analysis) whether the addition of other independent variables will improve the utility of the model, as Example 10.2 indicates.

EXAMPLE 10.2

Refer to Example 10.1, in which we modeled a plant's lost work-hours as a function of number of employees and the presence or absence of a safety program. Suppose a safety engineer believes that safety programs tend to help larger companies even more than smaller ones. Thus instead of a relationship like that shown in Figure 10.7(a), in which the rate of increase in mean lost work-hours per employee is the same for companies with and without safety programs, the engineer believes that the relationship is like that shown in Figure 10.7(b). Note that although the mean number of lost work-hours is smaller for companies with safety programs no matter how many employees the company has, the magnitude of the difference becomes greater as the number of employees increases. When the slope of the relationship between $E(Y)$ and one independent variable (x_1) depends on the value of a second independent variable (x_2), as is the case here, we say that x_1 and x_2 *interact*. The model for mean lost work-hours that includes interaction is written

$$E(Y) = \beta_0 + \beta_1 x_1 + \beta_2 x_2 + \beta_3 x_1 x_2$$

Note that the increase in mean lost work-hours $E(Y)$ for each person increase in number of employees x_1 is no longer given by the constant β_1, but is now $\beta_1 + \beta_3 x_2$. That is, the amount that $E(Y)$ increases for each 1-unit increase in x_1 is dependent on whether the company has a safety program. Thus the two variables x_1 and x_2 interact to affect Y.

The forty data points listed in Table 10.2 were used to fit the model with interaction. A portion of the computer printout is shown in Figure 10.8. Test the hypothesis that the effect of safety programs on mean lost work-hours is greater for larger companies.

FIGURE 10.7
Examples of No Interaction and Interaction Models

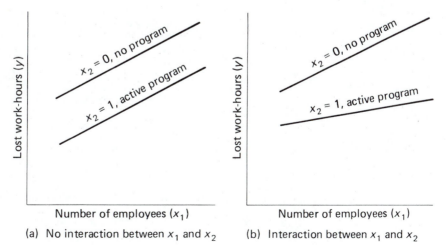

(a) No interaction between x_1 and x_2 (b) Interaction between x_1 and x_2

FIGURE 10.8
Computer Printout for the Safety Program Example

The regression equation is
$$y = -0.74 + 0.0197 x1 + 5.4 x2 - 0.0107 x1*x2$$

Predictor	Coef	Stdev	t-ratio	p
Constant	−0.739	7.927	−0.09	0.926
x1	0.019696	0.001219	16.16	0.000
x2	5.37	11.08	0.48	0.631
x1*x2	−0.010737	0.001715	−6.26	0.000

$s = 14.01$ R-sq = 93.3%   R-sq(adj) = 92.7%

Analysis of Variance

SOURCE	DF	SS	MS	F	p
Regression	3	97749	32583	166.11	0.000
Error	36	7062	196		
Total	39	104811			

FIGURE 10.9
Residuals and Stem-and-Leaf Plot for Model Without Interaction (Example 10.1)

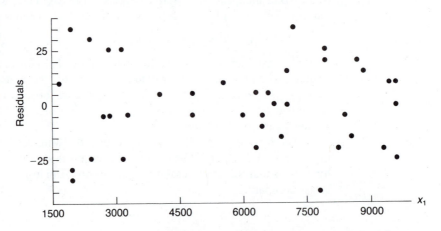

Stem-and-leaf of residuals, N = 40
Leaf unit = 1.0

```
  2     -3  87
  2     -3
  4     -2  95
  7     -2  431
 11     -1  9775
 11     -1
 16     -0  87665
 19     -0  433
 (4)     0  0013
 17      0  55778
 12      1  123
  9      1  6
  8      2  02
  6      2  5668
  2      3  3
  1      3  5
```

Solution   The interaction model is

$$E(Y) = \beta_0 + \beta_1 x_1 + \beta_2 x_2 + \beta_3 x_1 x_2$$

so the slope of $E(Y)$ versus x_1 is β_1 for companies with no safety programs and $\beta_1 + \beta_3$ for those with safety programs. Our research hypothesis is that the slope is smaller for companies with safety programs, so we will test

$$H_0: \beta_3 = 0 \qquad H_a: \beta_3 < 0$$

Test statistic: $t = \dfrac{\hat{\beta}_3}{s_{\hat{\beta}_3}}$

Rejection region: for $\alpha = 0.05$,

$$t < -t_{0.05}[n - (k + 1)] \quad \text{or} \quad t < -t_{0.05}(36) = -1.645$$

The t value corresponding to the test for β_3 is indicated in Figure 10.8. The value $t = -6.26$ is less than -1.645 and therefore falls in the rejection region. Thus the safety engineer can conclude that the larger the company, the greater the benefit of the safety program in terms of reducing the mean number of lost work-hours.   ∎

FIGURE 10.10
Residuals and Stem-and-Leaf Plot for Model with Interaction (Example 10.2)

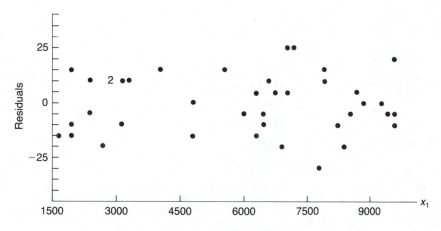

Stem-and-leaf of residuals, $N = 40$
Leaf unit = 1.0

1	-2	8
3	-2	10
6	-1	975
10	-1	3220
16	-0	888665
(5)	-0	43210
19	0	0
18	0	5567999
11	1	0011444
4	1	6
3	2	02
1	2	7

In comparing the results for Examples 10.1 and 10.2, we see that the model with interaction fits the data better than the model without an interaction term. The r^2 is greater for the model with interaction, and the interaction coefficient is significantly different from zero. How else might we compare how well these two models fit the data? Thinking back, for a moment, to Chapter 9, we might want to look at residuals as we did for the simple straight-line models.

Figure 10.9 shows the residuals plotted against x_1 for the model *without* interaction, and also shows the stem-and-leaf plot for these residuals. Figure 10.10 shows the residuals plotted against x_1 and the stem-and-leaf plot of residuals for the model *with* interaction. Note that the plot in Figure 10.10 seems to show more "randomness" in the points and less spread along the vertical axis. The stem-and-leaf plot in Figure 10.10 is more compact and more symmetrically mound shaped than the one in Figure 10.9. All of these features indicate that the interaction model fits the data better than the model without interaction.

EXERCISES

10.1   Suppose that you fit the model

$$Y = \beta_0 + \beta_1 x_1 + \beta_2 x_2 + \beta_3 x_1 x_2 + \beta_4 x_1^2 + \beta_5 x_2^2 + \varepsilon$$

to $n = 30$ data points and that

$$\text{SSE} = 0.37 \qquad R^2 = 0.89$$

(a)   Do the values of SSE and R^2 suggest that the model provides a good fit to the data? Explain.

(b)   Is the model of any use in predicting y? Test the null hypothesis that

$$E(Y) = \beta_0$$

that is,

$$H_0: \beta_1 = \beta_2 = \cdots = \beta_5 = 0$$

against the alternative hypothesis that

$$H_a: \text{At least one of the parameters } \beta_1, \beta_2, \ldots, \beta_5 \text{ is nonzero}$$

Use $\alpha = 0.05$.

10.2   In hopes of increasing the company's share of the fine food market, researchers for a meat-processing firm that prepares meats for exclusive restaurants are working to improve the quality of its hickory-smoked hams. One of their studies concerns the effect of time spent in the smokehouse on the flavor of the ham. Hams that were in the smokehouse for varying amounts of time were each subjected to a taste test by a panel of ten food experts. The following model was thought by the researchers to be appropriate:

$$Y = \beta_0 + \beta_1 t + \beta_2 t^2 + \varepsilon$$

where

Y = mean of the taste scores for the ten experts

t = time in the smokehouse (hours)

Assume the least-squares model estimated using a sample of twenty hams is

$$\hat{y} = 20.3 + 5.2t - 0.0025t^2$$

and that $s_{\beta_2} = 0.0011$. The coefficient of determination is $R^2 = 0.79$.

(a) Is there evidence to indicate that the overall model is useful? Test at $\alpha = 0.05$.

(b) Is there evidence to indicate the quadratic term is important in this model? Test at $\alpha = 0.05$.

10.3 Because the coefficient of determination R^2 always increases when a new independent variable is added to the model, it is tempting to include many variables in a model to force R^2 to be near 1. However, doing so reduces the degrees of freedom available for estimating σ^2, which adversely affects our ability to make reliable inferences. As an example suppose you want to use eighteen economic indicators to predict next year's GNP. You fit the model

$$Y = \beta_0 + \beta_1 x_1 + \beta_2 x_2 + \cdots + \beta_{17} x_{17} + \beta_{18} x_{18} + \varepsilon$$

where $Y = $ GNP and $x_1, x_2, \ldots, x_{18}$ are indicators. Only 20 years of data ($n = 20$) are used to fit the model, and you obtain $R^2 = 0.95$. Test to see whether this impressive-looking R is large enough for you to infer that this model is useful, that is, that at least one term in the model is important for predicting GNP. Use $\alpha = 0.05$.

10.4 A utility company of a major city gave the average utility bills listed in the table for a standard-size home during the last year.

(a) Plot the points in a scattergram.

(b) Use the methods of Chapter 9 to fit the model

$$Y = \beta_0 + \beta_1 x + \varepsilon$$

What do you conclude about the utility of this model?

(c) Hypothesize another model that might better describe the relationship between the average utility bill and average temperature. If you have access to a computer package, fit the model and test its utility.

Month	Average Monthly Temperature $x,°F$	Average Utility Bill y, dollars
January	38	99
February	45	91
March	49	78
April	57	61
May	69	55
June	78	63
July	84	80
August	89	95
September	79	65
October	64	56
November	54	74
December	41	93

10.5 To run a manufacturing operation efficiently, it is necessary to know how long it takes employees to manufacture the product. Without such information the cost of making the product cannot be determined. Furthermore management would not be able to establish an effective incentive plan for its employees because it would not know how to set work standards (Chase & Aquilano, *Production and Operations Management*, Irwin, 1979). Estimates of production time are frequently obtained by using time studies. The data in the table were obtained from a recent time study of a sample of fifteen employees on an automobile assembly line.

Time to Complete Task y (minutes)	Months of Experience x
10	24
20	1
15	10
11	15
11	17
19	3
11	20
13	9
17	3
18	1
16	7
16	9
17	7
18	5
10	20

(a) The computer printout for fitting the model $y = \beta_0 + \beta_1 x + \beta_2 x^2 + \varepsilon$ is shown here. Find the least-squares prediction equation.

Predictor	Coef	Stdev	t-ratio	p
Constant	20.0911	0.7247	27.72	0.000
x	−0.6705	0.1547	−4.33	0.001
x*x	0.009535	0.006326	1.51	0.158

$s = 1.091$ R-sq $= 91.6\%$ R-sq(adj) $= 90.2\%$

Analysis of Variance

SOURCE	DF	SS	MS	F	p
Regression	2	156.119	78.060	65.59	0.000
Error	12	14.281	1.190		
Total	14	170.400			

(b) Plot the fitted equation on a scattergram of the data. Is there sufficient evidence to support the inclusion of the quadratic term in the model? Explain.

(c) Test the null hypothesis that $\beta_2 = 0$ against the alternative that $\beta_2 \neq 0$. Use $\alpha = 0.01$. Does the quadratic term make an important contribution to the model?

(d) Your conclusion in part (c) should have been to drop the quadratic term from the model. Do so and fit the "reduced model" $y = \beta_0 + \beta_1 x + \varepsilon$ to the data.

(e) Define β_1 in the context of this exercise. Find a 90% confidence interval for β_1 in the reduced model of part (d).

10.6 A researcher wished to investigate the effects of several factors on production line supervisors' attitudes toward handicapped workers. A study was conducted involving forty randomly selected supervisors. The response Y, a supervisor's attitude toward handicapped workers, was measured with a standardized attitude scale. Independent variables used in the study were

$$x_1 = \begin{cases} 1 & \text{if the supervisor is female} \\ 0 & \text{if the supervisor is male} \end{cases}$$

x_2 = number of years of experience in a supervisory job

The researcher fit the model,

$$Y = \beta_0 + \beta_1 x_1 + \beta_2 x_2 + \beta_3 x_2^2 + \varepsilon$$

to the data with the following results:

$$\hat{y} = 50 + 5x_1 + 5x_2 - 0.1x_2^2$$

$$s_{\beta_3} = 0.03$$

(a) Do these data provide sufficient evidence to indicate that the quadratic term in years of experience x_2^2 is useful for predicting attitude score? Use $\alpha = 0.05$.

(b) Sketch the predicted attitude score $\hat{y}$ as a function of the number of years of experience x_2 for male supervisors ($x_1 = 0$). Next substitute $x_1 = 1$ into the least-squares equation and thereby obtain a plot of the prediction equation for female supervisors. [*Note:* For both males and females, plotting $\hat{y}$ for $x_2 = 0, 2, 4, 6, 8,$ and 10 will produce a good picture of the prediction equations. The vertical distance between the males' and females' prediction curves is the same for all values of x_2.]

10.7 To project personnel needs for the Christmas shopping season, a department store wants to project sales for the season. The sales for the previous Christmas season are an indication of what to expect for the current season. However, the projection should also reflect the current economic environment by taking into consideration sales for a more recent period. The following model might be appropriate:

$$y = \beta_0 + \beta_1 x_1 + \beta_2 x_2 + \varepsilon$$

where

x_1 = previous Christmas sales

x_2 = sales for August of current year

y = sales for upcoming Christmas

(All units are in thousands of dollars.) Data for 10 previous years were used to fit the prediction equation, and the following were calculated:

$$\hat{\beta}_1 = 0.62 \qquad s_{\hat{\beta}_1} = 0.273$$
$$\hat{\beta}_2 = 0.55 \qquad s_{\hat{\beta}_2} = 0.181$$

Use these results to determine whether there is evidence to indicate that the mean sales this Christmas are related to this year's August sales in the proposed model.

10.8 Suppose you fit the second-order model,

$$Y = \beta_0 + \beta_1 x + \beta_2 x^2 + \varepsilon$$

to $n = 30$ data points. Your estimate of β_2 is $\hat{\beta}_2 = 0.35$, and the standard error of the estimate is $s_{\hat{\beta}_2} = 0.13$.

  (a) Test the null hypothesis that the mean value of Y is related to x by the (first-order) linear model

$$E(Y) = \beta_0 + \beta_1 x$$

  Test $H_0: \beta_2 = 0$ against the alternative hypothesis that the true relationship is given by the quadratic model (a second-order linear model),

$$E(Y) = \beta_0 + \beta_1 x + \beta_2 x^2$$

  ($H_a: \beta_2 \neq 0$.) Use $\alpha = 0.05$.
  (b) Suppose you wanted only to determine whether the quadratic curve opens upward; that is, as x increases, the slope of the curve increases. Give the test statistic and the rejection region for the test for $\alpha = 0.05$. Do the data support the theory that the slope of the curve increases as x increases? Explain.
  (c) What is the value of the F statistic for testing the null hypothesis that $\beta_2 = 0$?
  (d) Could the F statistic in part (c) be used to conduct the tests in parts (a) and (b)? Explain.

10.9 How is the number of degrees of freedom available for estimating σ^2 (the variance of ε) related to the number of independent variables in a regression model?

10.10 An employer has found that factory workers who are with the company longer tend to invest more in a company investment program per year than workers with less time with the company. The following model is believed to be adequate in modeling the relationship of annual amount invested Y to years working for the company x:

$$Y = \beta_0 + \beta_1 x + \beta_2 x^2 + \varepsilon$$

The employer checks the records for a sample of fifty factory employees for a previous year, and fits the above model to get $\hat{\beta}_2 = 0.0015$ and $s_{\hat{\beta}_2} = 0.00712$. The basic shape of a quadratic model depends upon whether $\beta_2 < 0$ or $\beta_2 > 0$. Test to see whether the employer can conclude that $\beta_2 > 0$. Use $\alpha = 0.05$.

10.11 Automobile accidents result in a tragic loss of life and, in addition, they represent a serious dollar loss to the nation's economy. Shown in the table are the number of highway deaths (to the nearest hundred) and the number of licensed vehicles (in hundreds of thousands) for the years 1950–1979. (The years are coded 1–30 for convenience.) During the years 1974–1979 (years 25–30 in the table) the nationwide 55-mile-per-hour speed limit was in effect.

Year	Deaths y	Number of Vehicles x_1	Year	Deaths y	Number of Vehicles x_1
1	34.8	49.2	16	49.1	91.8
2	37.0	51.9	17	53.0	95.9
3	37.8	53.3	18	52.9	98.9
4	38.0	56.3	19	54.9	103.1
5	35.6	58.6	20	55.8	107.4
6	38.4	62.8	21	54.6	111.2
7	39.6	65.2	22	54.3	116.3
8	38.7	67.6	23	56.3	122.3
9	37.0	68.8	24	55.5	129.8
10	37.9	72.1	25	46.4	134.9
11	38.1	74.5	26	45.9	137.9
12	38.1	76.4	27	47.0	143.5
13	40.8	79.7	28	49.5	148.8
14	43.6	83.5	29	51.5	153.6
15	47.7	87.3	30	51.9	159.4

(a) Write a second-order model relating the number y of highway deaths for a year to the number x_1 of licensed vehicles.

(b) The computer printout for fitting the model to the data is shown. Is there sufficient evidence to indicate that the model provides information for the prediction of the number of annual highway deaths? Test by using $\alpha = 0.05$.

Predictor	Coef	Stdev	t-ratio	p
Constant	−1.408	6.895	−0.20	0.840
x_1	0.8455	0.1462	5.78	0.000
$x_1{}^*x_1$	−0.0033220	0.0007093	−4.68	0.000

$s = 3.706$ R-sq = 76.7%   R-sq(adj) = 75.0%

Analysis of Variance

SOURCE	DF	SS	MS	F	p
Regression	2	1222.16	611.08	44.50	0.000
Error	27	370.79	13.73		
Total	29	1592.95			

(c) Give the p-value for the test of part (b) and interpret it.

(d) Does the second-order term contribute information for the prediction of y? Test by using $\alpha = 0.05$.

(e) Give the p value for the test of part (d) and interpret it.

(f) Plot the residuals from the fitted model against x_1. Do you think this is the best model to use here?

10.12 If producers (providers) of goods (services) are able to reduce the unit cost of their goods (services) by increasing the scale of their operation, they are the beneficiaries of an economic force known as *economies of scale*. Economies of scale cause the firm's long-run average costs to decline (Ferguson and Maurice, *Economics Analysis*, Irwin, 1971).

The question of whether economies of scale, diseconomies of scale, or neither (i.e., constant economies of scale) exist in the U.S. motor freight, common carrier industry has been debated for years. In an effort to settle the debate within a specific subsection of the trucking industry, Sugrue, Ledford, and Glaskowsky (*Transportation Journal*, pp. 27–41, 1982) used regression analysis to model the relationship between each of a number of profitability/cost measures and the size of the operation. In one case they modeled expense per vehicle mile Y_1 as a function of the firm's total revenue X. In another case they modeled expense per ton mile Y_2 as a function of X. Data were collected from 264 firms, and the following least-squares results were obtained:

Dependent Variable	$\hat{\beta}_0$	$\hat{\beta}_1$	r	F
Expense per vehicle mile	2.279	−0.00000069	−0.0783	1.616
Expense per ton mile	0.1680	−0.000000066	−0.0902	2.148

(a) Investigate the usefulness of the two models estimated by Sugrue, Ledford, and Glaskowsky. Use $\alpha = 0.05$. Draw appropriate conclusions in the context of the problem.

(b) Are the observed significance levels of the hypothesis tests you conducted in part (a) greater than 0.10 or less than 0.10? Explain.

(c) What do your hypothesis tests of part (a) suggest about economies of scale in the subsection of the trucking industry investigated by Sugrue, Ledford, and Glaskowsky—the long-haul, heavy-load, intercity general-freight common carrier section of the industry? Explain.

10.6 REGRESSION SOFTWARE

A number of different statistical program packages feature regression software. Some of the most popular are BMD, Minitab, SAS, and SPSS. More recently many regression packages have been offered for microcomputers. You probably have access to one or more of these packages.

The multiple regression computer programs for these packages may differ in what they are programmed to do, how they do it, and the appearance of their computer printouts, but all of them print the basic statistics needed for a regression analysis. For example, some will compute confidence intervals for $E(Y)$ and prediction intervals for Y; others will not. Some test the null hypotheses that the individual β parameters equal zero, using Student's t tests, while others use F tests.[2] But all give the least-squares estimates, the values of SSE, s^2, and so forth.

To illustrate, the Minitab, SAS, and SPSS regression analysis computer printouts for Example 10.2 are shown in Figure 10.11. For that example we fit the model

$$y = \beta_0 + \beta_1 x_1 + \beta_2 x_2 + \beta_3 x_1 x_2 + \varepsilon$$

[2] A two-tailed Student's t test based on v df is equivalent to an F test where the F statistic possesses 1 df in the numerator and v df in the denominator.

to $n = 40$ data points. The variables in the model were

y = lost hours due to accidents

x_1 = number of employees

x_2 = safety program (1 if active, 0 if no program)

Notice that the Minitab printout gives the prediction equation at the top of the printout. The independent variables, shown in the prediction equation and listed at the left side of the printout, are x_1, x_2, and x_3. Thus Minitab treats the

FIGURE 10.11
Computer Printouts
for Example 10.2

The regression equation is
y = −0.74 + 0.0197 x1 + 5.4 x2 − 0.0107 x1*x2

Predictor	Coef	Stdev	t-ratio	p
Constant	−0.739	7.927	−0.09	0.926
x1	0.019696	0.001219	16.16	0.000
x2	5.37	11.08	0.48	0.631
x1*x2	−0.010737	0.001715	−6.26	0.000

s = 14.01   R-sq = 93.3%   R-sq(adj) = 92.7%

Analysis of Variance

SOURCE	DF	SS	MS	F	p
Regression	3	97749	32583	166.11	0.000
Error	36	7062	196		
Total	39	104811			

a. Minitab Regression Printout

SAS

GENERAL LINEAR MODELS PROCEDURE

DEPENDENT VARIABLE, Y

SOURCE	DF	SUM OF SQUARES	MEAN SQUARE	F VALUE
MODEL	3	97749.09184154	32583.03061385	166.11
ERROR	36	7061.68315846	196.15786551	PR > F
CORRECTED TOTAL	39	104810.77500000		0.0001

R-SQUARE	C.V.	ROOT MSE	Y MEAN
0.932624	16.0846	14.00563692	87.07500000

PARAMETER	ESTIMATE	T FOR H0, PARAMETER = 0	PR > |T|	STD ERROR OF ESTIMATE
INTERCEPT	−0.73908475	−0.09	0.9262	7.92730641
X1	0.01969560	16.16	0.0001	0.00121874
X2	5.36689032	0.48	0.6310	11.07978453
X1*X2	−0.01073708	−6.26	0.0001	0.00171477

b. SAS Regression Printout

FIGURE10.11 SSPS
(Cont) ****** MULTIPLE REGRESSION ******
LISTWISE DELETION OF MISSING DATA
EQUATION NUMBER 1   DEPENDENT VARIABLE.. Y

BEGINNING BLOCK NUMBER 1. METHOD, ENTER
VARIABLE(S) ENTERED ON STEP NUMBER 1 . .   X1X2
                                     2 . . X1
                                     3 . . X2

ANALYSIS OF VARIANCE

	DF	SUM OF SQUARES	MEAN SQUARE
REGRESSION	3	97749.09184	32583.03061
RESIDUAL	36	7061.68316	196.15787

F= 166.10616   SIGNIF F= .0000

MULTIPLE R	.96572
R SQUARE	.93262
ADJUSTED R SQUARE	.92701
STANDARD ERROR	14.00564

**** VARIABLES IN THE EQUATION ****

VARIABLE	B	SE B	BETA	T	SIG T
X1X2	−.010737	.001715	−.725998	−6.262	.0000
X1	.019696	.001219	.994050	16.161	.0000
X2	5.366890	11.079785	.052423	.484	.6310
(CONSTANT)	−.739085	7.927306		−.093	.9262

END BLOCK NUMBER 1 ALL REQUESTED VARIABLES ENTERED.

c. SPSS Regression Printout

product $x_1 x_2$ as a third independent variable x_3, which must be computed before the fitting commences. For this reason the Minitab prediction equation will always appear on the printout as first-order even though some of the independent variables shown in the prediction equation may actually be the squares or cross products of other independent variables. The inclusion of the squares or cross products of independent variables is treated in the same manner in the SPSS program shown in Figure 10.11(c). The SAS program is the only one of these three that can be instructed to include these terms automatically, and they appear in the printout with a star (*) that indicates multiplication. Thus in the SAS printout $x_1 x_2$ is printed as X1 * X2.

The estimates of the regression coefficients appear opposite the identifying variable in the Minitab column titled COEF, in the SAS column titled ESTIMATE, and in the SPSS column titled B. Compare the estimates given in these three columns. Note that the Minitab printout gives the estimates with a much lesser degree of precision (fewer decimal places) than the SAS and SPSS. (Ignore the column titled BETA in the SPSS printout. These are standardized estimates and will not be discussed in this text.)

The estimated standard errors of the estimates are given in the Minitab column titled STDEV, in the SAS column titled STD ERROR OF ESTIMATE, and in the SPSS column titled SE B.

The values of the test statistics for testing $H_0: \beta_i = 0$, where $i = 1, 2, 3$, are shown in the Minitab column titled T-RATIO and in the SAS column titled T FOR HO: PARAMETER = 0. Note that the computed t values shown in the Minitab and SAS columns are identical. Minitab gives the observed significance level in the column headed P, the SAS printout gives the observed significance level for each t test in the column titled PR > |T|, and SPSS provides it in the column headed SIG T. Note that these observed significance levels have been computed by assuming that the tests are two-tailed. The observed significance levels for one-tailed tests would equal half of these values.

The Minitab printout gives the value SSE = 7062 under the ANALYSIS OF VARIANCE column headed SS and in the row identified as ERROR. The value of $s^2 = 196$ is shown in the same row under the column headed MS, and the degrees of freedom DF appears in the same row as 36. The corresponding values are shown at the top of the SAS printout in the row labeled ERROR and in the columns designated as SUM OF SQUARES, MEAN SQUARE, and DF, respectively. These quantities appear with similar headings in the SPSS printout, except that ERROR is termed RESIDUAL.

The value of R^2, as defined in Section 10.4, is given in the Minitab printout as 93.3 PERCENT (we defined this quantity as a ratio where $0 \leqslant R^2 \leqslant 1$). It is given on the left of the SAS printout as 0.932624 and it is shown in the left column of the SPSS printout as 0.93262. (Ignore the quantities shown in the Minitab as R-SQ(ADJ) and in the SPSS printout as ADJUSTED R SQUARE. These quantities are adjusted for the degrees of freedom associated with the total SS and SSE and are not used or discussed in this test).

The F statistic for testing the utility of the model (Section 10.4)—that is, testing the null hypothesis that all model parameters (except β_0) equal zero—is shown under the title F VALUE as 166.11 at the top right of the SAS printout. In addition, the SAS printout gives the observed significance level of this F test under PR > F as 0.0001. This F value 166.10616 is also printed at the left side of the SPSS printout with the observed significance level given. The F statistic for testing the utility of the model is given in the Minitab printout as 166.11, with a p value of 0.000.

You can also compute the value of the F statistic directly from the mean square entries given in the ANALYSIS OF VARIANCE table. Thus an equivalent form of the F statistic for testing

$$H_0: \beta_1 = \beta_2 = \cdots = \beta_k = 0$$

is

$$F = \frac{\text{mean square for regression}}{\text{mean square for error (or residuals)}}$$

$$= \frac{\text{mean square for regression}}{s^2}$$

These quantities are given in the Minitab printout under the column marked MS. Thus

$$F = \frac{32,583}{196} = 166$$

a value that agrees (to three significant digits) with the values given in the SAS and SPSS printouts. The logic behind this test and other tests of hypotheses concerning sets of the β parameters is presented in Sections 10.4, 10.5, and 10.7.

We will not comment on the merits or demerits of the various packages because you will have to use the package(s) available at your computer center and become familiar with that output. Most of the computer printouts are similar, and it is relatively easy to learn how to read one output after you have become familiar with another.

10.7 MODEL BUILDING: TESTING PORTIONS OF A MODEL

In Section 10.4 we discussed testing all the parameters in a multiple regression model using the coefficient of determination R^2. Then in Section 10.5 a t test for individual model parameters was presented. We will now develop a test of sets of β parameters representing a *portion* of the model. In doing so, we will also present some useful techniques in constructing multiple regression models.

An example will best demonstrate the need for such a test for portions of a regression model, as well as the flexibility of multiple regression models. Suppose a construction firm wishes to compare the performance of its three sales engineers, using the mean profit per sales dollar. The sales engineers bid on jobs in two states, so that the true mean profit per sales dollar is to be considered a function of two factors: sales engineer and state. The six means are symbolically represented by μ_{ij}, with the i subscript used for sales engineer ($i = 1, 2, 3$) and the j subscript for state ($j = 1, 2$). The mean values are presented in Table 10.3.

TABLE 10.3
Mean Profit per Sales Dollar for Six Sales Engineer/State Combinations

		State	
		S_1	S_2
	E_1	μ_{11}	μ_{12}
Sales Engineer	E_2	μ_{21}	μ_{22}
	E_3	μ_{31}	μ_{32}

Since both sales engineer and state are *qualitative* factors, dummy variables will be used to represent them in the multiple regression model (see Example 10.1). Thus we define

$$x_1 = \begin{cases} 1 & \text{if state} = S_2 \\ 0 & \text{if state} = S_1 \end{cases}$$

to represent the state effect in the model, and

$$x_2 = \begin{cases} 1 & \text{if sales engineer} = E_2 \\ 0 & \text{if sales engineer} = E_1 \text{ or } E_3 \end{cases}$$

$$x_3 = \begin{cases} 1 & \text{if sales engineer} = E_3 \\ 0 & \text{if sales engineer} = E_1 \text{ or } E_2 \end{cases}$$

to represent the sales engineer effect. Note that two dummy variables are defined to represent the three sales engineers. The reason for this will be apparent when we write the model, which, using $Y = $ profit per sales dollar, is

$$E(Y) = \beta_0 + \overbrace{\beta_1 x_1}^{\text{State}} + \overbrace{\beta_2 x_2 + \beta_3 x_3}^{\substack{\text{Sales} \\ \text{Engineer}}} + \overbrace{\beta_4 x_1 x_2 + \beta_5 x_1 x_3}^{\substack{\text{State} \times \text{Sales} \\ \text{Engineer} \\ \text{Interaction}}}$$

Note that there are six parameters in the model: β_0, β_1, β_2, β_3, β_4, and β_5—one corresponding to each mean in Table 10.3. The correspondence is shown in Table 10.4. Other definitions of dummy variables can be used, which will generate a different correspondence between the means μ_{ij} and the β parameters of the model.

TABLE 10.4
Correspondence Between Six Sales Engineer/State Means (μ_{ij}) and Model Parameters (βs)

$$\mu_{11} = E(Y|E_1, S_1) = E(Y|x_1 = x_2 = x_3 = 0) = \beta_0$$
$$\mu_{12} = E(Y|E_1, S_2) = E(Y|x_1 = 1, x_2 = x_3 = 0) = \beta_0 + \beta_1$$
$$\mu_{21} = E(Y|E_2, S_1) = E(Y|x_1 = 0, x_2 = 1, x_3 = 0) = \beta_0 + \beta_2$$
$$\mu_{22} = E(Y|E_2, S_2) = E(Y|x_1 = 1, x_2 = 1, x_3 = 0) = \beta_0 + \beta_1 + \beta_2 + \beta_4$$
$$\mu_{31} = E(Y|E_3, S_1) = E(Y|x_1 = x_2 = 0, x_3 = 1) = \beta_0 + \beta_3$$
$$\mu_{32} = E(Y|E_3, S_2) = E(Y|x_1 = 1, x_2 = 0, x_3 = 1) = \beta_0 + \beta_1 + \beta_3 + \beta_5$$

However, the use of one fewer dummy variables than the number of factor levels (that is, one dummy variable for the 2 states, and two dummy variables for the 3 sales engineers) will yield an exact correspondence between the *number* of means and the *number* of model parameters. Fewer dummy variables will result in too few model parameters, and more dummy variables will yield more model parameters than necessary to generate the six mean profit values.

Now suppose the construction firm wishes to test whether the relative performance of the three sales engineers is the same in both states, as shown in Figure 10.12. Thus although the level of profit may shift between states, the shift is

FIGURE 10.12
Relative Performance of Sales Engineers Is the Same in Both States: Effects Are Additive

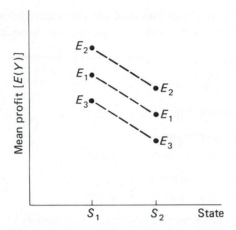

the same for all three sales engineers. Therefore the *relative* position of the engineers' mean profits is the same. This phenomenon is referred to as *additivity* of the state and sales engineer effects, and results in a simplification of the model. The interaction terms $\beta_4 x_1 x_2$ and $\beta_5 x_1 x_3$ are now unnecessary, since only β_1 is needed to represent the additive effect of state, as shown in Table 10.5. Note that the difference between the state means for each sales engineer is the same, β_1.

TABLE 10.5
Representation of Mean Profit in the Additive Model: $E(Y) = \beta_0 + \beta_1 x_1 + \beta_2 x_2 + \beta_3 x_3$

Sales Engineer	State	Mean
E_1	S_1	$\mu_{11} = \beta_0$
E_1	S_2	$\mu_{12} = \beta_0 + \beta_1$
E_2	S_1	$\mu_{21} = \beta_0 + \beta_2$
E_2	S_2	$\mu_{22} = \beta_0 + \beta_1 + \beta_2$
E_3	S_1	$\mu_{31} = \beta_0 + \beta_3$
E_3	S_2	$\mu_{32} = \beta_0 + \beta_1 + \beta_3$

If the sales engineers perform differently in the two states, their mean profits will not maintain the same relative positions. An example of this *nonadditive* relationship between state and sales engineer is shown in Figure 10.13. This more complex relationship requires all six model parameters

$$E(Y) = \beta_0 + \beta_1 x_1 + \beta_2 x_2 + \beta_3 x_3 + \beta_4 x_1 x_2 + \beta_5 x_1 x_3$$

to describe it. Thus you can see that the difference between an additive and a nonadditive relationship is the interaction terms $\beta_4 x_1 x_2$ and $\beta_5 x_1 x_3$. If β_4 and β_5 are zero, the effects are additive; otherwise, the State/Sales Engineer effects are nonadditive.

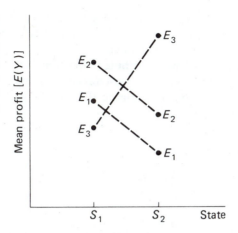

FIGURE 10.13
Relative Performance of Sales Engineers Is Not the Same in Two States: Effects Are Nonadditive

We will call the interaction model the *complete model*, and the reduced model the *main effects model*. Now suppose we wanted to use some data to test which of these models is more appropriate. This can be done by testing the hypothesis that the β parameters for the interaction terms equal zero:

H_0: $\beta_4 = \beta_5 = 0$

H_a: At least one interaction β parameter differs from zero

In Section 10.5 we presented the t test for a single coefficient, and in Section 10.4 we gave the F test for *all* the β parameters (except β_0) in the model. Now we need a test for *some* of the β parameters in the model. The test procedure is intuitive: first, we use the method of least squares to fit the main effects model, and calculate the corresponding sum of squares for error SSE_1 (the sum of squares of the deviations between observed and predicted y values). Next, we fit the interaction model, and calculate its sum of squares for error SSE_2. Then, we compare SSE_1 to SSE_2 by calculating the difference $SSE_1 - SSE_2$. If the interaction terms contribute to the model, then SSE_2 should be much smaller than SSE_1, and the difference $SSE_1 - SSE_2$ will be large. That is, the larger the difference, the greater the weight of evidence that the variables, sales engineer and state, interact to affect the mean profit per sales dollar in the construction job.

The sum of squares for error will always decrease when new terms are added to the model. The question is whether this decrease is large enough to conclude that it is due to more than just an increase in the number of model terms and to chance. To test the null hypothesis that the interaction terms β_4 and β_5 simultaneously equal zero, we use an F statistic calculated as follows:

$$F = \frac{(SSE_1 - SSE_2)/2}{SSE_2/[n - (5 + 1)]}$$

$$= \frac{\text{drop in SSE/number of } \beta \text{ parameters being tested}}{s^2 \text{ for complete model}}$$

When the assumptions listed in Section 10.3 about the error term ε are satisfied and the β parameters for interaction are all zero (H_0 is true), this F statistic has an F

distribution with $v_1 = 2$ and $v_2 = n - 6$ degrees of freedom. Note that v_1 is the number of β parameters being tested, and v_2 is the number of degrees of freedom associated with s^2 in the complete model.

If the interaction terms *do* contribute to the model (H_a is true), we expect the F statistic to be large. Thus we use a one-tailed test and reject H_0 when F exceeds some critical value F_α, as shown in Figure 10.14.

FIGURE 10.14
Rejection Region for the F Test H_o:
$\beta_4 = \beta_5 = 0$

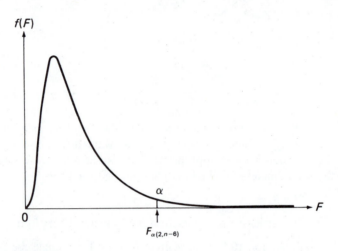

EXAMPLE 10.3

The profit per sales dollar Y for the six combinations of sales engineers and states is shown in Table 10.6. Note that the number of construction jobs per combination varies from one for levels (E_1, S_2) to three for levels (E_1, S_1). A total of twelve jobs are sampled.

TABLE 10.6
Profit Data for Combinations of Sales Engineers and States

Sales Engineer		State S_1	State S_2
	E_1	$\$0.065$ 0.073 0.068	$\$0.036$
	E_2	0.078 0.082	0.050 0.043
	E_3	0.048 0.046	0.061 0.062

(a) Assume the interaction between E and S is negligible. Fit the model for $E(Y)$ with interaction terms omitted.

(b) Fit the complete model for $E(Y)$, allowing for the fact that interactions might occur.

(c) Test the hypothesis that the interaction terms do not contribute to the model.

Solution

(a) The computer printout for the main effects model

$$E(Y) = \beta_0 + \overbrace{\beta_1 x_1}^{\substack{S \\ \text{Main} \\ \text{Effect}}} + \overbrace{\beta_2 x_2 + \beta_3 x_3}^{\substack{E \\ \text{Main Effect}}}$$

is given in Figure 10.15. The least-squares prediction equation is

$$\hat{y} = 0.0645 - 0.0158 x_1 + 0.067 x_2 - 0.00230 x_3$$

FIGURE 10.15
Printout for Main Effects Model of Example 10.3

SOURCE	DF	SUM OF SQUARES	MEAN SQUARE	F VALUE	PR>F
MODEL	3	0.00085826	0.00028609	1.51	0.2838
ERROR	8	0.00151241	0.00018905		
CORRECTED TOTAL	11	0.00237067		R-SQUARE	STD DEV
				0.362032	0.01374959

PARAMETER	ESTIMATE	T FOR H0: PARAMETER = 0	PR>|T|	STD ERROR OF ESTIMATE
INTERCEPT	0.06445455	8.98	0.0001	0.00718049
X1	−0.01581818	−1.91	0.0928	0.00829131
X2	0.00670455	0.67	0.5190	0.00994093
X3	−0.00229545	−0.23	0.8232	0.00994093

(b) The complete model printout is given in Figure 10.16. Recall that the complete model is

$$E(Y) = \beta_0 + \beta_1 x_1 + \beta_2 x_2 + \beta_3 x_3 + \beta_4 x_1 x_2 + \beta_5 x_1 x_3$$

FIGURE 10.16
Printout for Complete Model (Includes Interaction) of Example 10.3

SOURCE	DF	SUM OF SQUARES	MEAN SQUARE	F VALUE	PR>F
MODEL	5	0.00230300	0.00046060	40.84	0.0001
ERROR	6	0.00006767	0.00001128		
CORRECTED TOTAL	11	0.00237067		R-SQUARE	STD DEV
				0.971457	0.00335824

PARAMETER	ESTIMATE	T FOR H0: PARAMETER = 0	PR > |T|	STD ERROR OF ESTIMATE
INTERCEPT	0.06866667	35.42	0.0001	0.00193888
X1	−0.03266667	−8.42	0.0002	0.00387776
X2	0.01133333	3.70	0.0101	0.00306564
X3	−0.02166667	−7.07	0.0004	0.00306564
X1 • X2	−0.00083333	−0.16	0.8763	0.00512980
X1 • X3	0.04716667	9.19	0.0001	0.00512980

The least-squares prediction equation is

$$\hat{y} = 0.0687 - 0.0327x_1 + 0.0113x_2 - 0.0217x_3 - 0.0008x_1x_2$$

$$+ 0.0472x_1x_3$$

(c) Referring to the printouts shown in Figures 10.15 and 10.16, we find the following:

Main effects model: $SSE_1 = 0.00151241$

Interaction model: $SSE_2 = 0.00006767$

The test statistic is

$$F = \frac{(SSE_1 - SSE_2)/2}{SSE_2/(12 - 6)} = \frac{(0.00151241 - 0.00006767)/2}{0.00006767/6}$$

$$= \frac{0.00072237}{0.00001128} = 64.04$$

The critical value of F for $\alpha = 0.05$, $v_1 = 2$, and $v_2 = 6$ is found in Table 7 of the Appendix to be

$$F_{0.05}(2, 6) = 5.14$$

Since the calculated $F = 64.04$ greatly exceeds 5.14, we are quite confident in concluding that the interaction terms contribute to the prediction of Y, profit per sales dollar. They should be retained in the model.   ∎

The F test can be used to determine whether *any* set of terms should be included in a model by testing the null hypothesis that a particular set of β parameters simultaneously equal zero. For example, we may want to test to determine whether a set of quadratic terms for quantitative variables or a set of main effect terms for a qualitative variable should be included in a model. The F test appropriate for testing the null hypothesis that all of a set of β parameters are equal to zero is summarized below.

F TEST FOR TESTING THE NULL HYPOTHESIS:
SET OF β PARAMETERS EQUAL ZERO

Reduced model: $E(Y) = \beta_0 + \beta_1x_1 + \cdots + \beta_gx_g$

Complete model: $E(Y) = \beta_0 + \beta_1x_1 + \cdots + \beta_gx_g$
$$+ \beta_{g+1}x_{g+1} + \cdots + \beta_kx_k$$

H_0: $\beta_{g+1} = \beta_{g+2} = \cdots = \beta_k = 0$

H_a: At least one of the β parameters under test is nonzero

Test statistic: $F = \dfrac{(SSE_1 - SSE_2)/(k - g)}{SSE_2/[n - (k + 1)]}$

where

SSE_1 = sum of squared errors for the reduced model

SSE_2 = sum of squared errors for the complete model

$k - g$ = number of β parameters specified in H_0

$k + 1$ = number of β parameters in the complete model (including β_0)

n = total sample size

Rejection region: $F > F_\alpha(v_1, v_2)$

where

$v_1 = k - g$ = degrees of freedom for the numerator

$v_2 = n - (k + 1)$ = degrees of freedom for the denominator

EXERCISES

10.13 Suppose you fit the regression model

$$Y = \beta_0 + \beta_1 x_1 + \beta_2 x_2 + \beta_3 x_3 + \beta_4 x_4 + \varepsilon$$

to $n = 25$ data points and you wish to test the null hypothesis $\beta_1 = \beta_2 = 0$.
(a) Explain how you would find the quantities necessary for the F statistic.
(b) How many degrees of freedom would be associated with F?

10.14 An insurance company is experimenting with three different training programs, A, B, and C, for its salespeople. The following main effects model is proposed:

$$E(Y) = \beta_0 + \beta_1 x_1 + \beta_2 x_2 + \beta_3 x_3$$

where

Y = monthly sales (in thousands of dollars)

x_1 = number of months experience

$x_2 = \begin{cases} 1 & \text{if training program } B \text{ was used} \\ 0 & \text{otherwise} \end{cases}$

$x_3 = \begin{cases} 1 & \text{if training program } C \text{ was used} \\ 0 & \text{otherwise} \end{cases}$

Training program A is the base level.
(a) What hypothesis would you test to determine whether the mean monthly sales differ for salespeople trained by the three programs?
(b) After experimenting with fifty salespeople over a 5-year period, the complete

model is fit, with the result

$$\hat{y} = 10 + 0.5x_1 + 1.2x_2 - 0.4x_3 \qquad SSE = 140.5$$

Then the reduced model $E(Y) = \beta_0 + \beta_1 x_1$ is fit to the same data, with the result

$$\hat{y} = 11.4 + 0.4x_1 \qquad SSE = 183.2$$

Test the hypothesis you formulated in part (a). Use $\alpha = 0.05$.

10.15 A large research and development company rates the performance of each of the members of its technical staff once a year. Each person is rated on a scale of 0 to 100 by his or her immediate supervisor, and this merit rating is used to help determine the size of the person's pay raise for the coming year. The company's personnel department is interested in developing a regression model to help them forecast the merit rating that an applicant for a technical position will receive after he or she has been with the company 3 years. The company proposes to use the following model to forecast the merit ratings of applicants who have just completed their graduate studies and have no prior related job experience:

$$E(Y) = \beta_0 + \beta_1 x_1 + \beta_2 x_2 + \beta_3 x_1 x_2 + \beta_4 x_1^2 + \beta_5 x_2^2$$

where

$Y =$ applicant's merit rating after 3 years

$x_1 =$ applicant's grade-point average (GPA) in graduate school

$x_2 =$ applicant's verbal score on the Graduate Record Examination (percentile)

A random sample of $n = 40$ employees who have been with the company more than 3 years was selected. Each employee's merit rating after 3 years, his or her graduate GPA, and the percentile in which the verbal Graduate Record Exam score fell were recorded. The above model was fit to these data. Below is a portion of the resulting computer printout:

SOURCE	DF	SUM OF SQUARES	MEAN SQUARE
MODEL	5	4911.56	982.31
ERROR	34	1830.44	53.84
TOTAL	39	6742.00	R-SQUARE
			0.729

The reduced model $E(Y) = \beta_0 + \beta_1 x_1 + \beta_2 x_2$ was fit to the same data, and the resulting computer printout is partially reproduced below:

SOURCE	DF	SUMS OF SQUARES	MEAN SQUARE
MODEL	2	3544.84	1772.42
ERROR	37	3197.16	86.41
TOTAL	39	6742.00	R-SQUARE
			0.526

(a) Identify the null and alternative hypotheses for a test to determine whether the complete model contributes information for the prediction of Y.

(b) Identify the null and alternative hypotheses for a test to determine whether a second-order model contributes more information than a first-order model for the prediction of Y.

(c)   Conduct the hypothesis test you described in part (a). Test using $\alpha = 0.05$. Draw the appropriate conclusions in the context of the problem.

(d)   Conduct the hypothesis test you described in part (b). Test using $\alpha = 0.05$. Draw the appropriate conclusions in the context of the problem.

10.16   In an attempt to reduce the number of work-hours lost due to accidents, a company tested three safety programs, A, B, and C, each at three of the company's nine factories. The proposed complete model is

$$E(Y) = \beta_0 + \beta_1 x_1 + \beta_2 x_2 + \beta_3 x_3$$

where

$Y = $ total work-hours lost due to accidents for a 1-year period beginning 6 months after the plan is instituted

$x_1 = $ total work-hours lost due to accidents during the year before the plan was instituted

$$x_2 = \begin{cases} 1 & \text{if program } B \text{ is in effect} \\ 0 & \text{otherwise} \end{cases}$$

$$x_3 = \begin{cases} 1 & \text{if program } C \text{ is in effect} \\ 0 & \text{otherwise} \end{cases}$$

After the programs have been in effect for 18 months, the complete model is fit to the $n = 9$ data points, with the result

$$\hat{y} = -2.1 + 0.88x_1 - 150x_2 + 35x_3 \qquad \text{SSE} = 1527.27$$

Then the reduced model $E(Y) = \beta_0 + \beta_1 x_1$ is fit, with the result

$$\hat{y} = 15.3 + 0.84x_1 \qquad \text{SSE} = 3113.14$$

Test to see whether the mean work-hours lost differ for the three programs. Use $\alpha = 0.05$.

10.17   The following model was proposed for testing salary discrimination against women in a state university system:

$$E(Y) = \beta_0 + \beta_1 x_1 + \beta_2 x_2 + \beta_3 x_1 x_2 + \beta_4 x_2^2$$

where

$Y = $ annual salary (in thousands of dollars)

$$x_1 = \begin{cases} 1 & \text{if female} \\ 0 & \text{if male} \end{cases}$$

$x_2 = $ experience (years)

Below is a portion of the computer printout that results from fitting this model to a sample of 200 faculty members in the university system:

SOURCE	DF	SUM OF SQUARES	MEAN SQUARE
MODEL	4	2351.70	587.92
ERROR	195	783.90	4.02
TOTAL	199	3135.60	R-SQUARE
			0.7500

The reduced model $E(Y) = \beta_0 + \beta_2 x_2 + \beta_4 x_2^2$ is fit to the same data, and the resulting computer printout is partially reproduced below:

SOURCE	DF	SUM OF SQUARES	MEAN SQUARE
MODEL	2	2340.37	1170.185
ERROR	197	795.23	4.04
TOTAL	199	3135.60	R-SQUARE
			0.7464

Do these data provide sufficient evidence to support the claim that the mean salary of faculty members is dependent upon sex? Use $\alpha = 0.05$.

10.8 USING THE MODEL FOR ESTIMATION AND PREDICTION

In Section 9.8 we discussed the use of the least-squares line for estimating the mean value of Y, $E(Y)$, for some value of x, say $x = x_p$. We also showed how to use the same fitted model to predict, when $x = x_p$, some value of Y to be observed in the future. Recall that the least-squares line yielded the same value for both the estimate of $E(Y)$ and the prediction of some future value of Y. That is, both are the result of substituting x_p into the prediction equation $\hat{y}_p = \hat{\beta}_0 + \hat{\beta}_1 x_p$ and calculating $\hat{y}_p$. There the equivalence ends. The confidence interval for the mean $E(Y)$ was narrower than the prediction interval for Y, because of the additional uncertainty attributable to the random error ε when predicting some future value of Y.

These same concepts carry over to the multiple regression model. For example, suppose we want to estimate the mean peak power load for a given daily high temperature $x_p = 90$ degrees. Assuming the quadratic model represents the true relationship between peak power load and maximum temperature, we want to estimate

$$E(Y) = \beta_0 + \beta_1 x_p + \beta_2 x_p^2$$
$$= \beta_0 + \beta_1 90 + \beta_2 (90)^2$$

Substituting into the least-squares prediction equation, the estimate of $E(Y)$ is

$$\hat{y} = \hat{\beta}_0 + \hat{\beta}_1 90 + \hat{\beta}_2 (90)^2$$
$$= 1784.188 - 42.3862(90) + 0.27216(90)^2$$
$$= 173.93$$

To form a confidence interval for the mean, we need to know the standard deviation of the sampling distribution for the estimator $\hat{y}$. For multiple regression models the form of this standard deviation is rather complex. However, most regression computer programs allow us to obtain the confidence intervals for mean values of Y for any given combination of values of the independent variables. This portion of the computer output for the peak power load examples is shown in Figure 10.17. The mean value and corresponding 95% confidence interval for

	ESTIMATED	LOWER 95% CL	UPPER 95% CL
x	MEAN VALUE	FOR MEAN	FOR MEAN
90	173.92791187	159.14624576	188.70957799

FIGURE 10.17
*Printout for
Estimated Value and
Corresponding
Confidence Interval
for x = 90*

$x = 90$ are shown in the columns labeled ESTIMATED MEAN VALUE, LOWER 95% CL FOR MEAN, and UPPER 95% CL FOR MEAN. Note that

$$\hat{y} = 173.93$$

which agrees with our earlier calculation. The 95% confidence interval for the true mean of Y is shown to be 159.15 to 188.71 (see Figure 10.18).

FIGURE 10.18
*Confidence Interval
for Mean Peak
Power Load*

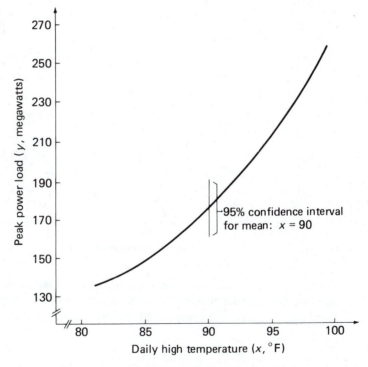

If we were interested in predicting the electrical usage for a particular day on which the high temperature is 90 degrees, $\hat{y} = 173.93$ would be used as the predicted value. However, the prediction interval for a particular value of Y will be wider than the confidence interval for the mean value. This is reflected by the printout shown in Figure 10.19, which gives the predicted value of Y and corresponding 95% prediction interval when $x = 90$. The prediction interval for $x = 90$ is 139.91 to 207.94 (see Figure 10.20).

Unfortunately not all computer packages have the capability to produce confidence intervals for means and prediction intervals for particular Y values. This

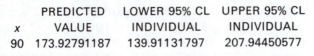

FIGURE 10.19
Printout for
Predicted Value and
Corresponding
Prediction Interval
for x=90

FIGURE 10.20
Prediction Interval
for Electrical Usage

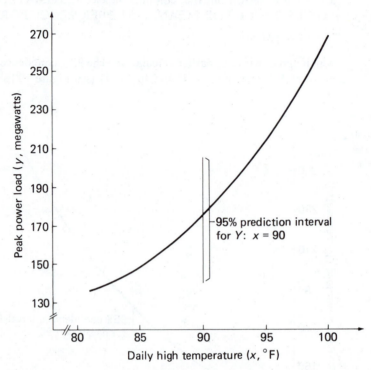

is a rather serious oversight, since the estimation of mean values and the prediction of particular values represent the culmination of our model-building efforts: using the model to make inferences about the dependent variable *Y*.

10.9 MULTIPLE REGRESSION: AN EXAMPLE

Many companies manufacture products that are at least partially chemically produced (e.g., steel, paint, gasoline). In many instances the quality of the finished product is a function of the temperature and pressure at which the chemical reactions take place.

Suppose you wanted to model the quality *Y* of a product as a function of the temperature x_1 and the pressure x_2 at which it is produced. Four inspectors independently assign a quality score between 0 and 100 to each product, and then the quality y is calculated by averaging the four scores. An experiment is conducted by varying temperature between 80 and 100°F and pressure between 50 and 60 pounds per square inch. The resulting data ($n = 27$) are given in Table 10.7.

TABLE 10.7
Temperature, Pressure, and Quality of the Finished Product

x_1, °F	x_2, pounds per square inch	y	x_1, °F	x_2, pounds per square inch	y	x_1, °F	x_2, pounds per square inch	y
80	50	50.8	90	50	63.4	100	50	46.6
80	50	50.7	90	50	61.6	100	50	49.1
80	50	49.4	90	50	63.4	100	50	46.4
80	55	93.7	90	55	93.8	100	55	69.8
80	55	90.9	90	55	92.1	100	55	72.5
80	55	90.9	90	55	97.4	100	55	73.2
80	60	74.5	90	60	70.9	100	60	38.7
80	60	73.0	90	60	68.8	100	60	42.5
80	60	71.2	90	60	71.3	100	60	41.4

Step 1

The first step is to hypothesize a model relating product quality to the temperature and pressure at which it was manufactured. A model that will allow us to find the setting of temperature and pressure that maximize quality is the equation for a *paraboloid*. The visualization of the paraboloid appropriate for this application is an inverted bowl-shaped surface, and the corresponding mathematical model for the mean quality at any temperature/pressure setting is

$$E(Y) = \beta_0 + \beta_1 x_1 + \beta_2 x_2 + \beta_3 x_1^2 + \beta_4 x_2^2 + \beta_5 x_1 x_2$$

This model is also referred to as a *complete second-order model* because it contains all first- and second-order terms in x_1 and x_2.[3] Note that a model with only first-order terms (a plane) would have no curvature, so that the mean quality could not reach a maximum within the experimental region of temperature/pressure even if the data indicated the probable existence of such a value. The inclusion of second-order terms allows curvature in the three-dimensional response surface traced by mean quality, so a maximum mean quality can be reached if the experimental data indicate one exists.

Step 2

Next we use the least-squares technique to estimate the model coefficients β_0, β_1, ..., β_5 of the paraboloid, using the data in Table 10.7. The least-squares model (see

---

[3] Recall from your analytic geometry that the order of a term is the sum of the exponents of the variables in the term. Thus $\beta_5 x_1 x_2$ is a second-order term, as is $\beta_3 x_1^2$. The term $\beta_i x_1^2 x_2$ is a third-order term.

Figure 10.21) is

$$\hat{y} = -5127.90 + 31.10x_1 + 139.75x_2 - 0.133x_1^2 - 1.14x_2^2 - 0.146x_1x_2$$

A three-dimensional graph of this model is shown in Figure 10.21.

FIGURE 10.21
*Plot of Second-
Order Least-
Squares Model for
Product Quality
Example*

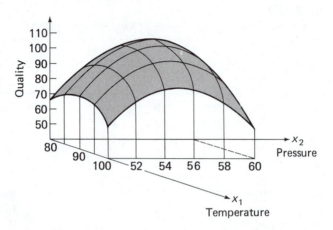

Step 3
The next step is to specify the probability distribution of ε, the random error component. We assume that ε is normally distributed, with a mean of zero and a constant variance σ^2. Furthermore we assume that the errors are independent. The estimate of the variance σ^2 is given in the printout (Figure 10.22) as

$$s^2 = \text{MSE} = \frac{\text{SSE}}{n - (k + 1)} = \frac{\text{SSE}}{27 - (5 + 1)} = 2.818$$

FIGURE 10.22
*Printout for the
Product Quality
Example*

SOURCE	DF	SUM OF SQUARES	MEAN SQUARE	F VALUE	PR>F
MODEL	5	8402.26453714	1680.45290743	596.32	0.0001
ERROR	21	59.17842582	2.81802028		
CORRECTED TOTAL	26	8461.44296296		R-SQUARE	STD DEV
				0.993006	1.67869601

PARAMETER	ESTIMATE	T FOR H0: PARAMETER=0	PR>T	STD ERROR OF ESTIMATE
INTERCEPT	-5127.89907417	-46.49	0.0001	110.29601483
X1	31.09638889	23.13	0.0001	1.34441322
X2	139.74722222	44.50	0.0001	3.14005411
X1 · X1	-0.13338889	-19.46	0.0001	0.00685325
X2 · X2	-1.14422222	-41.74	0.0001	0.02741299
X1 · X2	-0.14550000	-15.01	0.0001	0.00969196

Step 4
Next we want to evaluate the adequacy of the model. First note that $R^2 = 0.993$. This implies that 99.3% of the variation in y, observed quality ratings for the 27 experiments, is accounted for by the model. The statistical significance of this can

be tested:

$$H_0: \beta_1 = \beta_2 = \beta_3 = \beta_4 = \beta_5 = 0$$

H_a: At least one model coefficient is nonzero

Test statistic: $$F = \frac{R^2/k}{(1 - R^2)/[n - (k + 1)]}$$

Rejection region: For $\alpha = 0.05$,

$$F > F_\alpha[k, n - (k + 1)] = F_{0.05}(5, 21)$$

that is

$$F > 2.57$$

The test statistic is given on the printout of Figure 10.22. Since $F = 596.32$ greatly exceeds the tabulated value, we conclude that the model does contribute information about product quality.

Step 5
The culmination of the modeling effort is to use the model for estimation and/or prediction. In this example suppose the manufacturer is interested in estimating the mean product quality for the setting of temperature and pressure at which the estimated model reaches a maximum. To find this setting, we solve the equations

$$\frac{\partial \hat{y}}{\partial x_1} = \hat{\beta}_1 + 2\hat{\beta}_3 x_1 + \hat{\beta}_5 x_2 = 0$$

$$\frac{\partial \hat{y}}{\partial x_2} = \hat{\beta}_2 + 2\hat{\beta}_4 x_2 + \hat{\beta}_5 x_1 = 0$$

for x_1 and x_2, obtaining $x_1 = 86.25°$ and $x_2 = 55.58$ pounds per square inch. The fact that $\partial^2 \hat{y}/\partial x_1^2 = \hat{\beta}_3 < 0$ and $\partial^2 \hat{y}/\partial x_2^2 = \hat{\beta}_4 < 0$ ensures that this setting of x_1 and x_2 corresponds to a maximum value of $\hat{y}$.

To obtain an estimated mean quality for this temperature/pressure combination, we use the least-squares model:

$$\hat{y} = \hat{\beta}_0 + \hat{\beta}_1(86.25) + \hat{\beta}_2(55.58) + \hat{\beta}_3(86.25)^2 + \hat{\beta}_4(55.58)^2$$
$$+ \hat{\beta}_5(86.25)(55.58)$$

The estimated mean value is given in Figure 10.23, a partial reproduction of the regression printout for this example. The estimated mean quality is 96.9, and a 95% confidence interval is 95.5 to 98.3. Thus we are confident that the mean quality rating will be between 95.5 and 98.3 when the product is manufactured at 86.25°F and 55.58 pounds per square inch.

FIGURE 10.23
Partial Printout for
Product Quality
Example

X1	X2	PREDICTED VALUE	LOWER 95% CL FOR MEAN	UPPER 95% CL FOR MEAN
86.25	55.58	96.87401860	95.46463236	98.28340483

10.10   SOME COMMENTS ON THE ASSUMPTIONS

When we apply a regression analysis to a set of data, we never know for certain that the assumptions of Section 10.3 are satisfied. How far can we deviate from the assumptions and still expect a multiple regression analysis to yield results that will possess the reliability stated in this chapter? How can we detect departures (if they exist) from the assumptions of Section 10.3, and what can we do about them? We will provide some partial answers to these questions in this section and direct you to further discussion in succeeding chapters.

Remember (from Sections 10.1 and 10.3) that

$$Y = E(Y) + \varepsilon$$

where the expected value $E(Y)$ for a given set of values of $x_1, x_2, \ldots, x_k$ is

$$E(Y) = \beta_0 + \beta_1 x_1 + \beta_2 x_2 + \cdots + \beta_k x_k$$

and ε is a random error. The first assumption we made was that the mean value of the random error for *any* given set of values of $x_1, x_2, \ldots, x_k$ is $E(\varepsilon) = 0$.

One consequence of this assumption is that the mean $E(Y)$ for a specific set of values of $x_1, x_2, \ldots, x_k$ is

$$E(Y) = \beta_0 + \beta_1 x_1 + \beta_2 x_2 + \cdots + \beta_k x_k$$

That is,

$$
Y = \underbrace{E(Y)}_{\substack{\text{Mean value of } y \\ \text{for specific values} \\ \text{of } x_1, x_2, \ldots, x_k}} + \underbrace{\varepsilon}_{\substack{\text{Random} \\ \text{error}}}
$$

The second consequence of the assumption is that the least-squares estimators of the model parameters $\beta_0, \beta_1, \beta_2, \ldots, \beta_k$ will be unbiased regardless of the remaining assumptions that we attribute to the random errors and their probability distributions.

The properties of the sampling distributions of the parameter estimators $\hat{\beta}_0, \hat{\beta}_1, \ldots, \hat{\beta}_k$ will depend on the remaining assumptions that we specify concerning the probability distributions of the random errors. You will recall that we assumed that for any given set of values of $x_1, x_2, \ldots, x_k$, ε has a normal probability distribution with mean equal to zero and variance equal to σ^2. Also we assumed that the random errors are independent (in a probabilistic sense).

It is unlikely that the assumptions stated above are satisfied exactly for many practical situations. If departures from the assumptions are not too great, experience has shown that a least-squares regression analysis produces estimates—predictions and statistical test results—that possess, for all practical purposes, the properties specified in this chapter. If the observations are likely to be correlated (as in the case of data collected over time), we must check for correlation between the random errors and may have to modify our methodology if correlation exists. A test for correlation of the random errors can be found in Mendenhall and McClave (*A Second Course in Business Statistics: Regression Analysis*, Dellen, 1981). If the variance of the random error ε changes from one setting of the independent x variables to another, we can sometimes transform the data so that the standard least-squares methodology will be appropriate. Techniques for detecting non-homogeneous variances of the random errors (a condition called *heteroscedasticity*) and some methods for treating this type of data are also discussed in Mendenhall and McClave, as well as many other texts that focus on regression analysis.

Frequently the data $(y, x_1, x_2, \ldots, x_k)$ are observational; that is, we just observe an experimental unit and record values for $y, x_1, x_2, \ldots, x_k$. Do these data violate the assumption that $x_1, x_2, \ldots, x_k$ are fixed? For this particular case if we can assume that $x_1, x_2, \ldots, x_k$ are *measured without error*, the mean value $E(Y)$ can be viewed as a conditional mean. That is, it gives the mean value of Y, *given* that the x variables assume a specific set of values. With this modification in our thinking, the least-squares regression analysis is applicable to observational data.

To conclude, remember that when you perform a regression analysis, the reliability you can place in your inferences is dependent on the satisfaction of the assumptions prescribed in Section 10.3. Although the random errors will rarely satisfy these assumptions exactly, the reliability specified by a regression analysis will hold approximately for many types of data encountered in practice.

10.11 SOME PITFALLS: ESTIMABILITY, MULTICOLLINEARITY, AND EXTRAPOLATION

There are several problems you should be aware of when constructing a prediction model for some response Y. A few of the most important will be discussed in this section.

Problem 1: Parameter Estimability. Suppose you want to fit a model relating a firm's monthly profit Y to the advertising expenditure x. We propose the first-order model

$$E(Y) = \beta_0 + \beta_1 x$$

Now suppose we have 3 months of data, and the firm spent \$1,000 on advertising during each month. The data are shown in Figure 10.24. You can see the problem: The parameters of the line cannot be estimated when all the data are concentrated at a single x value. Recall that it takes two points (x values) to fit a straight line. Thus the parameters are not estimable when only one x value is observed.

A similar problem would occur if we attempted to fit the second-order model

$$E(Y) = \beta_0 + \beta_1 x + \beta_2 x^2$$

to a set of data for which only one *or two* different x values were observed (see Figure 10.25). At least three different x values must be observed before a second-

FIGURE 10.24
*Profit and
Advertising
Expenditure Data: 3
Months*

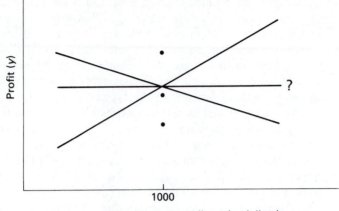

FIGURE 10.25
*Only Two x Values
Observed—the
Second-Order
Model Is Not
Estimable*

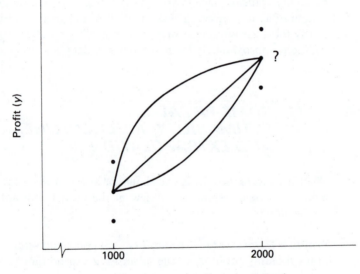

order model can be fit to a set of data (that is, before all three parameters are estimable). In general, the number of levels of x must be at least one more than the order of the polynomial in x that you want to fit.

Since many variables are not controlled by the researcher, the independent variables will almost always be observed at a sufficient number of levels to permit estimation of the model parameters. However, when the computer program you use suddenly refuses to fit a model, the problem is probably inestimable parameters.

Problem 2: Multicollinearity.   Often two or more of the independent variables used in the model for $E(Y)$ will contribute redundant information because they are correlated with each other. For example, suppose we want to construct a model to predict the gasoline mileage rating of a truck as a function of its load x_1 and the horsepower x_2 of its engine. In general you would expect heavy loads to require greater horsepower and to result in lower mileage ratings. Thus although both x_1 and x_2 contribute information for the prediction of mileage rating, some of the information is overlapping because x_1 and x_2 are correlated.

If the model

$$E(Y) = \beta_0 + \beta_1 x_1 + \beta_2 x_2$$

were fit to a set of data, we might find that the t values for both $\hat{\beta}_1$ and $\hat{\beta}_2$ (the least-squares estimates) are nonsignificant. However, the F test for $H_0: \beta_1 = \beta_2 = 0$ would probably be highly significant. The tests may seem to be contradictory, but really they are not. The t tests indicate that the contribution of one variable, say $x_1 =$ Load, is not significant after the effect of $x_2 =$ Horsepower has been discounted (because x_2 is also in the model). The significant F test, on the other hand, tells us that at least one of the two variables is making a contribution to the prediction of y (i.e., β_1, β_2, or both differ from zero). In fact both are probably contributing, but the contribution of one overlaps that of the other.

When highly correlated independent variables are present in a regression model, the results may be confusing. The researcher may want to include only one of the variables in the final model.

Problem 3: Prediction Outside the Experimental Region.   By the late 1960s many research economists had developed highly technical models to relate the state of the economy to various economic indices and other independent variables. Many of these models were multiple regression models where, for example, the

FIGURE 10.26
Using a Regression Model Outside the Experimental Region

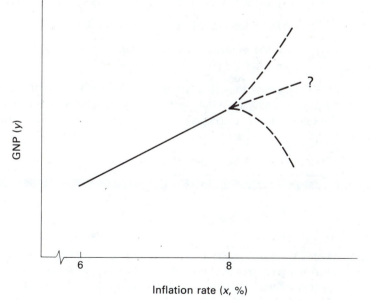

dependent variable Y might be next year's growth in GNP, and the independent variables might include this year's rate of inflation, this year's Consumer Price Index, and other factors. In other words the model might be constructed to predict next year's economy using this year's knowledge.

Unfortunately these models were almost unanimously unsuccessful in predicting the recession in the early 1970s. What went wrong? One of the problems was that the regression models were used to predict Y for values of the independent variables that were outside the region in which the model was developed. For example the inflation rate in the late 1960s, when the models were developed, ranged from 6% to 8%. When the double-digit inflation of the early 1970s became a reality, some researchers attempted to use the same models to predict future growth in GNP. As you can see in Figure 10.26, the model may be very accurate for predicting Y when x is in the range of experimentation, but the use of the model outside that range is a dangerous practice.

10.12   CONCLUSION

We have discussed some of the methodology of *multiple regression analysis*, a technique for modeling a dependent variable Y as a function of several independent variables $x_1, x_2, \ldots, x_k$. The steps we follow in constructing and using multiple regression models are much the same as those for the simple straight-line models:

1. The form of the probabilistic model is hypothesized.

2. The model coefficients are estimated by using least squares.

3. The probability distribution of ε is specified and σ^2 is estimated.

4. The adequacy of the model is checked.

5. If the model is deemed useful, it may be used to make estimates and to predict values of Y to be observed in the future.

We stress that this is not intended to be a complete coverage of multiple regression analysis. Whole texts have been devoted to this topic. However, we have presented the core necessary for a basic understanding of multiple regression.

SUPPLEMENTARY EXERCISES

[*Note:* Starred (*) exercises require the use of a computer.]

10.18   After a regression model is fit to a set of data, a confidence interval for the mean value of Y at a given setting of the independent variables will *always* be narrower than the corresponding prediction interval for a particular value of Y at the same setting of the independent variables. Why?

10.19 Before accepting a job, a computer at a major university estimates the cost of running the job to see if the user's account contains enough money to cover the cost. As part of the job submission, the user must specify estimated values for two variables: central processing unit (CPU) time and number of lines printed. While the CPU time required and the number of lines printed do not account for the complete cost of the run, it is thought that knowledge of their values should allow a good prediction of job cost. The following model is proposed to explain the relationship of CPU times and lines printed to job cost:

$$E(Y) = \beta_0 + \beta_1 x_1 + \beta_2 x_2 + \beta_3 x_1 x_2$$

where

Y = job cost (in dollars)

x_1 = number of lines printed

x_2 = CPU time

Records from twenty previous runs were used to fit this model. The SAS printout is shown.

Portion of SAS Printout for Exercise 10.19

SOURCE	DF	SUM OF SQUARES	MEAN SQUARE	F VALUE	PR > F
MODEL	3	43.25090461	14.41696820	84.96	0.0001
ERROR	16	2.71515039	0.16969690		
CORRECTED TOTAL	19	45.9660550		R-SQUARE	STD DEV
				0.940931	0.41194283

PARAMETER	ESTIMATE	T FOR H0: PARAMETER = 0	PR > |T|	STD ERROR OF ESTIMATE
INTERCEPT	0.04564705	0.22	0.8313	0.21082636
X1	0.00078505	5.80	0.0001	0.00013537
X2	0.23737262	7.50	0.0001	0.03163301
X1 · X2	−0.00003809	−2.99	0.0086	0.00001273

X1	X2	PREDICTED VALUE	LOWER 95% CL FOR MEAN	UPPER 95% CL FOR MEAN
2000	42	8.38574865	7.32284845	9.44864885

(a) Identify the least-squares model that was fit to the data.
(b) What are the values of SSE and s^2 (estimate of σ^2) for the data?
(c) What do we mean by the statement: This value of SSE [see part (b)] is minimum?

10.20 Refer to Exercise 10.19 and the portion of the SAS printout shown
(a) Is there evidence that the model is useful (as a whole) for predicting job cost? Test at $\alpha = 0.05$.
(b) Is there evidence that the variables x_1 and x_2 interact to affect Y? Test at $\alpha = 0.01$.
(c) What assumptions are necessary for the validity of the tests conducted in parts (a) and (b)?

10.21 Refer to Exercise 10.19 and the portion of the SAS printout shown. Use a 95% confidence interval to estimate the mean cost of computer jobs that require 42 seconds of CPU time and print 2000 lines.

10.22 Several states now require all high school seniors to pass an achievement test before they can graduate. On the test the seniors must demonstrate their familiarity with basic verbal and mathematical skills. Suppose that the educational testing company that creates and administers these exams wants to model the score Y on one of their exams as a function of the student's IQ x_1 and socioeconomic status (SES). The SES is a categorical (or *qualitative*) variable with three levels: low, medium, and high.

$$x_2 = \begin{cases} 1 & \text{if SES is medium} \\ 0 & \text{if SES is low or high} \end{cases}$$

$$x_3 = \begin{cases} 1 & \text{if SES is high} \\ 0 & \text{if SES is low or medium} \end{cases}$$

Data were collected for a random sample of sixty seniors who have taken the test, and the model

$$E(Y) = \beta_0 + \beta_1 x_1 + \beta_2 x_2 + \beta_3 x_3$$

was fit to these data, with the results shown in the SAS printout.

SOURCE	DF	SUM OF SQUARES	MEAN SQUARE	F VALUE	PR > F
MODEL	3	12268.56439492	4089.52146497	188.33	0.0001
ERROR	56	1216.01893841	21.71462390		STD DEV
CORRECTED TOTAL	59	13484.58333333		R-SQUARE	4.65989527
				0.909822	

PARAMETER	ESTIMATE	T FOR H0: PARAMETER = 0	PR > |T|	STD ERROR OF ESTIMATE
INTERCEPT	−13.06166081	−3.21	0.0022	4.07101383
X1	0.74193946	17.56	0.0001	0.04224805
X2	18.60320572	12.49	0.0001	1.48895324
X3	13.40965415	8.97	0.0001	1.49417069

(a) Identify the least-squares equation.
(b) Interpret the value of R^2 and test to determine whether the data provide sufficient evidence to indicate that this model is useful for predicting achievement test scores.
(c) Sketch the relationship between predicted achievement test score and IQ for the three levels of SES. [*Note*: Three graphs of $\hat{y}$ versus x_1 must be drawn: the first for the low-SES model ($x_2 = x_3 = 0$), the second for the medium-SES model ($x_2 = 1$, $x_3 = 0$), and the third for the high-SES model ($x_2 = 0$, $x_3 = 1$). Note that increase in predicted achievement test score per unit increase in IQ is the same for all three levels of SES; that is, all three lines are parallel.]

10.23 Refer to Exercise 10.22. We now use the same data to fit the model

$$E(Y) = \beta_0 + \beta_1 x_1 + \beta_2 x_2 + \beta_3 x_3 + \beta_4 x_1 x_2 + \beta_5 x_1 x_3$$

Thus we now add the interaction between IQ and SES to the model. The SAS printout for this model is shown.
(a) Identify the least-squares prediction equation.
(b) Interpret the value of R^2 and test to determine whether the data provide sufficient evidence to indicate that this model is useful for predicting achievement test scores.
(c) Sketch the relationship between predicted achievement test score and IQ for the three levels of SES.

(d) Test to determine whether there is evidence that the mean increase in achievement test score per unit increase in IQ differs for the three levels of SES.

SOURCE	DF	SUM OF SQUARES	MEAN SQUARE	F VALUE	PR > F
MODEL	5	12515.10021009	2503.02004202	139.42	0.0001
ERROR	54	969.48312324	17.95339117		
CORRECTED TOTAL	59	13484.58333333			

				R-SQUARE	STD DEV
				0.928104	4.23714422

PARAMETER	ESTIMATE	T FOR H0: PARAMETER = 0	PR > |T|	STD ERROR OF ESTIMATE
INTERCEPT	0.60129643	0.11	0.9096	5.26818519
X1	0.59526252	10.70	0.0001	0.05563379
X2	−3.72536406	−0.37	0.7115	10.01967496
X3	−16.23196444	−1.90	0.0631	8.55429931
X1 · X2	0.23492147	2.29	0.0260	0.10263908
X1 · X3	0.30807756	3.53	0.0009	0.08739554

10.24 The EPA wants to model the gas mileage ratings Y of automobiles as a function of their engine size x. A quadratic model

$$Y = \beta_0 + \beta_1 x + \beta_2 x^2$$

is proposed. A sample of fifty engines of varying sizes is selected, and the miles per gallon rating of each is determined. The least-squares model is

$$\hat{y} = 51.3 - 10.1x + 0.15x^2$$

The size x of the engine is measured in hundreds of cubic inches. Also $s_{\hat{\beta}_2} = 0.0037$ and $R^2 = 0.93$.

(a) Sketch this model between $x = 1$ and $x = 4$.
(b) Is there evidence that the quadratic term in the model is contributing to the prediction of the miles per gallon rating Y? Use $\alpha = 0.05$.
(c) Use the model to estimate the mean miles per gallon rating for all cars with 350-cubic-inch engines ($x = 3.5$).
(d) Suppose a 95% confidence interval for the quantity estimated in part (c) is (17.2, 18.4). Interpret this interval.
(e) Suppose you purchase an automobile with a 350-cubic-inch engine and determine that the miles per gallon rating is 14.7. Is the fact that this value lies outside the confidence interval given in part (d) surprising? Explain.

10.25 To increase the motivation and productivity of workers, an electronics manufacturer decides to experiment with a new pay incentive structure at one of two plants. The experimental plan will be tried at plant A for 6 months, while workers at plant B will remain on the original pay plan. To evaluate the effectiveness of the new plan, the average assembly time for part of an electronic system was measured for employees at both plants at the beginning and end of the 6-month period. Suppose the following model was proposed:

$$Y = \beta_0 + \beta_1 x_1 + \beta_2 x_2 + \varepsilon$$

where

Y = assembly time (hours) at end of 6-month period

x_1 = assembly time (hours) at beginning of 6-month period

$$x_2 = \begin{cases} 1 & \text{if plant A} \\ 0 & \text{if plant B} \end{cases} \quad \text{(dummy variable)}$$

A sample of $n = 42$ observations yielded

$$\hat{y} = 0.11 + 0.98x_1 - 0.53x_2$$

where

$$s_{\hat{\beta}_1} = 0.231 \qquad s_{\hat{\beta}_2} = 0.48$$

Test to see whether, after allowing for the effect of initial assembly time, plant A had a lower mean assembly time than plant B. Use $\alpha = 0.01$. [*Note:* When the (0, 1) coding is used to define a dummy variable, the coefficient of the variable represents the difference between the mean response at the two levels represented by the variable. Thus the coefficient β_2 is the difference in mean assembly time between plant A and plant B at the end of the 6-month period, and $\hat{\beta}_2$ is the sample estimator of that difference.]

10.26 One fact that must be considered in developing a shipping system that is beneficial to both the customer and the seller is time of delivery. A manufacturer of farm equipment can ship its products by either rail or truck. Quadratic models are thought to be adequate in relating time of delivery to distance traveled for both modes of transportation. Consequently it has been suggested that the following model be fit:

$$E(Y) = \beta_0 + \beta_1 x_1 + \beta_2 x_2 + \beta_3 x_1 x_2 + \beta_4 x_2^2$$

where

Y = shipping time

$$x_1 = \begin{cases} 1 & \text{if rail} \\ 0 & \text{if truck} \end{cases}$$

x_2 = distance to be shipped

(a) What hypothesis would you test to determine whether the data indicate that the quadratic distance term is useful in the model, that is, whether curvature is present in the relationship between mean delivery time and distance?

(b) What hypothesis would you test to determine whether there is a difference in mean delivery time by rail and by truck?

10.27 Refer to Exercise 10.26. Suppose the proposed second-order model is fit to a total of fifty observations on delivery time. The sum of squared errors is SSE = 226.12. Then the reduced model

$$E(Y) = \beta_0 + \beta_2 x_2 + \beta_4 x_2^2$$

is fit to the same data, and SSE = 259.34. Test to see whether the data indicate that the mean delivery time differs for rail and truck deliveries.

10.28 *Operations management* is concerned with planning and controlling those organizational functions and systems that produce goods and services (Schroeder, *Operations Management: Decision Making in the Operations Function*, McGraw-Hill, 1981). One concern of the operations manager of a production process is the level of productivity of the process. An operations manager at a large manufacturing plant is interested in predicting the level of productivity of assembly line A next year (i.e., the number of units that will be produced by the assembly line next year). To do so, she has decided to use regression analysis to model the level of productivity Y as a function of time x. The number of units produced by assembly line A was determined for each of the past 15 years ($x = 1, 2, \ldots, 15$). The model

$$E(Y) = \beta_0 + \beta_1 x + \beta_2 x^2$$

was fit to these data, using Minitab. The results shown in the printout were obtained.

THE REGRESSION EQUATION IS
Y = −1187 − 1333 X1 − 45.6 X2

COLUMN		COEFFICIENT	ST. DEV. OF COEF.	T-RATIO = COEF./S.D.
	—	−1187	446	−2.66
X1	C2	1333	128	10.38
X2	C3	−45.59	7.80	−5.84

THE ST. DEV. OF Y ABOUT REGRESSION LINE IS
S = 501

WITH (15−3) = 12 DEGREES OF FREEDOM
R-SQUARED = 97.3 PERCENT
R-SQUARED = 96.9 PERCENT. ADJUSTED FOR D.F.

ANALYSIS OF VARIANCE

DUE TO	DF	SS	MS = SS/DF
REGRESSION	2	110578719	55289359
RESIDUAL	12	3013365	251114
TOTAL	14	113592083	

(a) Identify the least-squares prediction equation.
(b) Find R^2 and interpret its value in the context of this problem.
(c) Is there sufficient evidence to indicate that the model is useful for predicting the productivity of assembly line A? Test by using $\alpha = 0.05$.
(d) Test the null hypothesis $H_0: \beta_2 = 0$ against the alternative hypothesis H_a: $\beta_2 \neq 0$ by using $\alpha = 0.05$. Interpret the results of your test in the context of this problem.
(e) Which (if any) of the assumptions we make about ε in regression analysis are likely to be violated in this problem? Explain.

10.29 Many companies must accurately estimate their costs before a job is begun in order to acquire a contract and make a profit. For example, a heating and plumbing contractor may base cost estimates for new homes on the total area of the house, the number of baths in the plans, and whether central air conditioning is to be installed.

(a) Write a first-order model relating the mean cost of material and labor $E(Y)$ to the area, number of baths, and central air conditioning variables.
(b) Write a complete second-order model for the mean cost as a function of the same three variables.
(c) How would you test the research hypothesis that the second-order terms are useful for predicting mean cost?

10.30 Refer to Exercise 10.29. The contractor samples twenty-five recent jobs and fits both the complete second-order model [part (b)] and the reduced main effects model in part (a), so that a test can be conducted to determine whether the additional complexity of the second-order model is necessary. The resulting SSE and R^2 are given in the table.

	SSE	R^2
First-order	8.548	0.950
Second-order	6.133	0.964

(a) Is there sufficient evidence to conclude that the second-order terms are important for predicting the mean cost?

(b) Suppose the contractor decides to use the main effects model to predict costs. Use the global F test to determine whether the main effects model is useful for predicting costs.

*10.31 A company that services two brands of microcomputers would like to be able to predict the amount of time it takes a service person to perform preventive maintenance on each brand. They believe the following predictive model is appropriate:

$$Y = \beta_0 + \beta_1 x_1 + \beta_2 x_2 + \varepsilon$$

where

Y = maintenance time

$$x_1 = \begin{cases} 1 & \text{if brand A} \\ 0 & \text{if brand I} \end{cases}$$

x_2 = service person's number of months of experience in preventive maintenance

Ten different service people were randomly selected, and each was randomly assigned to perform preventive maintenance on either a brand A or brand I microcomputer. The following data were obtained.

Maintenance Time (hours)	Brand	Experience (months)
2.0	1	2
1.8	1	4
0.8	0	12
1.1	1	12
1.0	0	8
1.5	0	2
1.7	1	6
1.2	0	5
1.4	1	9
1.2	0	7

(a) Fit the model to the data.

(b) Investigate whether the overall model is useful. Test by using $\alpha = 0.05$.

(c) Find R^2 for the fitted model. Does the value of R^2 support your findings in part (b)? Explain.

(d) Find a 90% confidence interval for β_2. Interpret your result in the context of this exercise.

(e) Use the fitted model to predict how long it will take a person with 6 months of experience to service a brand I microcomputer.

(f) How long would it take the person referred to in part (e) to service ten brand I microcomputers? List any assumptions you made in reaching your prediction.

(g) Find a 95% prediction interval for the time required to perform preventive maintenance on a brand A microcomputer by a person with 4 months of experience.

*10.32 Many colleges and universities develop regression models for predicting the grade-point average (GPA) of incoming freshmen. This predicted GPA can then be used to help make admission decisions. Although most models use many independent variables to predict GPA, we will illustrate by choosing two variables:

x_1 = verbal score on college entrance examination (percentile)

x_2 = mathematics score on college entrance examination (percentile)

The data in the table are obtained for a random sample of forty freshmen at one college.

Verbal x_1	Mathematics x_2	GPA Y	Verbal x_1	Mathematics x_2	GPA Y
81	87	3.49	79	75	3.45
68	99	2.89	81	62	2.76
57	86	2.73	50	69	1.90
100	49	1.54	72	70	3.01
54	83	2.56	54	52	1.48
82	86	3.43	65	79	2.98
75	74	3.59	56	78	2.58
58	98	2.86	98	67	2.73
55	54	1.46	97	80	3.27
49	81	2.11	77	90	3.47
64	76	2.69	49	54	1.30
66	59	2.16	39	81	1.22
80	61	2.60	87	69	3.23
100	85	3.30	70	95	3.82
83	76	3.75	57	89	2.93
64	66	2.70	74	67	2.83
83	72	3.15	87	93	3.84
93	54	2.28	90	65	3.01
74	59	2.92	81	76	3.33
51	75	2.48	84	69	3.06

(a) Fit the first-order model (no quadratic and no interaction terms)

$$Y = \beta_0 + \beta_1 x_1 + \beta_2 x_2 + \varepsilon$$

Interpret the value of R^2, and test whether these data indicate that the terms in the model are useful for predicting freshman GPA. Use $\alpha = 0.05$.

(b) Sketch the relationship between predicted GPA $\hat{y}$ and verbal score x_1 for the following mathematics scores: $x_2 = 60$, 75, and 90.

*10.33 Refer to Exercise 10.32. Now fit the following second-order model to the data:

$$Y = \beta_0 + \beta_1 x_1 + \beta_2 x_2 + \beta_3 x_1^2 + \beta_4 x_2^2 + \beta_5 x_1 x_2 + \varepsilon$$

(a) Interpret the value of R^2, and test whether the data indicate that this model is useful for predicting freshman GPA. Use $\alpha = 0.05$.

(b) Sketch the relationship between predicted GPA $\hat{y}$ and the verbal score x_1 for the following mathematics scores: $x_2 = 60$, 75, and 90. Compare these graphs with those for the first-order model in Exercise 10.32.

(c) Test to see whether the interaction term $\beta_5 x_1 x_2$ is important for the prediction of GPA. Use $\alpha = 0.10$. Note that this term permits the distance between three mathematics score curves for GPA versus verbal score to change as the verbal score changes.

10.34 Plastics made under different environmental conditions are known to have differing strengths. A scientist would like to know which combination of temperature and pressure yields a plastic with a high breaking strength. A small preliminary experiment was run at two pressure levels and two temperature levels. The following model was proposed:

$$E(Y) = \beta_0 + \beta_1 x_1 + \beta_2 x_2 + \beta_3 x_1 x_2$$

where

Y = breaking strength (pounds)

x_1 = temperature (°F)

x_2 = pressure (pounds per square inch)

A sample of $n = 16$ observations yielded

$$\hat{y} = 226.8 + 4.9x_1 + 1.2x_2 - 0.7x_1 x_2$$

with

$$s_{\beta_1} = 1.11 \qquad s_{\beta_2} = 0.27 \qquad s_{\beta_3} = 0.34$$

Do the data indicate there is an interaction between temperature and pressure? Test by using $\alpha = 0.05$.

*10.35 Air pollution regulations for power plants are often written so that the maximum amount of pollutant that can be omitted is increased as the plant's output increases. Suppose the following data are collected over a period of time:

Output x (megawatts)	Sulfur Dioxide Emission Y (parts per million)
525	143
452	110
626	173
573	161
422	105
712	240
600	165
555	140
675	210

(a) Plot the data in a scattergram.

(b) Use the least-squares method to fit a straight line relating sulfur dioxide emission to output. Plot the least-squares line on the scattergram.

(c) Use a computer program to fit a second-order (quadratic) model relating sulfur dioxide emission to output.

(d) Conduct a test to determine whether the quadratic model provides a better description of the relationship between sulfur dioxide and output than the linear model.

(e) Use the quadratic model to estimate the mean sulfur dioxide emission when the power output is 500 megawatts. Use a 95% confidence interval.

10.36 The recent expansion of U.S. grain exports has intensified the importance of the linkage between the domestic grain transportation system and international transportation. As a first step in evaluating the economies of this interface, Martin and Clement (*Transportation Journal*, 1982) used multiple regression to estimate ocean transport rates for grain shipped from the Lower Columbia River international ports. These ports include Portland, Oregon; and Vancouver, Longview, and Kalama, Washington. Rates per long ton Y were modeled as a function of the following independent variables:

$x_1 = $ shipment size in long tons

$x_2 = $ distance to destination port in miles

$x_3 = $ bunker fuel price in dollars per barrel

$x_4 = \begin{cases} 1 & \text{if American flagship} \\ 0 & \text{if foreign flagship} \end{cases}$

$x_5 = $ size of the port as measured by the U.S. Defense Mapping Agency's Standards

$x_6 = $ quantity of grain exported from the region during the year of interest

The method of least squares was used to fit the model to 140 observations from the period 1978 through 1980. The following results were obtained:

$$\hat{Y} = -18.469 - 0.367x_1 + 6.434x_2 - 0.2692x_2^2 + 1.7992x_3 + 50.292x_4$$
$$\phantom{\hat{Y} =} (-2.76) \quad (-5.62) \quad (3.64) \quad (-2.25) \quad (12.96) \quad (19.14)$$

$$+ 2.275x_5 - 0.018x_6$$
$$ (1.17) \quad (-2.69)$$

$$R^2 = 0.8979$$

$$F = 130.665$$

The numbers in parentheses are the t statistics associated with the $\hat{\beta}$ values above them.

(a) Test $H_0: \beta_1 = \beta_2 = \beta_3 = \beta_4 = \beta_5 = \beta_6 = \beta_7 = 0$. Use $\alpha = 0.01$. Interpret the results of your test in the context of the problem.

(b) Binkley and Harrer (*American Journal of Agricultural Economics*, pp. 47–51, 1979) estimated a similar rate function by using multiple regression, but used different independent variables. The coefficient of determination for their model was 0.46. Compare the explanatory power of the Binkley and Harrer model with that of Martin and Clement.

(c) According to the least-squares model, do freight charges increase with distance? Do they increase at an increasing rate? Explain.

10.37 An economist has proposed the following model to describe the relationship between the number of items produced per day (output) and the number of hours of labor expended per day (input) in a particular production process:

$$Y = \beta_0 + \beta_1 x + \beta_2 x^2 + \varepsilon$$

where

Y = number of items produced per day

x = number of hours of labor per day

A portion of the computer printout that results from fitting this model to a sample of 25 weeks of production data is shown. Test the hypothesis that, as the amount of input increases, the amount of output also increases, but at a decreasing rate. Do the data provide sufficient evidence to indicate that the *rate* of increase in output per unit increase of input decreases as the input increases? Test by using $\alpha = 0.05$.

```
THE REGRESSION EQUATION IS
Y = -6.17 + 2.04 X1 - 0.0323 X2
```

COLUMN		COEFFICIENT	ST. DEV. OF COEF.	T-RATIO. = COEF./S.D.
	—	-6.173	1.666	-3.71
X1	C2	2.036	0.185	11.02
X2	C3	-0.03231	0.00489	-6.60

```
THE ST. DEV. OF Y ABOUT REGRESSION LINE IS
S = 1.243
WITH (25 - 3) = 22 DEGREES OF FREEDOM
R-SQUARED = 95.5 PERCENT
R-SQUARED = 95.1 PERCENT, ADJUSTED FOR D.F.
```

ANALYSIS OF VARIANCE

DUE TO	DF	SS	MS = SS/DF
REGRESSION	2	718.168	359.084
RESIDUAL	22	33.992	1.545
TOTAL	24	752.160	

10.38 An operations engineer is interested in modeling $E(Y)$, the expected length of time per month (in hours) that a machine will be shut down for repairs, as a function of the type of machine (001 or 002) and the age of the machine (in years). He has proposed the following model:

$$E(Y) = \beta_0 + \beta_1 x_1 + \beta_2 x_1^2 + \beta_3 x_2$$

where

$x_1 = $ age of machine

$$x_2 = \begin{cases} 1 & \text{if machine type 001} \\ 0 & \text{if machine type 002} \end{cases}$$

Data were obtained on $n = 20$ machine breakdowns and were used to estimate the parameters of the above model. A portion of the regression analysis computer printout is shown below:

SOURCE	DF	SUM OF SQUARES	MEAN SQUARE
MODEL	3	2396.364	798.788
ERROR	16	128.586	8.037
TOTAL	19	2524.950	R-SQUARE
			0.949

The reduced model $E(Y) = \beta_0 + \beta_1 x_1 + \beta_2 x_2$ was fit to the same data. The regression analysis computer printout is partially reproduced below.

SOURCE	DF	SUM OF SQUARES	MEAN SQUARE
MODEL	2	2342.42	1171.21
ERROR	17	182.53	10.74
TOTAL	19	2524.95	R-SQUARE
			0.928

Do these data provide sufficient evidence to conclude that the second-order (x_1^2) term in the model proposed by the operations engineer is necessary? Test by using $\alpha = 0.05$.

*10.39 Refer to Exercise 10.38. The data that were used to fit the operations engineer's complete and reduced models are displayed in the table.

Downtime per Month (hours)	Machine Type	Machine Age (years)
10	001	1.0
20	001	2.0
30	001	2.7
40	001	4.1
9	001	1.2
25	001	2.5
19	001	1.9
41	001	5.0
22	001	2.1
12	001	1.1

Downtime per Month (hours)	Machine Type	Machine Age (years)
10	002	2.0
20	002	4.0
30	002	5.0
44	002	8.0
9	002	2.4
25	002	5.1
20	002	3.5
42	002	7.0
20	002	4.0
13	002	2.1

(a) Use these data to test the null hypothesis that $\beta_1 = \beta_2 = 0$ in the complete model. Test by using $\alpha = 0.10$.

(b) Carefully interpret the results of the test in the context of the problem.

10.40 Refer to Exercise 10.11, where we presented data on the number of highway deaths and the number of licensed vehicles on the road for the years 1950–1979. We mentioned that the number of deaths Y may also have been affected by the existence of the national 55-mile-per-hour speed limit during the years 1974–1979 (i.e., years 25–30). Define the dummy variable

$$x_2 = \begin{cases} 1 & \text{if 55-mile-per-hour speed limit was in effect} \\ 0 & \text{if not} \end{cases}$$

(a) Introduce the variable x_2 into the second-order model of Exercise 10.11 to account for the presence or absence of the 55-mile-per-hour speed limit in a given year. Include terms involving the interaction between x_2 and x_1.

(b) Refer to your model for part (a). Sketch on a single piece of graph paper your visualization of the two response curves, the second-order curves relating Y to x_1 before and after the imposition of the 55-mile-per-hour speed limit.

(c) Suppose that x_1 and x_2 do not interact. How would that affect the graphs of the two response curves of part (b)?

(d) Refer to part (c). Suppose that x_1 and x_2 do interact. How would this affect the graphs of the two response curves?

10.41 In Exercise 10.11 we fit a second-order model to data relating the number Y of U.S. highway deaths per year to the number x_1 of licensed vehicles on the road. In Exercise 10.40 we added a qualitative variable x_2 to account for the presence or absence of the 55-mile-per-hour national speed limit. The accompanying SAS computer printout gives the results of fitting the model

$$E(Y) = \beta_0 + \beta_1 x_1 + \beta_2 x_1^2 + \beta_3 x_2 + \beta_4 x_1 x_2 + \beta_5 x_1^2 x_2$$

to the data. Use this printout and the printout for Exercise 10.11 to determine whether the data provide sufficient evidence to indicate that the qualitative variable (speed limit) contributes information for the prediction of the annual number of highway deaths. Test by using $\alpha = 0.05$. Discuss the practical implications of your test results.

DEPENDENT VARIABLE: DEATHS

SOURCE	DF	SUM OF SQUARES	MEAN SQUARE	F VALUE
MODEL	5	1428.02852498	285.60570500	41.56
ERROR	24	164.91847502	6.87160313	PR > F
CORRECTED TOTAL	29	1592.94700000		0.0001

R-SQUARE	C.V.	ROOT MSE	DEATHS MEAN
0.896470	5.7752	2.62137428	45.39000000

PARAMETER	ESTIMATE	T FOR H0: PARAMETER=0	PR > |T|	STD ERROR OF ESTIMATE
INTERCEPT	16.94054815	2.21	0.0372	7.68054270
VEHICLES	0.34942808	1.89	0.0715	0.18528901
VEHICLES • VEHICLES	−0.00017135	−0.16	0.8724	0.00105538
LIMIT	41.38712279	0.11	0.9155	385.80676597
VEHICLES • LIMIT	−0.75118159	−0.14	0.8878	5.26806364
VEHICLE • VEHICLE • LIMIT	0.00245917	0.14	0.8921	0.01794470

*10.42   *Productivity* has been defined as the relationship between inputs and outputs of a productive system. To manage a system's productivity, it is necessary to measure it. Productivity is typically measured by dividing a measure of system output by a measure of the inputs to the system. Some examples of productivity measures are sales/salespeople, yards of carpet laid/number of carpet layers, shipments/[(direct labor)+(indirect labor)+(materials)]. Notice that productivity can be improved by producing greater output with the same inputs, or producing the same output with fewer inputs. In manufacturing operations, productivity ratios like the third example above generally vary with the volume of output produced (Schroeder, 1981). The production data below have been collected for a random sample of months for the three regional plants of a particular manufacturing firm. Each plant manufactures the same product.

North Plant		South Plant		West Plant	
Productivity Ratio	No. of Units Produced	Productivity Ratio	No. of Units Produced	Productivity Ratio	No. of Units Produced
1.30	1000	1.43	1015	1.61	501
0.90	400	1.50	925	0.74	140
1.21	650	0.91	150	1.19	303
0.75	200	0.99	222	1.88	930
1.32	850	1.33	545	1.72	776
1.29	600	1.15	402	1.39	400
1.18	756	1.51	709	1.86	810
1.10	500	1.01	176	0.99	220
1.26	925	1.24	392	0.79	160
0.93	300	1.49	699	1.59	626
0.81	258	1.37	800	1.82	640
1.12	590	1.39	660	0.91	190

(a) Constuct a scattergram for these data. Plot the data from the North plant using dots, from the South plant using small circles, and from the West plant using small triangles.

(b) By eye, fit each plant's response curve to the scattergram.

(c) Based on the results of part (b), propose a second-order regression model that could be used to estimate the relationship between productivity and volume for the three plants.

(d) Fit the model you proposed in part (c) to the data.

(e) Do the data provide sufficient evidence to conclude that the productivity response curves for the three plants differ? Test by using $\alpha = 0.05$.

(f) Do the data provide sufficient evidence to conclude that your second-order model contributes more information for the prediction of productivity than does a first-order model? Test by using $\alpha = 0.05$.

(g) Next month 890 units are scheduled to be produced at the West plant. Use the model you developed in part (d) to predict next month's productivity ratio at the West plant.

11

The Analysis of Variance

ABOUT THIS CHAPTER

We have now developed the techniques necessary to extend the two-sample hypothesis-testing results of Chapter 8 to cases involving more than two samples. More specifically, we will now construct tests of the equality of two or more means for sampling situations analogous to both the unpaired and paired samples of Chapter 8. The tools of multiple regression analysis will be used in this development.

CONTENTS

11.1 INTRODUCTION

We learned how to compare two means in Chapter 8, and we now want to extend the hypothesis-testing ideas developed there to the cases in which we might have three more means to compare. For example, a reliability engineer may wish to compare the mean lifelengths of three brands of capacitors, or a chemical engineer may desire to compare the average yields of four processes designed to produce an industrial chemical.

In this chapter we present a method for testing the equality of two or more means. We then show how this analysis can be handled through the linear regression techniques developed in Chapters 9 and 10. The remainder of the chapter deals with procedures for estimating individual means and differences between means, and with testing and estimation procedures for more complex experimental designs.

11.2 ANALYSIS OF VARIANCE FOR THE COMPLETELY RANDOMIZED DESIGN

The two-sample *t* test of Section 8.3 was designed to test the hypothesis that two population means are equal versus the alternative that they differ. Recall that, in order to use that test, the experiment must have resulted in two independent random samples, one sample from each of the populations under study. As an illustration, Example 8.10 reports results from a random sample of nine measurements on stamping times for a standard machine and an independently selected random sample of nine measurements from a new machine. This is a simple example of a completely randomized design.

DEFINITION 11.1	A **completely randomized design** is a plan for collecting data in which a random sample is selected from each population of interest, and the samples are independent.

The stamping time example consists of only two populations, but Definition 11.1 allows any finite number of populations. For example, suppose mean tensile strengths are to be compared for steel specimens coming from three processes, each involving a different percentage of carbon. If random samples of tensile strength measurements are taken on specimens from each process, and the samples are independent, then the resulting experiment would be a completely randomized design.

The populations of interest in experimental design problems are generally referred to as *treatments*. Thus the stamping time problem has two treatments (standard machine and new machine), whereas the tensile strength problem has three treatments (one for each percentage of carbon in the steel).

The notation we will employ for the k-population (or k-treatment) problem is summarized in Table 11.1.

TABLE 11.1
Notation for a Completely Randomized Design

Populations (Treatments)

	1	2	3	$\cdots$	k
Mean	μ_1	μ_2	μ_3	$\cdots$	μ_k
Variance	σ_1^2	σ_2^2	σ_3^2	$\cdots$	σ_k^2

Independent Random Samples

Sample Size	n_1	n_2	n_3	$\cdots$	n_k
Sample Totals	T_1	T_2	T_3	$\cdots$	T_k
Sample Means	$\bar{y}_1$	$\bar{y}_2$	$\bar{y}_3$	$\cdots$	$\bar{y}_k$

Total sample size $= n = n_1 + n_2 + n_3 + \cdots + n_k$
Overall sample total $= \sum y = T_1 + T_2 + \cdots + T_k$
Overall sample mean $= \bar{y} = \sum y/n$
Sum of squares of all n measurements $= \sum y^2$

The method we will present to analyze data from a completely randomized design (as well as many other designs) is called the *analysis of variance*. The basic inference problem for which analysis of variance provides an answer is the test of the null hypothesis

$$H_0: \mu_1 = \mu_2 = \cdots = \mu_k$$

versus the alternative

H_a: At least two treatment means differ

To see how we might formulate a test statistic for the null hypothesis indicated above, let us once again look at the two-sample problem involving the stamping times. Example 8.10 provides the following data:

$n_1 = 9$ $\qquad\qquad$ $n_2 = 9$

$\bar{y}_1 = 35.22$ seconds $\qquad$ $\bar{y}_2 = 31.56$ seconds

$(n_1 - 1)s_1^2 = 195.50$ $\qquad$ $(n_2 - 1)s_2^2 = 160.22$

If we want to test $H_0: \mu_1 = \mu_2$ versus $H_a: \mu_1 \neq \mu_2$, the t test can be used, with the statistic calculated as follows:

$$t = \frac{(\bar{y}_1 - \bar{y}_2)}{s_p \sqrt{\dfrac{1}{n_1} + \dfrac{1}{n_2}}}$$

The s_p here is the square root of the "pooled" s_p^2 given by

$$s_p^2 = \frac{(n_1 - 1)s_1^2 + (n_2 - 1)s_2^2}{n_1 + n_2 - 2}$$

The value of this statistic is calculated in Example 8.10 to be $t = 1.65$. The critical value with $\alpha = 0.05$ and 16 degrees of freedom is $t_{0.025} = 2.120$, and hence we cannot reject H_0.

The t test as presented above cannot be extended to a test of the equality of more than two means. However, if we square the t statistic, we obtain, after some algebra,

$$t^2 = \frac{(\bar{y}_1 - \bar{y}_2)^2}{s_p^2\left(\dfrac{1}{n_1} + \dfrac{1}{n_2}\right)} = \frac{\displaystyle\sum_{i=1}^{2} n_i(\bar{y}_i - \bar{y})^2}{s_p^2}$$

In repeated sampling this statistic can be shown to have an F distribution with $v_1 = 1$ and $v_2 = n_1 + n_2 - 2$ degrees of freedom in the numerator and denominator, respectively. Thus we could have tested the above hypothesis by calculating

$$\frac{\displaystyle\sum_{i=1}^{2} n_i(\bar{y}_i - \bar{y})^2}{s_p^2} = \frac{n_1(\bar{y}_1 - \bar{y})^2 + n_2(\bar{y}_2 - \bar{y})^2}{s_p^2} = \frac{60.28}{22.23} = 2.71$$

and comparing the value to $F_{0.05}(1, 16) = 4.49$. Again we would not reject H_0.

Now the second approach can be generalized to k populations quite easily. In general the numerator must have a divisor of $(k - 1)$, so that for testing

$$H_0: \mu_1 = \mu_2 = \cdots = \mu_k$$

versus

$$H_a: \text{At least two treatment means differ}$$

we use the test statistic

$$F = \frac{\displaystyle\sum_{i=1}^{k} n_i(\bar{y}_i - \bar{y})^2/(k - 1)}{s_p^2}$$

where

$$s_p^2 = \frac{\displaystyle\sum_{i=1}^{k} (n_i - 1)s_i^2}{n - k}$$

By inspecting this ratio, we can see that large deviations among the sample means will cause the numerator to be large, and thus may provide evidence for rejection of the null hypothesis. The denominator is an average within-sample variance and is not affected by differences among the sample means.

The usual analysis of variance notation is given as follows:

SST = Sum of Squares for Treatments

$$= \sum_{i=1}^{k} n_i(\bar{y}_i - \bar{y})^2$$

SSE = Sum of Squares for Error

$$= s_p^2(n - k)$$

TSS = Total Sum of Squares

$$= \text{SST} + \text{SSE}$$

$$\text{MST} = \text{Mean Square for Treatments} = \frac{\text{SST}}{k - 1}$$

and

$$\text{MSE} = \text{Mean Square for Error} = \frac{\text{SSE}}{n - k} = s_p^2$$

Thus the F ratio for testing H_0: $\mu_1 = \mu_2 = \cdots = \mu_k$ is given by

$$F = \frac{\text{MST}}{\text{MSE}}$$

Table 11.2 gives formulas for SST and TSS that minimize computational difficulties, but which are equivalent to those given above.

TABLE 11.2
Calculation Formulas for the Completely Randomized Design

$\text{TSS} = \sum y^2 - \dfrac{(\sum y)^2}{n}$	$\text{SST} = \displaystyle\sum_{i=1}^{k} \dfrac{T_i^2}{n_i} - \dfrac{(\sum y)^2}{n}$	$\text{SSE} = \text{TSS} - \text{SST}$

As in the two-sample t test the population probability distributions must be normal and the population variances must be equal in order for $F = \text{MST}/\text{MSE}$ to have an F distribution. But the F test works reasonably well if the populations are only approximately normal (mound shaped) with similar sized variances. The basic analysis of variance F test is summarized below.

TEST TO COMPARE k TREATMENT MEANS
FOR A COMPLETELY RANDOMIZED DESIGN

H_0: $\mu_1 = \mu_2 = \cdots = \mu_k$

H_a: At least two treatment means differ

Test statistic: $F = \dfrac{\text{MST}}{\text{MSE}}$

Assumptions:

1. Each population has a normal probability distribution.
2. The k population variances are equal.

Rejection region: $F > F_\alpha(k - 1, n - k)$

The analysis of variance (ANOVA) summary table is usually written as suggested in Table 11.3.

TABLE 11.3
ANOVA for Completely Randomized Design

Source	df	SS	MS	F ratio
Treatments	$k - 1$	SST	MST	MST/MSE
Error	$n - k$	SSE	MSE	
Total	$n - 1$	TSS		

EXAMPLE 11.1

In the past ten years computers and computer training have become integral parts of the curricula of both secondary schools and universities. As a result younger business professionals tend to be more comfortable with computers than their more senior counterparts.

"The older the person is, the worse it is," said Arnold S. Kahn of the American Psychological Association. "The older they are and the longer they wait before learning, the more dissatisfied they will become. The computer's unrelenting march into offices and factories is often cited as a chief cause of work-related stress. As computers and robots come into the work place, many workers fear they'll never master the new skills required." (Aplin-Brownlee, 1984).

Gary W. Dickson (personal communication, 1984), Professor of Management Information Systems at the University of Minnesota, investigated the computer literacy of middle managers with 10 years or more management experience. As part of his study Dickson designed a questionnaire that he hoped would measure managers' technical knowledge of computers. If the questionnaire were properly designed, the scores received by managers could be used as predictors of their knowledge of computers, with higher scores indicating more knowledge. To check the design of the questionnaire (i.e., its validity), nineteen middle managers from the Minneapolis–St. Paul metropolitan area were randomly sampled and asked to complete the questionnaire. Their scores appear in the following table. (The highest possible score on the questionnaire was 169.)

Questionnaire Scores of Middle Managers

Manager I.D.	Level of Technical Expertise	Score
1	A	82
2	A	114
3	A	90
4	A	80
5	B	128
6	B	90
7	C	156
8	A	88
9	A	93
10	B	130
11	A	80
12	A	105
13	B	110
14	B	133
15	C	128
16	B	130
17	B	104
18	C	151
19	C	140

Prior to completing the questionnaire, the managers were asked to describe their knowledge of and experience with computers. This information was used to classify the managers as possessing a low (A), medium (B), or high (C) level of technical computer expertise. These data also appear in the table.

Is there sufficient evidence to conclude that the mean score differs for the three groups of managers?

Solution   First, we group the data according to the "treatment" (level of technical expertise) and compute treatment totals.

Level:	A	B	C
	82	128	156
	114	90	128
	90	130	151
	80	110	140
	88	133	
	93	130	
	80	104	
	105		
Totals:	732	825	575

Then, using the calculation formulas for the completely randomized design, we find

$$TSS = \sum y^2 - \frac{(\sum y)^2}{n} = 250,048 - \frac{(2,132)^2}{19}$$

$$= 10,815.16$$

$$SST = \sum_{i=1}^{3} \frac{T_i^2}{n_i} - \frac{(\sum y)^2}{n}$$

$$= \frac{(732)^2}{8} + \frac{(825)^2}{7} + \frac{(575)^2}{4} - \frac{(2,132)^2}{19}$$

$$= 7,633.55$$

and

$$SSE = TSS - SST = 3,181.61$$

The test statistic is

$$F = \frac{MST}{SSE} = \frac{SST/(k-1)}{SSE/(n-k)} = \frac{7633.55/2}{3181.61/16}$$

$$= 19.19$$

Since $F_{0.05}(2, 16) = 3.63$, we can reject the hypothesis that the population mean scores for the three groups of managers are equal.

The ANOVA table is as follows:

Source	df	SS	MS	F Ratio
Treatments	2	7,633.55	3,816.78	19.19
Error	16	3,181.61	198.85	
Total	18	10,815.16		

Any standard statistical computing software package will produce an analysis of variance table similar to the one given above. The Minitab printout for this problem is shown below.

```
ANALYSIS OF VARIANCE
SOURCE   DF     SS     MS     F      p
FACTOR    2    7634   3817  19.19  0.000
ERROR    16    3182    199
TOTAL    18   10815

LEVEL     N   MEAN   STDEV
A         8   91.50  12.31
B         7  117.86  16.62
C         4  143.75  12.45

POOLED STDEV =   14.10
```

Note that "Treatments" is replaced by "Factor," unless specified otherwise in the computer analysis. The printout also shows each treatment mean and its standard deviation. With Minitab it is easy to show a dot plot of the original data, as seen below.

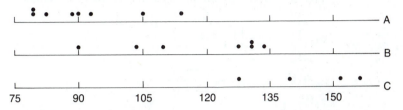

The dot plot shows that the three samples each display about the same amount of variability, but the sample sizes are too small to show the normality (or lack of it) in a definitive way.

If the equal variance assumption required by the analysis of variance procedure cannot be met, sometimes a transformation of the data is in order. (We could conduct a test for equality of more than two variances, but we will forgo that in favor of a graphical approach.) We saw how to choose transformations to stabilize variability in Chapter 1, and now we will use that technique in an analysis of variance setting.

EXAMPLE 11.2

Porosity of metal, an important determinant of strength and other properties, can be measured by looking at cross sections of the metal under a microscope. The *pore* (void) is dark, the metal is light, and the boundary between these two phases is often clearly delineated. If a grid is laid on the microscopic field, the number of intersections between the grid lines and the pore boundaries is proportional to the length of that boundary per unit area (and hence is proportional to the pore surface area per unit volume of metal).

Such counts, denoted by N, were made for samples of antimony, Linde copper, and electrolytic copper. (These data came from the Materials Science Department, University of Florida).

	Antimony				Linde Copper					Electrolytic Copper				
10	9	9	9	10	14	12	15	14	10	42	46	44	39	50
11	14	11	8	11	17	16	11	13	14	34	42	40	36	37
7	6	9	7	8	15	11	16	12	6	46	42	43	50	32
10	12	14	9	8	13	20	17	10	16	41	37	49	28	34
8	9	8	7	10	10	13	12	17	13	34	38	43	39	42

	Antimony	Linde Copper	Electrolytic Copper
Mean	9.36	13.48	40.32
Variance	4.07	9.01	31.58

Do the mean counts show significant differences?

Solution Obviously the variation is not constant across the three samples. In fact the variances are nearly proportional to the means, which indicates that a square root transformation is in order. The box plots in Figure 11.1 show the original data and the square roots for each metal. Observe how the square root transformation does make the variability about the same for all three samples.

FIGURE 11.1
Box Plots for
Example 10.2

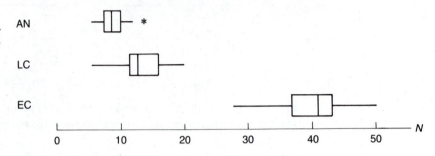

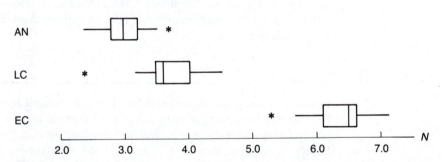

From looking at the data, it is obvious that the electrolytic copper has a much higher mean count, and this is borne out by the very high *F* value (475.16) in the analysis of variance, as shown in the printout.

ANALYSIS OF VARIANCE

SOURCE	DF	SS	MS	F	p
TYPE	2	153.498	76.749	475.16	0.000
ERROR	72	11.630	0.162		
TOTAL	74	165.128			

LEVEL	*N*	MEAN	STDEV
1	25	3.0430	0.3229
2	25	3.6479	0.4247
3	25	6.3347	0.4472

What is not so obvious is that the mean for Linde copper is very significantly higher than the mean count for antimony, as seen in the analysis of variance for only these two sets of samples.

```
ANALYSIS OF VARIANCE
SOURCE  DF    SS      MS      F      p
FACTOR   1   4.573   4.573  32.14  0.000
ERROR   48   6.830   0.142
TOTAL   49  11.403

LEVEL    N   MEAN   STDEV
  1      25  3.0430  0.3229
  2      25  3.6479  0.4247
```

So the mean counts do differ, which is not surprising, but the analysis must be done on transformed data because of the lack of homogeneity of variance.

We will show in Section 11.3 that the analysis just conducted can be performed by using regression models. Then we will show how to compare the pairs of treatment means in Section 11.4. ∎

11.1 Independent random samples were selected from three normally distributed populations with common (but unknown) variance σ^2. The data are shown below:

Sample 1	Sample 2	Sample 3
3.1	5.4	1.1
4.3	3.6	0.2
1.2	4.0	3.0
	2.9	

(a) Compute the appropriate sums of squares and mean squares and fill in the appropriate entries in the analysis of variance table shown below:

ANOVA Table

Source	df	SS	MS	F
Treatments				
Error				
Total				

(b) Test the hypothesis that the population means are equal (that is, $\mu_1 = \mu_2 = \mu_3$) against the alternative hypothesis that at least one mean is different from the other two. Test by using $\alpha = 0.05$.

11.2 A partially completed ANOVA table for a completely randomized design is shown next:

Source	df	SS	MS	F
Treatments	4	24.7		
Error				
Total	34	62.4		

    (a) Complete the ANOVA table.
    (b) How many treatments are involved in the experiment?
    (c) Do the data provide sufficient evidence to indicate a difference among the population means? Test by using $\alpha = 0.01$.

11.3   Some varieties of *nematodes* (round worms that live in the soil and frequently are so small they are invisible to the naked eye) feed upon the roots of lawn grasses and other plants. This pest, which is particularly troublesome in warm climates, can be treated by the application of nematicides. Data collected on the percentage of kill for nematodes for four particular rates of application (dosages given in pounds of active ingredient per acre) are as follows:

Rate of Application			
2	3	5	7
86	87	94	90
82	93	99	85
76	89	97	86
		91	

Do the data provide sufficient evidence to indicate a difference in the mean percentage of kill for the four different rates of application of nematicide? Use $\alpha = 0.05$.

11.4   It has been hypothesized that treatment (after casting) of a plastic used in optic lenses will improve wear. Four different treatments are to be tested. To determine whether any differences in mean wear exist among treatments, twenty-eight castings from a single formulation of the plastic were made and seven castings were randomly assigned to each of the treatments. Wear was determined by measuring the increase in "haze" after 200 cycles of abrasion (better wear being indicated by small increases).

Treatment			
A	B	C	D
9.16	11.95	11.47	11.35
13.29	15.15	9.54	8.73
12.07	14.75	11.26	10.00
11.97	14.79	13.66	9.75
13.31	15.48	11.18	11.71
12.32	13.47	15.03	12.45
11.78	13.06	14.86	12.38

(a) Is there evidence of a difference in mean wear among the four treatments? Use $\alpha = 0.05$.

(b) Estimate the mean difference in haze increase between treatments B and C, using a 99% confidence interval. [*Hint*: Use the two-sample t statistic.]

(c) Find a 90% confidence interval for the mean wear for lenses receiving treatment A. [*Hint*: Use the one-sample t statistic.]

11.5 The application of *management by objectives* (MBO), a method of performance appraisal, was the object of a study by Y. K. Shetty and H. M. Carlisle of Utah State University ("Organizational Correlates of a Management by Objectives Program," *Academy of Management Journal*, 1974, *17*). The study dealt with the reactions of a university faculty to an MBO program. One hundred nine faculty members were asked to comment on whether they thought the MBO program was successful in improving their performance within their respective departments and the university. Each response was assigned a score from 1 (significant improvement) to 5 (significant decrease). The table shows the sample sizes, sample totals, mean scores, and sum of squares of deviations *within* each sample for samples of scores corresponding to the four academic ranks. Assume that the four samples in the table can be viewed as independent random samples of scores selected from among the four academic ranks.

		ACADEMIC RANK		
	Instructor	Assistant Professor	Associate Professor	Professor
Sample Size	15	41	29	24
Sample Total	42.960	145.222	92.249	73.224
Sample Mean	2.864	3.542	3.181	3.051
Within-Sample Sum of Squared Deviations	2.0859	14.0186	7.9247	5.6812

(a) Perform an analysis of variance for the data.

(b) Arrange the results in an analysis of variance table.

(c) Do the data provide sufficient evidence to conclude there is a difference in mean scores among the four academic ranks? Test by using $\alpha = 0.05$.

(d) Find a 95% confidence interval for the difference in mean scores between instructors and professors.

(e) Do the data provide sufficient evidence to indicate a difference in mean scores between nontenured faculty members (instructors and assistant professors) and tenured faculty members?

11.6 The concentration of a catalyst used in producing grouted sand is thought to affect its strength. An experiment designed to investigate the effects of three different concentrations of the catalyst utilized five specimens of grout per concentration. The strength of a grouted sand was determined by placing the test specimen in a press and applying pressure until the specimen broke. The pressures required to break the specimens, expressed in pounds per square inch, are shown on the next page.

Concentration of Catalyst		
35%	40%	45%
5.9	6.8	9.9
8.1	7.9	9.0
5.6	8.4	8.6
6.3	9.3	7.9
7.7	8.2	8.7

Do the data provide sufficient evidence to indicate a difference in mean strength of the grouted sand among the three concentrations of catalyst? Test by using $\alpha = 0.05$.

11.7 Several companies are experimenting with the concept of paying production workers (generally paid by the hour) on a salary basis. It is believed that absenteeism and tardiness will increase under this plan, yet some companies feel that the working environment and overall productivity will improve. Fifty production workers under the salary plan are monitored at Company A, and likewise, fifty under the hourly plan at Company B. The number of work-hours missed due to tardiness or absenteeism over a 1-year period is recorded for each worker. The results are partially summarized in the table.

Source	df	SS	MS	F
Company		3,237.2		
Error		16,167.7		
Total	99			

(a) Fill in the missing information above.
(b) Is there evidence at the $\alpha = 0.05$ level of significance that the mean number of hours missed differs for employees of the two companies?

11.8 One of the selling points of golf balls is their durability. An independent testing laboratory is commissioned to compare the durability of three different brands of golf balls. Balls of each type will be put into a machine that hits the balls with the same force that a golfer does on the course. The number of hits required until the outer covering cracks is recorded for each ball, with the results given in the table below. Ten balls from each manufacturer are randomly selected for testing.

Brand		
A	B	C
310	261	233
235	219	289
279	263	301
306	247	264
237	288	273
284	197	208

(*continued*)

	Brand	
A	B	C
259	207	245
273	221	271
219	244	298
301	228	276

Is there evidence that the mean durabilities of the three brands differ? Use $\alpha = 0.05$.

11.9 Eight independent observations on percent copper content were taken on each of four castings of bronze. The sample means for each casting are as follows:

Casting:	1	2	3	4
Means:	80	81	86	90

SSE = 700

Is there sufficient evidence to say that there are differences among the mean percentages of copper for the four castings? Use $\alpha = 0.01$.

11.10 Three thermometers are used regularly in a certain laboratory. To check the relative accuracies of the thermometers, they are randomly and independently placed in a cell kept at zero degrees Celsius. Each thermometer is placed in the cell four times, with the following results:

Thermometer	1	2	3
Reading	0.10	−0.20	0.90
(°C)	0.90	0.80	0.20
	−0.80	−0.30	0.30
	−0.20	0.60	−0.30

Are there significant differences among the means for the three thermometers? Use $\alpha = 0.05$.

11.11 Casts of aluminum were subjected to four standard heat treatments and their tensile strengths measured. Five measurements were taken on each treatment, with the following results (in 1000 psi):

Treatment:	A	B	C	D
	35	41	42	31
	31	40	49	32
	40	43	45	30
	36	39	47	32
	32	45	48	34

Perform an analysis of variance. Do the mean tensile strengths differ from treatment to treatment at the 5% significance level?

11.12 How does flextime, which allows workers to set their individual work schedules, affect worker job satisfaction? Researchers recently conducted a study to compare a measure of job satisfaction for workers, using three types of work scheduling: flextime, staggered starting hours, and fixed hours. Workers in each group worked according to their specified work scheduling system for 4 months. Although each worker filled out job satisfaction questionnaires both before and after the 4-month test period, we will examine only the posttest-period scores. The sample sizes, means, and standard deviations of the scores for the three groups are shown in the table.

	Group		
	Flextime	Staggered	Fixed
Sample Size	27	59	24
Mean	35.22	31.05	28.71
Standard Deviation	10.22	7.22	9.28

(a) Assume that the data were collected according to a completely randomized design. Use the information in the table to calculate the treatment totals and SST.
(b) Use the values of the sample standard deviations to calculate the sum of squares of deviations *within* each of the three samples. Then calculate SSE, the sum of these quantities.
(c) Construct an analysis of variance table for the data.
(d) Do the data provide sufficient evidence to indicate differences in mean job satisfaction scores among the three groups? Test by using $\alpha = 0.05$.
(e) Find a 90% confidence interval for the difference in mean job satisfaction scores between workers on flextime and those on fixed schedules.
(f) Do the data provide sufficient evidence to indicate a difference in mean scores between workers on flextime and those using staggered starting hours? Test by using $\alpha = 0.05$.

11.3 A LINEAR MODEL FOR THE COMPLETELY RANDOMIZED DESIGN

The analysis of variance, as presented in Section 11.2, can be produced through the regression techniques given in Chapters 9 and 10. This approach is particularly beneficial if a computer program for multiple regression is available.

Consider the experiment of Example 11.1 involving three treatments. If we let *Y* denote the response variable for the test score of one manager, we can model *Y* as

$$Y = \beta_0 + \beta_1 x_1 + \beta_2 x_2 + \varepsilon$$

where

$$x_1 = \begin{cases} 1 & \text{if the response is from treatment B} \\ 0 & \text{otherwise} \end{cases}$$

$$x_2 = \begin{cases} 1 & \text{if the response is from treatment C} \\ 0 & \text{otherwise} \end{cases}$$

and ε is the random error, with $E(\varepsilon) = 0$ and $V(\varepsilon) = \sigma^2$. For a response Y_A from treatment A, $x_1 = 0$ and $x_2 = 0$, and hence

$$E(Y_A) = \beta_0 = \mu_A$$

For a response Y_B from treatment B, $x_1 = 1$ and $x_2 = 0$, so

$$E(Y_B) = \beta_0 + \beta_1 = \mu_B$$

It follows that

$$\beta_1 = \mu_B - \mu_A$$

In like manner

$$E(Y_C) = \beta_0 + \beta_2 = \mu_C$$

and thus

$$\beta_2 = \mu_C - \mu_A$$

The null hypothesis of interest in the analysis of variance, namely H_0: $\mu_A = \mu_B = \mu_C$, is now equivalent to H_0: $\beta_1 = \beta_2 = 0$.

For fitting the above model by the method of least squares, the data would be arrayed as follows:

y	x_1	x_2	
82	0	0	
114	0	0	
90	0	0	
80	0	0	Treatment A
88	0	0	
93	0	0	
80	0	0	
105	0	0	
128	1	0	
90	1	0	
130	1	0	
110	1	0	Treatment B
133	1	0	
130	1	0	
104	1	0	
156	0	1	
128	0	1	Treatment C
151	0	1	
140	0	1	

The method of least squares will give

$$\hat{\beta}_0 = \bar{y}_A$$
$$\hat{\beta}_1 = \bar{y}_B - \bar{y}_A$$
$$\hat{\beta}_2 = \bar{y}_C - \bar{y}_A$$

The error sum of squares for this model, denoted by SSE_2, turns out to be

$$SSE_2 = TSS - SST = 3181.61$$

Under $H_0: \beta_1 = \beta_2 = 0$, the reduced model becomes $Y = \beta_0 + \varepsilon$. The error sum of squares, when this model is fit by the method of least squares, is denoted by SSE_1 and is computed to be

$$SSE_1 = TSS = 10,815.2$$

The F test discussed in Section 10.6 for testing $H_0: \beta_1 = \beta_2 = 0$ has the form

$$F = \frac{(SSE_1 - SSE_2)/2}{SSE_2/(n-3)}$$

which is equivalent to

$$F = \frac{[TSS - (TSS - SST)]/2}{SSE_2/(n-3)} = \frac{SST/2}{SSE_2/(n-3)} = \frac{MST}{MSE}$$

Thus the F test arising from the regression formulation is equivalent to the analysis of variance F test as given in Section 11.2.

All of the analysis of variance problems discussed in this chapter can be formulated in terms of regression problems, as shown in Sections 11.6 and 11.8, and we show in Figure 11.2 the Minitab printout for the case described above.

FIGURE 11.2
Linear Model Analysis for Example 11.1

The regression equation is
y = 91.5 + 26.4 x1 + 52.3 x2

Predictor	Coef	Stdev	t-ratio	p
Constant	91.500	4.986	18.35	0.000
x1	26.357	7.298	3.61	0.002
x2	52.250	8.635	6.05	0.000

s = 14.10   R-sq = 70.6%   R-sq(adj) = 66.9%

Analysis of Variance

SOURCE	DF	SS	MS	F	p
Regression	2	7633.6	3816.8	19.19	0.000
Error	16	3181.6	198.9		
Total	18	10815.2			

Note that SSE_2 is the error sum of squares for this model, and SSE_1 is the total sum of squares. Also

$$\hat{\beta}_0 = \bar{y}_A = 91.5$$
$$\hat{\beta}_1 = \bar{y}_B - \bar{y}_A = 117.9 - 91.5 = 26.4$$

and

$$\hat{\beta}_2 = \bar{y}_C - \bar{y}_A = 143.8 - 91.5 = 52.3$$

The regression model approach allows easy access to the residuals for further analysis and checking of assumptions. Figure 11.3(a) shows the residuals plotted against the normal scores. The strong linear trend in this plot indicates that the normality assumption may be satisfied. The stem-and-leaf plot in Figure 11.3(b) shows a fairly compact display of residuals, somewhat symmetric about zero. Figure 11.3(c) shows the residuals for each treatment. Again they are somewhat

FIGURE 11.3
Residual Plots for Example 11.1

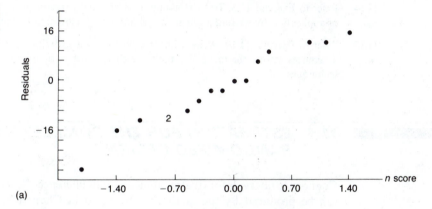

(a)

Stem-and-leaf of residuals N = 19
Leaf unit = 1.0

```
  1      -2   7
  1      -2
  2      -1   5
  5      -1   311
  7      -0   97
 (3)     -0   331
  9       0   1
  8       0   7
  7       1   02223
  2       1   5
  1       2   2
```

(b)

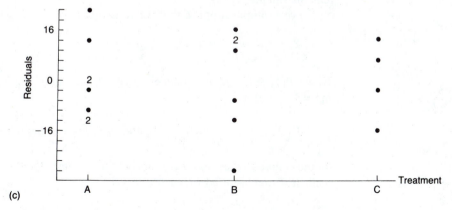

(c)

symmetric about zero and show about the same amount of variability for each of the three treatments. Thus the assumptions underlying the analysis of variance (or linear regression) seem to be met satisfactorily.

EXERCISES

11.13 Refer to Exercise 11.1. Answer part (b) by writing the appropriate linear model and using regression techniques.

11.14 Refer to Exercise 11.3. Test the assumption of equal percentages of kill for the four rates of application by writing a linear model and using regression techniques.

11.15 Refer to Exercise 11.10. Write a linear model for this experiment. Test for significant differences among the means for the three thermometers by making use of regression techniques.

11.4 ESTIMATION FOR THE COMPLETELY RANDOMIZED DESIGN

Confidence intervals for treatment means and differences between treatment means can be produced by the methods introduced in Chapter 8. Recall that we are assuming all treatment populations to be normally distributed with a common variance. Thus the confidence intervals based upon the t distribution can be employed. The common population variance σ^2 is estimated by the pooled sample variance $s^2 = \text{MSE}$.

Since we have k means in an analysis of variance problem, we may want to construct a number of confidence intervals based on the same set of experimental data. For example we may want to construct confidence intervals for all k means individually or for all possible differences between pairs of means. If such multiple intervals are to be used, we must use extreme care in selecting the confidence coefficients for the individual intervals so that the *experimentwise* error rate remains small.

For example suppose that we construct two confidence intervals,

$$\bar{Y}_A \pm t_{\alpha/2} \frac{s}{\sqrt{n}} \quad \text{and} \quad \bar{Y}_B \pm t_{\alpha/2} \frac{s}{\sqrt{n}}$$

Now $P(\bar{Y}_A \pm t_{\alpha/2} s/\sqrt{n} \text{ includes } \mu_A) = 1 - \alpha$ and $P(\bar{Y}_B \pm t_{\alpha/2} s/\sqrt{n} \text{ includes } \mu_B) = 1 - \alpha$. But

$$P\left(\bar{Y}_A \pm t_{\alpha/2} \frac{s}{\sqrt{n}} \text{ includes } \mu_A \text{ and } \bar{Y}_B \pm t_{\alpha/2} \frac{s}{\sqrt{n}} \text{ includes } \mu_B \right) < 1 - \alpha$$

so the simultaneous coverage probability is *less than* the $(1 - \alpha)$ confidence coefficient that we used on each interval.

If c intervals are to be constructed on one set of experimental data, then one method of keeping the simultaneous coverage probability (or confidence coefficient) at a value of *at least* $1 - \alpha$ is to make the individual confidence coefficients as close as possible to $1 - (\alpha/c)$. This technique for multiple confidence intervals is outlined in the following summary.

CONFIDENCE INTERVALS FOR MEANS
IN THE COMPLETELY RANDOMIZED DESIGN

> Suppose c intervals are to be constructed from one set of data.
>
> Single treatment mean μ_i: $\bar{y}_i \pm t_{\alpha/2c} \dfrac{s}{\sqrt{n_i}}$
>
> Difference between two treatment means $\mu_i - \mu_j$:[1]
>
> $$(\bar{y}_i - \bar{y}_j) \pm t_{\alpha/2c} s \sqrt{\frac{1}{n_i} + \frac{1}{n_j}}$$
>
> Note that $s = \sqrt{\text{MSE}}$ and all t values depend on $n - k$ degrees of freedom.

EXAMPLE 11.3

Using the data of Example 11.1, construct confidence intervals for all three possible differences between pairs of treatment means so that the simultaneous confidence coefficient is at least 0.95.

Solution Since there are three intervals to construct ($c = 3$), each interval should be of the form

$$(\bar{y}_i - \bar{y}_j) \pm t_{0.05/2(3)} s \sqrt{\frac{1}{n_i} + \frac{1}{n_j}}$$

Now $t_{0.05/2(3)} = t_{0.05/6}$ is approximately $t_{0.01}$. (This is as close as we can get with Table 5 in the Appendix.) With $n - k = 16$ degrees of freedom the tabled value is $t_{0.01} = 2.583$.

The three intervals are then constructed as follows:

$$\mu_A - \mu_B: (\bar{y}_A - \bar{y}_B) \pm t_{0.01} s \sqrt{\frac{1}{n_A} + \frac{1}{n_B}}$$

$$(91.5 - 117.9) \pm 2.583\sqrt{198.85}\sqrt{\frac{1}{8} + \frac{1}{7}} = -26.4 \pm 18.9$$

$$\mu_A - \mu_C: (\bar{y}_A - \bar{y}_C) \pm t_{0.01} s \sqrt{\frac{1}{n_A} + \frac{1}{n_C}}$$

---

[1]This is the *Bonferroni* method for conducting multiple comparison of means.

$$(91.5 - 143.8) \pm 2.583\sqrt{198.85}\sqrt{\frac{1}{8} + \frac{1}{4}} = -52.3 \pm 22.3$$

$$\mu_B - \mu_C: (\bar{y}_B - \bar{y}_C) \pm t_{0.01}s\sqrt{\frac{1}{n_B} + \frac{1}{n_C}}$$

$$(117.9 - 143.8) \pm 2.583\sqrt{198.85}\sqrt{\frac{1}{7} + \frac{1}{4}} = -25.9 \pm 22.8$$

In interpreting these results, we would be inclined to conclude that all pairs of means differ, because none of the three observed confidence intervals contains zero. Furthermore inspection of the means leads to the inference that $\mu_C > \mu_B > \mu_A$. The questionnaire appears to be a valid predictor of the level of the manager's technical knowledge of computers. Our combined confidence level for all three confidence intervals is at least 0.95.

If the $t_{0.025}$ value (with 6 degrees of freedom) had been used in place of the $t_{0.01}$ value, all the intervals would have been narrower. However, our combined confidence level is reduced to something less than 0.95. ■

EXERCISES

11.16 Refer to Exercise 11.1.
  (a) Find a 90% confidence interval for $(\mu_2 - \mu_3)$. Interpret the interval.
  (b) What would happen to the width of the confidence interval in part (a) if you quadrupled the number of observations in the two samples?
  (c) Find a 95% confidence interval for μ_2.
  (d) Approximately how many observations would be required if you wished to be able to estimate a population mean correct to within 0.4 with probability equal to 0.95?

11.17 Refer to Exercise 11.2.
  (a) Suppose that $\bar{y}_1 = 3.7$ and $\bar{y}_2 = 4.1$. Do the data provide sufficient evidence to indicate a difference between μ_1 and μ_2? Assume that there are seven observations for each treatment. Test by using $\alpha = 0.10$.
  (b) Refer to part (a). Find a 90% confidence interval for $(\mu_1 - \mu_2)$.
  (c) Refer to part (a). Find a 90% confidence interval for μ_1.

11.18 Refer to Exercise 11.3.
  (a) Estimate the true difference in mean percentage of kill between rate 2 and rate 5. Use a 90% confidence interval.
  (b) Construct confidence intervals for the six possible differences between treatment means, with simultaneous confidence close to 0.90.

11.19 Refer to Exercise 11.6.
  (a) Find a 95% confidence interval for the difference in mean strength for specimens produced with a 35% concentration of catalyst versus those containing a 45% concentration of catalyst.
  (b) Construct confidence intervals for the three treatment means, with simultaneous confidence coefficient approximately 0.90.

11.20 Refer to Exercise 11.9. Casting 1 is a standard and 2, 3, and 4 involve slight modifications to the process. Compare the standard with each of the modified processes by producing three confidence intervals with simultaneous confidence coefficient approximately 0.90.

11.5 ANALYSIS OF VARIANCE FOR THE RANDOMIZED BLOCK DESIGN

In the completely randomized design only one source of variation, the treatment-to-treatment variation, is specifically considered in the design and analyzed after the data are collected. That is, observations are taken from each of k treatments, and then a test and confidence intervals are constructed to analyze how the treatment means may differ.

Most often, however, the responses of interest are subject to sources of variation in addition to the treatments under study. Suppose, for example, that an engineer is studying the gas mileage resulting from four brands of gasoline. If more than one automobile is used in the study, then automobiles would form another important source of variation. To control for this additional variation, it would be crucially important to run each brand of gasoline at least once in each automobile. In this type of experiment the brands of gasoline are the *treatments* and the automobiles are called the *blocks*. If the order of running the brands in each automobile is randomized, the resulting design is called a *randomized block design*. Letting the brands be denoted by A, B, C, and D and the automobiles by I, II, and III, the structure of the design might look like Figure 11.4.

FIGURE 11.4
Typical Randomized
Block Design

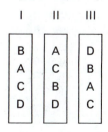

For auto I brand B was run first, followed in order by A, C, and D.

DEFINITION 11.2

A **randomized block design** is a plan for collecting data in which each of k treatments is measured once in each of b blocks. The order of the treatments within the blocks is random.

The blocking helps the experimenter control a source of variation in responses so that any true treatment differences are more likely to show up in the analysis. Suppose that the gasoline mileage study did, in fact, use more than one automobile, but that we had not blocked on automobiles as indicated above. Assuming that we still take three measurements on each treatment, as indicated in Figure 11.5, two undesirable conditions might result.

FIGURE 11.5
A Completely Randomized Design for the Gasoline Mileage Study

A   B   C   D

— — — —

— — — —

— — — —

First, the auto-to-auto variation would add to the variance of the measurements within each sample and hence inflate the MSE. The *F* test could then turn out to be nonsignificant even when differences do exist among the true treatment means. Second, all of the A responses could turn out to be from automobile I and all of the B responses from automobile II. If the mean for A and the mean for B are significantly different, we still don't know how to interpret the result. Are the brands really different, or are the automobiles different with respect to gas mileage? A randomized block design will help us avoid both of these difficulties.

After the measurements are completed in a randomized block design, as shown in Figure 11.4, they can be rearranged into a two-way table, as given in Table 11.4.

TABLE 11.4
Notation for the Results of Randomized Block Experiment

			Treatments			
		1	2	$\cdots$	k	Totals
	1	y_{11}	y_{12}	$\cdots$	y_{1k}	B_1
Blocks	2	y_{21}	y_{22}	$\cdots$	y_{2k}	B_2
	$\vdots$	$\vdots$	$\vdots$	$\cdots$	$\vdots$	$\vdots$
	b	y_{b1}	y_{b2}	$\cdots$	y_{bk}	B_b
Totals		T_1	T_2	$\cdots$	T_k	

Total sample size $= n = bk$
Overall sample total $= \sum y = T_1 + T_2 + \cdots + T_k$
$$= B_1 + B_2 + \cdots + B_b$$
Overall sample mean $= \sum y/n$
Sum of squares of all *n* measurements $= \sum y^2$
Note that there are *k* treatments and *b* blocks.

The total sum of squares (TSS) can now be partitioned into a treatment sum of squares (SST), a block sum of squares (SSB), and an error sum of squares (SSE). The computation formulas are given in Table 11.5.

The test of the null hypothesis that the *k* treatment means $\mu_1, \mu_2, \ldots, \mu_k$ are equal is, once again, an *F* test constructed as the ratio of MST to MSE. In an

TABLE 11.5
Calculation Formulas for the Randomized Block Design

$$\text{TSS} = \sum y^2 - \frac{(\sum y)^2}{n}$$

$$\text{SST} = \frac{1}{b} \sum_{i=1}^{k} T_i^2 - \frac{(\sum y)^2}{n}$$

$$\text{SSB} = \frac{1}{k} \sum_{i=1}^{b} B_i^2 - \frac{(\sum y^2)}{n}$$

$$\text{SSE} = \text{TSS} - \text{SST} - \text{SSB}$$

$$\text{MST} = \frac{\text{SST}}{k-1}$$

$$\text{MSB} = \frac{\text{SSB}}{b-1}$$

$$\text{MSE} = \frac{\text{SSE}}{n-k-b+1} = \frac{\text{SSE}}{(b-1)(k-1)}$$

analogous fashion we could construct a test of the hypothesis that the block means are equal. The resulting F test for this case would be the ratio of MSB to MSE.

TEST TO COMPARE k TREATMENT MEANS
FOR A RANDOMIZED BLOCK DESIGN

$$H_0: \mu_1 = \mu_2 = \cdots = \mu_k$$

H_a: At least two treatment means differ

Test statistic: $F = \dfrac{\text{MST}}{\text{MSE}}$

Assumptions:
1. Each population (treatment/block combination) has an independent normal probability distribution.
2. The variances of the probability distributions are equal.

Rejection region: $F > F_\alpha[(k-1), (b-1)(k-1)]$

The analysis of variance summary table for a randomized block design is given in Table 11.6.

TABLE 11.6
ANOVA for a Randomized Block Design

Source	df	SS	MS	F ratio
Treatments	$k - 1$	SST	MST	MST/MSE
Blocks	$b - 1$	SSB	MSB	MSB/MSE
Errors	$(b - 1)(k - 1)$	SSE	MSE	
Total	$bk - 1$	TSS		

EXAMPLE 11.4

Four chemical treatments for fabric are to be compared with regard to their ability to resist stains. Two different types of fabric are available for the experiment, so it is decided to apply each chemical to a sample of each type of fabric. The result is a randomized block design with four treatments and two blocks. The measurements are as follows:

		Block (Fabric) 1	2	Totals
	1	5	9	14
Treatment	2	3	8	11
(Chemicals)	3	8	13	21
	4	4	6	10
	Totals	20	36	56

Is there evidence of significant differences among the treatment means? Use $\alpha = 0.05$.

Solution For these data the computations (see Table 11.5) are as follows:

$$\text{TSS} = 464 - \frac{(56)^2}{8} = 464 - 392 = 72$$

$$\text{SST} = \frac{1}{2} [(14)^2 + (11)^2 + (21)^2 + (10)^2] - \frac{(56)^2}{8}$$

$$= 429 - 392 = 37$$

$$\text{SSB} = \frac{1}{4} [(20)^2 + (36)^2] - \frac{(56)^2}{8} = 424 - 392 = 32$$

$$\text{SSE} = \text{TSS} - \text{SST} - \text{SSB} = 72 - 37 - 32 = 3$$

The ANOVA summary table then becomes

Source	df	SS	MS	F ratio
Treatments	3	37	$\dfrac{37}{3} = 12.33$	$\dfrac{12.33}{1} = 12.33$
Blocks	1	32	$\dfrac{32}{1} = 32$	$\dfrac{32}{1} = 32$
Error	3	3	$\dfrac{3}{3} = 1$	
Total	7			

Since $F_{0.05}(3, 3) = 9.28$ and our observed F ratio for treatments is 12.33, we reject $H_0: \mu_1 = \mu_2 = \mu_3 = \mu_4$ and conclude that there is significant evidence of at least one treatment difference.

We also see in this summary table that the F ratio for blocks is 32. Since $F_{0.05}(1, 3) = 10.13$, we conclude that there is evidence of a difference between the block (fabric) means. That is, the fabrics seem to differ with respect to their ability to resist stains when treated with these chemicals. The decision to "block out" the variability between fabrics appears to have been a good one. ■

The Minitab printout for this analysis is shown below. Note that the treatment means are provided along with the standard analysis of variance table. Block means can be easily produced, as well.

ANALYSIS OF VARIANCE

SOURCE	DF	SS	MS
TREATMT	3	37.00	12.33
BLOCK	1	32.00	32.00
ERROR	3	3.00	1.00
TOTAL	7	72.00	

TREATMT	Mean
1	7.00
2	5.50
3	10.50
4	5.00

Next we turn to a brief discussion of how the randomized block experiment can be analyzed by a linear model, and then consider estimation of means for this design.

11.21 A randomized block design was conducted to compare the mean responses for three treatments, A, B, and C, in four blocks. The data are shown below:

Treatment	Block 1	2	3	4
A	3	6	1	2
B	5	7	4	6
C	2	3	2	2

(a) Compute the appropriate sums of squares and mean squares and fill in the entries in the analysis of variance table shown below:

ANOVA Table

Source	df	SS	MS	F
Treatment				
Block				
Error				
Total				

(b) Do the data provide sufficient evidence to indicate a difference among treatment means? Test by using $\alpha = 0.05$.

(c) Do the data provide sufficient evidence to indicate that blocking was effective in reducing the experimental error? Test by using $\alpha = 0.05$.

11.22 The analysis of variance for a randomized block design produced the ANOVA table entries shown below:

ANOVA Table

Source	df	SS	MS	F
Treatment	3	27.1		
Block	5		14.90	
Error		33.4		
Total				

(a) Complete the ANOVA table.

(b) Do the data provide sufficient evidence to indicate a difference among the treatment means? Test by using $\alpha = 0.01$.

(c) Do the data provide sufficient evidence to indicate that blocking was a useful design strategy to employ for this experiment? Explain.

11.23 An evaluation of diffusion bonding of zircaloy components is performed. The main objective is to determine which of three elements—nickel, iron, or copper—is the best bonding agent. A series of zircaloy components are bonded, using each of the possible bonding agents. Since there is a great deal of variation in components machined from different ingots, a randomized block design is used, blocking on the ingots. A pair of components from each ingot are bonded together, using each of the three agents, and the pressure (in units of 1000 pounds per square inch) required to separate the bonded components is measured. The following data are obtained.

| | Bonding Agent | | |
Ingot	Nickel	Iron	Copper
1	67.0	71.9	72.2
2	67.5	68.8	66.4
3	76.0	82.6	74.5
4	72.7	78.1	67.3
5	73.1	74.2	73.2
6	65.8	70.8	68.7
7	75.6	84.9	69.0

Is there evidence of a difference in pressure required to separate the components among the three bonding agents? Use $\alpha = 0.05$.

11.24 A construction firm employs three cost estimators. Usually only one estimator works on each potential job, but it is advantageous to the company if the estimators are consistent enough so that it does not matter which of the three estimators is assigned to a particular job. To check on the consistency of the estimators, several jobs are selected and all three estimators are asked to make estimates. The estimates for each job by each estimator are given in the table.

Estimates (in thousands of dollars)

| | Estimator | | |
Job	A	B	C
1	27.3	26.5	28.2
2	66.7	67.3	65.9
3	104.8	102.1	100.8
4	87.6	85.6	86.5
5	54.5	55.6	55.9
6	58.7	59.2	60.1

(a) Do these estimates provide sufficient evidence that the means for the estimators differ? Use $\alpha = 0.05$.
(b) Present the complete ANOVA summary table for this experiment.

11.25 A power plant, which uses water from the surrounding bay for cooling its condensors, is required by EPA to determine whether discharging its heated water into the bay has a detrimental effect on the flora (plant life) in the water. The EPA requests that the power plant makes its investigation at three strategically chosen locations, called *stations*. Stations 1 and 2 are located near the plant's discharge tubes, while Station 3 is located farther out in the bay. During one randomly selected day in each of four months, a diver sent down to each of the stations randomly samples a square meter area of the bottom and counts the number of blades of the different types of grasses present. The results are as follows for one important grass type:

Month	Station 1	2	3
May	28	31	53
June	25	22	61
July	37	30	56
August	20	26	48

    (a) Is there sufficient evidence to indicate that the mean number of blades found per square meter per month differs for the three stations? Use $\alpha = 0.05$.

    (b) Is there sufficient evidence to indicate that the mean number of blades found per square meter differs across the 4 months? Use $\alpha = 0.05$.

11.26   The Bell Telephone Company's long-distance phone charges may appear to be exorbitant when compared with some of its competitors, but this is because a comparison of charges between competing companies is often analogous to comparing apples and eggs. Bell's charges for individuals are on a per-call basis. In contrast, its competitors often charge a monthly minimum long-distance fee, reduce the charges as the usage rises, or both. Shown in the table is a sampling of long-distance charges from Orlando, Florida, to twelve cities for three non-Bell companies offering long-distance service. The data were contained in an advertisement in the *Orlando Sentinel*, March 19, 1984. A note in fine print below the advertisement states that the rates are based on "30 hours of usage" for each of the servicing companies.

    The data in the table are pertinent for companies making phone calls to large cities. Therefore assume that the cities receiving the calls were randomly selected from among all large cities in the United States.

From Orlando to	Time	Length of Call (minutes)	Company 1	2	3
New York	Day	2	$0.77	$0.79	$0.66
Chicago	Evening	3	0.69	0.71	0.59
Los Angeles	Day	2	0.87	0.88	0.66
Atlanta	Evening	1	0.22	0.23	0.20
Boston	Day	3	1.15	1.19	0.99
Phoenix	Day	5	1.92	1.98	1.65
West Palm Beach	Evening	2	0.49	0.42	0.40
Miami	Day	3	1.12	1.05	0.99
Denver	Day	10	3.85	3.96	3.30
Houston	Evening	1	0.22	0.23	0.20
Tampa	Day	3	1.06	1.00	0.99
Jacksonville	Day	3	1.06	1.00	0.99

(a) What type of design was used for the data collection?

(b) Perform an analysis of variance for the data. Present the results in an ANOVA table.

(c) Do the data provide sufficient evidence to indicate differences in mean charges among the three companies? Test by using $\alpha = 0.05$.

(d) Company 3 placed the advertisement, so it might be more relevant to compare the charges for companies 1 and 2. Do the data provide sufficient evidence to incidate a difference in mean changes for these two companies? Test by using $\alpha = 0.05$.

11.27   From time to time one branch office of a company must make shipments to a certain branch office in another state. There are three package delivery services between the two cities where the branch offices are located. Since the price structures for the three delivery services are quite similar, the company wants to compare the delivery times. The company plans to make several different types of shipments to its branch office. To compare the carriers, each shipment will be sent in triplicate, one with each carrier. The results listed in the table are the delivery times in hours.

		Carrier	
Shipment	I	II	III
1	15.2	16.9	17.1
2	14.3	16.4	16.1
3	14.7	15.9	15.7
4	15.1	16.7	17.0
5	14.0	15.6	15.5

Is there evidence of a difference in mean delivery times among the three carriers? Use $\alpha = 0.05$.

11.28   Due to increased energy shortages and costs, utility companies are stressing ways in which home and apartment utility bills can be cut. One utility company reached an agreement with the owner of a new apartment complex to conduct a test of energy-saving plans for apartments. The tests were to be conducted before the apartments were rented. Four apartments were chosen that were identical in size, amount of shade, and direction faced. Four plans were to be tested, one on each apartment. The thermostat was set at 75°F in each apartment, and the monthly utility bill was recorded for each of the 3 summer months. The results are listed in the table.

		Treatment		
Month	1	2	3	4
June	$74.44	$68.75	$71.34	$65.47
July	89.96	73.47	83.62	72.33
August	82.00	71.23	79.98	70.87

Treatment 1:   No insulation
Treatment 2:   Insulation in walls and ceilings

Treatment 3:   No insulation; awnings for windows
Treatment 4:   Insulation and awnings for windows

    (a)   Is there evidence that the mean monthly utility bills differ for the four treatments? Use $\alpha = 0.01$.

    (b)   Is there evidence that blocking is important, that is, that the mean bills differ for the 3 months? Use $\alpha = 0.05$.

11.29   A chemist runs an experiment to study the effect of four treatments on the glass transition temperature of a particular polymer compound. Raw material used to make this polymer is bought in small batches. The material is thought to be fairly uniform within a batch but variable between batches. Therefore each treatment was run on samples from each batch with the following results:

Temperature (in K)

Batch	I	II	III	IV
		Treatment		
1	576	584	562	543
2	515	563	522	536
3	562	555	550	530

    (a)   Do the data provide sufficient evidence to indicate a difference in mean temperature among the four treatments? Use $\alpha = 0.05$.

    (b)   Is there sufficient evidence to indicate a difference in mean temperature among the three batches? Use $\alpha = 0.05$.

    (c)   If the experiment is conducted again in the future, would you recommend any changes in its design?

11.30   The table lists the number of strikes that occurred per year in five U.S. manufacturing industries over the period from 1976 to 1981. Only work stoppages that continued for at least 1 day and involved six or more workers were counted.

Year	Food and Kindred Products	Primary Metal Industry	Electrical Equipment and Supplies	Fabricated Metal Products	Chemicals and Allied Products
1976	227	197	204	309	129
1977	221	239	199	354	111
1978	171	187	190	360	113
1979	178	202	195	352	143
1980	155	175	140	280	89
1981	109	114	106	203	60

Source: *Statistical Abstract of the United States.*

    (a)   Perform an analysis of variance and determine whether there is sufficient evidence to conclude that the mean number of strikes per year differs among the five industries. Test by using $\alpha = 0.05$.

    (b)   Construct the appropriate ANOVA table.

11.6 A LINEAR MODEL FOR THE RANDOMIZED BLOCK DESIGN

Section 11.3 contains a discussion of the regression approach to the analysis of a completely randomized design. A similar linear model can be written for the randomized block design.

Letting Y denote a response from the randomized block design of Example 11.4, we can write

$$Y = \beta_0 + \beta_1 x_1 + \beta_2 x_2 + \beta_3 x_3 + \beta_4 x_4 + \varepsilon$$

where

$$x_1 = \begin{cases} 1 & \text{if the response is from block 2} \\ 0 & \text{otherwise} \end{cases}$$

$$x_2 = \begin{cases} 1 & \text{if the response is from treatment 2} \\ 0 & \text{otherwise} \end{cases}$$

$$x_3 = \begin{cases} 1 & \text{if the response is from treatment 3} \\ 0 & \text{otherwise} \end{cases}$$

$$x_4 = \begin{cases} 1 & \text{if the response is from treatment 4} \\ 0 & \text{otherwise} \end{cases}$$

and ε is the random error term. The error sum of squares for this model is denoted by SSE_2. Testing the null hypothesis that the four treatment means are equal is now equivalent to testing $H_0: \beta_2 = \beta_3 = \beta_4 = 0$. The reduced model is then

$$Y = \beta_0 + \beta_1 x_1 + \varepsilon$$

which, when fit by the method of least squares, will produce an error sum of squares denoted by SSE_1.

The F test for $H_0: \beta_2 = \beta_3 = \beta_4 = 0$ versus the alternative that at least one β_i, $i = 2, 3, 4$, is different from zero then has the form

$$F = \frac{(SSE_1 - SSE_2)(k - 1)}{SSE_2/(b - 1)(k - 1)}$$

$$= \frac{SST/(k - 1)}{SSE_2/(b - 1)(k - 1)} = \frac{MST}{MSE}$$

The test is equivalent to the F test for equality of treatment means given in Section 11.5.

The regression printouts for the two models are shown in Figure 11.6. The sum of squared errors (SSE) for the complete and reduced model are boxed:

Reduced model: $SSE_1 = 40$
Complete model: $SSE_2 = 3$

FIGURE 11.6
Printouts for
Complete and
Reduced Models,
Example 11.4

The regression equation is
$$y = 5.00 + 4.00\ x1 - 1.50\ x2 + 3.50\ x3 - 2.00\ x4$$

Predictor	Coef	Stdev	t-ratio	p
Constant	5.0000	0.7906	6.32	0.008
x1	4.0000	0.7071	5.66	0.011
x2	-1.500	1.000	-1.50	0.231
x3	3.500	1.000	3.50	0.039
x4	-2.000	1.000	-2.00	0.139

$s = 1.000$ R-sq $= 95.8\%$ R-sq(adj) $= 90.3\%$

Analysis of Variance

SOURCE	DF	SS	MS	F	p
Regression	4	69.000	17.250	17.25	0.021
Error	3	3.000	1.000		
Total	7	72.000			

(a)   Complete Model

The regression equation is
$$y = 5.00 + 4.00\ x1$$

Predictor	Coef	Stdev	t-ratio	p
Constant	5.000	1.291	3.87	0.008
x1	4.000	1.826	2.19	0.071

$s = 2.582$ R-sq $= 44.4\%$ R-sq(adj) $= 35.2\%$

Analysis of Variance

SOURCE	DF	SS	MS	F	p
Regression	1	32.000	32.000	4.80	0.071
Error	6	40.000	6.667		
Total	7	72.000			

(b)   Reduced Model

Then

$$F = \frac{(40 - 3)/3}{3/3} = 12.33$$

which is exactly the same F value calculated in Example 11.4. The regression approach leads us to the same conclusion as did the ANOVA approach: there is evidence that the mean stain resistances of the four chemical differ.

The plot of residuals against their normal scores (Figure 11.7) shows a very straight line, indicating that the normality assumption is adequately satisfied for these data.

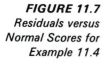

FIGURE 11.7
Residuals versus
Normal Scores for
Example 11.4

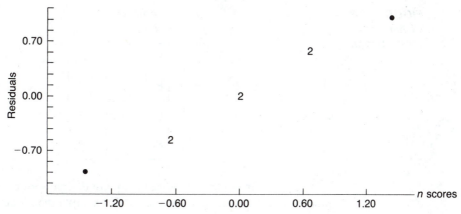

The randomized block design is only one of many types of block designs. When there are two sources of nuisance variation, it is necessary to block on both sources to eliminate this unwanted variability. Blocking in two directions can be accomplished by using a Latin square design.

To evaluate the toxicity of certain compounds in water, samples must be preserved for long periods of time. Suppose an experiment is being conducted to evaluate the Maximum Holding Time (MHT) for four different preservatives used to treat a mercury-base compound. The MHT is defined as the time that elapses before the solution loses 10% of its initial concentration. Both the level of the initial concentration and the analyst who measures the MHT are sources of variation, so an experimental design that blocks on both is necessary to allow the difference in mean MHT between preservatives to be more accurately estimated. A Latin square design is constructed wherein each preservative is applied exactly *once* to each initial concentration, and is analyzed *exactly once* by each analyst. You can see why the design must be "square," since the single application of each treatment level at each block level requires that the number of levels of each block *equal* the number of treatment levels. Thus to apply the Latin square design, using four different preservatives, we must employ four initial concentrations and four analysts. The design would be constructed as shown in Figure 11.8, where P_i = Preservative $i (i = 1, 2, 3, 4)$. Note that each preservative appears exactly once in each row (initial concentration) and each column (analyst). The resulting design is a 4×4 Latin square. A Latin square design for three treatments will require a 3×3 configuration and, in general, p treatments will require a $p \times p$ array of experimental units. If more observations are desired per treatment, the experimenter would utilize several Latin square configurations in one experiment. In the example above it would be necessary to run two Latin squares to obtain eight observations per treatment.

A comparison of mean MHT for any pair of preservatives would eliminate the variability due to initial concentration and to analysts, because each preservative was applied with equal frequency (once) at each level of both blocking factors. Consequently the block effects would be canceled when comparing mean MHT for any pair of preservatives.

FIGURE 11.8
A Latin Square Design

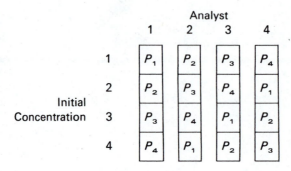

The linear model for the Latin square design is an extension of that for the randomized block design. For the MHT example the model is

$$Y = \beta_0 + \overbrace{\beta_1 x_1 + \beta_2 x_2 + \beta_3 x_3}^{\text{Initial Concentration}} + \overbrace{\beta_4 x_4 + \beta_5 x_5 + \beta_6 x_6}^{\text{Analyst}}$$

$$+ \overbrace{\beta_7 x_7 + \beta_8 x_8 + \beta_9 x_9}^{\text{Preservative}} + \varepsilon$$

where $Y = $ MHT.

$$x_1 = \begin{cases} 1 & \text{if Initial Concentration 1} \\ 0 & \text{otherwise} \end{cases} \qquad x_6 = \begin{cases} 1 & \text{if Analyst 3} \\ 0 & \text{otherwise} \end{cases}$$

$$x_2 = \begin{cases} 1 & \text{if Initial Concentration 2} \\ 0 & \text{otherwise} \end{cases} \qquad x_7 = \begin{cases} 1 & \text{if Preservative 1} \\ 0 & \text{otherwise} \end{cases}$$

$$x_3 = \begin{cases} 1 & \text{if Initial Concentration 3} \\ 0 & \text{otherwise} \end{cases} \qquad x_8 = \begin{cases} 1 & \text{if Preservative 2} \\ 0 & \text{otherwise} \end{cases}$$

$$x_4 = \begin{cases} 1 & \text{if Analyst 1} \\ 0 & \text{otherwise} \end{cases} \qquad x_9 = \begin{cases} 1 & \text{if Preservative 3} \\ 0 & \text{otherwise} \end{cases}$$

$$x_5 = \begin{cases} 1 & \text{if Analyst 2} \\ 0 & \text{otherwise} \end{cases}$$

Thus dummy variables are used to represent the block effects, as well as the treatments. To test the null hypothesis of no treatment differences, we test $H_0: \beta_7 = \beta_8 = \beta_9 = 0$ by fitting complete and reduced models, just as we did for the randomized block design. There are equivalent ANOVA formulas (not presented here) similar to those presented in Section 11.5 for the randomized block design for calculating the F test statistic for treatment (and block) differences.

In summary, there are many types of block designs for comparing treatment means while controlling sources of nuisance variation. As with the randomized block and Latin square designs, other block designs can be most easily analyzed by using the linear model approach, with dummy variables representing both block and treatment effects. Then the complete and reduced model approach is used to test for treatment and block effects on the mean response.

Note that, in order for the F tests given above to be valid, we must still make the usual assumptions about the independence and normality of population measurements with equal variances across all treatment/block combinations.

EXERCISES

11.31 Refer to Exercise 11.21. Answer parts (b) and (c) by fitting a linear model to the data and using regression techniques.

11.32 Refer to Exercise 11.25. Answer part (a) by fitting a linear model to the data and using regression techniques.

11.33 Suppose that three automobile engine designs are being compared to determine differences in mean time between breakdowns.
   (a) Show how three test automobiles and three different test drivers could be used in a 3×3 Latin square design aimed at comparing the engine designs.
   (b) Write the linear model for the design in part (a).
   (c) What null hypothesis would you test to determine whether the mean time between breakdowns differs for the engine designs?

11.7 ESTIMATION FOR THE RANDOMIZED BLOCK DESIGN

As in the case of the completely randomized design, we may want to estimate the mean for a particular treatment, or the difference between the means of two treatments, after conducting the initial F test. Since we will generally be interested in c intervals simultaneously, we will want to set the confidence coefficient of each individual interval at $(1 - \alpha/c)$. This will guarantee that the probability of all c intervals simultaneously covering the parameters being estimated is at least $1 - \alpha$. A summary of the estimation procedure is given below.

CONFIDENCE INTERVALS FOR MEANS IN THE RANDOMIZED BLOCK DESIGN

Suppose c intervals are to be constructed from one set of data.

Single treatment mean μ_i: $\bar{y}_i \pm t_{\alpha/2c}s\sqrt{\dfrac{1}{b}}$

Difference between two treatment means $\mu_i - \mu_j$:[2]

$$(\bar{y}_i - \bar{y}_j) \pm t_{\alpha/2c}s\sqrt{\dfrac{1}{b} + \dfrac{1}{b}}$$

---

[2]This is again the *Bonferroni* method for conducting multiple comparisons of means.

> Note that $s = \sqrt{\text{MSE}}$ and all t values depend upon $(k - 1)(b - 1)$ degrees of freedom, where k is the number of treatments and b the number of blocks.

EXAMPLE 11.5

Refer to the four chemical treatments and two blocks (fabrics) of Example 11.4. It is of interest to estimate simultaneously all possible differences between treatment means. Construct confidence intervals for these differences with a simultaneous confidence coefficient of at least 0.90, approximately.

Solution   Since there are four treatment means, μ_1, μ_2, μ_3, and μ_4, there will be $c = 6$ differences of the form $\mu_i - \mu_j$. Since the simultaneous confidence coefficient is to be $1 - \alpha = 0.90$, $\alpha/2c$ becomes $0.10/2(6) = 0.008$. The closest tabled t value in Table 5 of the Appendix has a tail area of 0.01, so we will use $t_{0.01}$ as an approximation to $t_{0.008}$.

For the data of Example 11.4 the four sample treatment means are

$$\bar{y}_1 = 7.0, \qquad \bar{y}_2 = 5.5, \qquad \bar{y}_3 = 10.5, \qquad \bar{y}_4 = 5.0$$

Also $s = \sqrt{\text{MSE}} = \sqrt{1} = 1$ and $b = 2$. Thus any interval of the form

$$(\bar{y}_i - \bar{y}_j) \pm t_{0.01} s \sqrt{\frac{1}{b} + \frac{1}{b}}$$

will become

$$(\bar{y}_i - \bar{y}_j) \pm (4.541)(1) \sqrt{\frac{1}{2} + \frac{1}{2}}$$

or

$$(\bar{y}_i - \bar{y}_j) \pm 4.541$$

since the degrees of freedom are $(b - 1)(k - 1) = 3$. The six confidence intervals are as follows:

Parameter	Interval
$\mu_1 - \mu_2$	$(\bar{y}_1 - \bar{y}_2) \pm 4.541$ or 1.5 ± 4.541
$\mu_1 - \mu_3$	$(\bar{y}_1 - \bar{y}_3) \pm 4.451$ or -3.5 ± 4.541
$\mu_1 - \mu_4$	$(\bar{y}_1 - \bar{y}_4) \pm 4.541$ or 2.0 ± 4.541
$\mu_2 - \mu_3$	$(\bar{y}_2 - \bar{y}_3) \pm 4.541$ or -5.0 ± 4.541
$\mu_2 - \mu_4$	$(\bar{y}_2 - \bar{y}_4) \pm 4.541$ or 0.5 ± 4.541
$\mu_3 - \mu_4$	$(\bar{y}_3 - \bar{y}_4) \pm 4.541$ or 5.5 ± 4.541

The sample data would suggest that μ_2 and μ_3 are significantly different, and μ_3 and μ_4 are significantly different, since these intervals do not overlap zero. ■

EXERCISES

11.34 Refer to Exercise 11.23.
  (a) Form a 95% confidence interval to estimate the true mean difference in pressure between nickel and iron. Interpret this interval.
  (b) Form confidence intervals on the three possible differences between the means for the bonding agents, using a simultaneous confidence coefficient of approximately 0.90.

11.35 Refer to Exercise 11.24. Use a 90% confidence interval to estimate the true difference between the mean responses given by estimators B and C.

11.36 Refer to Exercise 11.25. What are the significant differences among the station means? (Use a simultaneous confidence coefficient of 0.90.)

11.37 Refer to Exercise 11.26. Compare the three telephone companies' mean rates by using a simultaneous confidence coefficient of approximately 0.90.

11.38 Refer to Exercise 11.27.
  (a) Use a 99% confidence interval to estimate the difference between the mean delivery time for carriers I and II.
  (b) What assumptions are necessary for the validity of the procedures you used in part (a)?

*11.8 THE FACTORIAL EXPERIMENT

The response of interest Y is, in many experimental situations, related to one or more other variables that can be controlled by the experimenter. For example the yield Y in an experimental study of the process of manufacturing a chemical may be related to the temperature x_1 and pressure x_2 at which the experiment is run. The variables, temperature and pressure, can be controlled by the experimenter. In a study of heat loss through ceilings of houses, the amount of heat loss Y will be related to the thickness of insulation x_1 and the temperature differential x_2 between the inside and outside of the house. Again the thickness of insulation and the temperature differential can be controlled by the experimenter as he or she designs the study. The strength Y of concrete may be related to the amount of aggregate x_1, the mixing time x_2, and the drying time x_3.

The variables that are thought to affect the response of interest and are controlled by the experimenter are called *factors*. The various settings of these factors in an experiment are called *levels*. A factor-level combination then defines a *treatment*. In the chemical yield example the temperatures of interest may be 90°, 100°, and 110°C. These are the three experimental levels of the factor "temperature." The pressure settings (levels) of interest may be 400 and 450 psi. Setting temperature at 90° and pressure at 450 psi would define a particular treatment, and observations on yield could then be obtained for this combination of levels. Note

*Optional section.

that three levels of temperature and two of pressure would result in $2 \times 3 = 6$ different treatments, if all factor-level combinations are to be used.

An experiment in which treatments are defined by specified factor-level combinations is referred to as a *factorial experiment*. In this section we assume that *r* observations on the response of interest are realized from each treatment, with each observation being independently selected. (That is, the factorial experiment is run in a completely randomized design with *r* observations per treatment.) The objective of the analysis is to decide if, and to what extent, the various factors affect the response variable. Does the yield of chemical increase as temperature increases? Does the strength of concrete decrease as less and less aggregate is used? These are the types of questions we will be able to answer with the results of this section.

We illustrate the concepts and calculations involved in the analysis of variance for a factorial experiment by considering an example consisting of two factors each at two levels. In the production of a certain industrial chemical the yield Y depends upon the cooking time x_1 and the cooling time x_2. Two cooking times and two cooling times are of interest. Only the relative magnitudes of the levels are important in the analysis of factorial experiments, and thus we can call the two levels of cooking time 0 and 1. That is, x_1 can take on the value 0 or 1. Similarly we can call the levels of cooling time 0 and 1; that is, x_2 can take on the values 0 or 1. For convenience we will refer to cooking time as factor A and cooling time as factor B. Schematically we then have the arrangement of four treatments seen in Figure 11.9, and we will assume that r observations will be available from each.

FIGURE 11.9
A 2 × 2 Factorial Experiment (r observations per treatment)

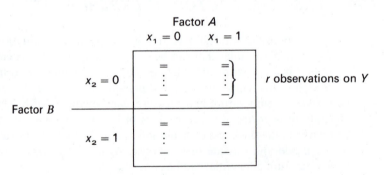

The analysis of this factorial experiment could be accomplished by writing a linear regression model for the response Y as a function of x_1 and x_2, and then employing the theory of Chapters 9 and 10. We will proceed by writing the models for illustrative purposes, but we will then present simple calculation formulas that make the actual fitting of the models by least squares unnecessary.

Since Y depends, supposedly, on x_1 and x_2, we could start with the simple model

$$E(Y) = \beta_0 + \beta_1 x_1 + \beta_2 x_2$$

On Figure 11.10 we see that this model implies that the rate of change on $E(Y)$ as x_1 goes from 0 to 1 is the same (β_1) for each value of x_2. This is referred to as the "no-interaction" case.

FIGURE 11.10
E(Y) for a 2 × 2
Factorial Experiment
with No Interaction

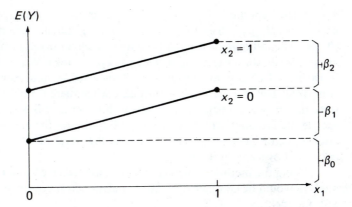

It is quite likely that the rate of change on $E(Y)$, as x_1 goes from 0 to 1, could be different for different values of x_2. To account for this *interaction*, we write the model as

$$E(Y) = \beta_0 + \beta_1 x_1 + \beta_2 x_2 + \beta_3 x_1 x_2$$

$\beta_3 x_1 x_2$ is referred to as the interaction term. Now if $x_2 = 0$, we have

$$E(Y) = \beta_0 + \beta_1 x_1$$

and if $x_2 = 1$, we have

$$E(Y) = (\beta_0 + \beta_2) + (\beta_1 + \beta_3)x_1$$

The slope (rate of change) when $x_2 = 0$ is β_1, but the slope when $x_2 = 1$ changes to $\beta_1 + \beta_3$. Figure 11.11 depicts a possible "interaction" case.

FIGURE 11.11
E(Y) for a 2 × 2
Factorial Experiment
with Interaction

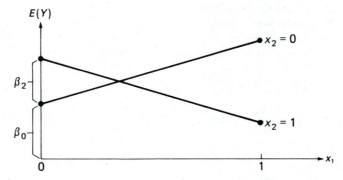

One way to proceed with an analysis of this 2×2 factorial experiment is as follows. First test the hypothesis $H_0: \beta_3 = 0$. If you reject this hypothesis, thus establishing that there is evidence of interaction, *do not* proceed with tests on β_1 and β_2. The significance of β_3 is enough to establish the fact that there are some differences among the treatment means. It may then be best simply to estimate the individual treatment means, or differences between means, by the methods given in Section 11.4.

If the hypothesis $H_0: \beta_3 = 0$ is not rejected, it is then appropriate to test the hypotheses $H_0: \beta_2 = 0$ and $H_0: \beta_1 = 0$. These are often referred to as tests of "main effects." The test of $\beta_2 = 0$ actually compares the observed mean of all observations at $x_2 = 1$ with the corresponding mean at $x_2 = 0$, regardless of the value of x_1. That is, each level of B is averaged over all levels of A, and then the levels of B are compared. This is a reasonable procedure when there is no evidence of interaction, since the change in response from low to high level of B is essentially the same for each level of A. Similarly the test of $\beta_1 = 0$ compares the mean response at the high level of factor $A(x_1 = 1)$ with that at the low level of $A(x_1 = 0)$ with the means computed over all levels of factor B.

Using the method of fitting complete and reduced models (Section 10.7), the analysis would be completed by fitting the models given below and calculating the SSE for each:

1. $Y = \beta_0 + \beta_1 x_1 + \beta_2 x_2 + \beta_3 x_1 x_2 + \varepsilon, \quad \text{SSE}_1$

2. $Y = \beta_0 + \beta_1 x_1 + \beta_2 x_2 + \varepsilon, \quad \text{SSE}_2$

3. $Y = \beta_0 + \beta_1 x_1 + \varepsilon, \quad \text{SSE}_3$

4. $Y = \beta_0 + \varepsilon, \quad \text{SSE}_4$

Now

$$\text{SSE}_2 - \text{SSE}_1 = \text{SS}(A \times B)$$

the sum of squares for the $A \times B$ interaction. Also

$$\text{SSE}_3 - \text{SSE}_2 = \text{SS}(B)$$

TABLE 11.7
Calculation Formulas for a Two-Factor Factorial Experiment with r Observations per Cell

$$n = abr = \text{Total number of observations}$$

$$\text{TSS} = \sum y^2 - \frac{(\sum y)^2}{n}$$

$$\text{SST} = \sum_{ij} \frac{T_{ij}^2}{r} - \frac{(\sum y)^2}{n}$$

$$\text{SS}(A) = \sum_{i=1}^{a} \frac{A_i^2}{br} - \frac{(\sum y)^2}{n}$$

$$\text{SS}(B) = \sum_{i=1}^{b} \frac{B_j^2}{ar} - \frac{(\sum y)^2}{n}$$

$$\text{SS}(A \times B) = \text{SST} - \text{SS}(A) - \text{SS}(B)$$

$$\text{SSE} = \text{TSS} - \text{SST}$$

the sum of squares due to factor B, and

$$SSE_4 - SSE_3 = SS(A)$$

the sum of squares due to factor A. This approach works well if a computer is available for fitting the indicated models.

The sums of squares shown above can be calculated by the direct formulas given in Table 11.7. Table 11.7 shows the general formulas for any two-factor factorial experiment with r observations per cell, assuming that there are a levels of factor A and b levels of factor B. That is, the data must follow the format given in Figure 11.12.

FIGURE 11.12
A General Two-Factor Factorial Experiment with r Observations per Cell

		Factor A					
		1	2	3	$\cdots$	a	Totals
	1	— — — T_{11}	— — — T_{12}	— — — T_{13}		— — T_{1a}	B_1
	2	— — — T_{21}	— — — T_{22}	— — — T_{23}		— — T_{2a}	B_2
Factor B	3	— — — T_{31}	— — — T_{32}	— — — T_{33}		— — T_{3a}	B_3
	$\vdots$						
		— — — T_{b1}	— — — T_{b2}	— — — T_{b3}		— — T_{ba}	B_b
Totals		A_1	A_2	A_3		A_a	$\sum y$

T_{ij} = Total of observations in row i and column j.

Notice that the treatment sum of squares is now partitioned into a sum of squares for A, a sum of squares for B, and an interaction sum of squares. The degrees of freedom for the latter three sums of squares are $(a - 1)$, $(b - 1)$, and $(a - 1)(b - 1)$, respectively. These facts are shown on the analysis of variance table given in Table 11.8.

We illustrate the calculations and F tests in the following example.

TABLE 11.8
ANOVA for a Two-Factor Factorial Experiment in a Completely Randomized Design

Source	df	SS	MS	F ratio
Treatments	$ab - 1$	SST	$MST = \dfrac{SST}{ab - 1}$	
Factor A	$a - 1$	SS(A)	$MS(A) = \dfrac{SS(A)}{a - 1}$	MS(A)/MSE
Factor B	$b - 1$	SS(B)	$MS(B) = \dfrac{SS(B)}{b - 1}$	MS(B)/MSE
$A \times B$ Interaction	$(a - 1)(b - 1)$	SS($A \times B$)	$MS(A \times B) = \dfrac{SS(A + B)}{(a - 1)(b - 1)}$	MS($A \times B$)/MSE
Error	$ab(r - 1)$	SSE	$MSE = \dfrac{SSE}{ab(r - 1)}$	
Total	$n - 1$	TSS		

Assumption: All factor combinations have independent normal distributions with equal variances

EXAMPLE 11.6

For the chemical experiment with two cooking times (factor A) and two cooling times (factor B), the yields are as given below, with $r = 2$ observations per treatment.

$$
\begin{array}{c|c|c|c}
 & \multicolumn{2}{c}{A} & \\
 & x_1 = 0 & x_1 = 1 & \\
\hline
x_2 = 0 & \begin{array}{c} 9 \\ 8 \\ \hline 17 \end{array} & \begin{array}{c} 5 \\ 6 \\ \hline 11 \end{array} & 28 \\
\hline
x_2 = 1 & \begin{array}{c} 8 \\ 7 \\ \hline 15 \end{array} & \begin{array}{c} 3 \\ 4 \\ \hline 7 \end{array} & 22 \\
\hline
 & 32 & 18 & 50
\end{array}
$$

B

Perform an analysis of variance.

Solution   Using the computation formulas of Table 11.7, with $a = 2$, $b = 2$, and $r = 2$, we have

$$TSS = 9^2 + 8^2 + \cdots + 3^2 + 4^2 - \frac{(50)^2}{8} = 31.5$$

$$SST = \frac{1}{2}[(17)^2 + (11)^2 + (15)^2 + (7)^2] - \frac{(50)^2}{8} = 29.5$$

$$SS(A) = \frac{1}{4}[(32)^2 + (18)^2] - \frac{(50)^2}{8} = 24.5$$

$$SS(B) = \frac{1}{4}[(28)^2 + (22)^2] - \frac{(50)^2}{8} = 4.5$$

$$SSE(A \times B) = SST - SS(A) - SS(B) = 0.5$$

and

$$SSE = TSS - SST = 2.0$$

The analysis of variance, using the format of Table 11.8, proceeds as follows:

Source	df	SS	MS	F ratio
Treatments	3	29.5		
A	1	24.5	24.5	$\frac{24.5}{0.5} = 49.0$
B	1	4.5	4.5	$\frac{4.5}{0.5} = 9.0$
$A \times B$	1	0.5	0.5	$\frac{0.5}{0.5} = 1.0$
Error	4	2.0	0.5	
Total	7	31.5		

Since $F_{0.05}(1, 4) = 7.71$, the interaction effect is not significant. We can then proceed with "main effect" tests for factors A and B. The F ratio of 49.0 for factor A is highly significant. Thus there is a significant difference between the mean response at $x_1 = 0$ and that at $x_1 = 1$. Looking at the data, we see that the yield falls off as cooking time goes from the low level to the high level.

The F ratio of 9.0 for factor B is also significant. The mean yield also seems to decrease as cooling time is changed from the low to the high level. ∎

In Example 11.7 we show an analysis in which the interaction term is highly significant.

EXAMPLE 11.7

Suppose a chemical experiment like the one of Example 11.6 (two factors each at two levels) gave the following responses:

$$A$$

	$x_1 = 0$	$x_1 = 1$	
$x_2 = 0$	9 8	5 6	28
$x_2 = 1$	3 4	8 7	22
	24	26	50

B (label to the left spanning both rows)

Perform an analysis of variance. If the interaction is significant, construct confidence intervals for the six possible differences between treatment means.

Solution   The data involve the same responses as in Example 11.6, with the observations in the lower left and lower right cells interchanged. Thus we still have

$$\text{TSS} = 9^2 + 8^2 + \cdots + 8^2 + 7^2 - \frac{(50)^2}{8} = 31.5$$

$$\text{SST} = \frac{1}{2}[(17)^2 + (11)^2 + (7)^2 + (15)^2] - \frac{(50)^2}{8} = 29.5$$

and

$$\text{SSE} = \text{TSS} - \text{SST} = 2.0$$

Now

$$\text{SS}(A) = \frac{1}{4}[(24)^2 + (26)^2] - \frac{(50)^2}{8} = 0.5$$

$$\text{SS}(B) = \frac{1}{4}[(28)^2 + (22)^2] - \frac{(50)^2}{8} = 4.5$$

and

$$\text{SS}(A \times B) = \text{SST} - \text{SS}(A) - \text{SS}(B) = 24.5$$

The analysis of variance table is as follows:

Source	df	SS	MS	F ratio
Treatments	3	29.5		
A	1	0.5	0.5	
B	1	4.5	4.5	
$A \times B$	1	24.5	24.5	49.0
Error	4	2.0	0.5	
Total	7	31.5		

At the 5% level the interaction term is highly significant. Hence we will not make any "main effect" tests, but instead will place confidence intervals on all possible differences between treatment means. For convenience we will identify the $(x_1 = 0, x_2 = 0)$ combination as treatment 1 with mean μ_1, $(x_1 = 1, x_2 = 0)$ as treatment 2 with mean μ_2, $(x_1 = 0, x_2 = 1)$ as treatment 3 with mean μ_3, and $(x_1 = 1, x_2 = 1)$ as treatment 4 with mean μ_4. To form a confidence interval on $\mu_i - \mu_j$, we follow the format of Section 11.4, which gives the interval of the form

$$(\bar{y}_i - \bar{y}_j) \pm t_{\alpha/2c} s \sqrt{\frac{1}{n_i} + \frac{1}{n_j}}$$

In this problem $c = 6$ and $n_i = n_j = 2$. Also $s = \sqrt{MSE}$ and is based upon 4 degrees of freedom. If we settle for $1 - \alpha = 0.90$, then $\alpha/2c = 0.10/12 \approx 0.01$. From Table 5 in the Appendix, $t_{0.01} = 3.747$, with 4 degrees of freedom. Thus all six intervals will be of the form

$$(\bar{y}_i - \bar{y}_j) \pm 3.747\sqrt{0.5} \sqrt{\frac{1}{2} + \frac{1}{2}}$$

or

$$(\bar{y}_i - \bar{y}_j) \pm 2.65$$

The sample means are given by

$$\bar{y}_1 = 8.5, \qquad \bar{y}_2 = 5.5, \qquad \bar{y}_3 = 4.5, \qquad \bar{y}_4 = 7.5$$

and the confidence intervals would then result in

$$(\bar{y}_1 - \bar{y}_2) \pm 2.65 \quad \text{or} \quad 3.0 \pm 2.65^*$$

$$(\bar{y}_1 - \bar{y}_3) \pm 2.65 \quad \text{or} \quad 5.0 \pm 2.65^*$$

$$(\bar{y}_1 - \bar{y}_4) \pm 2.65 \quad \text{or} \quad 1.0 \pm 2.65$$

$$(\bar{y}_2 - \bar{y}_3) \pm 2.65 \quad \text{or} \quad 2.0 \pm 2.65$$

$$(\bar{y}_2 - \bar{y}_4) \pm 2.65 \quad \text{or} \quad -2.0 \pm 2.65$$

$$(\bar{y}_3 - \bar{y}_4) \pm 2.65 \quad \text{or} \quad -4.0 \pm 2.65^*$$

The significant differences are then between μ_1 and μ_2, μ_1 and μ_3, and μ_3 and μ_4 (see the intervals marked with an asterisk). In practical terms the yield is reduced significantly as x_1 goes from 0 to 1 and x_2 remains at 0. Also the yield is increased as x_1 goes from 0 to 1 and x_2 remains at 1. The yield decreases as x_2 goes from 0 to 1 with x_1 at 0, but there is no significant change in mean yield between x_2 at 0 and x_2 at 1 with x_1 at 1. Since μ_1 and μ_4 do not appear to differ, the best choices for maximizing yield appear to be either $(x_1 = 0, x_2 = 0)$ or $(x_1 = 1, x_2 = 1)$.   ■

Suppose that the chemical example under discussion had three cooking times (levels of factor *A*) and two cooling times (levels of factor *B*). If the three cooking times were equally spaced (such as 20 minutes, 30 minutes, and 40 minutes), then the levels of *A* could be coded as $x_1 = -1$, $x_1 = 0$, and $x_1 = 1$. Again only the *relative* magnitudes of the levels are important in the analysis. Since there are now three levels of factor A, a quadratic term in x_1 can be added to the model. Thus the complete model would have the form

$$E(Y) = \beta_1 + \beta_1 x_1 + \beta_2 x_1^2 + \beta_3 x_2 + \beta_4 x_1 x_2 + \beta_5 x_1^2 x_2$$

Note that if $x_2 = 0$, we simply have a quadratic model in x_1, given by

$$E(Y) = \beta_0 + \beta_1 x_1 + \beta_2 x_1^2$$

If $x_2 = 1$, we have another quadratic model in x_1, but the coefficients have changed, and

$$E(Y) = (\beta_0 + \beta_3) + (\beta_1 + \beta_4)x_1 + (\beta_2 + \beta_5)x_1^2$$

If $\beta_4 = \beta_5 = 0$, the two curves will have the same shape, but if either β_4 or β_5 differs from zero, the curves will differ in shape. Thus β_4 and β_5 are both components of interaction. (See Figure 11.13.)

FIGURE 11.13 *E(Y) for a 3×2 Factorial Experiment, (a) without Interaction, (b) with Interaction*

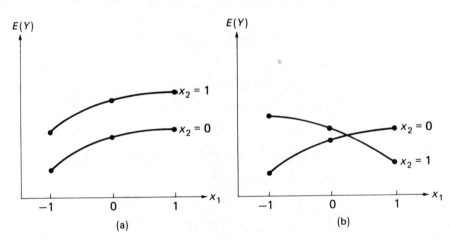

In the analysis first we test $H_0: \beta_4 = \beta_5 = 0$ (no interaction). If we reject this hypothesis, then we do no further F tests but may compare treatment means. If we do not reject the hypothesis of no interaction, we proceed with tests of $H_0: \beta_3 = 0$ (no main effect for factor B) and $H_0: \beta_1 = \beta_2 = 0$ (no main effect for factor A). The test of $H_0: \beta_3 = 0$ merely compares the mean responses at the high and low levels of factor B. The test of $H_0: \beta_1 = \beta_2 = 0$ looks for both linear and quadratic trends among the three means for the levels of factor A, averaged across the levels of factor B.

The analysis could be conducted by fitting complete and reduced models, and using the results of Section 10.7, or by using the computational formulas of Table 11.7. We illustrate both approaches with the following example.

EXAMPLE 11.8

In manufacturing a certain beverage, an important measurement is the percentage of impurities present in the final product. The following data show the percentage of impurities present in samples taken from products manufactured at three different temperatures (factor A), and two sterilization times (factor B). The three levels of *A* were actually 75°, 100°, and 125°C. The two levels of *B* were actually 15 minutes and 20 minutes. The data are as follows:

			A			
		75°	100°	125°		
		14.05	10.55	7.55		
		14.93	9.48	6.59		
	15 min.	28.98	20.03	14.14	63.15	
B						
		16.56	13.63	9.23		
		15.85	11.75	8.78		
	20 min.	32.41	25.38	18.01	75.80	
		61.39	45.41	32.15	138.95	

Perform an analysis of variance.

Using the formulas of Table 11.7, with $a = 3$, $b = 2$, and $r = 2$, we have

$$\text{TSS} = (14.05)^2 + \cdots + (8.78)^2 - \frac{(138.95)^2}{12} = 124.56$$

$$\text{SST} = \frac{1}{2}[(28.98)^2 + (20.03)^2 + \cdots (18.01)^2] - \frac{(138.95)^2}{12} = 121.02$$

$$\text{SS}(A) = \frac{1}{4}[(61.39)^2 + (45.41)^2 + (32.15)^2] - \frac{(138.95)^2}{12} = 107.18$$

$$\text{SS}(B) = \frac{1}{6}[(63.15)^2 + (75.80)^2] - \frac{(138.95)^2}{12} = 13.34$$

$$\text{SS}(A \times B) = \text{SST} - \text{SS}(A) - \text{SS}(B) = 0.50$$

and

$$\text{SSE} = \text{TSS} - \text{SST} = 3.54$$

The analysis of variance table then has the following form:

Source	df	SS	MS	*F* ratio
Treatments	5	121.02		
A	2	107.18	53.59	90.83
B	1	13.34	13.34	22.59
A × *B*	2	0.50	0.25	0.42
Error	6	3.54	0.59	
Total	11	124.56		

Since $F_{0.05}(2, 5) = 5.14$ and $F_{0.05}(1, 6) = 5.99$, it is clear that the interaction is not significant and that the main effects for both factor A and B are highly significant. The means for factor A tend to decrease as the temperature goes from low to high, and this decreasing trend is of approximately the same degree for both the low and high levels of factor B.   ■

For the regression approach we begin by fitting the complete model, with interactions, as shown in the following printout.

The regression equation is
$y = 11.4 - 3.66\,x1 + 1.34\,x2 + 0.340\,x1{**}2 + 0.055\,x1x2 - 0.425\,x1{**}2x2$

Predictor	Coef	Stdev	t-ratio	p
Constant	11.3525	0.3841	29.56	0.000
x1	−3.6550	0.2716	−13.46	0.000
x2	1.3375	0.3841	3.48	0.013
x1**2	0.3400	0.4704	0.72	0.497
x1x2	0.0550	0.2716	0.20	0.846
x1**2x2	−0.4250	0.4704	−0.90	0.401

$s = 0.7682$ R-sq $= 97.2\%$ R-sq(adj) $= 94.8\%$

Analysis of Variance

SOURCE	DF	SS	MS	F	p
Regression	5	121.022	24.204	41.01	0.000
Error	6	3.541	0.590		
Total	11	124.562			

Thus $SSE_1 = 3.541$.

To test for significant interaction ($H_0: \beta_4 = \beta_5 = 0$), we fit the reduced model as follows.

The regression equation is
$y = 11.4 - 3.66\,x1 + 1.05\,x2 + 0.340\,x1{**}2$

Predictor	Coef	Stdev	t-ratio	p
Constant	11.3525	0.3556	31.92	0.000
x1	−3.6550	0.2515	−14.54	0.000
x2	1.0542	0.2053	5.13	0.000
x1**2	0.3400	0.4355	0.78	0.457

$s = 0.7112$ R-sq $= 96.8\%$ R-sq(adj) $= 95.5\%$

Analysis of Variance

SOURCE	DF	SS	MS	F	p
Regression	3	120.516	40.172	79.41	0.000
Error	8	4.047	0.506		
Total	11	124.563			

We now have $SSE_2 = 4.047$, and the F statistic becomes

$$\frac{(4.047 - 3.541)/2}{3.541/6} = \frac{0.25}{0.59} = 0.42$$

the same as the analysis of variance F ratio for interaction.

To test for significant main effect for factor B, we must reduce the model even further by letting the coefficient of x_2 equal zero.

The regression equation is
$y = 11.4 - 3.66\,x1 + 0.340\,x1**2$

Predictor	Coef	Stdev	t-ratio	p
Constant	11.3525	0.6949	16.34	0.000
X1	−3.6550	0.4913	−7.44	0.000
X1**2	0.3400	0.8510	0.40	0.699

$s = 1.390$ R-sq $= 86.0\%$ R-sq(adj) $= 82.9\%$

Analysis of Variance

SOURCE	DF	SS	MS	F	p
Regression	2	107.180	53.590	27.75	0.000
Error	9	17.382	1.931		
Total	11	124.563			

Now

$$SSE_3 = 17.382$$

and

$$SS(B) = SSE_3 - SSE_2$$
$$= 17.382 - 4.047 = 13.335$$

Also

$$SSE_4 = SST = 124.563$$

so that

$$SS(A) = SSE_4 - SSE_3$$
$$= 124.563 - 17.382 = 107.181$$

Note that $SS(B)$ and $SS(A)$ as calculated from the regression models agree with the calculations on the analysis of variance table.

Since the regression model allows easy checking of residuals, we plot these residuals (from the main effects model, with no interaction) against their normal scores, as shown in Figure 11.14. The nearly straight-line plot indicates that the normality assumption is reasonable.

Two-level factorial experiments (2×2 or, in general, $k \times 2$ if there are k factors of interest) are often used as screening experiments in preliminary investigations. If

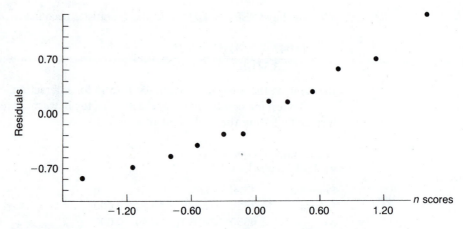

a process depends upon many factors, the two-level experiments are run to determine which factors seem to be affecting the response most dramatically. A higher-level experiment can be run subsequently to find optimal settings for the levels of those factors. Since two-level factorial experiments are so common, we will present more details on their analysis, focusing on graphical techniques.

Going back to the data of Example 11.6 for a 2×2 factorial experiment, we first observe that the numerical values of x_1 and x_2 have no effect on the analysis. Thus we will simply refer to the low levels as $(-)$ and the high levels as $(+)$. We then have 2 observations on yield at $A^- B^-$ (low cooking time, low cooling time), 2 at $A^- B^+$, and so on. A convenient way to display the data is on back-to-back dot plots, as in Figure 11.15. We can immediately see a large shift from high to low values as we move from A^- to A^+, and a less dramatic shift downward as we move from B^- to B^+.

FIGURE 11.15
Dot Plots for
Factors A and B

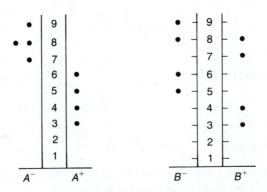

In a similar way interaction terms can be plotted. The upper left and lower right cells of the data table make positive $(+)$ contributions to the interaction, and the other two cells make negative $(-)$ contributions. Thus the interaction dot plot looks like Figure 11.16. Although the $(AB)^+$ values are spread out more, there is no shift of location as we move from $(AB)^-$ to $(AB)^+$. Thus there is little interaction effect.

FIGURE 11.16
Dot Plot for
Interaction

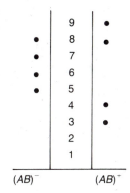

The shifts seen in Figures 11.15 and 11.16 can be quantified more precisely in terms of averages. The average of all 8 yields is 6.25, while the average of A^+ yields is 4.5 and the average of A^- yields is 8.0. Similar averages can be calculated for B^+ and B^-. Figure 11.17 shows the overall average as the center line and the factor-level (cell) averages plotted relative to this center. This is a convenient way to show all averages on one plot.

FIGURE 11.17
Plot of Factor-Level
Means

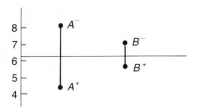

Notice that the distance between A^+ and the overall mean is the same as the distance between A^- and the overall mean, with a similar result holding for factor B. This distance,

$$\frac{1}{2}(A^+ \text{ average} - A^- \text{ average}) = -1.75$$

is called the *main effect* due to A. It tells us that, on the average, the high cooking time lowers the yield by 1.75 units compared to the experimentwide average yield. Similarly the *main effect* for B, -0.75, tells us that the high cooling time lowers the average yield by 0.75 units.

The *interaction effect* is measured by

$$\frac{1}{2}[(AB)^+ \text{ average} - (AB)^- \text{ average}] = -0.25$$

and turns out to be small compared to the main effects in this case.

There is a simple and elegant graphical interpretation of main effects and interaction, as developed in Figure 11.18. First we plot the A^- and A^+ averages and connect them with a dashed line. Since we can think of A^- as being at -1 and

FIGURE 11.18
*Plot of Main Effect
and Interaction
Lines*

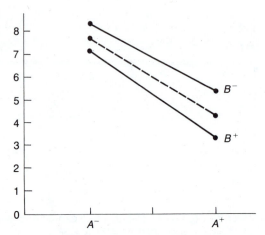

A^+ as being at $+1$ on the horizontal axis, these points are two units apart. This makes the slope of the dashed line equal to the main effect for A, -1.75 in this case.

We then plot the A^-B^- and A^+B^- averages, and connect them with the line labeled B^-. Repeating the procedure for A^-B^+ and A^+B^+ gives the B^+ line.

The slope of the B^+ line is

$$\frac{1}{2}(A^+B^+ \text{ average} - A^-B^+ \text{ average}) = \frac{1}{2}(3.5 - 7.5) = -2.0$$

The difference between the slopes of the dashed line (main effect for A) and the B^+ line is

$$[-2.0 - (-1.75)] = -0.25$$

which is the interaction effect. Whereas a main effect measures the difference between an overall average and a factor-level average, an interaction effect measures the difference between the slope of a main effect line and the slope for that factor plotted at a fixed level of a second factor. To see what happens to these lines for a situation with a large interaction, make a similar plot for the data in Example 11.7.

EXAMPLE 11.9

An experiment was conducted to compare the strengths of two brands of facial tissues, the famous brands X and Y. The two brands (two "levels" of factor A) were each tested under both dry and damp conditions (the two levels of factor B). Thus the four treatments tested are

A^-B^- (brand X, dry)

A^+B^- (brand Y, dry)

A^-B^+ (brand X, damp)

A^+B^+ (brand Y, damp)

Four tissues were tested under each treatment. The strength measurement was produced as follows. A tissue was stretched over the opening of a plastic cup and held by a rubber band. Then a marble was dropped onto the stretched tissue. The lowest height (in inches) from which a marble that went through the tissue was dropped is the recorded strength measurement. The data are as shown on the accompanying table. (One of the authors actually conducted the experiment!)

	A −	A +
B −	14, 14 12, 11	10, 11 11, 12
B +	4, 4 3, 5	5, 5 6, 6

The "14" in the upper left corner means that a tissue of brand X that was dry broke first from a marble dropped from 14 inches about its surface. Analyze the data for significant main effects and interaction.

Solution   Since these are two-digit measurements, stem-and-leaf plots can be used in place of dot plots. The main effect and interaction plots are shown below.

```
      |1|                      |1|                    |1|
4421  |1| 0112      44221110   |1|           2110     |1| 1244
   5  |0| 5566                 |0| 55566         5     |0| 5566
 443  |0|                      |0| 344         443     |0|
 ─────────────────          ──────────────      ─────────────────
  A⁻        A⁺              B⁻          B+       (AB)⁻     (AB)⁺
```

These plots show little shift as we move from A^- to A^+, although A^- values are more spread out than A^+ values. A dramatic shift is seen as we move from B^- to B^+, indicating a large main effect for moisture content. There is some slight shift to larger values as we move from $(AB)^-$ to $(AB)^+$, signaling a possible interaction.

The main effect for A (brand) is

$$\frac{1}{2}(A^+ \text{ average} - A^- \text{ average}) = \frac{1}{2}(8.250 - 8.375) = -0.0625$$

which shows little difference in strength between brands.

The main effect for B (dry versus damp) is

$$\frac{1}{2}(B^+ \text{ average} - B^- \text{ average}) = \frac{1}{2}(4.750 - 11.875) = -3.5625$$

a rather large value in comparison to the sizes of the measurements. It should not be surprising, however, that damp tissues have less strength than dry tissues. We do not have to experiment very carefully to see this result!

The interaction effect is

$$\frac{1}{2}[(AB)^+ \text{ average} - (AB)^- \text{ average}] = \frac{1}{2}[9.125 - 7.500] = 0.8125$$

This is much larger than the main effect for A and indicates that the interaction is obscuring the effect of brands. A more careful look at the data suggests that brand X tissues tend to be stronger than brand Y tissues when dry, but brand Y tissues are slightly stronger when damp. When do you want tissues to be strong? Do you think you could state a preference for either brand?

The main effect plot here could be misleading because of the interaction, so we show only the interaction plot in Figure 11.19. Note that the two solid lines have obviously different slopes, indicating some interaction.

FIGURE 11.19
Interaction Plot for
Example 11.9

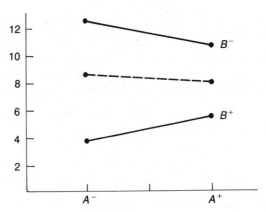

The analysis of variance for these data shows a large interaction mean square and a very large mean square due to factor B, just as we expected. Both of these effects are highly significant.

ANALYSIS OF VARIANCE

SOURCE	DF	SS	MS
A	1	0.063	0.063
B	1	203.062	203.062
INTERACTION	1	10.563	10.563
ERROR	12	11.750	0.979
TOTAL	15	225.437	

The analyses of factorial experiments with more than two factors proceed along similar lines. However, we will not present those results here. The interested reader should consult a text on experimental design.

EXERCISES

11.39 In pressure sintering of alumina, two important variables controlled by the experimenter are the pressure and time of sintering. An experiment involving two pressures and two times, with three specimens tested on each pressure/time combination, showed the following densities (in g/cc):

		Pressure	
		100 psi	200 psi
Time	10 min	3.6, 3.5, 3.3	3.8, 3.9, 3.8
	20 min	3.4, 3.7, 3.7	4.1, 3.9, 4.2

(a) Perform an analysis of variance, constructing appropriate tests for interaction and main effects. Use $\alpha = 0.05$.

(b) Estimate the difference between the true densities for specimens sintered for 20 minutes and those sintered for 10 minutes. Use a 95% confidence coefficient.

11.40 The yield percentage Y of a chemical process depends upon the temperature at which the process is run and the length of time the process is active. For two levels of temperature and two lengths of time, the yields were as follows (with two observations per treatment):

		Temperature	
		Low	High
	Low	24	28
		25	30
Time			
	High	26	23
		28	22

(a) Perform an analysis of variance, testing first for interaction. Use $\alpha = 0.05$.

(b) If interaction is significant, construct confidence intervals on the six possible differences between treatment means, using a simultaneous confidence coefficient of approximately 0.90.

11.41 The yield percentage Y of a certain precipitate depends upon the concentration of the reactant and the rate of addition of diammonium hydrogen phosphate. Experiments were run at three different concentration levels and three addition rates. The yield percentages, with two observations per treatment, were as follows:

| | Concentration of Reactant | | |
	−1	0	1
−1	90.1	92.4	96.4
	90.3	91.8	96.8
Addition Rate 0	91.2	94.3	98.2
	92.3	93.9	97.6
1	92.4	96.1	99.0
	92.5	95.8	98.9

(a)   Perform an analysis of variance, constructing all appropriate tests at the 5% significance level.

(b)   Estimate the average yield percentage, in a 95% confidence interval, for the treatment at the middle level of both concentration and addition rate.

11.42   How do women compare with men in their ability to perform laborious tasks that require strength? Some information on this question is provided in a study, by M. D. Phillips and R. L. Pepper, of the firefighting ability of men and women ("Shipboard Fire-Fighting Performance of Females and Males," *Human Factors*, 1982, *24*). Phillips and Pepper conducted a 2×2 factorial experiment to investigate the effect of the factor Sex (male or female) and the factor Weight (light or heavy) on the length of time required for a person to perform a particular firefighting task. Eight persons were selected for each of the $2 \times 2 = 4$ Sex/Weight categories of the 2×2 factorial experiment, and the length of time needed to complete the task was recorded for each of the thirty-two persons. The means and standard deviations of the four samples are shown in the table.

| | Light | | Heavy | |
	Mean	Standard deviation	Mean	Standard deviation
Female	18.30	6.81	14.50	2.93
Male	13.00	5.04	12.25	5.70

(a)   Calculate the total of the $n = 8$ time measurements for each of the four categories of the 2×2 factorial experiment.

(b)   Use the result of part (a) to calculate the sums of squares for Sex, Weight, and the Sex/Weight interaction.

(c)   Calculate each sample variance. Then calculate the sum of squares of deviations *within* each sample for each of the four samples.

(d)   Calculate SSE. [*Hint*: SSE is the pooled sum of squares of the deviations calculated in part (c).]

(e)   Now that you know SS(Sex), SS(Weight), SS(Sex × Weight), and SSE, find SS(Total).

(f) Summarize the calculations in an analysis of variance table.

(g) Explain the practical significance of the presence (or absence) of Sex/Weight interaction. Do the data provide evidence of a Sex/Weight interaction?

(h) Do the data provide sufficient evidence to indicate a difference in time required to complete the task between light men and women? Test by using $\alpha = 0.05$.

(i) Do the data provide sufficient evidence to indicate a difference in time to complete the task between heavy men and women? Test by using $\alpha = 0.05$.

11.43 Refer to Exercise 11.42. Phillips and Pepper (1982) give data on another 2×2 factorial experiment utilizing twenty males and twenty females. The experiment involved the same treatments with ten persons assigned to each Sex/Weight category. The response measured for each person was the pulling force the person was able to exert on the starter cord of a P-250 fire pump. The means and standard deviations of the four samples (corresponding to the $2 \times 2 = 4$ categories of the experiment) are shown in the table.

	Light		Heavy	
	Mean	Standard deviation	Mean	Standard deviation
Females	46.26	14.23	62.72	13.97
Males	88.07	8.32	86.29	12.45

(a) Use the procedures outlined in Exercise 11.42 to perform an analysis of variance for the experiment. Display your results in an ANOVA table.

(b) Explain the practical significance of the presence (or absence) of Sex/Weight interaction. Do the data provide sufficient evidence of a Sex/Weight interaction?

(c) Do the data provide sufficient evidence to indicate a difference in force exerted between light men and women? Test by using $\alpha = 0.05$.

(d) Do the data provide sufficient evidence to indicate a difference in force exerted between heavy men and women? Test by using $\alpha = 0.05$.

11.44 In analyzing coal samples for ash content, two types of crucibles and three temperatures were used in a complete factorial arrangement, with the following results:

		Crucible			
		Steel		Silica	
	825	8.7	7.2	9.3	9.1
Temperature	875	9.4	9.6	9.7	9.8
	925	10.1	10.2	10.4	10.7

(Two independent observations were taken on each treatment.) Note that "crucible" is not a quantitative factor, but it can still be considered to have two levels even though the levels cannot be ordered as to high and low.

Perform an analysis of variance, conducting appropriate tests at the 5% significance level. Does there appear to be a difference between the two types of crucible, with respect to average ash content?

11.45 Four different types of heads were tested on each of two sealing machines. Four independent measurements of strain were then made on each head/machine combination. The data are as follows:

	Head			
	1	2	3	4
Machine A	3	2	6	3
	0	1	5	0
	4	1	8	1
	1	3	8	1
Machine B	6	7	2	4
	8	6	0	7
	6	3	1	6
	5	4	2	7

(a) Perform an analysis of variance, making all appropriate tests at the 5% level of significance.

(b) If you were to use machine A, which head type would you recommend for maximum strain resistance in the seal? Why?

11.46 An experiment was conducted to determine the effects of two alloying elements (carbon and manganese) on the ductility of specimens of metal. Two specimens from each treatment were measured, and the data (in work required to break a specimen of standard dimension) are as follows:

		Carbon	
		0.2%	0.5%
Manganese	0.5%	34.5	38.2
		37.5	39.4
	1.0%	36.4	42.8
		37.1	43.4

(a) Perform an analysis of variance, conducting all appropriate tests at the 5% significance level.

(b) Which treatment would you recommend to maximize average breaking strength? Why?

11.9 CONCLUSION

Before measurements are actually made in any experimental investigation, careful attention should be paid to the *design* of the experiment. If the experimental objective is the comparison of k treatment means, and no other major source of variation is present, then a completely randomized design will be adequate. If a second source of variation is present, it can often be controlled through the use of a randomized block design. These two designs serve as an introduction to the topic of design of experiments, but many more complex designs, such as the Latin square, can be considered.

The analysis of variance for either design considered in this chapter can be carried out through a linear model (regression) approach or through the use of direct calculation formulas. The reader would benefit by trying both approaches on some examples to gain familiarity with the concepts and techniques.

The factorial experiment arises in cases for which treatments are defined by various combinations of factor levels. Notice that the factorial arrangement defines the treatments of interest, but is *not* in itself a design. Factorial experiments can be run in completely randomized, randomized block, or other designs, but we considered only the completely randomized design in this chapter. Many more topics dealing with factorial experiments can be found in texts that focus on experimental design and analysis of variance.

SUPPLEMENTARY EXERCISES

11.47   Three methods have been devised to reduce the time spent in transferring materials from one location to another. With no previous information available on the effectiveness of these three approaches, a study is performed. Each approach is tried several times, and the amount of time to completion (in hours) is recorded in the table.

Method		
A	B	C
8.2	7.9	7.1
7.1	8.1	7.4
7.8	8.3	6.9
8.9	8.5	6.8
8.8	7.6	
	8.5	

(a)   What type of experimental design was used?

(b)   Is there evidence that the mean time to completion of the task differs for the three methods? Use $\alpha = 0.01$.

(c)   Form a 95% confidence interval for the mean time to completion for method B.

11.48   One important consideration in determining which location is best for a new retail business is the amount of traffic that passes the location each business day. Counters are placed at each of four locations on the 5 weekdays, and the number of cars passing each location is recorded in the following table.

		Location		
Day	I	II	III	IV
1	453	482	444	395
2	500	605	505	490
3	392	400	383	390
4	441	450	429	405
5	427	431	440	430

(a)   What type of design does this represent?
(b)   Is there evidence of a difference in the mean number of cars per day at the four locations?
(c)   Estimate the difference between the mean numbers of cars that pass locations I and III each weekday.

11.49   Mileage tests were performed to compare three different brands of regular gas. Four different automobiles were used in the experiment, and each brand of gas was used in each car until the mileage was determined. The results are shown in the table.

Miles per Gallon

		Automobile		
Brand	1	2	3	4
A	20.2	18.7	19.7	17.9
B	19.7	19.0	20.3	19.0
C	18.3	18.5	17.9	21.1

(a)   Is there evidence of a difference in the mean mileage rating among the three brands of gasoline? Use $\alpha = 0.05$.
(b)   Construct the ANOVA summary table for this experiment.
(c)   Is there evidence of a difference in the mean mileage for the four models, that is, is blocking important in this type of experiment? Use $\alpha = 0.05$.
(d)   Form a 99% confidence interval for the difference between the mileage ratings of brands B and C.
(e)   Form confidence intervals for the three possible differences between the means for the brands, with a simultaneous confidence coefficient of approximately 0.90.

11.50   The table shows the partially completed analysis of variance for a two-factor factorial experiment.

Source	df	SS	MS	FC
A	3	2.6		
B	5	9.2		
AB			3.1	
Error		18.7		
Total	47			

(a) Complete the analysis of variance table.

(b) Give the number of levels for each factor and the number of observations per factor-level combination.

(c) Do the data provide sufficient evidence to indicate an interaction between the factors? Test by using $\alpha = 0.05$.

(d) State the practical implications of your test result in part (c).

11.51 The data shown in the table are for a 4×3 factorial experiment with two observations per factor-level combination.

		Level of B		
		1	2	3
Level of A	1	2 4	5 6	1 3
	2	5 4	2 2	10 9
	3	7 10	1 0	5 3
	4	8 7	12 11	7 4

(a) Perform an analysis of variance for the data and display the results in an analysis of variance table.

(b) Do the data provide sufficient information to indicate an interaction between the factors? Test by using $\alpha = 0.05$.

(c) Suppose the objective of the experiment is to select the factor-level combination with the largest mean. Based on the data and using a simultaneous confidence coefficent of approximately 0.90, which pairs of means appear to differ?

11.52 The set of activities and decisions through which a firm moves from its initial awareness of an innovative industrial procedure to its final adoption or rejection of the innovation is referred to as the *industrial adoption process*. The process can be described as having five stages: (1) awareness, (2) interest (additional information requested), (3) evaluation (advantages and disadvantages compared), (4) trial (innovation tested), and (5) adoption. As part of a study of the industrial adoption process, Ozanne and Churchill (1971) hypothesized that firms use a greater number of informational inputs to the process (e.g., visits by salespersons) in the later stages than in the earlier stages. In particular they tested the hypotheses that a greater number of informational inputs are used in the interest stage than in the awareness stage, and that a greater number are used in the evaluation stage than in the interest stage. Ozanne and Churchill collected the information given in the table on the number of informational inputs used by a sample of thirty-seven industrial firms that had recently adopted a particular new automatic machine tool.

	Number of Information Sources Used				Number of Information Sources Used		
Company	Awareness Stage	Interest Stage	Evaluation Stage	Company	Awareness Stage	Interest Stage	Evaluation Stage
1	2	2	3	20	1	1	1
2	1	1	2	21	1	2	1
3	3	2	3	22	1	4	3
4	2	1	2	23	2	3	3
5	3	2	4	24	1	1	1
6	3	4	6	25	3	1	4
7	1	1	2	26	1	1	5
8	1	3	4	27	1	3	1
9	2	3	3	28	1	1	1
10	3	2	4	29	3	2	2
11	1	2	4	30	4	2	3
12	3	3	3	31	1	2	3
13	4	4	4	32	1	3	2
14	4	2	7	33	1	2	2
15	3	2	2	34	2	3	2
16	4	5	4	35	1	2	4
17	2	1	1	36	2	2	3
18	1	1	3	37	2	3	2
19	1	3	2				

(Source: Ozanne and Churchill, *Marketing Models, Behavioral Science Applications*, pp. 249–265. Intext Educational Publishers, 1971.)

(a) Do the data provide sufficient evidence to indicate that differences exist in the mean number of informational inputs of the three stages of the industrial adoption process studied by Ozanne and Churchill? Test by using $\alpha = 0.05$.

(b) Find a 90% confidence interval for the difference in the mean number of informational inputs between the evaluation and awareness stages.

11.53 England has experimented with different 40-hour workweeks to maximize production and minimize expenses. A factory tested a 5-day week (8 hours per day), a 4-day week (10 hours per day), and a $3\frac{1}{3}$-day week (12 hours per day), with the weekly production results shown in the table (in thousands of dollars worth of items produced).

8-Hour Day	10-Hour Day	12-Hour Day
87	75	95
96	82	76
75	90	87
90	80	82
72	73	65
86		

(a) What type of experimental design was employed here?
(b) Construct an ANOVA summary table for this experiment.
(c) Is there evidence of a difference in the mean productivities for the three lengths of workdays?
(d) Form a 90% confidence interval for the mean weekly productivity when 12-hour workdays are used.

11.54 To compare the preferences of technicians for three brands of calculators, each technician was required to perform an identical series of calculations on each of the three calculators, A, B, and C. To avoid the possibility of fatigue, a suitable time period separated each set of calculations, and the calculators were used in random order by each technician. A preference rating, based on a 0–100 scale, was recorded for each machine/technician combination. These data are shown here:

	Calculator Brand		
Technician	A	B	C
1	85	90	95
2	70	70	75
3	65	60	80

(a) Do the data provide sufficient evidence to indicate a difference in technician preference among the three brands? Use $\alpha = 0.05$.
(b) Why did the experimenter have each technician test all three calculators? Why not randomly assign three different technicians to each calculator?

11.55 Sixteen workers were randomly selected to participate in an experiment to determine the effects of work scheduling and method of payment on attitude toward the job. Two types of scheduling were employed, the standard 8–5 workday and a modification whereby the worker was permitted to start the day at either 7 or 8 A.M. and to vary the starting time as desired; in addition, the worker was allowed to choose, on a daily basis, either a $\frac{1}{2}$-hour or 1-hour lunch period. The two methods of payment were a standard hourly rate and a reduced hourly rate with an added piece rate based on the worker's production. Four workers were randomly assigned to each of the four scheduling/payment combinations, and each completed an attitude test after 1 month on the job. The test scores are shown in the table.

		Payment	
		Hourly Rate	Hourly and Piece Rate
Scheduling	8–5	54, 68 55, 63	89, 75 71, 83
	Worker-Modified Schedule	79, 65 62, 74	83, 94 91, 86

(a) Construct an analysis of variance table for the data.
(b) Do the data provide sufficient information to indicate a factor interaction? Test by using $\alpha = 0.05$. Explain the practical implications of your test.
(c) Do the data indicate that any of the scheduling/payment combinations produce a mean attitude score that is clearly higher than the other three? Test by using a simultaneous Type I error rate of approximately $\alpha = 0.05$, and interpret your results.

11.56 An experiment was conducted to compare the yields of orange juice for six different juice extractors. Because of a possibility of a variation in the amount of juice per orange from one truckload of oranges to another, equal weights of oranges from a single truckload were assigned to each extractor, and this process was repeated for fifteen loads. The amount of juice recorded for each extractor for each truckload produced the following sums of squares:

Source	df	SS	MS	F
Extractor		84.71		
Truckload		159.29		
Error		94.33		
Total		339.33		

(a) Complete the ANOVA table.
(b) Do the data provide sufficient evidence to indicate a difference in mean amount of juice extracted by the six extractors? Use $\alpha = 0.05$.

11.57 A farmer wants to determine the effect of five different concentrations of lime on the pH (acidity) of the soil on a farm. Fifteen soil samples are to be used in the experiment, five from each of the three different locations. The five soil samples from each location are then randomly assigned to the five concentrations of lime, and 1 week after the lime is applied, the pH of the soil is measured. The data are shown in the following table:

Location	Lime Concentration				
	0	1	2	3	4
I	3.2	3.6	3.9	4.0	4.1
II	3.6	3.7	4.2	4.3	4.3
III	3.5	3.9	4.0	3.9	4.2

    (a) What type of experimental design was used here?

    (b) Do the data provide sufficient evidence to indicate that the five concentrations of lime have different mean soil pH levels? Use $\alpha = 0.05$.

    (c) Is there evidence of a difference in soil pH levels among locations? Test by using $\alpha = 0.05$.

11.58   In the hope of attracting more riders, a city transit company plans to have express bus service from a suburban terminal to the downtown business district. These buses should save travel time. The city decides to perform a study of the effect of four different plans (such as a special bus lane and traffic signal progression) on the travel time for the buses. Travel times (in minutes) are measured for several weekdays during a morning rush-hour trip while each plan is in effect. The results are recorded in the table.

	Plan		
1	2	3	4
27	25	34	30
25	28	29	33
29	30	32	31
26	27	31	
	24	36	

    (a) What type of experimental design was employed?

    (b) Is there evidence of a difference in the mean travel times for the four plans? Use $\alpha = 0.01$.

    (c) Form a 95% confidence interval for the difference between plan 1 (express lane) and plan 3 (a control: no special travel arrangements).

11.59   Five sheets of writing paper are randomly selected from each of three batches produced by a certain company. A measure of brightness is obtained for each sheet, with the following results:

	Batch	
1	2	3
28	34	27
32	36	25
25	32	29
27	38	31
26	39	21

    (a) Do the mean brightness measurements seem to differ among the three batches? Use $\alpha = 0.05$.

    (b) If you were to select the batch with the largest mean brightness, which would you select? Why?

11.60 To be able to provide its clients with comparative information on two large suburban residential communities, a realtor wants to know the average home value in each community. Eight homes are selected at random within each community and are appraised by the realtor. The appraisals are given in the table (in thousands of dollars). Can you conclude that the average home value is different in the two communitities? You have three ways of analyzing this problem.

Community	
A	B
43.5	73.5
49.5	62.0
38.0	47.5
66.5	36.5
57.5	44.5
32.0	56.0
67.5	68.0
71.5	63.5

(a) Use the two-sample t statistic to test $H_0: \mu_A = \mu_B$.

(b) Consider the regression model

$$y = \beta_0 + \beta_1 x + \varepsilon$$

where

$$x = \begin{cases} 1 & \text{if community B} \\ 0 & \text{if community A} \end{cases} \qquad y = \text{appraised price}$$

Since $\beta_1 = \mu_B - \mu_A$, testing $H_0: \beta_1 = 0$ is equivalent to testing $H_0: \mu_A = \mu_B$. Use the partial reproduction of the regression printout shown here to test $H_0: \beta_1 = 0$. Use $\alpha = 0.05$.

SOURCE	df	SUM OF SQUARES	MEAN SQUARE	F VALUE	PR > F
MODEL	1	40.64062500	40.64062500	0.21	0.6501
ERROR	14	2648.71875000	189.19419643		
CORRECTED TOTAL	15	2689.35937500			

				R-SQUARE	ROOT MSE
				0.015112	13.75478813

PARAMETER	ESTIMATE	T FOR H0: PARAMETER=0	PR > |T|	STD ERROR OF ESTIMATE
INTERCEPT	53.25000000	10.95	0.0001	4.86305198
X	3.18750000	0.46	0.6501	6.87739406

(c) Use the ANOVA method to test $H_0: \mu_A = \mu_B$. Use $\alpha = 0.05$.

(d) Using the results of the three tests in parts (a)–(c), verify that the tests are equivalent (for this special case, $k = 2$) for the completely randomized design in terms of the test-statistic value and rejection region. For the three methods used what are the advantages and disadvantages (limitations) of using each in analyzing results for this type of experimental design?

12

Nonparametric Statistics

ABOUT THIS CHAPTER

Many of the inferential techniques of Chapters 7 through 11 required some distributional assumptions on the population of measurements under study. Most often the population was assumed to be normally distributed. Obviously this assumption is not met for certain populations of interest. Also there are occasions when exact numerical measurements are impossible to obtain, as in the case of ranking the appearance of four designs of a building. In these cases we must resort to *nonparametric* statistical techniques. Five of these techniques are presented in this chapter.

CONTENTS

12.1 INTRODUCTION

Chapters 8 and 11 presented techniques for making inferences about the mean of a single population and for comparing the means of two or more populations. Most of the techniques discussed in those chapters were based on the assumption that the sampled populations have probability distributions that are approximately normal with equal variances. But how can you analyze data that are sampled from populations that do *not* satisfy these assumptions? Or how can you make comparisons between populations when you cannot assign specific numerical values to your observations?

In this chapter we present statistical techniques for comparing two or more populations that are based on an ordering of the sample measurements according to their relative magnitudes. These techniques, which require fewer or less stringent assumptions concerning the nature of the probability distributions of the populations, are called *nonparametric statistical methods*. The statistical tests presented in this chapter apply to the same experimental designs as those covered in the introduction to an analysis of variance (Chapter 11). Thus we will present nonparametric statistical techniques for comparing two or more populations, using either a completely randomized or a randomized block design.

The t and F tests for comparing the means of two or more populations (Chapters 8 and 11) are unsuitable for analyzing some types of experimental data. For these tests to be appropriate we assumed that the random variables being measured had normal probability distributions with equal variances. Yet in practice the observations from one population may exhibit much greater variability than those from another, or the probability distributions may be decidedly nonnormal. For example the distribution might be very flat, peaked, or strongly skewed to the right or left. When any of the assumptions required for the t and F tests are seriously violated, the computed t and F statistics may not follow the standard t and F distributions. If this is true, the tabulated values of t and F (Tables 5, 7, and 8 in the Appendix) are not applicable, the correct value of α for the test is unknown, and the t and F tests are of dubious value.

Often data occur in practical situations for which the t and F tests are inappropriate. Such data are not susceptible to a meaningful numerical measurement, but can be ranked in order of magnitude. For example if we want to compare the teaching ability of two college instructors based on subjective evaluations of students, we cannot assign a number that exactly measures the teaching ability of a single instructor. But we can compare two instructors by ranking them according to a subjective evaluation of their teaching abilities. If instructors A and B are evaluated by each of ten students, we have the standard problem of comparing the probability distributions for two populations of ratings: one for instructor A and one for B. However, the t test of Chapter 8 would be inappropriate, because the only data that can be recorded are the preference statements of the ten students; that is, each student decides that either A is better than B or vice versa.

Another situation in which the assumption of normality is not generally appropriate involves the measurement of extreme values, like breaking strengths of cables or maximum annual water levels in streams. These sets of data tend to possess a few very large measurements, and the symmetric normal distribution would not provide a good model for the respective populations.

The nonparametric counterparts of the t and F tests compare the probability distributions of the sampled populations, rather than specific parameters of these populations (such as the means or variances). For example nonparametric tests can be used to compare the probability distribution of the strength measurements for a new cable to the probability distributions of the strength measurements for a cable in current use. If it can be inferred that the distribution from the new cable lies above (to the right of) the other (see Figure 12.1), the implication is that the new cable tends to be stronger than the one in current use. Such an inference might lead to a decision to replace the old cable.

FIGURE 12.1
Probability Distributions of Strength Measurements for Cables (New cable tends to be stronger.)

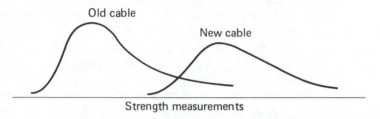

Many nonparametric methods use the relative ranks of the sample observations, rather than their actual numerical values. These tests are particularly valuable when we are unable to obtain numerical measurements of some phenomena but are able to rank them in comparison to each other. Statistics based on ranks of measurements are called *rank statistics*. In Sections 12.2 and 12.4 we present rank statistics for comparing two probability distributions, using independent samples. In Sections 12.3 and 12.5 the matched-pairs and randomized block design are used to make nonparametric comparisons of populations. Finally in Section 12.6 we present a nonparametric measure of correlation between two variables: Spearman's rank correlation coefficient.

12.2 COMPARING TWO POPULATIONS: WILCOXON RANK SUM TEST FOR INDEPENDENT SAMPLES

Suppose two independent random samples are to be used to compare two populations and the t test of Chapter 8 is inappropriate for making the comparison. We may be unwilling to make assumptions about the form of the underlying population probability distributions, or we may be unable to obtain exact values of the sample measurements. For either of these situations, if the data can be ranked in order of magnitude, the Wilcoxon rank sum test (developed by Frank Wilcoxon) can be used to test the hypothesis that the probability distributions associated with the two populations are equivalent.

For example suppose an engineer wants to compare completion times for technicians on task A to those on a similar task B. Prior experience has shown that populations of completion time measurements often possess probability distributions that are skewed to the right, as shown in Figure 12.2. Consequently a *t* test should not be used to compare the mean completion times for the two tasks because the normality assumption that is required for the *t* test may not be valid (particularly if the sample sizes are small).

FIGURE 12.2
Typical Probability Distribution of Completion Times

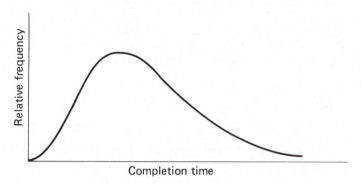

The engineer randomly assigns fourteen technicians to each of two groups, one group of seven to perform task A and the other to perform task B. The completion time for each technician is measured. These data (with the exception of the measurement for one technician in group A who was eliminated from the experiment for personal reasons) are shown in Table 12.1.

TABLE 12.1
Completion Times for Technicians on Task A or B

Task A Completion Time (seconds)	Rank	Task B Completion Time (seconds)	Rank
1.96	4	2.11	6
2.24	7	2.43	9
1.71	2	2.07	5
2.41	8	2.71	11
1.62	1	2.50	10
1.93	3	2.84	12
		2.88	13

The population of completion times for either of the tasks, say task A, is that which could conceptually be obtained by giving task A to all possible technicians. To compare the probability distributions for populations A and B, first we rank the sample observations as though they were all drawn from the same population. This is, we pool the measurements from both samples, and then rank the measurements from the smallest (a rank of 1) to the largest (a rank of 13). The results of this ranking process are shown in Table 12.1.

If the two populations were identical, we would expect the ranks to be randomly mixed between the two samples. On the other hand if one population tends to have larger completion times than the other, we would expect the larger ranks to be mostly in one sample and the smaller ranks mostly in the other. Thus the test statistic for the Wilcoxon test is based on the totals of the ranks for each of the two samples, that is, on the rank sums. For example, when the sample sizes are equal, the greater the difference in the rank sums, the greater will be the weight of evidence to indicate a difference between the probability distributions for populations A and B. In the reaction times example we denote the rank sum for task A by T_A and that for task B by T_B. Then

$$T_A = 4 + 7 + 2 + 8 + 1 + 3 = 25$$

and

$$T_B = 6 + 9 + 5 + 11 + 10 + 12 + 13 = 66$$

The sum of T_A and T_B will always equal $n(n + 1)/2$, where $n = n_1 + n_2$. So for this example $n_1 = 6$, $n_2 = 7$, and

$$T_A + T_B = \frac{13(13 + 1)}{2} = 91$$

Since $T_A + T_B$ is fixed, a small value for T_A implies a large value for T_B (and vice versa) and a large difference between T_A and T_B. Therefore the smaller the value of one of the rank sums, the greater will be the evidence to indicate that the samples were selected from different populations.

Values that locate the rejection region for the rank sum associated with the smaller sample are given in Table 9 in the Appendix. The columns of the table represent n_1, the first sample size, and the rows represent n_2, the second sample size. The T_L and T_U entries in the table are the boundaries of the lower and upper regions, respectively, for the rank sum associated with the sample that has fewer measurements. If the sample sizes n_1 and n_2 are the same, either rank sum may be used as the test statistic. To illustrate, suppose $n_1 = 8$ and $n_2 = 10$. For a two-tailed test with $\alpha = 0.05$ we consult the table and find that the null hypothesis will be rejected if the rank sum of sample 1 (the sample with fewer measurements), T_A, is less than or equal to $T_L = 54$ or greater than or equal to $T_U = 98$. The Wilcoxon rank sum test is summarized here.

WILCOXON RANK SUM TEST: INDEPENDENT SAMPLES[1]

One-Tailed Test	Two-Tailed Test
H_0: Two sampled populations have identical probability distributions	H_0: Two sampled populations have identical probability distributions

[1]Another statistic used for comparing two populations based on independent random samples is the Mann-Whitney U statistic. The U statistic is a simple function of the rank sums. It can be shown that the Wilcoxon rank sum test and the Mann-Whitney U test are equivalent.

H_a: The distribution for population A is shifted to the right of that for B

H_a: The probability distribution for population A is shifted to the left *or* to the right of that for B

Test statistic:

The rank sum T associated with the sample with fewer measurements (if sample sizes are equal, either rank sum can be used)

Test statistic:

The rank sum T associated with the sample with fewer measurements (if the sample sizes are equal, either rank sum can be used)

Rejection region:

Assuming the smaller sample size is associated with distribution A (or, if sample sizes are equal, we use the rank sum T_A), we reject if

$$T_A \geqslant T_U$$

where T_U is the upper value given by Table 10 of the Appendix for the chosen *one-tailed* α value.

[*Note*: if the one-sided alternative is that the probability distribution for A is shifted to the *left* of B (and T_A is the test statistic), we reject if $T_A \leqslant T_L$.]

Rejection region:

$T \leqslant T_L$ or $T \geqslant T_U$, where T_L is the lower value given by Table 9 in the Appendix for the chosen *two-tailed* α value, and T_U is the upper value from Table 10

Assumptions:
1. The two samples are random and independent.
2. The observations obtained can be ranked in order of magnitude.
   [*Note*: No assumptions have to be made about the shape of the population probability distributions.]

EXAMPLE 12.1

Do the data given in Table 12.1 provide sufficient evidence to indicate a shift in the probability distributions for tasks A and B? That is, can we say that the probability distribution corresponding to task A lies either to the right or left of the probability distribution corresponding to task B? Test at the 0.05 level of significance.

Solution For this problem we have

H_0: The two populations of completion times corresponding to task A and task B have the same probability distribution.

H_a: The probability distribution for task A is shifted to the right or left of the probability distribution corresponding to task B.

Test statistic: Since task A has fewer subjects than task B, the test statistic is T_A, the rank sum of task A's completion times.

Rejection region: Since the test is two-sided, we consult part (a) of Table 9 for the rejection region corresponding to $\alpha = 0.05$. We will reject H_0 for $T_A \leqslant T_L$ or $T_A \geqslant T_U$. Thus we will reject H_0 if $T_A \leqslant 28$ or $T_A \geqslant 56$.

Since T_A, the rank sum of task A's completion times in Table 12.1, is 25, it is in the rejection region (see Figure 12.3).[2] Therefore we can conclude that the probability distributions for tasks A and B are not identical. In fact it appears that task B tends to be associated with completion times that are larger than those associated with task A (because T_A fell in the lower tail of the rejection region).

■

FIGURE 12.3
Alternative Hypothesis and Rejection Region for Example 12.1

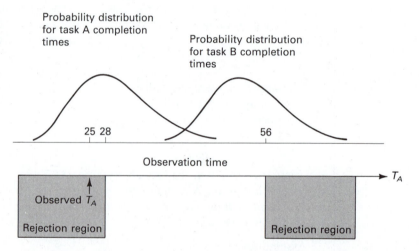

When you apply the Wilcoxon rank sum test in a practical situation, you may encounter one or more ties in the observations. A tie occurs when two of the sample observations are equal. The Wilcoxon rank sum test is still valid if the number of ties is small in comparison with the number of sample measurements, and if you assign to each tied observation the average of the ranks the two observations would have received if the observations had not been tied. For example, suppose the fourth and fifth smallest observations are tied. Since these observations would have received the ranks 4 and 5, you should assign the average of these ranks, 4.5, to both of them and then proceed with the test in the usual manner.

Table 9 of the Appendix gives values of T_L and T_U for the sample sizes n_1 and n_2 less than or equal to 10. When both sample sizes are 10 or larger, the sampling distribution of T_A can be approximated by a normal distribution with mean and variance

---

[2]The figure depicts only one side of the two-sided alternative hypothesis. The other would show distribution A shifted to the right of distribution B.

$$E(T_A) = \frac{n_1(n_1 + n_2 + 1)}{2} \quad \text{and} \quad \sigma_{T_A}^2 = \frac{n_1 n_2 (n_1 + n_2 + 1)}{12}$$

Therefore for $n_1 \geqslant 10$ and $n_2 \geqslant 10$ we can conduct the Wilcoxon rank sum test using a standard normal (z) test statistic, as summarized in the box.

WILCOXON RANK SUM TEST:
LARGE INDEPENDENT SAMPLES

One-Tailed Test	Two-Tailed Test
H_0: Two sampled populations have identical probability distributions	H_0: Two sampled populations have identical probability distributions
H_a: The probability distribution for population A is shifted to the right of that for B	H_a: The probability distribution for population A is shifted to the left *or* to the right of that for B

Test statistic:

$$Z = \frac{T_A - \dfrac{n_1(n_1 + n_2 + 1)}{2}}{\sqrt{\dfrac{n_1 n_2 (n_1 + n_2 + 1)}{12}}}$$

Rejection region: $z > z_\alpha$

Test statistic:

$$Z = \frac{T_A - \dfrac{n_1(n_1 + n_2 + 1)}{2}}{\sqrt{\dfrac{n_1 n_2 (n_1 + n_2 + 1)}{12}}}$$

Rejection region:

$$z < -z_{\alpha/2} \text{ or } z > z_{\alpha/2}$$

Assumptions: $n_1 \geqslant 10$ and $n_2 \geqslant 10$
Independent, random samples

EXERCISES

12.1 Suppose you wish to compare two treatments, A and B, and you want to determine whether the distribution of the population of B measurements is shifted to the right of the distribution of the population of A measurements. If $n_1 = 7$, $n_2 = 5$, and $\alpha = 0.05$, give the rejection region for the test.

12.2 Refer to Exercise 12.1. Suppose you wish to detect a shift in the distributions, either A to the right of B or vice versa. Locate the rejection region for the test (assume $n_1 = 7$, $n_2 = 5$, and $\alpha = 0.05$).

12.3 An industrial psychologist claims that the order in which test questions are presented affects a prospective employee's chance of answering correctly. To investigate this assertion, the psychologist randomly divides thirteen applicants into two groups, seven in one and six in the other. The test questions are arranged in order of increasing difficulty on test A, but the order is reversed on test B. One group of applicants is given test A and the other test B. The resulting scores are as shown below:

Test A: 90, 71, 83, 82, 75, 91, 65

Test B: 66, 78, 50, 68, 80, 60

Do the data provide sufficient evidence to indicate a difference between the two tests? Test by using $\alpha = 0.05$.

12.4 A major razor blade manufacturer advertises that its twin-blade disposable razor will "get you a lot more shaves" than any single-blade disposable razor on the market. A rival blade company, which has been very successful in selling single-blade razors, wishes to test this claim. Independent random samples of eight single-blade shavers and eight twin-blade shavers are taken, and the number of shaves that each gets before the razor is disposed of is recorded. The results are shown below:

Number of Shaves	
Twin Blades	Single Blades
8	10
17	6
9	3
11	7
15	13
10	14
6	5
12	7

(a) Do the data support the twin-blade manufacturer's claim? Use $\alpha = 0.05$.

(b) Do you think that this experiment was designed in the best possible way? If not, what design might have been better?

12.5 Six specimens were independently selected from each of two processes for producing plastic. The ultimate strength measurements (in 1000 psi) for the specimens were as follows:

Plastic A	Plastic B
15.8	18.3
16.3	22.5
18.9	19.6
17.6	21.3
21.4	20.9
16.9	19.8

Do the ultimate strengths for the plastics appear to differ in location? Use $\alpha = 0.05$. (Note that these are *maximum* strength measurements, and such measurements usually do not follow the normal distribution.)

12.6 Counts on the number of defects per square yard for fabrics woven on two different machines were recorded for independently selected samples from each machine. The data are as follows:

Machine A: 3,   0,   4,   9,   25,   10,   8,   14

Machine B: 12,   21,   2,   3,   7

(a) Is there sufficient evidence to suggest a difference in location for the defect populations of the two machines, at the 10% significance level?

(b) Discuss a possible parametric analysis designed to answer the question in (a).

12.7 Electronic components of similar type but from two different suppliers, A and B, are operating in an industrial system. Five new components from each supplier are put into the system at the same time. As the components fail and are replaced, over time, their supplier is duly noted. The sequence of failures, ordered in time, is:

ABBABBBBAAA

Is there evidence to suggest a difference between the two suppliers? Use $\alpha = 0.10$.

12.8 A *management information system* (MIS) is a computer-based information processing system that is designed to support the operations, management, and decision functions of an organization. The development of an MIS involves three stages: definition of the system, physical design of the system, and implementation of the system (Davis, *Management Information Systems*, McGraw-Hill, 1974). Steven Alter and Michael Ginzberg ("Managing Uncertainty in MIS Implementation," *Sloan Management Review*, pp. 23–31, Fall 1978) have shown that the successful implementation of an MIS is related to the quality of the entire development process. The implementation of an MIS could fail for many reasons. For example failure could be due to inadequate planning by and negotiating between the designers and the future users of the system prior to the construction of the system. Or it could fail simply because members of the organization were improperly trained to use the system effectively.

Thirty firms that recently implemented an MIS were surveyed; 16 were satisfied with the implementation results, 14 were not. These firms were asked to rate the quality of the planning and negotiation stages of the development process. Quality was to be rated on a scale of 0 to 100 with higher numbers indicating better quality. In particular a score of 100 indicates all the problems that occurred in the planning and negotiation stages appeared to have been resolved. The following results were obtained:

Firms with Successful MIS's	Firms with Unsuccessful MIS's	
52	60	
70	50	
40	55	
80	70	
82	41	
65	40	*(continued)*

Firms with Successful MIS's	Firms with Unsuccessful MIS's
59	55
60	65
90	55
75	70
80	90
95	85
90	80
86	90
95	
93	

Source: Based on Alter and Ginzberg (1978).

(a) Ginzberg used the Mann-Whitney U test, a procedure equivalent to the Wilcoxon rank sum test, on data similar to those above to investigate differences in the quality of the development processes of successfully and unsuccessfully implemented MIS's. Use the Wilcoxon rank sum test (the large-sample procedure) to determine whether the distribution of quality scores for successfully implemented systems lies above the distribution of scores for unsuccessfully implemented systems. Test by using $\alpha = 0.05$.

(b) Under what circumstances could you employ the two-sample t test of Chapter 8 to conduct a similar test comparing the population means?

12.3 COMPARING TWO POPULATIONS: WILCOXON SIGNED RANK TEST FOR THE PAIRED-DIFFERENCE EXPERIMENT

Nonparametric techniques may also be employed to compare two probability distributions when a paired-difference design is used. For example consumer preferences for two competing products are often compared by having each of a sample of consumers rate both products. Thus the ratings have been paired on each consumer. Here is an example of this type of experiment.

For some paper products softness of the paper is an important consideration in determining consumer acceptance. One method of determining softness is to have judges give a softness rating to a sample of the products. Suppose each of ten judges is given a sample of two products that a company wants to compare. Each judge rates the softness of each product on a scale from 1 to 10, with higher ratings implying a softer product. The results of the experiment are shown in Table 12.2.

Since this is a paired-difference experiment, we should analyze the differences between the measurements. The paired-difference t test of Section 8.3 requires that the differences represent a random sample from a normal distribution, an assumption that is probably unwarranted when the measurements are ratings or

TABLE 12.2
Softness Ratings of Paper

Judge	Product A	B	Difference (A − B)	Absolute Value of Difference	Rank Absolute Value
1	6	4	2	2	5
2	8	5	3	3	7.5
3	4	5	−1	1	2
4	9	8	1	1	2
5	4	1	3	3	7.5
6	7	9	−2	2	5
7	6	2	4	4	9
8	5	3	2	2	5
9	6	7	−1	1	2
10	8	2	6	6	10

$$T_+ = \text{sum of positive ranks} = 46$$
$$T_- = \text{sum of negative ranks} = 9$$

ranks. A nonparametric approach is to calculate the ranks of the absolute values of the differences between the measurements, that is, the ranks of the differences after removing any minus signs. Note that tied absolute differences are assigned the average of the ranks they would receive if they were unequal but successive measurements. After the absolute differences are ranked, the sum of the ranks of the positive differences T_+ and the sum of the ranks of the negative differences T_- are computed.

Now we are prepared to test the nonparametric hypothesis:

H_0: The probability distributions of the ratings for products A and B are identical.

H_a: The probability distributions of the ratings differ (in location) for the two products. (Note that this is a two-sided alternative and therefore that it implies a two-tailed test.)

Test statistic: $T =$ Smaller of the positive and negative rank sums T_+ and T_-.

The smaller the value of T, the greater will be the evidence to indicate that the two probability distributions differ in location. The rejection region for T can be determined by consulting Table 10 in the Appendix. This table gives a value T_0 for both one- and two-tailed tests for each value of n, the number of matched pairs. For a two-tailed test with $\alpha = 0.05$ we will reject H_0 if $T \leqslant T_0$. You can see from the table that the value of T_0 that locates the rejection region for the judges' ratings for $\alpha = 0.05$ and $n = 10$ pairs of observations is 8. Therefore the rejection region for the test (see Figure 12.4) is

Rejection region: $T \leqslant 8$ for $\alpha = 0.05$

Since the smaller rank sum for the paper data $T_- = 9$ does not fall within the rejection region, the experiment has not provided sufficient evidence to indicate

FIGURE 12.4
*Rejection
Region for
Paired-Difference
Experiment*

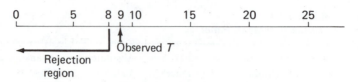

that the two paper products differ with respect to their softness ratings at the $\alpha = 0.05$ level.

Note that if a significance level of $\alpha = 0.10$ had been used, the rejection region would have been $T \leqslant 11$, and we would have rejected H_0. In other words the samples do provide evidence that the probability distributions of the softness ratings differ at the $\alpha = 0.10$ significance level.

When the Wilcoxon signed rank test is applied to a set of data, it is possible that one (or more) of the paired differences may equal 0. The Wilcoxon test will continue to be valid if the number of zeros is small in comparison to the number of pairs, but to perform the test you must delete the zeros and reduce the number of differences accordingly. For example if you have $n = 12$ pairs and two of the differences equal 0, you should delete these pairs, rank the remaining $n = 10$ differences, and use $n = 10$ when locating T_0 in Table 10. Ties in ranks are treated in the same manner as for the Wilcoxon rank sum test for a completely randomized design. Assign to each of the tied ranks the average of the ranks the two observations would have received if the observations had not been tied.

The Wilcoxon signed rank test for a paired-difference experiment is summarized in the box below.

WILCOXON SIGNED RANK TEST FOR A ————
PAIRED-DIFFERENCE EXPERIMENT

One-Tailed Test	Two-Tailed Test
H_0: Two sampled populations have identical probability distributions	H_0: Two sampled populations have identical probability distributions
H_a: The probability distribution for population A is shifted to the right of that for population B	H_a: The probability distribution for population A is shifted to the right *or* to the left of that for population B
Test statistic: T_-, the negative rank sum (we assume the differences are computed by subtracting each paired B measurement from the corresponding A measurement)	Test statistic: T, the smaller of the positive and negative rank sums T_+ and T_-

Rejection region:
$T_- \leqslant T_0$, where T_0 is found in Table 10 in the Appendix for the one-tailed significance level α and the number of untied pairs n.

Rejection region:
$T \leqslant T_0$, where T_0 is found in Table 10 in the Appendix for the two-tailed significance level α and the number of untied pairs n.

[*Note*: If the alternative hypothesis is that the probability distribution for A is shifted to the left of B, we use T_+ as the test statistic and reject H_0 if $T_+ \leqslant T_0$.]

Assumptions:
1. A random sample of pairs of observations has been taken.
2. The absolute differences in the paired observations can be ranked. [*Note*: No assumptions have to be made about the shape of the population probability distributions.]

EXAMPLE 12.2

The president of a corporation must choose between two plans for improving employee safety: plan A and plan B. To aid in reaching a decision, both plans are examined by ten safety experts, each of whom is asked to rate the plans on a scale from one to ten (high ratings imply a better plan). The corporation will adopt plan B, which is more expensive, only if the data provide evidence that the safety experts rate plan B higher than plan A.

The results of the study are shown in Table 12.3. Do the data provide evidence at the $\alpha = 0.05$ level that the distribution of ratings for plan B lies above that for plan A?

Solution The null and alternative hypotheses are

H_0: The two probability distributions of ratings are identical

H_a: The ratings of the more expensive plan (B) tend to exceed those of plan A

Observe that the alternative hypothesis is one-sided (that is, we wish to detect a shift only in the distribution of the B ratings to the right of the distribution of A ratings), and therefore it implies a one-tailed test of the null hypothesis (see Figure 12.5). When the alternative hypothesis is true, the B ratings will tend to be larger than their paired A ratings, more negative differences in pairs will occur, T_- will be large, and T_+ will be small. Because Table 10 is constructed to give lower-tail values of T_0, we will use T_+ as the test statistic and reject H_0 for $T_+ \leqslant T_0$. The differences in ratings for the pairs (A − B) are shown in Table 12.3. Note that one of

TABLE 12.3
Ratings by Ten Qualified Safety Experts

Safety Expert	Plan A	Plan B	Difference $(A - B)$	Rank of Absolute Difference
1	7	9	−2	4.5
2	4	5	−1	2
3	8	8	0	(Eliminated)
4	9	8	1	2
5	3	6	−3	6
6	6	10	−4	7.5
7	8	9	−1	2
8	10	8	2	4.5
9	9	4	5	9
10	5	9	−4	7.5

Positive rank sum $= T_+ = 15.5$

FIGURE 12.5
The Research Hypothesis for Example 12.2: We Expect T_+ to be Small

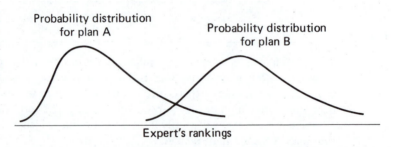

Probability distribution for plan A

Probability distribution for plan B

Expert's rankings

the differences equals 0. Consequently we eliminate this pair from the ranking and reduce the number of pairs to $n = 9$. Using this value to enter Table 10 in the Appendix, you see that for a one-tailed test with $\alpha = 0.05$ and $n = 9$, $T_0 = 8$. Therefore the test statistic and rejection region for the test are

Test statistic: T_+, the positive rank sum

Rejection region: $T_+ \leqslant 8$

Summing the ranks of the positive differences from Table 12.3, we find $T_+ = 15.5$. Since this value exceeds the critical value $T_0 = 8$, we conclude that this sample provides insufficient evidence at the $\alpha = 0.05$ level to support the research hypothesis. The president cannot conclude that plan B is rated higher than plan A. ∎

As is the case for the rank sum test for independent samples, the sampling distribution of the signed rank statistic can be approximated by a normal distribution when the number n of paired observations is large (say, $n \geqslant 25$). The large-sample z test is summarized as follows.

WILCOXON SIGNED RANK TEST FOR A
PAIRED-DIFFERENCE EXPERIMENT: LARGE SAMPLE

One-Tailed Test	Two-Tailed Test

H_0: Two sampled populations have identical probability distributions

H_a: The probability distribution for population A is shifted to the right of that for population B

Test statistic:

$$z = \frac{T_+ - \dfrac{n(n+1)}{4}}{\sqrt{\dfrac{n(n+1)(2n+1)}{24}}}$$

Rejection Region: $z > z_\alpha$

Assumptions: $n \geqslant 25$

H_0: Two sampled populations have identical probability distributions

H_a: The probability distribution for population A is shifted to the right *or* to the left of that for population B

Test statistic:

$$z = \frac{T_+ - \dfrac{n(n+1)}{4}}{\sqrt{\dfrac{n(n+1)(2n+1)}{24}}}$$

Rejection Region: $z < -z_{\alpha/2}$ or $z > z_{\alpha/2}$

EXERCISES

12.9 Suppose you wish to test a hypothesis that two treatments, A and B, are equivalent against the alternative that the responses for A tend to be larger than those for B. If $n = 9$ and $\alpha = 0.01$, give the rejection region for a Wilcoxon signed rank test.

12.10 Refer to Exercise 12.9. Suppose you wish to detect a difference in the locations of the distributions of the responses for A and B. If $n = 7$ and $\alpha = 0.10$, give the rejection region for the Wilcoxon signed rank test.

12.11 Economic indices provide measures of economic change. The January 23, 1984, issue of *U.S. News & World Report* listed the following indices for the first week of January 1984 and the first week of January 1983. By comparing these two sets of indices, you can obtain information regarding changes in the economy that occurred during 1983.

	January 1984	January 1983
Steel Production	66.3	47.2
Automobile Production	107.8	66.4
Crude Petroleum Production	97.8	98.3
Lumber Production	59.7	91.4
Freight Car Loadings	57.2	50.8
Electric Power Production	217.1	190.8

(a)   Conduct a paired-difference *t* test to compare the mean values of these indices for January 1983 and January 1984. Use $\alpha = 0.05$. What assumptions are necessary for the validity of this procedure? Why might these assumptions be doubtful?

(b)   Use the Wilcoxon signed rank test to determine whether these data provide evidence that the probability distribution of the economic indices has changed. Use $\alpha = 0.05$.

12.12   On clear, cold nights in the central Florida citrus region, the precise location of the below-freezing temperatures is important since the methods of protecting trees from freezing conditions are very expensive. One method of locating likely cold spots is by relating temperature to elevation. It is conjectured that, on calm nights, the cold spots will be at low elevation. The highest and lowest spots in a particular grove showed the following minimum temperatures for ten cold nights in a recent winter.

Night	1	2	3	4	5	6	7	8	9	10
High Elevation	32.9	33.2	32.0	33.1	33.5	34.6	32.1	33.1	30.2	29.1
Low Elevation	31.8	31.9	29.2	33.2	33.0	33.9	31.0	32.5	28.9	28.0

(a)   Is there sufficient evidence to support the conjecture that low elevations tend to be colder?

(b)   Would it be reasonable to use a *t* test on the above data? Why or why not?

12.13   The following data set lists the number of industrial accidents in twelve manufacturing plants for one-week periods before and after an intensive promotion on safety:

Plant	1	2	3	4	5	6	7	8	9	10	11	12
Before	3	4	6	3	4	5	5	3	2	4	4	5
After	2	1	3	5	4	2	3	3	0	3	1	2

(a)   Do the data support the claim that the campaign was successful in reducing accidents? Use $\alpha = 0.05$.

(b)   Discuss the problems associated with a parametric analysis designed to answer the question in (a).

12.14   Dental researchers have developed a new material for preventing cavities, a plastic sealant, which is applied to the chewing surfaces of teeth. To determine whether the sealant is effective, it was applied in half of the teeth of each of twelve school-age children. After 5 years, the number of cavities in the sealant-coated teeth and untreated teeth were counted. The results are given in the accompanying table. Is there sufficient evidence to indicate that sealant-coated teeth are less prone to cavities than are untreated teeth? Test by using $\alpha = 0.05$.

Child	Sealant-Coated	Untreated	Child	Sealant-Coated	Untreated
1	3	3	7	1	5
2	1	3	8	2	0
3	0	2	9	1	6
4	4	5	10	0	0
5	1	0	11	0	3
6	0	1	12	4	3

12.15 A food vending company currently uses vending machines made by two different manufacturers. Before purchasing new machines, the company wants to compare the two types in terms of reliability. Records for 7 weeks are given in the accompanying table; the data indicate the number of breakdowns per week for each type of machine. The company has the same number of machines of each type. Do the data present sufficient evidence to indicate that one of the machine types is less prone to breakdowns than the other? Test by using $\alpha = 0.05$.

Week	Machine Type A	B
1	14	12
2	17	13
3	10	14
4	15	12
5	14	9
6	9	11
7	12	11

12.16 According to the American Bar Association, in 1981 there were 498,249 lawyers in the United States. About 73% of these lawyers were in private practice, about 15% worked in government as judges, prosecutors, legislators, and so forth, and about 10% worked for businesses. Because of mushrooming government regulation, high outside legal fees, and complex litigation, the number of corporate lawyers has been growing at a rapid pace. The following salary data were collected for a sample of ten U.S. cities.

Average Salary for Lawyer with 8 Years' Experience

	Corporate Lawyers	Lawyers with Law Firms
Atlanta	$35,000	$34,000
Boston	$32,500	$35,500
Cincinnati	$34,000	$30,500
Des Moines	$32,000	$39,000
Houston	$36,000	$35,500
Los Angeles	$45,000	$45,500

(continued)

Average Salary for Lawyer with 8 Years' Experience (Cont.)

	Corporate Lawyers	Lawyers with Law Firms
Milwaukee	$36,500	$37,000
New York	$42,000	$50,000
Pittsburgh	$34,500	$34,000
San Francisco	$31,500	$35,500

Source: *The American Almanac of Jobs and Salaries*, 1982, pp. 379–389.

(a) Use the Wilcoxon signed rank test to determine whether the data provide sufficient evidence to conclude that the salaries of corporate lawyers differ from those of lawyers working for law firms. Test by using $\alpha = 0.05$.

(b) Under what circumstances would it be appropriate to conduct the test in part (a) by using the paired-difference t test described in Chapter 8?

12.17 Traditionally jobs in the United States have required employees to perform their work during a fixed 8-hour workday. A recent job scheduling innovation that is helping managers to overcome the motivation and absenteeism problems associated with the fixed workday is a concept called *flextime*. Flextime is a flexible-working-hours program that permits employees to design their own 40-hour workweek (Certo, 1980). The management of a large manufacturing firm is considering adopting a flextime program for its hourly employees and has decided to base the decision on the success or failure of a pilot flextime program. Ten employees were randomly selected and given a questionnaire designed to measure their attitude toward their job. These same people were then permitted to design and follow a flextime workday. After 6 months attitudes toward their jobs were again measured. The resulting attitude scores are displayed in the table. The higher the score, the more favorable is the employee's attitude toward his or her work. Use a nonparametric test procedure to evaluate the success of the pilot flextime program. Test by using $\alpha = 0.05$.

Employee	Before Flextime	After Flextime
1	54	68
2	25	42
3	80	80
4	76	91
5	63	70
6	82	88
7	94	90
8	72	81
9	33	39
10	90	93

Source: Certo, *Principles of Modern Management*, Wm. C. Brown, 1980.

12.4 KRUSKAL-WALLIS H TEST FOR A COMPLETELY RANDOMIZED DESIGN

In Chapter 11 we used an analysis of variance and the F test to compare the means of k populations based on random sampling from populations that were normally distributed with a common variance σ^2. Now we present a nonparametric technique for comparing the populations that requires no assumptions concerning the shape of the population probability distributions.

A quality-control engineer in an electronics plant has sampled the output of three assembly lines and recorded the number of defects observed. The samples involve the entire output of the three lines for ten randomly selected hours from a given week. (Different hours were selected for each line.) The number of defects observed could be quite large since a specific inspected component could have many different defects (see Table 12.4). It is quite unlikely that the data would follow a normal distribution, and thus a nonparametric analysis is in order. Our comparison is based on the rank sums for the three sets of sample data.

TABLE 12.4
Number of Defects

Line 1		Line 2		Line 3	
Defects	Rank	Defects	Rank	Defects	Rank
6	5	34	25	13	9.5
38	27	28	19	35	26
3	2	42	30	19	15
17	13	13	9.5	4	3
11	8	40	29	29	20
30	21	31	22	0	1
15	11	9	7	7	6
16	12	32	23	33	24
25	17	39	28	18	14
5	4	27	18	24	16
	$R_1 = 120$		$R_2 = 210.5$		$R_3 = 134.5$

Just as with two independent samples (Section 12.2), the ranks are computed for each observation according to the relative magnitude of the measurements when the data for all the samples are combined (see Table 12.4). Ties are treated as they were for the Wilcoxon rank sum and signed rank tests by assigning the average value of the ranks to each of the tied observations. We test

H_0: The probability distributions of the number of defects are the same for all three assembly lines.

H_a: At least two of the probability distributions differ in location.

If we denote the three sample rank sums by R_1, R_2, R_3, the test statistic is given by

$$H = \frac{12}{n(n+1)} \sum \frac{R_j^2}{n_j} - 3(n+1)$$

where n_j is the number of measurements in the jth sample and n is the total sample size $(n = n_1 + n_2 + \cdots + n_k)$. For the data in Table 12.4, we have $n_1 = n_2 = n_3 = 10$ and $n = 30$. The rank sums are $R_1 = 120$, $R_2 = 210.5$, and $R_3 = 134.5$. Thus

$$H = \frac{12}{30(31)} \left[\frac{(120)^2}{10} + \frac{(210.5)^2}{10} + \frac{(134.5)^2}{10} \right] - 3(31)$$

$$= 99.097 - 93 = 6.097$$

If the null hypothesis is true, the distribution of H in repeated sampling is approximately a χ^2 (chi-square) distribution. This approximation for the sampling distribution of H is adequate as long as each of the k sample sizes exceeds five (see the references for more detail). The χ^2 probability distribution (see Section 5.5) is characterized by a single parameter, called the degrees of freedom associated with the distribution. Several χ^2 probability distributions with different degrees of freedom are shown in Figure 12.6. The degrees of freedom corresponding to the approximate sampling distribution of H will always be $(k-1)$, 1 less than the number of probability distributions being compared. Because large values of H support the research hypothesis that the populations have different probability distributions, the rejection region for the test will be located in the upper tail of the χ^2 distribution.

FIGURE 12.6
Several χ^2
Probability
Distributions

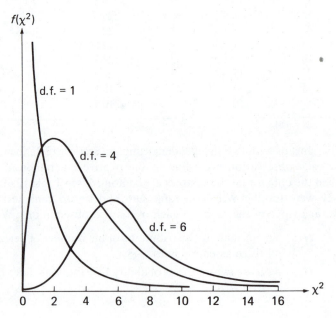

For the data of Table 12.4 the approximate distribution of the test statistic H is a χ^2 with $(k - 1) = 2$ df. To determine how large H must be before we reject the null hypothesis, we consult Table 6 in the Appendix. Entries in the table give an upper-tail value of χ^2, call it χ_α^2, such that $P(\chi^2 > \chi_\alpha^2) = \alpha$. The columns of the table identify the value of α associated with the tabulated value of χ_α^2, and the rows correspond to the degrees of freedom. Thus for $\alpha = 0.05$ and df $= 2$ we can reject the null hypothesis that the three probability distributions are the same if

$$H > \chi_{0.05}^2(2) \qquad \text{where } \chi_{0.05}^2(2) = 5.99147$$

The rejection region is pictured in Figure 12.7. Since the calculated $H = 6.097$ exceeds the critical value of 5.99147, we conclude that at least one of the three lines tends to have a larger number of defects than the others.

FIGURE 12.7
Rejection Region for the Comparison of Three Probability Distributions

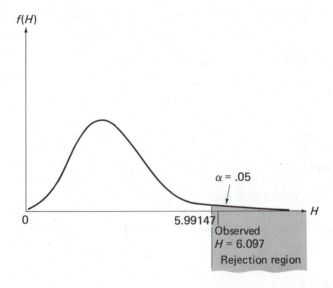

Note that prior to conducting the experiment, we might have decided to compare the defect rates for a specific pair of lines. The Wilcoxon rank sum test presented in Section 12.2 could be used for this purpose.

The Kruskal-Wallis H test for comparing more than two probability distributions is summarized as follows.

KRUSKAL-WALLIS H TEST FOR COMPARING k PROBABILITY DISTRIBUTIONS

H_0: The k probability distributions are identical

H_a: At least two of the k probability distributions differ in location

Test statistic: $H = \dfrac{12}{n(n + 1)} \sum \dfrac{R_j^2}{n_j} - 3(n + 1)$

where

n_j = number of measurements in sample j

R_j = rank sum for sample j, where the rank of each measurement is computed according to its relative magnitude in the totality of data for the k samples

n = total sample size = $n_1 + n_2 + \cdots + n_k$

Assumptions:

1. The k samples are random and independent.
2. There are five or more measurements in each sample.
3. The observations can be ranked.

[*Note*: No assumptions have to be made about the shape of the population probability distributions.]

Rejection region: $H > \chi^2_\alpha(k - 1)$

EXAMPLE 12.3
_____

An electrical engineer wants to compare the lifelengths of four brands of capacitors. He does so by randomly selecting four capacitors of each brand and placing them on test, under identical conditions. Since some of the lifelengths are extremely long, relative to the others, the normal distribution will not serve well as a population model. Thus a rank analysis is suggested, and the data in Table 12.5 are the ranks of the actual observations. Do the data provide sufficient evidence to indicate that at least one of the brands tends to have longer lifelengths than the others? Test at the 5% level of significance.

TABLE 12.5
Ranks for Lifelength Data

Brand 1	Brand 2	Brand 3	Brand 4
1	12	8	14
5	2	9	15
6	17	3	16
7	19	11	4
10	20	13	18
$R_1 = 29$	$R_2 = 70$	$R_3 = 44$	$R_4 = 67$

Solution The elements of the test are as follows:

H_0: The population probability distributions of lifelengths for the four brands are identical.

H_a: At least two of the brands have probability distributions with different locations.

Test statistic: $H = \dfrac{12}{n(n+1)} \sum \dfrac{R_j^2}{n_j} - 3(n+1)$

$$= \dfrac{12}{20(21)} \left[\dfrac{(29)^2}{5} + \dfrac{(70)^2}{5} + \dfrac{(44)^2}{5} + \dfrac{(67)^2}{5} \right] - 3(21)$$

$$= 69.5 - 63 = 6.5$$

Rejection region: Since we are comparing four probability distributions, there are $(4 - 1) = 3\,df$ associated with the test statistic. Thus we will reject H_0 if $H > \chi_{0.05}^2(3) = 7.81473$.

Since 6.5 is less than 7.81473, we have insufficient evidence to indicate that one or more of the brands tends to have longer lifelengths than the others. ∎

EXERCISES

12.18   To investigate possible differences among production rates for three production lines turning out similar items, independent random samples of total production figures were obtained for seven days for each line. The data are as follows:

Line 1	Line 2	Line 3
48	41	18
43	36	42
39	29	28
57	40	38
21	35	15
47	45	33
58	32	31

Do the data provide sufficient evidence to indicate any differences in location for the three sets of production figures at the 5% significance level?

12.19   Refer to Exercise 12.6. Suppose samples from a third machine produced the following data on number of defects per square yard:

Machine C: 22, 24, 26, 25, 28, 30

Is there evidence of significant differences in location among the three sets of defect measurements? Use $\alpha = 0.05$.

12.20   An experiment was conducted to compare the length of time it takes a human to recover from each of three types of influenza—Victoria A, Texas, and Russian. Twenty-one human subjects were selected at random from a group of volunteers and divided into three groups of seven each. Each group was randomly assigned a strain of the virus, and the influenza was induced in the subjects. All the subjects were then cared for under identical conditions, and the recovery time (in days) was recorded. The results are as follows:

Victoria A	Texas	Russian
12	9	7
6	10	3
13	5	7
10	4	5
8	9	6
11	8	4
7	11	8

(a) Do the data provide sufficient evidence to indicate that the recovery times for one (or more) type(s) of influenza tend(s) to be longer than for the other types? Test by using $\alpha = 0.05$.

(b) Do the data provide sufficient evidence to indicate a difference in locations of the distributions of recovery times for the Victoria A and Russian types? Test by using $\alpha = 0.05$.

12.21   The EPA wants to determine whether temperature changes in the ocean's water caused by a nuclear power plant will have a significant effect on the animal life in the region. Recently hatched specimens of a certain species of fish are randomly divided into four groups. The groups are placed in separate simulated ocean environments that are identical in every way except for water temperature. Six months later the specimens are weighed. The results (in ounces) are given in the table. Do the data provide sufficient evidence to indicate that one (or more) of the temperatures tend(s) to produce larger weight increases than the other temperatures? Test by using $\alpha = 0.10$.

Weights of Specimens

38°F	42°F	46°F	50°F
22	15	14	17
24	21	28	18
16	26	21	13
18	16	19	20
19	25	24	21
	17	23	

12.22   Three different brands of magnetron tubes (the key components in microwave ovens) were subjected to stressful testing, and the number of hours each operated without repair was recorded. Although these times do not represent typical lifelengths, they do indicate how well the tubes can withstand extreme stress:

Brand		
A	B	C
36	49	71
48	33	31
5	60	140
67	2	59
53	55	42

(a) Use the F test for a completely randomized design (Chapter 11) to test the hypothesis that the mean length of life under stress is the same for the three brands. Use $\alpha = 0.05$. What assumptions are necessary for the validity of this procedure? Is there any reason to doubt these assumptions?

(b) Use the Kruskal-Wallis H test to determine whether evidence exists to conclude that the brands of magnetron tubes tend to differ in length of life under stress. Test by using $\alpha = 0.05$.

12.23 A random sample of six senior computer systems analysts was selected from each of three industries: banking, federal government, and retail sales. Their salaries were determined and are recorded in the table. You have been hired to determine whether differences exist among the salary distributions for senior systems analysts in the three industries.

Banking	Federal Government	Retail Sales
$30,000	$28,000	$20,100
24,500	34,000	19,200
27,100	39,000	20,500
26,000	35,000	20,600
23,800	34,100	21,100
25,800	36,200	19,300

Source: Based on *The American Almanac of Jobs and Salaries*, 1982, p. 420.

(a) Under what circumstances would it be appropriate to use the F test for completely randomized design to perform the required analysis?

(b) Which assumption(s) required by the F test are likely to be violated in this problem? Explain.

(c) Use the Kruskal-Wallis H test to determine whether the salary distributions differ among the three industries. Specify your null and alternative hypotheses and state your conclusions in the context of the problem. Use $\alpha = 0.05$.

12.24 CRS, a national car rental company, was interested in comparing the quality of service at its three largest airport locations. Six business travelers who frequently rent from a competitor were randomly selected at each airport and asked to use (free of charge) a CRS car the next time they rented a car at that airport. The travelers were asked to rate the service they received in four categories: (1) timeliness of service, (2) friendliness of employees, (3) cleanliness of the car, and (4) mechanical performance of the car. They were to use a rating scale of 1 to 10 for each category, with higher numbers indicating better service. An average score was computed for each traveler. These averages are reported in the table.

New York Kennedy	Chicago O'Hare	Dallas Dallas-Fort Worth
7.50	7.25	9.25
6.25	6.75	8.75
7.50	5.75	9.00
4.75	8.00	9.25
6.25	5.75	8.75
7.00	6.00	8.75

(a) Use the Kruskal-Wallis H test to determine whether the levels of service ratings differ among the three car rental outlets. Use $\alpha = 0.10$.

(b) What experimental design was used by CRS?

12.5 THE FRIEDMAN F_r TEST FOR A RANDOMIZED BLOCK DESIGN

In Section 11.5 we employed an analysis of variance to compare k population means when the data were collected using a randomized block design. The Friedman F_r test[1] provides another method for testing to detect a shift in the locations of a set of k populations. Like other nonparametric tests it requires no assumptions concerning the nature of the populations other than that you be able to rank the individual observations.

In Section 12.2 we gave an example where a completely randomized design was used to compare the completion times of technicians assigned to one of two tasks. When the completion time varies greatly from person to person, it may be beneficial to employ a randomized block design. Using the technicians as blocks, we would hope to eliminate the variability among persons and thereby increase the amount of information in the experiment. Suppose that three tasks, A, B, and C, are to be compared by using a randomized block design. Each of the technicians is assigned to each task with suitable time lags between the three tasks. The order in which the tasks are completed is randomly determined for each technician. Thus one task would be completed by a technician, a completion time noted, and after a sufficient length of time the second task completed, and so on.

Suppose that six technicians are chosen and that the completion times for each task are shown in Table 12.6. To compare the three tasks, we rank the observations within each technician (block) and then compute the rank sums for each of the tasks (treatments). Tied observations within blocks are handled in the usual manner by assigning the average value of the rank to each of the tied observations.

[1] The Friedman F_r test is the product of the Nobel prize-winning economist, Milton Friedman.

TABLE 12.6
Completion Times for Three Tasks

Technician	Task A	Rank	Task B	Rank	Task C	Rank
1	1.21	1	1.48	2	1.56	3
2	1.63	1	1.85	2	2.01	3
3	1.42	1	2.06	3	1.70	2
4	2.43	2	1.98	1	2.64	3
5	1.16	1	1.27	2	1.48	3
6	1.94	1	2.44	2	2.81	3
		$R_1 = \overline{7}$		$R_2 = \overline{12}$		$R_3 = \overline{17}$

The null and alternative hypotheses are

H_0: The population of completion times are identically distributed for all three tasks

H_a: At least two of the tasks have probability distributions of completion times that differ in location

The Friedman F_r test statistic, which is based on the rank sums for each treatment, is

$$F_r = \frac{12}{bk(k+1)} \sum_{j=1}^{k} R_j^2 - 3b(k+1)$$

where b is the number of blocks, k is the number of treatments, and R_j is the jth rank sum. For the data in Table 12.6

$$F_r = \frac{12}{(6)(3)(4)} [(7)^2 + (12)^2 + (17)^2] - 3(6)(4)$$

$$= 80.33 - 72 = 8.33$$

As with the Kruskal-Wallis statistic, the Friedman F_r statistic has approximately a χ^2 sampling distribution with $(k-1)$ degrees of freedom. Empirical results show the approximation to be adequate if either b (the number of blocks) or k (the number of treatments) exceeds 5. For the task example we will use $\alpha = 0.05$ to form the rejection region,

$$F_r > \chi_{0.05}^2(2) = 5.99147$$

Consequently, because the observed value $F_r = 8.33$ exceeds 5.99147, we conclude that at least two of the three tasks have probability distributions of completion times that differ in location.

The Friedman F_r test for randomized block designs is summarized here.

FRIEDMAN F_r TEST FOR A RANDOMIZED BLOCK DESIGN

H_0: The probability distributions for the k treatments within each block are identical

H_a: At least two of the probability distributions differ in location

Test statistic: $F_r = \dfrac{12}{bk(k+1)} \displaystyle\sum_{j=1}^{k} R_j^2 - 3b(k+1)$

where
b = number of blocks
k = number of treatments
R_j = rank sum of the jth treatment, where the rank of each measurement is computed relative to its position *within its own block*

Assumptions:
1. The treatments are randomly assigned to experimental units within the blocks.
2. The measurements can be ranked within blocks.
3. Either the number of blocks (b) or the number of treatments (k) should exceed 5 for the χ^2 approximation to be adequate.

[*Note*: No assumptions have to be made about the shape of the population probability distributions.]

Rejection region: $F_r > \chi_\alpha^2(k-1)$

EXAMPLE 12.4

Suppose a marketing firm wants to compare the relative effectiveness of three different modes of advertising: direct-mail, newspaper, and magazine ads. For fifteen clients all three modes are used over a 1-year period, and the marketing firm records the year's percentage response to each type of advertising. That is, the firm divides the number of responses to a particular type of advertising by the total number of potential customers reached by the advertisements of that type. The results are shown in Table 12.7. Do these data provide sufficient evidence to indicate a difference in the locations of the probability distributions of response rates? Use the Friedman F_r test at the $\alpha = 0.10$ level of significance.

Solution The fifteen companies act as blocks in this experiment; thus we rank the observations within each company (block), and then compute the rank sums for each of the three types of advertising (treatments). The null and alternative hypotheses are

H_0: The probability distributions for the response rates are the same within each company for all three types of advertising

H_a: At least two of the probability distributions of response rates differ in location

TABLE 12.7
Percent Response to Three Types of Advertising for Fifteen Different Companies

Company	Direct-Mail	Rank	Newspaper	Rank	Magazine	Rank
1	7.3	1	15.7	3	10.1	2
2	9.4	2	18.3	3	8.2	1
3	4.3	1	11.2	3	5.1	2
4	11.3	2	19.1	3	6.5	1
5	3.3	1	9.2	3	8.7	2
6	4.2	1	10.5	3	6.0	2
7	5.9	1	8.7	2	12.3	3
8	6.2	1	14.3	3	11.1	2
9	4.3	2	3.1	1	6.0	3
10	10.0	1	18.8	3	12.1	2
11	2.2	1	5.7	2	6.3	3
12	6.3	2	20.2	3	4.3	1
13	8.0	1	14.1	3	9.1	2
14	7.4	2	6.2	1	18.1	3
15	3.2	1	8.9	3	5.0	2
		$R_1 = 20$		$R_2 = 39$		$R_3 = 31$

Friedman F_r test statistic:

$$F_r = \frac{12}{bk(k+1)} \sum_{j=1}^{k} R_j^2 - 3b(k+1)$$

$$= \frac{12}{(15)(3)(4)} (R_1^2 + R_2^2 + R_3^2) - (3)(15)(4)$$

$$= \frac{12}{(15)(3)(4)} [(20)^2 + (39)^2 + (31)^2] - (3)(15)(4)$$

$$= 192.13 - 180 = 12.13$$

Rejection region: $F_r > \chi^2_{0.10}(2) = 4.60517$

Since the calculated $F_r = 12.13$ exceeds the critical value of 4.60517 (see Figure 12.8), we conclude that the probability distributions of response rates differ in location for at least two of the three types of advertising.

Clearly the assumptions that the measurements are ranked within blocks and that the number of blocks (companies) is greater than 5 are satisfied. However, the experimenter must be sure that the treatments are randomly assigned to blocks. For the procedure to be valid we assume that the three modes of advertising are used in a random order by each company. Note that if this were not true, the difference in the response rates for the three advertising modes might be due to the order in which the modes are used.

FIGURE 12.8
Rejection Region
for the Advertising
Example

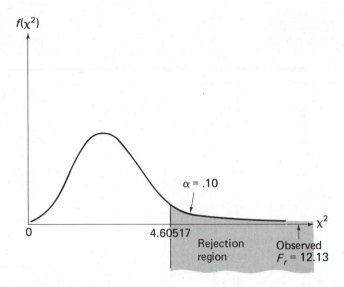

EXERCISES

12.25 Corrosion of different metals is a problem in many mechanical devices. Three sealers used to help retard the corrosion of metals were tested to see whether there were any differences among them. Samples of ten different metal compositions were treated with each of the three sealers, and the amount of corrosion was measured after exposure to the same environmental conditions for 1 month. The data are given in the table. Is there any evidence of a difference in the abilities of the sealers to prevent corrosion? Test by using $\alpha = 0.05$.

Metal	Sealer		
	I	II	III
1	4.6	4.2	4.9
2	7.2	6.4	7.0
3	3.4	3.5	3.4
4	6.2	5.3	5.9
5	8.4	6.8	7.8
6	5.6	4.8	5.7
7	3.7	3.7	4.1
8	6.1	6.2	6.4
9	4.9	4.1	4.2
10	5.2	5.0	5.1

12.26 A serious drought-related problem for farmers is the spread of aflatoxin, a highly toxic substance caused by mold, which contaminates field corn. In higher levels of contamination, aflatoxin is potentially hazardous to animal and possibly human health. (Officials of the FDA have set a maximum limit of 20 parts per billion aflatoxin as safe for interstate marketing.) Three sprays, A, B, and C, have been developed to control aflatoxin in field corn. To determine whether differences exist among the sprays, ten ears of corn are randomly chosen from a contaminated corn field, and each is divided into three pieces of equal size. The sprays are then randomly assigned to the pieces for each ear of corn, thus setting up a randomized block design. The table gives the amount (in parts per billion) of aflatoxin present in the corn samples after spraying. Use the Friedman F_r test to determine whether there are differences among the sprays for control of aflatoxin. Test at the $\alpha = 0.05$ level of significance.

		Spray	
Ear	A	B	C
1	21	23	15
2	29	30	21
3	16	19	18
4	20	19	18
5	13	10	14
6	5	12	6
7	18	18	12
8	26	32	21
9	17	20	9
10	4	10	2

12.27 In recent years domestic car manufacturers have devoted more attention to the small-car market. To compare the popularity of four domestic small cars within a city, a local trade organization obtained the information given in the accompanying table from four car dealers—one dealer for each of the four car makes. Is there evidence of differences in location among the probability distributions of the number of cars sold for each type? Use $\alpha = 0.10$.

Number of Small Cars Sold

		Make of Car		
Month	A	B	C	D
1	9	17	14	8
2	10	20	16	9
3	13	15	19	12
4	11	12	19	11
5	7	18	13	8

12.28 An experiment is conducted to investigate the toxic effect of three chemicals, A, B, and C, on the skin of rats. Three adjacent 1-inch squares are marked on the backs of eight rats, and each of the three chemicals is applied to each rat. The squares of skin are then scored from 0 to 10, depending on the degree of irritation. The data are given in the table. Is there sufficient evidence to support the research hypothesis that the probability distributions of skin irritation scores corresponding to the three chemicals differ in location? Use $\alpha = 0.01$.

		Chemical	
Rat	A	B	C
1	6	5	3
2	9	8	4
3	6	9	3
4	5	8	6
5	7	8	9
6	5	7	6
7	6	7	5
8	6	5	7

12.29 One of the by-products of inflationary times is increased replacement costs of materials. The June 20, 1977 issue of *Business Week* gave a table of ratios of current replacement costs to historical costs in four categories for three well-known pharmaceutical companies. Use the Friedman F_r test to determine whether evidence exists to indicate that the inflationary effects have been felt to different extents by the companies. Test by using $\alpha = 0.05$.[2]

	Gross Assets	Net Assets	Depreciation	Inventory
Bristol-Meyers	1.62	1.58	1.64	1.03
Eli-Lilly	1.74	1.41	1.57	1.00
Pfizer	1.84	1.85	1.67	1.16

12.30 An *optical mark reader* (OMR) is a machine that is able to "read" pencil marks that have been entered on a scannable form. When connected to a computer, such systems are able to read and analyze data in one step. As a result the keypunching of data into a machine-readable code can be eliminated. Eliminating this step reduces the possibility that the data will be contaminated by human error. OMRs are used by schools to grade exams and by survey research organizations to compile data from questionnaires (Oas, 1980). A manufacturer of OMRs believes its product can operate equally well in a variety of temperature and humidity environments. To determine whether operating data contradict this belief, the manufacturer asks a well-known industrial testing laboratory to test its product. Five recently produced OMRs were randomly selected, and each was operated in five different environments. The number of forms each was able to process in an hour was recorded and used as a measure of the OMR's operating efficiency. These data appear in the table. Use the Friedman F_r test to determine whether evidence exists to indicate that the probability distributions for the number of forms processed per hour differ in location for at least two of the environments. Test by using $\alpha = 0.10$.

[2] The true level of α for this test will only approximately equal 0.05, since the χ^2 distribution provides only a rough approximation when neither the number of blocks nor the number of treatments exceeds 5.

Machine	Environment				
Number	1	2	3	4	5
1	8,001	8,025	8,100	8,055	7,991
2	7,910	7,932	7,900	7,990	7,892
3	8,111	8,101	8,201	8,175	8,102
4	7,802	7,820	7,904	7,850	7,819
5	7,500	7,601	7,702	7,633	7,600

Source: Oas, "Processing Collection of Primary Data Simplified by Optimal Mark Reading," *Marketing News*, December 12, 1980.

12.31 In Exercise 11.30 the data below on the number of strikes per year in five U.S. manufacturing industries were treated as if they were generated by a randomized block design, with years used as the blocking variable. And analysis of variance F test was used to investigate whether the mean number of strikes per year differed among industries.

Manufacturing Industry

Year	Food and Kindred Products	Primary Metal Industry	Electrical Equipment and Supplies	Fabricated Metal Supplies	Chemicals and Allied Products
1976	227	197	204	309	129
1977	221	239	199	354	111
1978	171	187	190	360	113
1979	178	202	195	352	143
1980	155	175	140	280	89
1981	109	114	106	203	60

Source: *Statistical Abstract of the U.S.*

(a) What can be learned about the five industries by conducting a Friedman F_r test?

(b) Conduct a Friedman F_r test, using $\alpha = 0.05$. Specify the null and alternative hypotheses and state your conclusion in the context of the problem.

(c) Find the approximate p value for the test of part (b) and interpret its value.

12.6 SPEARMAN'S RANK CORRELATION COEFFICIENT

Suppose that ten applicants for a position as civil engineer are to be ranked from 1 (best) to 10 (worst) by each of two experienced recruiters, say A and B. We might want to determine whether or not the rankings by the two recruiters are related. If an applicant is rated high by recruiter A, is he also likely to be rated high by recruiter B? Or do high rankings by A correspond to low rankings by B? That is, we wish to determine whether the rankings by the two recruiters are correlated.

If the rankings are as shown in the "Percent Agreement" columns of Table 12.8, we immediately notice that the recruiters agree on the rank of every applicant. High ranks correspond to high ranks and low ranks to low ranks. This is an example of *perfect positive correlation* between the ranks.

TABLE 12.8
Rankings of Ten Applicants by Two Recruiters

Applicant	Perfect Agreement		Perfect Disagreement	
	Rec. A	Rec. B	Rec. A	Rec. B
1	4	4	9	2
2	1	1	3	8
3	7	7	5	6
4	5	5	1	10
5	2	2	2	9
6	6	6	10	1
7	8	8	6	5
8	3	3	4	7
9	10	10	8	3
10	9	9	7	4

In contrast, if the rankings appear as shown in the "Perfect Disagreement" columns of Table 12.8, high ranks by one applicant correspond to low ranks by the other. This is an example of *perfect negative correlation*.

In practice you will rarely see perfect positive or negative correlation between the ranks. In fact it is quite possible for the ranks to appear as shown in Table 12.9. You will note that these rankings indicate some agreement between the recruiters but not perfect agreement, thus indicating a need for a measure of rank correlation.

TABLE 12.9
Rankings of Applicants: Less Than Perfect Agreement

Applicant	Recruiter		Difference Between Rank 1 and Rank 2	d^2
	A	B		
1	4	5	-1	1
2	1	2	-1	1
3	9	10	-1	1
4	5	6	-1	1
5	2	1	1	1
6	10	9	1	1
7	7	7	0	0
8	3	3	0	0
9	6	4	2	4
10	8	8	0	0
				$\sum d^2 = \overline{10}$

Spearman's rank correlation coefficient r_s provides a measure of correlation between ranks. The formula for this measure of correlation is given below. We also give a formula that is identical to r_s when there are no ties in rankings; this provides a good approximation to r_s when the number of ties is small relative to the number of pairs.

SPEARMAN'S RANK CORRELATION COEFFICIENT ⎯⎯⎯⎯⎯⎯⎯

$$r_s = \frac{SS_{uv}}{\sqrt{SS_{uu}SS_{vv}}}$$

where

$$SS_{uv} = \sum (u_i - \bar{u})(v_i - \bar{v}) = \sum u_i v_i - \frac{(\sum u_i)(\sum v_i)}{n}$$

$$SS_{uu} = \sum (u_i - \bar{u})^2 = \sum u_i^2 - \frac{(\sum u_i)^2}{n}$$

$$SS_{vv} = \sum (v_i - \bar{v})^2 = \sum v_i^2 - \frac{(\sum v_i)^2}{n}$$

u_i = rank of the ith observation in sample 1
v_i = rank of the ith observation in sample 2
n = number of pairs of observations

Shortcut formula for r_s:

$$r_s = 1 - \frac{6 \sum d_i^2}{n(n^2 - 1)}$$

where

$d_i = u_i - v_i$ (difference in the ranks of the ith observations for samples 1 and 2)

Note that if the rankings by the two recruiters are identical, as in the second and third columns of Table 12.8, the differences between the ranks d will all be zero. Thus

$$r_s = 1 - \frac{6 \sum d^2}{n(n^2 - 1)} = 1 - \frac{6(0)}{10(99)} = 1$$

That is, perfect positive correlation between the pairs of ranks is characterized by a Spearman correlation coefficient of $r_s = 1$. When the ranks indicate perfect disagreement, as in the fourth and fifth columns of Table 12.8, $d_i^2 = 330$ and

$$r_s = 1 - \frac{6(330)}{10(99)} = -1$$

Thus perfect negative correlation is indicated by $r_s = -1$.

For the data of Table 12.9

$$r_s = 1 - \frac{6 \sum d^2}{n(n^2 - 1)} = 1 - \frac{6(10)}{10(99)} = 1 - \frac{6}{99} = 0.94$$

The fact that r_s is close to 1 indicates that recruiters A and B tend to agree, but the agreement is not perfect.

The value of r_s will always fall between -1 and $+1$, with $+1$ indicating the perfect positive correlation and -1 indicating perfect negative correlation. The closer r_s falls to $+1$ or -1, the greater the correlation between the ranks. Conversely the nearer r_s is to 0, the less the correlation. We summarize the properties of r_s.

PROPERTIES OF SPEARMAN'S RANK CORRELATION COEFFICIENT

1. The value of r_s is always between -1 and 1.
2. r_s positive: The ranks of the pairs of sample observations tend to increase together.
3. $r_s = 0$: The ranks are not correlated.
4. r_s negative: The ranks of one variable tend to decrease as the other variable's ranks increase.

You will note that the concept of correlation implies that two responses are obtained for each experimental unit. In the above example each applicant received two ranks (one by each recruiter), and the objective of the study was to determine the degree of positive correlation between the two rankings. Rank correlation methods can be used to measure the correlation between any pair of variables. If two variables are measured on each of n experimental units, we rank the measurements associated with each variable separately. Ties receive the average of the ranks of the tied observations. Then we calculate the value of r_s for the two rankings. This value will measure the rank correlation between the two variables. We illustrate the procedure with Example 12.5.

EXAMPLE 12.5

In a factory producing handcrafted items, certain workers produce many items per day while others produce few. It is hypothesized that those workers producing many items per day have a low quality rating on work produced, while those producing few items have a high quality rating. The data in Table 12.10 show the average number of items produced per day (over a one-month period) by fifteen randomly selected workers. In addition, Table 12.10 shows the average quality rating on inspected items for the same fifteen workers. (High values on the quality rating scale imply high-quality work.) Calculate and interpret Spearman's rank correlation coefficient for the data.

TABLE 12.10
Data and Calculations for Example 12.5

Worker	Items Produced Per Day	Rank	Quality Score	Rank	d	d^2
1	12	1	7.7	5	−4	16
2	15	2	8.1	9	−7	49
3	35	13	6.9	4	9	81
4	21	7	8.2	10	−3	9
5	20	5.5	8.6	13.5	−8	64
6	17	3	8.3	11.5	−8.5	72.25
7	19	4	9.4	15	−11	121
8	46	15	7.8	6	9	81
9	20	5.5	8.3	11.5	−6	36
10	25	8.5	5.2	1	7.5	56.25
11	39	14	6.4	3	11	121
12	25	8.5	7.9	7	1.5	2.25
13	30	12	8.0	8	4	16
14	27	10	6.1	2	8	64
15	29	11	8.6	13.5	−2.5	6.25
					Total =	795

Solution   First we rank the number of items produced per day, assigning a 1 to the smallest (12) and a 15 to the largest number (46). Note that the two ties received the averages of their respective ranks. Similarly we assign ranks to the fifteen quality scores. Since the number of ties is relatively small, we will use the shortcut formula to calculate r_s. The differences between the ranks of the quality scores and the ranks of the number of items produced per day are shown in Table 12.10. The squares of the differences are also given. Thus

$$r_s = 1 - \frac{6 \sum d_i^2}{n(n^2 - 1)} = 1 - \frac{6(795)}{15(15^2 - 1)} = 1 - 1.42 = -0.42$$

This negative correlation coefficient indicates that, in this sample, an increase in the number of items produced per day is associated with a decrease in the quality rating. Can this conclusion be generalized from the sample to the population? That is, can we infer that quality scores and the number of items produced per day are negatively correlated for the populations of observations for all workers?

If we define ρ_s as the population Spearman rank correlation coefficient, this question can be answered by conducting the test

$H_0: \rho_s = 0$ (no population correlation between ranks)

$H_a: \rho_s < 0$ (negative population correlation between ranks)

Test statistic: r_s, the sample Spearman rank correlation coefficient

To determine a rejection region, we consult Table 11 in the Appendix. Note that the left-hand column gives values of n, the number of pairs of observations. The entries in the table are values for an upper-tail rejection region, since only positive values are given. Thus for $n = 15$ and $\alpha = 0.05$, the value 0.441 is the boundary of the upper-tail rejection region, so that $P(r_s > 0.441) = 0.05$ when H_0: $\rho_s = 0$ is true.

Similarly, for negative values of r_s, $P(r_s < -0.441) = 0.05$ when $\rho_s = 0$. That is, we expect to see $r_s < -0.441$ only 5% of the time when there is really no relationship between the ranks of the variables. The lower-tailed rejection region is therefore

Rejection region: $\alpha = 0.05$ $r_s < -0.441$

Since the calculated $r_s = -0.42$ is not less than -0.441, we cannot reject H_0 at the $\alpha = 0.05$ level of significance. That is, this sample of fifteen workers provides insufficient evidence to conclude that a negative correlation exists between number of items produced and the quality scores of the populations of measurements corresponding to all workers. This does not, of course, mean that no relationship exists. A study using a larger sample of workers and taking other factors into account would be more likely to discover whether production rate and quality are related.   ∎

A summary of Spearman's nonparametric test for correlation is given here.

SPEARMAN'S NONPARAMETRIC TEST FOR RANK CORRELATION

One-Tailed Test	Two-Tailed Test
H_0: $\rho_s = 0$ H_a: $\rho_s > 0$ (or H_a: $\rho_s < 0$)	H_0: $\rho_s = 0$ H_a: $\rho_s \neq 0$
Test statistic: r_s, the sample rank correlation (formulas for calculating r_s are given on page 587)	Test statistic: r_s, the sample rank correlation (formulas for calculating r_s are given on page 587)
Rejection region: $\quad r_s > r_{s,\alpha}$ (or $r_s < -r_{s,\alpha}$ when H_a: $\rho_s < 0$) where $r_{s,\alpha}$ is the value from Table 11 in the appendix corresponding to the upper-tail area α and n pairs of observations.	Rejection region: $\quad r_s < -r_{s,\alpha/2}$ or $r_s > r_{s,\alpha/2}$ where $r_{s,\alpha/2}$ is the value from Table 11 in the Appendix corresponding to the upper-tail area $\alpha/2$ and n pairs of observations.

Assumption: The n pairs of observations are randomly sampled from the populations.

EXAMPLE 12.6

Manufacturers of perishable foods often use preservatives to retard spoilage. One concern is that using too much preservative will change the flavor of the food. Suppose an experiment is conducted, using samples of a food product with varying amounts of preservative added. The length of time until the food shows signs of spoiling and a taste rating are recorded for each sample. The taste rating is the average rating for three tasters, each of whom rates each sample on a scale from 1 (bad) to 5 (good). Twelve sample measurements are shown in Table 12.11. Use a nonparametric test to find whether the spoilage times and taste ratings are correlated. Use $\alpha = 0.05$. [*Note*: Tied measurements are assigned the average of the ranks that would be given the measurements if they were different but consecutive.]

TABLE 12.11
Data for Example 12.6

Sample	Time Until Spoilage (Days)	Rank	Taste Rating	Rank
1	30	2	4.3	11
2	47	5	3.6	7.5
3	26	1	4.5	12
4	94	11	2.8	3
5	67	7	3.3	6
6	83	10	2.7	2
7	36	3	4.2	10
8	77	9	3.9	9
9	43	4	3.6	7.5
10	109	12	2.2	1
11	56	6	3.1	5
12	70	8	2.9	4

Solution   The test is two-tailed, with

$$H_0: \rho_s = 0 \qquad H_a: \rho_s \neq 0$$

Test statistic: $r_s = 1 - \dfrac{6 \sum d_i^2}{n(n^2 - 1)}$

Rejection region: Since the test is two-tailed, we need to halve the α value before consulting Table 11 in the Appendix. For $\alpha = 0.05$ we calculate $\alpha/2 = 0.025$ and $n = 12$ pairs of measurements:

$$r_{s,\alpha/2} = r_{s,0.025} = 0.591$$

We will reject H_0 if

$$r_s < -0.591 \quad \text{or} \quad r_s > 0.591$$

The first step in the computation of r_s is to sum the squares of the differences between ranks:

$$\sum d_i^2 = (2 - 11)^2 + (5 - 7.5)^2 + \cdots + (8 - 4)^2 = 536.5$$

Then

$$r_s = 1 - \frac{6(536.5)}{12(144 - 1)} = -0.876$$

Since $-0.876 < -0.591$, we reject H_0 and conclude that the preservative does affect the taste of the food. The fact that r_s is negative suggests that the preservative has an average effect on the taste. ∎

EXERCISES

12.32   Suppose you wish to detect $\rho_s > 0$ and you have $n = 14$ and $\alpha = 0.01$. Give the rejection region for the test.

12.33   Suppose you wish to detect $\rho_s \neq 0$ and you have $n = 22$ and $\alpha = 0.05$. Give the rejection region for the test.

12.34   Two expert design engineers were asked to rank six designs for a new aircraft. The rankings are shown below:

Design	Engineer I	Engineer II
A	6	5
B	5	6
C	1	2
D	3	1
E	2	4
F	4	3

Do the data present sufficient evidence to indicate a positive correlation in the rankings of the two engineers?

12.35   For a certain factory job that requires great skill, it is thought that productivity on the job should increase with an increase in years of experience. Ten employees were randomly selected from among those who hold this type of job. Data on years of experience and a measure of productivity were found to be as follows:

Employee	Years of Experience	Productivity
1	4	80
2	6	82
3	10	88
4	2	81
5	12	92
6	6	85
7	5	83
8	10	86
9	13	91
10	9	90

Do the data support the conjecture that years of experience is positively correlated with productivity?

12.36 Many large businesses send representatives to college campuses to conduct job interviews. To aid the interviewer, one company decides to study the correlation between the strength of an applicant's references (the company requires three references) and the performance of the applicant on the job. Eight recently hired employees are sampled, and independent evaluations of both references and job performance are made on a scale from 1 to 20. The scores are given in the table.

Employee	References	Job Performance
1	18	20
2	14	13
3	19	16
4	13	9
5	16	15
6	11	18
7	20	15
8	9	12

(a) Compute Spearman's rank correlation coefficient for these data.
(b) Is there evidence that strength of references and job performance are positively correlated? Use $\alpha = 0.05$.

12.37 A large manufacturing firm wants to determine whether a relationship exists between the number of work-hours an employee misses per year and the employee's annual wages. A sample of fifteen employees produced the data in the table. Do these data provide evidence that the work-hours missed are related to annual wages? Use $\alpha = 0.05$.

Employee	Work-Hours Missed	Annual Wages (thousands of dollars)
1	49	12.8
2	36	14.5
3	127	8.3
4	91	10.2
5	72	10.0
6	34	11.5
7	155	8.8
8	11	17.2
9	191	7.8
10	6	15.8
11	63	10.8
12	79	9.7
13	43	12.1
14	57	21.2
15	82	10.9

12.38 The decision to build a new plant or move an existing plant to a new location involves the long-term commitment of both human and monetary resources. Accordingly such decisions should be made only after carefully considering the relevant factors associated with numerous alternative plant sites. G. Michael Epping ("Important factors in plant location in 1980," *Growth and Change, 13*, pp. 47–51, 1982) examined the relationship between the location factors deemed important by businesses that located in Arkansas and those that considered Arkansas, but located elsewhere. A questionnaire that asked manufacturers to rate the importance of thirteen general location factors on a nine-point scale was completed by 118 firms that had moved a plant to Arkansas in the period 1955–1977 and by 73 firms that had recently considered Arkansas, but located elsewhere. Epping averaged the importance ratings and arrived at the rankings shown in the table below. Calculate Spearman's rank correlation coefficient for these data and carefully interpret its value in the context of the problem.

Factor	Manufacturers Locating in Arkansas	Manufacturers Not Locating in Arkansas
Labor	1	1
Taxes	2	2
Industrial Site	3	4
Information Sources and Special Inducements	4	5
Legislative Laws and Structure	5	3
Utilities and Resources	6	7
Transportation Facilities	7	8
Raw Material Supplies	8	10
Community	9	6
Industrial Financing	10	9
Markets	11	12
Business Services	12	11
Personal Preferences	13	13

12.39 The table reports the 1950 and 1980 sales (in millions of dollars) of a sample of thirty major U.S. corporations.

No.	Name	1950 Sales ($ millions)	1980 Sales ($ millions)
1	Abbott Laboratories	74	2,038
2	Allis-Chalmers Corporation	344	2,064
3	American Cyanamid Company	322	3,454
4	Armstrong World Industries	187	1,323
5	Boeing Company	307	9,426
6	Borg-Warner Corporation	331	2,673
7	Bristol-Myers Company	52	3,158
8	Caterpiller Tractor Company	337	8,598
9	Celanese Corporation	233	3,348
10	Coca-Cola Company	215	5,913
11	Continental Group, Inc.	398	5,119
12	Corning Glass Works	117	1,530

No.	Name	1950 Sales ($ millions)	1980 Sales ($ millions)
13	Curtiss-Wright Corporation	136	228
14	Eaton Corporation	148	3,176
15	Fruehauf Corporation	128	2,082
16	General Electric Corporation	2,233	24,959
17	Gillette Company	99	2,315
18	Gulf Oil Corporation	1,150	26,483
19	IBM Corporation	215	26,213
20	International Paper Company	498	5,043
21	PepsiCo, Incorporated	40	5,975
22	Philip Morris, Incorporated	306	7,328
23	Philips Petroleum Company	533	13,377
24	RCA Corporation	584	8,011
25	R. J. Reynolds Industries, Inc.	758	8,449
26	Standard Brands, Incorporated	301	3,018
27	Sterling Drug Incorporated	139	1,701
28	The Timken Company	144	1,338
29	Union Carbide Corporation	758	9,994
30	Westinghouse Electric Corp.	1,020	8,514

Source: Dharan, "Empirical Identification Procedures for Earnings Models," *Journal of Accounting Education*, 21, pp. 256–270, 1983.

(a) Calculate Spearman's rank correlation coefficient for these data and interpret its value in the context of the problem.

(b) Test $H_0: \rho_s = 0$ against $H_a: \rho_s > 0$, using $\alpha = 0.05$. Interpret both the null and the alternative hypotheses in the context of the problem.

(c) Explain why Spearman's rank correlation coefficient can be more widely applied to make inferences about the correlation between two variables than Pearson's product moment correlation coefficient (Chapter 9).

12.7 CONCLUSION

We have presented several useful nonparametric techniques for comparing two or more populations. Nonparametric techniques are useful when the underlying assumptions for their parametric counterparts are not justified or when it is impossible to assign specific values to the observations. Rank sums are the primary tools of nonparametric statistics. The Wilcoxon rank sum statistic and the Wilcoxon signed rank statistic can be used to compare two populations for either an independent sampling experiment or a paired-difference experiment. The Kruskal-Wallis H test is applied when comparing k populations, using a completely randomized design. The Friedman F_r test is used to compare k populations when a randomized block design is conducted.

The strength of nonparametric statistics lies in their general applicability. They require few restrictive assumptions, and they may be used for observations that can be ranked but cannot be exactly measured. Therefore nonparametric tests provide very useful sets of statistical tests to use in conjunction with the parametric tests of Chapters 8 and 11.

SUPPLEMENTARY EXERCISES

12.40 A study was conducted to determine whether the installation of a traffic light was effective in reducing the number of accidents at a busy intersection. Samples taken 6 months before installation and 5 months after installation of the light yielded the following numbers of accidents per month:

Before	After
12	4
5	2
10	7
9	3
14	8
6	

(a) Is there sufficient evidence to conclude the traffic light aided in reducing the number of accidents? Test by using $\alpha = 0.025$.

(b) Explain why this type of data might or might not be suitable for analysis, using the t test of Chapter 8.

12.41 A manufacturer of household appliances is considering one of two department store chains to be the sales merchandiser for its product in a particular region of the United States. Before choosing one chain, the manufacturer wants to make a comparison of the product exposure that might be expected for the two chains. Eight locations are selected where both chains have stores and, on a specific day, the number of shoppers entering each store is recorded. The data are shown in the table. Do the data provide sufficient evidence to indicate that one of the chains tends to have more customers per day than the other? Test by using $\alpha = 0.05$.

Location	A	B
1	879	1,085
2	445	325
3	692	848
4	1,565	1,421
5	2,326	2,778
6	857	992
7	1,250	1,303
8	773	1,215

12.42 A drug company has synthesized two new compounds to be used in sleeping pills. The data in the table represent the additional hours of sleep gained by ten patients through the use of the two drugs. Do the data present sufficient evidence to indicate that one drug is more effective than the other in increasing the hours of sleep? Test by using $\alpha = 0.10$.

Patient	Drug A	Drug B
1	0.4	0.7
2	−0.7	−1.6
3	−0.4	−0.2
4	−1.4	−1.4
5	−1.6	−0.2
6	2.9	3.4
7	4.0	3.7
8	0.1	0.8
9	3.1	0.0
10	1.9	2.0

12.43 Weevils cause millions of dollars worth of damage each year to cotton crops. Three chemicals designed to control weevil populations are applied, one to each of three cotton fields. After 3 months ten plots of equal size are randomly selected within each field, and the percentage of cotton plants with weevil damage is recorded for each. Do the data in the table provide sufficient evidence to indicate a difference in location among the distributions of damage rates corresponding to the three treatments? Use $\alpha = 0.05$.

A	B	C
10.8	22.3	9.8
15.6	19.5	12.3
19.2	18.6	16.2
17.9	24.3	14.1
18.3	19.9	15.3
9.8	20.4	10.8
16.7	23.6	12.2
19.0	21.2	17.3
20.3	19.8	15.1
19.4	22.6	11.3

12.44 In Exercise 9.21 a calibration study undertaken by the Minnesota Department of Transportation to evaluate its newly installed weigh-in-motion scale was described. Pearson's product moment correlation coefficient was used to measure the strength of the relationship between the static weight of a truck and the truck's weight as measured by the weigh-in-motion equipment. The data are repeated below (in thousands of pounds):

Trial No.	Static Weight of Truck, x	Weigh-In-Motion Reading y_1 (prior to calibration adjustment)	Weigh-In-Motion Reading y_2 (after calibration adjustment)
1	27.9	26.0	27.8
2	29.1	29.9	29.1
3	38.0	39.5	37.8
4	27.0	25.1	27.1
5	30.3	31.6	30.6
6	34.5	36.2	34.3
7	27.8	25.1	26.9
8	29.6	31.0	29.6
9	33.1	35.6	33.0
10	35.5	40.2	35.0

Source: Adapted from data in Wright, Owen, and Pena, "Status of MN/DOT's Weigh-in-Motion Program," Minnesota DOT, 1983.

(a) Calculate Spearman's rank correlation coefficient for x and y_1 and for x and y_2. Interpret the values you obtain in the context of the problem. Compare your results to those of Exercise 9.21, part (c).

(b) In the context of this problem describe the circumstances that would result in Spearman's rank correlation coefficient being exactly 1; being exactly 0.

12.45 (a) Suppose a company wants to study how personality relates to leadership. Four supervisors with different types of personalities are selected. Several employees are then selected from the group supervised by each, and these employees are asked to rate the leader of their group on a scale from 1 to 20 (20 signifies highly favorable). The table shows the resulting data. Is there sufficient evidence to indicate that one or more of the supervisors tend to receive higher ratings than the others? Use $\alpha = 0.05$.

	Supervisor		
I	II	III	IV
20	17	16	8
19	11	15	12
20	13	13	10
18	15	18	14
17	14	11	9
	16		10

(b) Suppose the company is particularly interested in comparing the ratings of the personality types represented by supervisors I and III. Make this comparison by using $\alpha = 0.05$.

12.46 A manufacturer wants to determine whether the number of defectives produced by its employees tends to increase as the day progresses. Unknown to the employees, a complete inspection is made of every item produced on one day, and the hourly fraction defective is recorded. The table gives the resulting data. Do they provide evidence that the fraction defective increases as the day progresses? Test at the $\alpha = 0.05$ level.

Hour	Fraction Defective
1	0.02
2	0.05
3	0.03
4	0.08
5	0.06
6	0.09
7	0.11
8	0.10

12.47 An experiment is conducted to compare three calculators, A, B, and C, according to ease of operation. To make the comparison, six randomly selected students are assigned to perform the same sequence of arithmetic operations on each of the three calculators. The order of use of the calculators varies in a random manner from student to student. The times necessary for the completion of the sequence of tasks (in seconds) are recorded in the table. Do the data provide sufficient evidence to indicate that the calculators differ in ease of operation?

Student	Calculator Type A	B	C
1	306	330	300
2	260	265	285
3	281	290	277
4	288	301	305
5	301	309	319
6	262	245	240

12.48 A union wants to determine its members' preferences before negotiating with management. Ten union members are randomly selected, and each member completes an extensive questionnaire. The responses to the various aspects of the questionnaire will enable the union to rank in order of importance the items to be negotiated. The rankings are shown in the table. Is there sufficient evidence to indicate that one or more of the items are preferred to the others? Test by using $\alpha = 0.05$.

Person	More Pay	Job Stability	Fringe Benefits	Shorter Hours
1	2	1	3	4
2	1	2	3	4
3	4	3	2	1
4	1	4	2	3
5	1	2	3	4
6	1	3	4	2
7	2.5	1	2.5	4
8	3	1	4	2
9	1.5	1.5	3	4
10	2	3	1	4

12.49 A clothing manufacturer employs five inspectors who provide quality control of workmanship. Every item of clothing produced carries with it the number of the inspector who checked it. Thus the company can evaluate an inspector by keeping records of the number of complaints received about products bearing his or her inspection number. Records for 6 months are given in the table.

Number of Returns

Month	I	II	Inspector III	IV	V
1	8	10	7	6	9
2	5	7	4	12	12
3	5	8	6	10	6
4	9	6	8	10	13
5	4	13	3	7	15
6	4	8	2	6	9

(a) Do these data provide sufficient evidence to indicate a tendency to receive more complaints for one or more of the inspectors' products than for the others? Use $\alpha = 0.10$.

(b) Use the Wilcoxon signed rank test to determine whether evidence exists to indicate that the performances of inspectors I and IV differ. Use $\alpha = 0.05$.

12.50 Performance in a personal interview often determines whether a candidate is offered a job. Suppose the personnel director of a company interviewed six potential job applicants without knowing anything about their backgrounds, and then rated them on a scale from 1 to 10. Independently, the director's supervisor made an evaluation of the background qualifications of each candidate on the same scale. The results are shown in the table. Is there evidence that a candidate's qualification score is positively correlated with interview performance score? Use $\alpha = 0.05$.

Candidate	Qualifications	Interview Performance
1	10	8
2	8	9
3	9	10
4	4	5
5	5	3
6	6	6

12.51 David K. Campbell, James Gaertner, and Robert P. Vecchio ("Perceptions of Promotion and Tenure Criteria: A survey of accounting educators," *Journal of Accounting Education, 1,* pp. 83–92, 1983) investigated the perceptions of accounting professors with respect to the present and desired importance of various factors considered in promotion and tenure decisions at major universities. One hundred fifteen professors at universities with accredited doctoral programs responded to a mailed questionnaire. The questionnaire asked the professors to rate (1) the *current* importance placed on twenty factors in the promotion and tenure decisions at their universities and (2) how they believe the factors *should* be weighted. Responses were obtained on a five-point scale ranging from No Importance to Extreme Importance. The resulting ratings were averaged and converted to the ranks shown in the table. Calculate Spearman's rank correlation coefficient for these data and carefully interpret its value in the context of the problem.

Factor	Current Importance	Ideal Importance
I. *Teaching (and Related Items)*		
Teaching Performance	6	1
Advising and Counseling Students	19	15
Students' Complaints/Praises	14	17
II. *Research*		
Number of Journal Articles	1	6
Quality of Journal Articles	4	2
Referred Publications:		
A. Applied Studies	5	4
B. Theoretical Empirical Studies	2	3
C. Educationally Oriented	11	8
Papers at Professional Meetings	10	12
Journal Editor or Reviewer	9	10
Other (Textbooks, etc.)	7.5	11
III. *Service and Professional Interaction*		
Service to Profession	15	9
Professional/Academic Awards	7.5	6
Community Service	18	19
University Service	16	16
Collegiality/Cooperativeness	12	13
IV. *Other*		
Academic Degrees Attained	3	5
Professional Certification	17	14
Consulting Activities	20	20
Grantsmanship	13	18

12.52 With the emergence of Japan as a super industrial power, American businesses have begun taking a close look at Japanese management styles and philosophies. Some of the credit for the high quality of Japanese products has been attributed to the Japanese system of permanent employment for their workers. In the United States high job-turnover rates are common in many industries and are associated with high product-defect rates. High turnover rates mean U.S. plants are more highly populated with inexperienced workers who are unfamiliar with the company's product lines than is the case in Japan. In a study of the room air conditioner industry in Japan and the U.S., David Garvin ("Quality on the Line," *Harvard Business Review*, Sept.–Oct., 1983, pp. 65–75) reported that the difference in the average annual turnover rate of workers between U.S. plants and Japanese plants was 3.1%. In a different study five Japanese and five U.S. plants that manufacture room air conditioners were randomly sampled and their turnover rates were determined:

U.S. Plants	Japanese Plants
7.11%	3.52%
6.06%	2.02%
8.00%	4.91%
6.87%	3.22%
4.77%	1.92%

(a) Use an independent samples t test to compare the mean annual percentage turnover for American and Japanese plants. Use $\alpha = 0.05$. Be sure to specify all assumptions that must be met to ensure the validity of the test.

(b) You will recall that the variance of the binomial sample proportion $\hat{p}$ depends on the value of the population parameter p. As a consequence the variance of a sample percentage, $(100\hat{p})\%$, will also depend on p. The relevance of this fact is that if you conduct an unpaired t test [part (a)] to compare the means of two populations of percentages, you may be violating the assumption $\sigma_1^2 = \sigma_2^2$ upon which the t test is based. If the disparity in the variances is large, you will obtain more reliable test results by using the Wilcoxon rank sum test for independent samples. Compare the two populations of turnover rates by using the Wilcoxon rank sum test ($\alpha = 0.05$).

(c) Compare the results of the t test and Wilcoxon rank sum test by calculating the (approximate) observed significance level for each.

13

Applications to Quality Control

ABOUT THIS CHAPTER

An important use of statistics is in the area of industrial quality control. Quality of raw material and finished products is maintained by periodic sampling and inspection of representative items. Statistics calculated from these sample items are used in making decisions concerning acceptable or unacceptable quality levels. Many of the statistical techniques discussed earlier in this book are used in quality control, but some modifications are unique to this application.

CONTENTS

13.1 INTRODUCTION

Our theme throughout this text has been that many, if not most, experiments result in outcomes that are *not* purely deterministic. That is, a sequence of seemingly identical experiments will give slightly different results due to uncontrollable sources of variation. Because of this variability in outcomes of experiments, we have developed probabilistic models for such outcomes and statistical techniques for making decisions based on such outcomes.

Suppose our "experiment" now becomes the measurement of a quality trait on an item resulting from a production process. The measurement might be the strength of a section of steel cable, the weight of cereal dispensed by a filling machine, the percentage of copper in a casting of bronze, or the diameter of a machined rod for an engine. No matter how carefully a production process is controlled, these quality measurements will vary from item to item, and there will be a probability distribution associated with the population of such measurements.

If all important sources of variation are under control in a production process, then the slight variations among quality measurements usually cause no serious problems. Such a process should produce the same distribution of quality measurements no matter when it is sampled, and thus it can be referred to as a "stable system." For example a process for producing bronze castings may be regarded as stable as long as the percentages of copper in test samples lie between 80 and 82.

One objective of quality control is to develop a scheme for sampling a process, making the quality measurement of interest on the sample items, and then making a decision as to whether or not the process is in its stable state, or "in control." If the sample data suggest that the process is "out of control," a cause ("assignable cause") for the abnormality is sought. A common method for making these decisions involves the use of control charts. In the next four sections we discuss the $\bar{x}$-chart for checking central tendency, the r-chart for checking variability, the p-chart for checking fraction of defectives, and the c-chart for checking number of defects per item.

Before proceeding, some comments as to why these applications merit a separate chapter are in order. First, these are very important and widely used techniques in industry, and every engineer working in industry, even if not directly in quality control, should be aware of how the techniques work. Second, the statistical formulas and methodologies that have become standard in the area of quality control differ somewhat from similar methodologies presented earlier in the text for other purposes. We will develop the control-charting techniques using standard terminology wherever possible.

13.2 THE $\bar{x}$-CHART

Imagine an industrial process that is running continuously. A standard quality-control plan would require sampling one or more items from this process periodically, and making the appropriate quality measurements. Usually more than one item is measured at each time point to increase accuracy and measure variability. The objective of this section is to develop a technique that will help the quality-control engineer decide whether the center (or average or location) of the measurements has shifted up or down.

Suppose that n observations are to be made at each time point in which the process is checked. Let X_i denote the ith observation ($i = 1, \ldots, n$) at the specified time point, and $\bar{X}_j$ denote the average of the n observations at time point j. If $E(X_i) = \mu$ and $V(X_i) = \sigma^2$, for a process in control, then $\bar{X}_j$ should be approximately normally distributed with $E(\bar{X}_j) = \mu$ and $V(\bar{X}_j) = \sigma^2/n$. As long as the process remains in control, where will most of the $\bar{X}_j$ values lie? Since we know that $\bar{X}_j$ has approximately a normal distribution, we can find an interval that will have probability $1 - \alpha$ of containing $\bar{X}_j$. This interval is $\mu \pm z_{\alpha/2}(\sigma/\sqrt{n})$. The usual value for $z_{\alpha/2}$ in most control-charting problems is 3. Thus the interval $\mu \pm 3(\sigma/\sqrt{n})$ has probability 0.9973 of including $\bar{X}_j$, as long as the process is in control. If μ and σ were known or specified, we could use $\mu - 3(\sigma/\sqrt{n})$ as a lower control limit and $\mu + 3(\sigma/\sqrt{n})$ as an upper control limit. If a value of $\bar{X}_j$ was observed to fall outside of these limits, we would suspect that the process might be out of control (the mean may have shifted to a different value) since this event has a very small probability of occurring (0.0027) when the process is in control. Figure 13.1 shows schematically how the decision process would work.

FIGURE 13.1
Schematic Representation of x̄-Chart

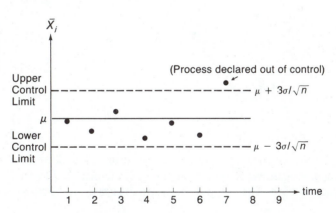

Note that if a process is declared to be out of control because $\bar{X}_j$ fell outside of the control limits, there is a positive probability of making an error. That is, we could declare that the mean of the quality measurements had shifted when, in fact, it had not. However, this probability of error is only 0.0027 for the 3-standard deviation limits used above. (This error is similar to the Type I error of a hypothesis-testing problem.)

If a control chart is being started for a new process, μ and σ will not be known and hence must be estimated from the data. For establishing control limits it is generally recommended that at least $k = 20$ time points be sampled before the control limits are calculated. We will now discuss the details of estimating $\mu \pm 3(\sigma/\sqrt{n})$ from k independent samples, each of size n.

For each of the k samples we will compute the mean $\bar{X}_j$, the sample variance S_j^2, and the range R_j of observations in the sample. (Recall that the range is simply the difference between the largest and smallest observations in the sample.) Traditionally quality-control engineers work with

$$S_j'^2 = \left(\frac{n-1}{n}\right) S_j^2$$

rather than S_j^2 itself. This is merely a matter of changing the denominator of the sample variance from $(n-1)$ to n.

To form an unbiased estimator of μ from $\bar{X}_1, \bar{X}_2, \ldots, \bar{X}_k$, we can simply calculate the average of the sample means $\bar{\bar{X}}$ where

$$\bar{\bar{X}} = \frac{1}{k} \sum_{j=1}^{k} \bar{X}_j$$

In a similar fashion we can calculate the average of the S_j' quantities to form

$$\bar{S}' = \frac{1}{k} \sum_{j=1}^{k} S_j'$$

Now $\bar{S}'$ is *not* an unbiased estimator of σ, but it can be made unbiased by dividing by a constant c_2 found in Table 12 of the Appendix. Thus $\bar{S}'/c_2$ is an unbiased estimator of σ. A good estimator of $\mu \pm 3(\sigma/\sqrt{n})$ then becomes

$$\bar{\bar{X}} \pm 3 \frac{\bar{S}'}{c_2\sqrt{n}}$$

or

$$\bar{\bar{X}} \pm A_1 \bar{S}'$$

where

$$A_1 = 3/c_2\sqrt{n}$$

For convenience, values of A_1 are also in Table 12 of the Appendix. In practice, then, we can use $\bar{\bar{X}} + A_1\bar{S}'$ as the upper control limit and $\bar{\bar{X}} - A_1\bar{S}'$ as the lower control limit. The computations are illustrated in Example 13.1.

EXAMPLE 13.1

We want to start a control chart for a new machine that fills boxes of cereal by weight. Five observations on amount of fill are taken every two hours until twenty such samples are obtained. The data are given in Table 13.1. Calculate upper and lower control limits on the mean. The weights are assumed to be normally distributed.

Solution   The sample means ($\bar{x}_j$) and adjusted standard deviations (s_j') are given on Table 13.1 for each of the twenty samples of size 5. Now we can calculate

$$\bar{\bar{x}} = \frac{1}{20} \sum_{j=1}^{20} \bar{x}_j = 16.32$$

TABLE 13.1
Amount of Cereal Dispensed by Filling Machines

Sample	Readings					$\bar{x}_j$	s'_j	r_j (range)
1	16.1	16.2	15.9	16.0	16.1	16.06	0.1020	0.3
2	16.2	16.4	15.8	16.1	16.2	16.14	0.1860	0.6
3	16.0	16.1	15.7	16.3	16.1	16.04	0.1960	0.6
4	16.1	16.2	15.9	16.4	16.6	16.24	0.2417	0.7
5	16.5	16.1	16.4	16.4	16.2	16.32	0.1470	0.4
6	16.8	15.9	16.1	16.3	16.4	16.30	0.3033	0.9
7	16.1	16.9	16.2	16.5	16.5	16.44	0.2800	0.8
8	15.9	16.2	16.8	16.1	16.4	16.28	0.3059	0.9
9	15.7	16.7	16.1	16.4	16.8	16.34	0.4030	1.1
10	16.2	16.9	16.1	17.0	16.4	16.52	0.3655	0.9
11	16.4	16.9	17.1	16.2	16.1	16.54	0.3929	1.0
12	16.5	16.9	17.2	16.1	16.4	16.62	0.3868	1.1
13	16.7	16.2	16.4	15.8	16.6	16.34	0.3200	0.9
14	17.1	16.2	17.0	16.9	16.1	16.66	0.4224	1.0
15	17.0	16.8	16.4	16.5	16.2	16.58	0.2856	0.8
16	16.2	16.7	16.6	16.2	17.0	16.54	0.3072	0.8
17	17.1	16.9	16.2	16.0	16.1	16.46	0.4499	1.1
18	15.8	16.2	17.1	16.9	16.2	16.44	0.4841	1.3
19	16.4	16.2	16.7	16.8	16.1	16.44	0.2728	0.7
20	15.4	15.1	15.0	15.2	14.9	15.12	0.1720	0.5

and

$$\bar{s}' = \frac{1}{20} \sum_{j=1}^{20} s'_j = 0.3017$$

Since $n = 5$, we have that $A_1 = 1.596$ from Table 12 of the Appendix. Thus the realization of the control limit is

$$\bar{x} \pm A_1 \bar{s}'$$

$$16.32 \pm 1.596(0.3017)$$

or

$$(15.84, 16.80)$$

Any sample mean below 15.84 or above 16.80, from a sample of size 5, would be declared out of control.

The control limits and data points are plotted in Figure 13.2.

Since the sample mean for sample 20 ($\bar{x}_{20} = 15.12$) falls below the lower control limit, we would declare the process to be out of control at that point. Now we might want to eliminate that sample and recompute the control limits from the remaining 19 samples. Doing these computations results in

$$\bar{x} = 16.38$$

$$\bar{s}' = 0.3085$$

FIGURE 13.2
An $\bar{x}$-Chart for the
Data in Table 13.1

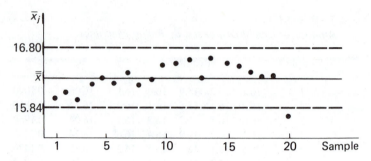

and

$$\bar{\bar{x}} \pm A_1 \bar{s}'$$

yielding

$$16.38 \pm 1.596(0.3085)$$

or

$$(15.89, 16.87)$$

All 19 sample means are now within these limits, and they can be used to check the means of future samples. As new data become available, the control limits should be recomputed from time to time. ∎

Computation of $\bar{s}'$ requires an effort that can often be avoided by using the ranges (R_j) rather than the adjusted standard deviations (S'_j). R_j denotes the range (difference between the largest and smallest observation) for sample j, $j = 1, \ldots, k$. We can average these ranges to produce

$$\bar{R} = \frac{1}{k} \sum_{j=1}^{k} R_j$$

Now the range (or average range) has some relationship to the variation of the data, but $\bar{R}$ is *not* an unbiased estimator of σ. However, it turns out that $\bar{R}/d_2$, where d_2 is found in Table 12 of the Appendix, *is* an unbiased estimator of σ, and thus $\mu \pm 3\sigma/\sqrt{n}$ can be estimated by

$$\bar{\bar{X}} \pm 3 \frac{\bar{R}}{d_2\sqrt{n}}$$

or

$$\bar{\bar{X}} \pm A_2 \bar{R}$$

where $A_2 = 3/d_2\sqrt{n}$. For convenience, values of A_2 are found in Table 12 of the Appendix.

EXAMPLE 13.2

Using the data in Table 13.1, construct control limits for the mean by using the sample ranges.

Solution Using all twenty samples, we have already seen that $\bar{x} = 16.32$. The sample realization of $\bar{R}$ is

$$\bar{r} = \frac{1}{20} \sum_{j=1}^{20} r_j = 0.82$$

Then $\bar{\bar{X}} \pm A_2\bar{R}$ takes on the value

$$\bar{x} \pm A_2\bar{r}$$

$$16.32 \pm (0.577)(0.82)$$

or

$$(15.85, 16.79)$$

The twentieth sample mean still falls outside of the control limits. Recomputing the limits on the 19 samples (excluding sample 20) results in

$$\bar{x} \pm A_2\bar{r}$$

$$16.38 \pm (0.577)(0.84)$$

or

$$(15.90, 16.86)$$

and all 19 means fall within the limits. ■

Note that the limits based on $\bar{R}$ are very close to those based on $\bar{S}'$, and the ones based on $\bar{R}$ require far fewer computations. In general $\bar{R}/d_2$ is not as good as $\bar{S}'/c_2$ as an estimator of σ, but the method employing $\bar{R}$ usually works very well if k is at least 20. The disadvantage of a slight loss in accuracy is offset by the advantage of easy computations.

EXERCISES

13.1 Production of ammeters is controlled for quality by periodically selecting an ammeter and obtaining four measurements on a test circuit designed to produce 15 amps. The following data were observed for 15 tested ammeters.
   (a) Construct control limits for the mean by using the sample standard deviations.
   (b) Do all observed sample means lie within the limits found in (a)? If not, recalculate the control limits after omitting those samples that are "out of control."

Ammeter	Readings
1	15.1 14.9 14.8 15.2
2	15.0 15.0 15.2 14.9
3	15.1 15.1 15.2 15.1
4	15.0 14.7 15.3 15.1
5	14.8 14.9 15.1 15.2
6	14.9 14.9 15.1 15.1
7	14.7 15.0 15.1 15.0
8	14.9 15.0 15.3 14.8
9	14.4 14.5 14.3 14.4
10	15.2 15.3 15.2 15.5
11	15.1 15.0 15.3 15.3
12	15.2 15.6 15.8 15.8
13	14.8 14.8 15.0 15.0
14	14.9 15.1 14.7 14.8
15	15.1 15.2 14.9 15.0

13.2 Refer to Exercise 13.1. Repeat both parts (a) and (b), using sample ranges instead of sample standard deviations.

13.3 Refer to Exercise 13.1, part (a). How would the control limits change if the probability of a sample mean falling outside of the limits, when the process is in control, is to be 0.01?

13.4 Refer to Exercises 13.1 and 13.2. The specifications for these ammeters require individual readings on the test circuit to be within 15.0 ± 0.4. Do the meters seem to be meeting this specification? [*Hint*: Construct an interval in which *individual* measurements will lie with high probability ($n = 1$).]

13.5 Bronze castings are controlled for copper content. Ten samples of five specimens each gave the following measurements on percentage of copper.

Sample	Percentages of Copper
1	82 84 80 86 83
2	90 91 94 90 89
3	86 84 87 83 80
4	92 91 89 91 90
5	84 82 83 81 84
6	82 81 83 84 81
7	80 80 79 83 82
8	79 83 84 82 82
9	81 84 85 79 86
10	81 92 94 79 80

(a) Construct control limits for the mean percentage of copper by using the sample ranges.

(b) Would you use the limits in (a) as control limits for future samples? Why?

13.3 THE r-CHART

In addition to deciding whether the center of the distribution of quality measurements has shifted up or down, it is frequently of interest to decide if the variability of the process measurements has significantly increased or decreased. A process that suddenly starts turning out highly variable products could cause severe problems in the operation of a plant.

Since we have already established the fact that there is a relationship between the ranges (R_j) and σ, it should seem natural to base our control chart for variability on the R_j's. A control chart for variability could be based on S_j also, but use of ranges provides nearly as much accuracy for much less computation.

Where will most of the range measurements lie if a process is in control? Using a similar argument to that employed for the mean, almost all R_j's should be within three standard deviations of the mean of R_j. Now

$$E(R_j) = d_2\sigma$$

and

$$V(R_j) = d_3^2\sigma^2$$

where d_2 and d_3 can be found in Table 12 of the Appendix.

The three-standard deviation interval about the mean of R_j then becomes

$$d_2\sigma \pm 3d_3\sigma$$

or

$$\sigma(d_2 \pm 3d_3)$$

For any sample there is a high probability (approximately 0.9973) that its range will fall inside this interval if the process is in control.

As in the case of the $\bar{X}$-chart, if σ is not specified, it must be estimated from the data. The best estimator of σ based on $\bar{R}$ is $\bar{R}/d_2$, and the estimator of $\sigma(d_2 \pm 3d_3)$ then becomes

$$\frac{\bar{R}}{d_2}(d_2 \pm 3d_3)$$

or

$$\bar{R}\left(1 \pm 3\frac{d_3}{d_2}\right)$$

Letting $1 - 3(d_3/d_2) = D_3$ and $1 + 3(d_3/d_2) = D_4$, the control limits take on the form $(\bar{R}D_3, \bar{R}D_4)$. D_3 and D_4 are given in Table 12 of the Appendix.

EXAMPLE 13.3

Use the data of Table 13.1 to construct control limits on the variability of the process by using ranges.

Solution From Table 13.1 the variable $\bar{R}$ has outcome $\bar{r} = 0.82$, using all twenty samples. From Table 12 of the Appendix we see that, for samples of size $n = 5$, $D_3 = 0$ and $D_4 = 2.115$. (The lower bound is set at zero since a range cannot be negative.) Hence the realization of the control limits is

$$(\bar{r}D_3, \bar{r}D_4)$$

$$((0.82)(0), (0.82)(2.115))$$

or

$$(0, 173)$$

All twenty sample ranges are well within these limits, so there appear to be no significant changes in the variability of the process over these twenty time points. The limits constructed above could now be used to check future samples.

The control limits are plotted in Figure 13.3. ■

FIGURE 13.3
An r-Chart for the Data in Table 13.1

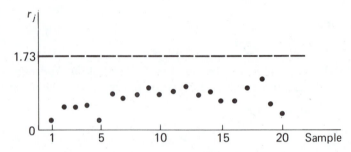

EXERCISES

13.6 Refer to Exercise 13.1. Construct control limits for variability. Does the process appear to be entirely in control with respect to variability?

13.7 Refer to Exercise 13.5. Construct control limits for variability of the percentages of copper. Based on all ten samples, would you use these limits as control limits for future samples? Why?

13.8 Refer to Exercise 13.6. Change the control limits so that a sample will be falsely declared to be out of control with only 0.01 probability.

13.4 THE p-CHART

The control-charting procedures outlined in the preceding sections of this chapter depend upon quality measurements that possess a continuous probability distribution. Sampling such measurements is commonly referred to as "sampling by variables." In many quality-control situations, however, we merely want to assess whether or not a certain item is defective. We then observe the number of defective items from a particular sample or series of samples. This is commonly referred to as "sampling by attributes."

As in previous control-charting problems, suppose a series of k independent samples, each of size n, is selected from a production process. Let p denote the proportion of defective items in the population (total production for a certain time period) for a process in control and let X_i denote the number of defectives observed in the ith sample. Then X_i has a binomial distribution, assuming random sampling from large lots, with $E(X_i) = np$ and $V(X_i) = np(1 - p)$. We will usually work with sample fractions of defectives (X_i/n) rather than the observed number of defectives. Where will these sample fractions tend to lie for a process in control? As argued previously, most sample fractions should lie within three standard deviations of their mean, or in the interval

$$p \pm 3 \sqrt{\frac{p(1 - p)}{n}}$$

since $E(X_i/n) = p$ and $V(X_i/n) = p(1 - p)/n$.

Since p is unknown, now we must estimate these control limits, using the data from all k samples. An unbiased estimator of p is

$$\bar{P} = \frac{\sum\limits_{i=1}^{k} X_i}{nk} = \frac{\text{total number of defectives observed}}{\text{total sample size}}$$

Thus estimated control limits are given by

$$\bar{P} \pm 3 \sqrt{\frac{\bar{P}(1 - \bar{P})}{n}}$$

The following example illustrates the calculation of these control limits.

EXAMPLE 13.4

A process that produces transistors is sampled every four hours. At each time point 50 transistors are randomly sampled, and the number of defectives x_i is observed. The data for 24 samples are given in Table 13.2. Construct a control chart based on these samples.

Solution   From the above data the observed value of $\bar{P}$ is

$$\bar{p} = \frac{\sum\limits_{i=1}^{k} x_i}{nk} = \frac{52}{50(24)} = 0.04$$

Thus

$$\bar{p} \pm 3 \sqrt{\frac{\bar{p}(1 - \bar{p})}{n}}$$

TABLE 13.2
Data for Example 13.4

Sample	x_i	x_i/n	Sample	x_i	x_i/n
1	3	0.06	13	1	0.02
2	1	0.02	14	2	0.04
3	4	0.08	15	0	0.00
4	2	0.04	16	3	0.06
5	0	0.00	17	2	0.04
6	2	0.04	18	2	0.04
7	3	0.06	19	4	0.08
8	3	0.06	20	1	0.02
9	5	0.10	21	3	0.06
10	4	0.08	22	0	0.00
11	1	0.02	23	2	0.04
12	1	0.02	24	3	0.06

becomes

$$0.04 \pm 3 \sqrt{\frac{(0.04)(0.96)}{50}}$$

$$0.04 \pm 0.08$$

This yields the interval $(-0.04, 0.12)$, but we set the lower control limit at zero since a fraction of defectives cannot be negative, and we obtain $(0, 0.12)$.

All of the observed sample fractions are within these limits, so we would feel comfortable in using them as control limits for future samples. The limits should be recalculated from time to time as new data become available.

The control limits and observed data points are plotted in Figure 13.4. ■

FIGURE 13.4
A p-Chart for the Data of Table 13.2

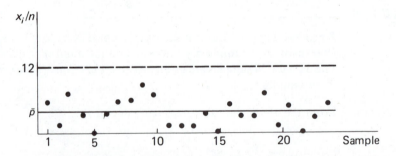

EXERCISES

13.9 Filled bottles of soft drink coming off a production line are checked for underfilling. One hundred bottles are sampled from each half-day's production. The results for twenty such samples are as follows:

Sample	No. of Defectives	Sample	No. of Defectives
1	10	11	15
2	12	12	6
3	8	13	3
4	14	14	2
5	3	15	0
6	7	16	1
7	12	17	1
8	10	18	4
9	9	19	3
10	11	20	6

(a) Construct control limits for the proportion of defectives.

(b) Do all the observed sample fractions fall within the control limits found in (a)? If not, adjust the limits for future use.

13.10 Refer to Exercise 13.9. Adjust the control limits so that a sample fraction will fall outside of the limits with probability 0.05, when the process is in control.

13.11 Bolts being produced in a certain plant are checked for tolerances. Those not meeting tolerance specifications are considered to be defective. Fifty bolts are gauged for tolerance every hour. The results for thirty hours of sampling are as follows:

Sample	No. Defective	Sample	No. Defective	Sample	No. Defective
1	5	11	3	21	6
2	6	12	0	22	4
3	4	13	1	23	9
4	0	14	1	24	3
5	8	15	2	25	0
6	3	16	5	26	6
7	10	17	5	27	4
8	2	18	4	28	2
9	9	19	3	29	1
10	7	20	1	30	2

(a) Construct control limits for the proportion of defectives based on these data.

(b) Would you use the control limits found in (a) for future samples? Why?

13.5 THE c-CHART

In many quality-control problems the particular items being subjected to inspection may have more than one defect, and so we may wish to count defects instead of merely classifying an item as to whether or not it is defective. If C_i denotes the number of defects observed on the ith inspected item, a good working model for many applications is to assume that C_i has a Poisson distribution. We will let the Poisson distribution have a mean of λ for a process in control.

Most of the C_i's should fall within three standard deviations of their mean if the process remains in control. Since $E(C_i) = \lambda$ and $V(C_i) = \lambda$, the control limits become

$$\lambda \pm 3\sqrt{\lambda}$$

If k items are inspected, then an unbiased estimator of λ is given by

$$\bar{C} = \frac{1}{k} \sum_{i=1}^{k} C_i$$

The estimated control limits then become

$$\bar{C} \pm 3\sqrt{\bar{C}}$$

The computations are illustrated in Example 13.5.

EXAMPLE 13.5

Twenty rebuilt pumps were sampled from the hydraulics shop of an aircraft rework facility. The numbers of defects recorded are given in Table 13.3. Use these data to construct control limits for defects per pump.

TABLE 13.3
Data for Example 13.5

Pump	No. of Defects (C_i)	Pump	No. of Defects (C_i)
1	6	11	4
2	3	12	3
3	4	13	2
4	0	14	2
5	2	15	6
6	7	16	5
7	3	17	0
8	1	18	7
9	0	19	2
10	0	20	1

Solution   Assuming that C_i, the number of defects observed on pump i, has a Poisson distribution, we estimate the mean number of defects per pump by $\bar{C}$ and use

$$\bar{C} \pm 3\sqrt{\bar{C}}$$

as the control limits. Using the data in Table 13.3, we have

$$\bar{c} = \frac{58}{20} = 2.9$$

and hence

$$\bar{c} \pm 3\sqrt{\bar{c}}$$

becomes

$$2.9 \pm 3\sqrt{2.9}$$

$$2.9 \pm 5.1$$

or

$$(0, 8.0)$$

Again the lower limit is raised to zero since a number of defects cannot be negative.

All of the observed counts are within the control limits, so we could use these limits for future samples. Again the limits should be recomputed from time to time as new data become available.

The control limits and data points are plotted in Figure 13.5.   ■

FIGURE 13.5
A c-Chart for the Data of Table 13.3

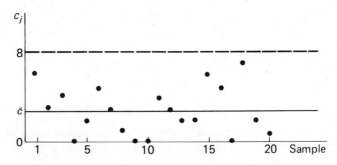

EXERCISES

13.12   Woven fabric from a certain loom is controlled for quality by periodically sampling one-square-meter specimens and counting the number of defects. For 15 such samples the observed number of defects was as follows:

Sample	No. of Defects
1	3
2	6
3	10
4	4
5	7
6	11
7	3
8	9
9	6
10	14
11	4
12	3
13	1
14	4
15	0

    (a)   Construct control limits for the average number of defects per square meter.

    (b)   Do all samples observed seem to be "in control?"

13.13   Refer to Exercise 13.12. Adjust the control limits so that a count for a process in control will fall outside the limits with an approximate probability of only 0.01.

13.14   Quality control in glass production involves counting the number of defects observed in samples of a fixed size periodically selected from the production process. For twenty samples the observed counts on numbers of defects are as follows:

Sample	No. of Defects	Sample	No. of Defects
1	3	11	7
2	0	12	3
3	2	13	0
4	5	14	1
5	3	15	4
6	0	16	2
7	4	17	2
8	1	18	1
9	2	19	0
10	2	20	2

    (a)   Construct control limits for defects per sample based on these data.

    (b)   Would you adjust the limits in (a) before using them on future samples?

13.15   Suppose variation among defect counts is important in a process like that of Exercise 13.14. Is it necessary to have a separate method to construct control limits for variation? Discuss.

13.6 ACCEPTANCE SAMPLING BY ATTRIBUTES

The preceding sections of this chapter dealt with statistical methods for controlling quality of products in an ongoing production process. In this section we discuss the problem of making a decision with regard to the proportion of defective items in a finite lot based on the number of defectives observed in a sample from that lot. Inspection in which a unit of product is classified simply as defective or nondefective is called *inspection by attributes*.

Lots of items, either raw materials or finished products, are sold by a producer to a consumer with some guarantee as to quality. In this section quality will be determined by the proportion p of defective items in the lot. To check on the quality characteristics of a lot, the consumer will sample some of the items, test them, and observe the number of defectives. Generally some defectives are allowable because defective-free lots may be too expensive for the consumer to purchase. But if the number of defectives is too large, the consumer will reject the lot and return it to the producer.

In the decision to accept or reject the lot, sampling is generally used because the cost of inspecting the entire lot may be too high or the inspection process may be destructive.

Before sampling inspection takes place for a lot, the consumer must have in mind a proportion p_0 of defectives that will be acceptable. Thus if the true proportion of defectives p is no greater than p_0, the consumer wants to accept the lot, and if p is greater than p_0, the consumer wants to reject the lot. This maximum proportion of defectives satisfactory to the consumer is called the *acceptable quality level* (AQL).

Note that we now have a hypothesis-testing problem in which we are testing $H_0: p \leq p_0$ versus $H_a: p > p_0$. A random sample of n items will be selected from the lot, and the number of defectives Y observed. The decision concerning rejection or nonrejection of H_0 (rejecting or accepting the lot) will be based on the observed value of Y.

The probability of a Type I error α in this problem is the probability that the lot will be rejected by the consumer when, in fact, the proportion of defectives is satisfactory to the consumer. This is referred to as the *producer's risk*. The value of α calculated for $p = p_0$ is the upper limit to the proportion of good lots rejected by the sampling plan being considered. The probability of a Type II error β calculated for some proportion of defectives p_1, where $p_1 > p_0$, represents the probability that an unsatisfactory lot will be accepted by the consumer. This is referred to as the *consumer's risk*. The value of β calculated for $p = p_1$ is the upper limit to the proportion of bad lots accepted by the sampling plan, for all $p \geq p_1$.

The null hypothesis $H_0: p \leq p_0$ will be rejected if the observed value of Y is larger than some constant a. Since the lot is accepted if $Y \leq a$, the constant a is called the *acceptance number*. In this context the significance level α is given by

$$\alpha = P(\text{Reject } H_0 \text{ when } p = p_0)$$

$$= P(Y > a \text{ when } p = p_0)$$

and

$$1 - \alpha = P(\text{Not rejecting } H_0 \text{ when } p = p_0)$$

$$= P(\text{Accepting the lot when } p = p_0)$$

$$= P(A)$$

Since p_0 may not be known precisely, it is of interest to see how $P(A)$ behaves as a function of the true p, for a given n and a. A plot of these probabilities is called an *operating characteristic curve*, abbreviated OC curve. We illustrate its calculation with the following numerical example.

EXAMPLE 13.6

One sampling inspection plan calls for $(n = 10, a = 1)$ while another calls for $(n = 25, a = 3)$. Plot the operating characteristic curves for both plans. If the plant using these plans can operate well on 30% defective raw materials, considering the price, but cannot operate efficiently if the proportion gets close to 40%, which plan should they choose?

Solution   We must calculate $P(A) = P(Y \leqslant a)$ for various values of p in order to plot the OC curves. For $(n = 10, a = 1)$ we have

$$p = 0, \qquad P(A) = 1$$
$$p = 0.1, \qquad P(A) = 0.736$$
$$p = 0.2, \qquad P(A) = 0.376$$
$$p = 0.3, \qquad P(A) = 0.149$$
$$p = 0.4, \qquad P(A) = 0.046$$
$$p = 0.6, \qquad P(A) = 0.002$$

using Table 2 of the Appendix. For $(n = 25, a = 3)$ we have

$$p = 0, \qquad P(A) = 1$$
$$p = 0.1, \qquad P(A) = 0.764$$
$$p = 0.2, \qquad P(A) = 0.234$$
$$p = 0.3, \qquad P(A) = 0.033$$
$$p = 0.4, \qquad P(A) = 0.002$$

These values are plotted in Figure 13.6. Note that the OC curve for the $(n = 10, a = 1)$ plan drops slowly and does not get appreciably small until p is in the neighborhood of 0.4. For p around 0.3 this plan still has a fairly high probability of accepting the lot. The other curve $(n = 25, a = 3)$ drops much more rapidly and

FIGURE 13.6
*Operating
Characteristic
Curves*

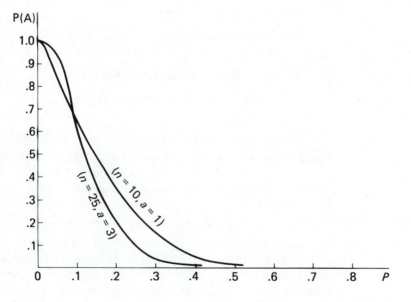

falls to a small $P(A)$ at $p = 0.3$. The latter plan would be better for the plant, if they can afford to sample $n = 25$ items out of each lot before making a decision. ■

Numerous handbooks give operating characteristic curves and related properties for a variety of sampling plans, when inspecting by attributes, but one of the most widely used sets of plans is Military Standard 105D (MIL-STD–105D).

Even though the probability of acceptance, as calculated in Example 13.6, generally does not depend upon the lot size, as long as the lot is large compared to the sample size, MIL-STD-105D is constructed so that sample size increases with lot size. The practical reasons for the increasing sample sizes are that small random samples are sometimes difficult to obtain from large lots and that discrimination between good and bad lots may be more important for large lots. The adjustments in sample sizes for large lots are based upon empirical evidence rather than strict probabilistic considerations.

MIL-STD-105D contains plans for three sampling levels, labeled I, II, and III. Generally, sampling or inspection level II is used. Level I gives slightly smaller samples and less discrimination whereas level III gives slightly larger samples and more discrimination. Table 13 in the Appendix gives the code letters for various lot or batch sizes and inspection levels I, II, and III. This code letter is used to enter either Table 14, 15, or 16 to find the sample size and acceptance number to be used for specified values of the AQL.

In addition to having three inspection levels, MIL-STD-105D also has sampling plans for normal, tightened, and reduced inspection. These plans are given in Tables 14, 15, and 16, respectively. Normal inspection is designed to protect the producer against the rejection of lots with percent defectives less than the AQL. Tightened inspection is designed to protect the consumer from accepting lots with percent defectives greater than the AQL. Reduced inspection is introduced as an economical plan to be used if the quality history of the lots is good. When sampling a series of lots, rules for switching among normal, tightened, and reduced sampling are given in MIL-STD-105D. When using MIL-STD-105D on a single lot, the OC curves for various plans, given in the handbook, should be studied carefully before deciding on the optimal plan for a particular problem.

EXAMPLE 13.7

A lot of size 1000 is to be sampled with normal inspection at level II. If the AQL of interest is 10%, find the appropriate sample size and acceptance number from MIL-STD-105D.

Solution　Entering Table 13 in the Appendix for a lot size 1000, we see that the code letter for level II is J. Since normal inspection is to be used, we enter Table 14 at row J. The sample size is 80 and, moving over to the column headed 10% AQL, we see that the acceptance number is 14. Thus we sample 80 items and accept the lot if the number of defectives is less than or equal to 14. This plan will have a high probability of accepting any lot with a true percentage of defectives below 10%. ■

13.7   ACCEPTANCE SAMPLING BY VARIABLES

In acceptance sampling by attributes, one merely records, for each sampled item, whether or not the item is defective. The decision to accept or reject the lot then depends upon the number of defectives observed in the sample. However, if the characteristic under study involves a measurement on each sampled item, one could base the decision to accept or reject on the measurements themselves, rather than simply on the number of items having measurements that do not meet a certain standard.

For example, suppose a lot of manufactured steel rods is checked for quality with the quality measurement being the diameter. Rods are not allowed to exceed ten centimeters in diameter, and a rod with a diameter in excess of ten centimeters is declared defective for this special case. In attribute sampling, the decision to accept or reject the lot is based on the number of defective rods observed in a sample. But this decision could be made on the basis of some function of the actual diameter observations themselves, say the mean of these observations. If the sample mean is close to or in excess of ten centimeters, one would expect the lot to contain a high percentage of defectives, whereas if the sample mean is much smaller than ten centimeters, one would expect the lot to contain relatively few defectives.

Using actual measurements of the quality characteristic as a basis for acceptance or rejection of a lot is called *acceptance sampling by variables*. Sampling by variables often has advantages over sampling by attributes since the variables method may contain more information for a fixed sample size than the attribute method. Also, when looking at the actual measurements obtained by the variables method, the experimenter may gain some insights into the degree of nonconformance and may be able quickly to suggest methods for improving the product.

For acceptance sampling by variables let

U = upper specification limit

and

L = lower specification limit

If X denotes the quality measurement being considered, then U is the maximum allowable value for X on an acceptable item. (U is equal to ten centimeters in the rod-diameter example.) Similarly L is the minimum allowable value for X on an acceptable item. Some problems may specify a U, others an L, and still others both U and L.

As in acceptance sampling by attributes, an AQL is specified for a lot, and a sampling plan is then determined that will give high probability of acceptance to lots with percent defectives lower than the AQL, and low probability of acceptance to lots with percent defectives higher than the AQL. In practice an *acceptability constant k* is determined such that a lot is accepted if

$$\frac{U - \bar{X}}{S} \geq k$$

or

$$\frac{\bar{X} - L}{S} \geq k$$

where $\bar{X}$ is the sample mean and S the sample standard deviation of the quality measurements.

As in the case of sampling by attributes, handbooks of sampling plans are available for acceptance sampling by variables. One of the most widely used of these handbooks is Military Standard 414 (MIL-STD-414). Tables 17 and 18 of the Appendix are necessary to illustrate how MIL-STD-414 is used. Table 17A shows which tabled AQL value to use if the problem calls for an AQL not specifically tabled. Table 17B gives the sample size code letter to use with any possible lot size. Sample sizes are tied to lot sizes, and different levels of sampling are used, as in MIL-STD-105D. Inspection level IV is recommended for general use.

Table 18 gives the sample size and value of k for a fixed code letter and AQL. Both normal and tightened inspection procedures can be found in this table.

The use of these tables is illustrated in Example 13.8.

EXAMPLE 13.8
_____

The tensile strengths of wires in a certain lot of size 400 are specified to exceed 5 kilograms. We want to set up an acceptance sampling plan with AQL = 1.4%. With inspection level IV and normal inspection, find the appropriate plan from MIL-STD-414.

Solution   Looking first at Table 17A, we see that a desired AQL of 1.4% would correspond to a tabled value of 1.5%. Finding the lot size of 400 for inspection level IV in Table 17B yields a code letter of I.

Now looking at Table 18 in the row labeled I, we find a sample size of 25. For that same row the column headed with an AQL of 1.5% gives $k = 1.72$. Thus we are to sample 25 wires from the lot and accept the lot if

$$\frac{\bar{X} - 5}{S} \geq 1.72$$

where $\bar{X}$ and S are the mean and standard deviation for the 25 tensile strength measurements.   ■

MIL-STD-414 also gives a table for reduced inspection, much as in MIL-STD-105D. The handbook provides OC curves for the sampling plans tabled there and, in addition, gives sampling plans based on sample ranges rather than standard deviations.

The theory connected with the development of the sampling plans for inspection by variables is more difficult than that for inspection by attributes, and cannot be developed fully here. However, we will outline the basic approach.

Suppose again that X is the quality measurement under study and U is the upper specification limit. The probability that an item is defective is then given by $P(X > U)$. (This probability also represents the proportion of defective items in the lot.) Now we want to accept only lots for which $P(X > U)$ is small, say less than or equal to a constant M. But

$$P(X > U) \leqslant M$$

is equivalent to

$$P\left(\frac{X - \mu}{\sigma} > \frac{U - \mu}{\sigma}\right) \leqslant M$$

where $\mu = E(X)$ and $\sigma^2 = V(X)$. If, in addition, X has a normal distribution, then the latter probability statement is equivalent to

$$P\left(Z > \frac{U - \mu}{\sigma}\right) \leqslant M$$

where Z has a standard normal distribution.

It follows that $P(X > U)$ will be "small" if and only if $(U - \mu)/\sigma$ is "large." Thus

$$P(X > U) \leqslant M$$

is equivalent to

$$\frac{U - \mu}{\sigma} \geqslant k^*$$

for some constant k^*. Since μ and σ are unknown, the actual acceptance criterion used in practice becomes

$$\frac{U - \bar{X}}{S} \geqslant k$$

where k is chosen so that $P(X > U) \leqslant M$.

For normally distributed X, the $P(X > U)$ can be estimated by making use of the sample information provided by $\bar{X}$ and S, and these estimates are tabled in MIL-STD-414. These estimates are used in the construction of OC curves. Also tables are given that show the values of M, in addition to the value of k, for sampling plans designed for specified AQLs.

EXERCISES

13.16 Construct operating characteristic curves for the following sampling plans, when sampling by attributes:

(a) $n = 10$, $a = 2$ (b) $n = 10$, $a = 4$
(c) $n = 20$, $a = 2$ (d) $n = 20$, $a = 4$

13.17 Suppose a lot of size 3000 is to be sampled by attributes. An AQL of 4% is specified for acceptable lots. Find the appropriate level II sampling plan under

(a) normal inspection.
(b) tightened inspection.
(c) reduced inspection.

13.18 Fuses of a certain type must not allow the current passing through them to exceed 20 amps. A lot of 1200 of these fuses is to be checked for quality by a lot-acceptance sampling plan with inspection by variables. The desired AQL is 2%. Find the

appropriate sampling plan for level IV sampling and
    (a)   normal inspection.   (b)   tightened inspection.

13.19   Refer to Exercise 13.18. If, under normal inspection, the sample mean turned out to be 19.2 amps and the sample standard deviation 0.3 amp, what would you conclude?

13.20   For a lot of 250 gallon cans filled with a certain industrial chemical, each gallon is specified to have a percentage of alcohol in excess of L. If the AQL is specified to be 10%, find an appropriate lot-acceptance sampling plan (level IV inspection) under
    (a)   normal inspection.   (b)   tightened inspection.

13.8 TOLERANCE LIMITS

The quality assurance techniques addressed thus far are concerned with estimating, and controlling, only one aspect of the distribution of a quality or productivity measurement, usually the mean or variance. In addition to information on these selected parameters, engineers and technicians often want to determine an interval that contains a certain (high) proportion of the measurements, with a known probability. For example, in the production of resistors it is helpful to have information of the form "we are 95% confident that 90% of the resistances produced are between 490 and 510 ohms." Such an interval is called a *tolerance interval* for 90% of the population with confidence coefficient $1 - \alpha = 0.95$. The lower end of the interval (490 in this case) is called the *lower tolerance limit*, and the upper end (510 in this case) is called the *upper tolerance limit*. We still use the term "confidence coefficient" to reflect the confidence we have in a particular method producing an interval that includes the prescribed portion of the population. That is, a method for constructing 95% tolerance intervals for 90% of a population should produce intervals that contain approximately 90% of the population measurements in 95 out of every 100 times it is used.

    Tolerance intervals are sometimes specified by a design engineer or production supervisor, in which case measurements must be taken from time to time to see if the specification is being met. More often, measurements on a process are taken first, and then tolerance limits are calculated to describe how the process is operating. These "natural" tolerance limits are often made available to customers.

    Suppose the key measurements (resistances, for example) on a process have been collected and studied over a long period of time and are known to have a normal distribution with mean μ and standard deviation σ. We can then easily construct a tolerance interval for 90% of the population since the interval

$$(\mu - 1.645\sigma, \mu + 1.645\sigma)$$

must contain precisely 90% of the population measurements. The confidence coefficient here is 1.0 since we have no doubt that exactly 90% of the measurements fall in the resulting interval. In most cases, however, μ and σ are not known and must be estimated by $\bar{x}$ and s, which are calculated from sample data. The interval

$$(\bar{x} - 1.645s, \bar{x} + 1.645s)$$

will no longer cover exactly 90% of the population because of the errors introduced by the estimators. However, it is possible to find a number K such that

$$(\bar{x} - Ks, \bar{x} + Ks)$$

has the property that it covers a proportion δ of the population with confidence coefficient $1 - \alpha$. Table 19 in the Appendix gives values of K for three values of δ and two values of $1 - \alpha$.

EXAMPLE 13.9

A random sample of 45 resistors was tested with a resulting average resistance produced of 498 ohms. The standard deviation of these resistances was 4 ohms. Find a 95% tolerance interval for 90% of the resistance measurements in the population from which this sample was selected. Assume a normal distribution for this population.

Solution For these data $n = 45$, $\bar{x} = 498$, and $s = 4$. From Table 19 with $\delta = 0.90$ and $1 - \alpha = 0.95$, we have $K = 2.021$. Thus

$$(\bar{x} - 2.021s, \bar{x} + 2.021s)$$

$$(498 - 2.021(4), 498 + 2.021(4))$$

or

$$(489.9, 506.1)$$

forms the realization of the 95% tolerance interval. We are 95% confident that 90% of the population resistances lie between 489.9 and 506.1 ohms. (In constructing such intervals, the lower tolerance limit should be rounded down and the upper tolerance limit rounded up.)

Notice that as n increases, the values of K tend toward the corresponding $z_{\alpha/2}$ values. Thus the tolerance intervals $\bar{x} \pm Ks$ tend toward the intervals $\mu \pm z_{\alpha/2}\sigma$ as n gets large. ∎

The tolerance interval formulation discussed in previous paragraphs requires an assumption of normality for the data on which the interval is based. We now present a simple nonparametric technique for constructing tolerance intervals for any continuous distribution. Suppose a random sample of measurements $X_1, X_2, \ldots, X_n$ is ordered from smallest to largest to produce an ordered set $Y_1, Y_2, \ldots, Y_n$. That is, $Y_1 = \min(X_1, \ldots, X_n)$ and $Y_n = \max(X_1, \ldots, X_n)$. It can then be shown that

$$P[(Y_1, Y_n) \text{ covers at least } \delta \text{ of the population}]$$

$$= 1 - n\delta^{n-1} + (n - 1)\delta^n$$

Thus the interval (Y_1, Y_n) is a tolerance interval for a proportion δ of the population with confidence coefficient $1 - n\delta^{n-1} + (n - 1)\delta^n$. For a specified δ the confidence coefficient is determined by the sample size. We illustrate the use of nonparametric tolerance intervals in the following examples.

EXAMPLE 13.10

A random sample of $n = 50$ measurements on the lifelength of a certain type of L.E.D. display showed the minimum to be $y_1 = 2150$ hours and the maximum to be $y_{50} = 2610$ hours. If (y_1, y_{50}) is used as a tolerance interval for 90% of the population measurements on lifelengths, find the confidence coefficient.

Solution   It is specified that $\delta = 0.90$ and $n = 50$. Thus the confidence coefficient for an interval of the form (Y_1, Y_{50}) is given by

$$1 - n\delta^{n-1} + (n-1)\delta^n = 1 - 50(0.9)^{49} + 49(0.9)^{50} = 0.97$$

We are 97% confident that the interval (2150, 2610) contains at least 90% of the lifelength measurements for the population under study.   ∎

EXAMPLE 13.11

In the setting of Example 13.10 suppose the experimenter wants to choose the sample size n so that (Y_1, Y_n) is a 90% tolerance interval for 90% of the population measurements. What value should he choose for n?

Solution   Here the confidence coefficient is 0.90 and $\delta = 0.90$. So we must solve the equation

$$1 - n(0.9)^{n-1} + (n-1)(0.9)^n = 0.90$$

This can be done quickly by trial and error, with a calculator. The value $n = 37$ gives a confidence coefficient of 0.896, and the value $n = 38$ gives a confidence coefficient of 0.905. Therefore $n = 38$ should be chosen to guarantee a confidence coefficient of at least 0.90. With $n = 38$ observations, (Y_1, Y_n) is a 90.5% tolerance interval for 90% of the population measurements.   ∎

EXERCISES

13.21   A bottle-filling machine is set to dispense 10.0 cc of liquid into each bottle. A random sample of 100 filled bottles gives a mean fill of 10.1 cc and a standard deviation of 0.02 cc. Assuming the amounts dispensed to be normally distributed,
　　(a)   Construct a 95% tolerance interval for 90% of the population measurements.
　　(b)   Construct a 99% tolerance interval for 90% of the population measurements.
　　(c)   Construct a 95% tolerance interval for 99% of the population measurements.

13.22   The average thickness of the plastic coating on electrical wire is an important variable in determining the wearing characteristics of the wire. Ten thickness measurements from randomly selected points along a wire gave the following (in thousandths of an inch):

　　　8, 7, 10, 5, 12, 9, 8, 7, 9, 10

Construct a 99% tolerance interval for 95% of the population measurements on coating thickness. What assumption is necessary for your answer to be valid?

13.23   A cutoff operation is supposed to cut dowel pins to 1.2 inches in length. A random sample of 60 such pins gave an average length of 1.1 inches and a standard deviation of 0.03 inch. Assuming normality of the pin lengths, find a tolerance interval for 90% of the pin lengths in the population from which this sample was selected. Use a confidence coefficient of 0.95.

13.24   In the setting of Exercise 13.23 construct the tolerance interval if the sample standard deviation is
　　(a)   0.015 inch.　　　　　　(b)   0.06 inch.

Compare these answers to that of Exercise 13.23. Does the size of s seem to have a large effect upon the length of the tolerance interval?

13.25 In the setting of Exercise 13.23 construct the tolerance interval if the sample size is
  (a) $n = 100$.   (b) $n = 200$.
Does the sample size seem to have a great effect upon the length of the tolerance interval, as long as it is "large"?

13.26 For the data of Exercise 13.22 and without the assumption of normality,
  (a) Find the confidence coefficient if (y_1, y_n) is to be used as a tolerance interval for 95% of the population.
  (b) Find the confidence coefficient if (y_1, y_n) is used as a tolerance interval for 80% of the population.

13.27 The interval (Y_1, Y_n) is to be used as a method of constructing tolerance intervals for a proportion δ of a population with confidence coefficient $1 - \alpha$. Find the minimum sample sizes needed to ensure that
  (a) $1 - \alpha = 0.90$ for $\delta = 0.95$.   (b) $1 - \alpha = 0.90$ for $\delta = 0.80$.
  (c) $1 - \alpha = 0.95$ for $\delta = 0.95$.   (d) $1 - \alpha = 0.95$ for $\delta = 0.80$.

13.9 CONCLUSION

Quality assurance often necessitates the use of control charts. The construction of control limits is based on standard statistical techniques, sometimes using the sample mean as a measure of central tendency and the sample variance as a measure of variability. More commonly, the range is used as a measure of variability. The normal distribution is employed as a basic model for control limits based on means, whereas the binomial or Poisson distribution is commonly used for counts.

Control limits are usually set so that the chance of stating that a process is out of control when, in fact, it is in control is very small, much smaller than the common 0.05 value for Type I errors in statistical tests. This is so because of the often serious and expensive consequences of falsely declaring a process to be out of control. Also the control chart uses a series of many "hypothesis tests," and a 1 in 20 error rate is too large.

There are many other adaptations of control-charting techniques. Some can be found in the suggested references.

If instead of controlling an ongoing process, one is interested in deciding whether to accept or reject a lot of manufactured items or raw materials, a lot-acceptance sampling plan is needed. Inspection for lot acceptance may be by attributes or by variables, and excellent military standard sampling plans are available for both types of inspection.

SUPPLEMENTARY EXERCISES

13.28 Casts of aluminum are checked for tensile strength by testing four specimens from each batch. The results for tests of ten batches are given in the table, with measurements in 1000 psi.

Batch	Tensile Strength Measurements
1	40 41 43 39
2	44 42 46 41
3	32 34 33 31
4	42 47 43 44
5	46 48 49 46
6	47 49 48 46
7	42 44 43 46
8	49 47 42 48
9	48 47 32 31
10	42 44 46 45

  (a) Construct control limits for the mean tensile strength, using ranges as a measure of variability.
  (b) Do all samples appear to come from a process "in control"?
  (c) How would you adjust the limits found in (a) for use in future sampling?

13.29 Use the data given in Exercise 13.28 to answer the following:
  (a) Construct control limits for variability.
  (b) Would you use these limits for future studies? If not, how would you adjust them for future use?

13.30 Refer to Exercise 13.28. Adjust the control limits so that a sample mean will fall outside the limits with probability approximately 0.02 when the process is in control.

13.31 Refer to Exercise 13.29. Adjust the control limits so that a sample range will fall outside the limits with probability approximately 0.01 when the process is in control.

13.32 Use the data given in Exercise 13.5 to answer the following. A specimen will be called "defective" if the percentage of copper is outside the interval (80, 90). Construct control limits for the percentage of specimens that are defective, using these data. Do all of the observed samples appear to be "in control"?

13.33 Occasionally, samples of equal size cannot be obtained at every time point. In that event the control limits must be computed separately for each sample size. Construct a three-standard deviation limit control chart for the mean, using the sample standard deviations as measures of variability, based on the following data on measures of pollutants in water samples.

Sample	Amount of Pollutant				
1	6.2	7.3	9.2	8.1	
2	5.4	6.8	3.2		
3	20.4	16.8	17.2	16.0	19.1
4	5.7	10.2	9.3		
5	6.8	10.1	7.2	8.4	
6	7.9	10.4	8.6	9.3	
7	17.1	16.8	17.3		
8	16.2	13.9	14.7	16.1	
9	10.2	11.4	13.2	10.9	
10	11.1	12.4	13.2		

[*Hint*: An unbiased estimate of μ can be found by using a weighted average of the sample means, with weights equal to the sample sizes. The best estimate of σ^2 involves a weighted average of the sample variances s_i^2.]

13.34 Discuss the construction of a p-chart for proportions when the sample sizes vary.

13.35 Discuss the construction of a c-chart when the sample sizes vary.

Experiences with Real Data

ABOUT THIS CHAPTER

Up to this point all of the exercises in this book were related to a particular section or, at least, to a particular chapter. Thus the techniques for solving the problem in question were readily available because they had just been discussed in the immediately preceding material. Little room was left for creativity on the part of the student in choosing an appropriate analysis. Now a set of problems is presented as exercises without guidance as to the most appropriate technique for solution. In fact there may be no one best technique for solving some of these problems, and it is up to the student to justify the assumptions for the chosen method of analysis.

The exercises that follow are all based upon real data or experimental situations. Part of the value in using real data derives from the fact that such data does not usually look like the neat textbook examples. Often the data analyst encounters outliers, missing observations, and data collected without the benefit of a carefully designed, balanced experimental plan. It is then up to the analyst to come up with an analysis that is meaningful to the experimenter but, at the same time, satisfies the assumptions upon which the analysis is based. This is part of the challenge that accompanies any good statistical analysis. For the problems that follow, an appropriate technique for data analysis can be found among those already presented in this book. So keep an open mind, think carefully about methods and their assumptions, and enjoy the challenge of being a statistical data analyst as you look for solutions to the following real problems.

EXERCISES

14.1   In building a model to study automobile fuel consumption, Biggs and Akcelik (*Journal of Transportation Engineering*, Vol. 113, No. 1, January 1987, pp. 101–106) begin by looking at the relationship between idle fuel consumption and engine capacity. Suppose the data are as follows:

Idle fuel consumption (mL/s)	Engine size (L)
0.18	1.2
0.21	1.2
0.17	1.2
0.31	1.8
0.34	1.8
0.29	1.8
0.42	2.5
0.39	2.5
0.45	2.5
0.52	3.4
0.61	3.4
0.44	3.4
0.62	4.2
0.65	4.2
0.59	4.2

(handwritten annotations: x, y; use it as + reorder highest to lowest; .186, .313, .42, .523, .62; same engine size; draw some kinds of conclusions)

What model do you think best explains the relationship between idle fuel consumption and engine size?

14.2   The Florida Game and Freshwater Fish Commission is interested in a model that will allow the accurate prediction of the weight of an alligator from more easily observed data on length. It is suggested that the model should have the form

$$w = al^b$$

for weight (w) and length (l). Using the data given below for 25 alligators, estimate a and b. What is the practical significance of b? Is the estimated value you obtained surprising?

Alligator	l	w
1	94	130
2	74	51
3	147	640
4	58	28
5	86	80
6	94	110
7	63	33
8	86	90
9	69	36

(continued)

Alligator	*l*	*w*
10	72	38
11	128	366
12	85	84
13	82	80
14	86	83
15	88	70
16	72	61
17	74	54
18	61	44
19	90	106
20	89	84
21	68	39
22	76	42
23	114	197
24	90	102
25	78	57

(Weight is in pounds; length in inches.)

14.3 Drying stresses in wood produce acoustic emissions (AE), and the rate of acoustic emission can be monitored as a check on drying conditions. The data shown below are the results of measuring peak values of acoustic emission over 10-second intervals for wood specimens with controlled diameter, thickness, temperature, and relative humidity. Which factors appear to affect the AE rates?

Test No.	Diameter	Thickness	Dry-Bulb Temp. ($°$C)	RH (%)	Peak Values of AE Event Count Rates (1/10s)
	---------(mm)---------				
1	180	20	40	40	780
2	180	20	50	50	695
3	180	20	60	30	1140
4	180	25	40	40	270
5	180	25	50	50	340
6	180	25	60	30	490
7	180	30	40	50	160
8	180	30	50	30	270
9	180	30	60	40	185
10	125	20	40	50	620
11	125	20	50	30	1395
12	125	20	60	40	1470
13	125	25	40	30	645
14	125	25	50	40	1020
15	125	25	60	50	755
16	125	30	40	30	545
17	125	30	50	40	635
18	125	30	60	50	335

(Source: Naguchi, Kitayama, Satoyoshi and Umetsu, *Forest Products Journal*, 37 (1), 1987, pp. 28–34.)

14.4   Is knowledge of right-of-way laws at four-legged intersections associated with similar knowledge for T intersections? Montgomery and Carstens (*Journal of Transportation Engineering*, Vol. 113, No. 3, May 1987, pp. 299–314) show results of a questionnaire designed to investigate this question (among others). The data are as follows:

		Response to T-Intersection Question		
		Correct	Incorrect	
Response to Four-Legged-Intersection Question	Correct	141	145	286
	Incorrect	13	188	201
		154	333	487

How would you answer the question posed above?

14.5   A study of volatile organic compounds (VOCs) in breath, personal air, and outdoor air was made for the Bayonne/Elizabeth, New Jersey area and for Los Angeles. One part of the study compared winter air to summer/spring air. From the samples taken in both time periods, rank correlation coefficients were calculated, as seen below.

Spearman Correlations Between Winter and Spring/Summer[a] in Bayonne/Elizabeth and Los Angeles

Compound	Breath Bayonne/ Elizabeth	Breath Los Angeles	Overnight personal air Bayonne/ Elizabeth	Overnight personal air Los Angeles	Daytime personal air Bayonne/ Elizabeth	Daytime personal air Los Angeles	Overnight outdoor air Bayonne/ Elizabeth	Overnight outdoor air Los Angeles	Daytime outdoor air Bayonne/ Elizabeth	Daytime outdoor air Los Angeles
Ethylbenzene	0.29	0.53[†]	0.11	0.37[†]	0.36[†]	0.27	−0.06	0.13	0.36	0.58[†]
o-Xylene	0.25	0.48[†]	0.17	0.29[†]	0.45[†]	0.10	0.02	0.14[·]	0.29	0.38
m,p-Xylene	0.32[†]	0.55[†]	0.23	0.27	0.43[†]	0.27	−0.12	0.29	0.42	0.42[†]
Benzene	0.02	0.57[†]	−0.13	0.08	0.22	0.36[†]	−0.77	0.29	0.26	0.51[†]
Tetrachloroethylene	0.25	0.42[†]	0.25	0.49[†]	0.02	0.21	0.14	0.16	0.22	0.67[†]
1,1,1-Trichloroethylene	0.02	0.44[†]	0.15	0.29[†]	0.23	0.26	−0.89[†]	0.18	−0.39	0.45[†]
m,p-Dichlorobenzene	0.40[†]	0.42[†]	0.33[†]	0.56[†]	0.40[†]	0.48[†]	−0.14	0.34	0.32	0.30
Styrene	0.27	0.24	0.16	0.13	0.25	0.29[†]	0.01	0.59[†]	0.00	0.08
Chloroform	−0.29	0.17	−0.17	0.60[†]	−0.07	0.17	0.31	0.05	−0.31	−0.09
Sample size	14–44	49–51	46–49	50–51	38–40	49–50	6–9	22–23	6–8	23–24

[a]Spring in Los Angeles, summer in Bayonne/Elizabeth.
[†]Correlations significantly different from zero at the 0.05 level.
(Source: Hartwell, Pellizzari, Perritt, Whitmore, Zelon, and Wallace, *Atmospheric Environment*, Vol. 21, No. 11, 1987, pp. 2413–2424.)

What explanations can you give to this analysis?

14.6   From data on measured alkalinity of certain streams and on the exposure of the water to certain rock types, a prediction model is to be constructed. It is desired to use the model to predict alkalinity of water flowing through certain types of rocks. Samples from seven streams and their surrounding basins produced the following data:

Stream	Rock Exposure (%)	Alkalinity (μeq/L)
1	30	474
2	49	558
3	40	518
4	27	199
5	39	377
6	42	613
7	24	285

(Source: DeWalle, Diricola and Sharpe, *Water Resources Bulletin*, Vol. 23, No. 1, 1987 pp. 29–35.)

Construct an appropriate prediction model from these data.

14.7 In a study of wood-laminating adhesives by Kreibich and Hemingway (*Forest Products Journal*, 37 (2), 1987 pp. 43–46), tannin is added to the usual phenol-resorcinol-formaldehyde (PRF) resin and tested for strength. For four mixtures of resin and tannin and two types of test, the resulting shear strengths of the wood specimens (in psi) were as follows:

PRF/Tannin Ratio	Test A	Test B
50/50	1,960	1,460
	2,267	140
40/60	1,450	1,620
	1,683	2,287
30/70	2,080	760
	1,750	130
23/77	1,493	650
	1,850	90

Test A: dry shear test
Test B: two-hour boil test

What can you conclude about the effects of the tannin mixtures on the strength of the laminate?

14.8 In an article on optimal maintenance decisions, Carnahan, Davis, Shahin, Keane, and Wu (*Journal of Transportation Engineering*, Vol. 113, No. 5, September 1987, pp. 554–572) present data on highway pavement condition as a function of age. Without actually doing an analysis, discuss how you would fit an appropriate linear model to these data.

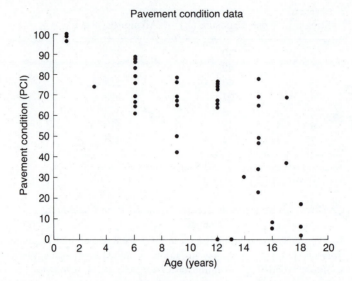

Pavement condition data

14.9 Sulfate concentrations in a shallow unconfined aquifer were measured quarterly for a period of seven years. Similarly chloride concentrations were measured quarterly for nine years at a different site. The data are shown below. Do these data sets show trends over time? Does either data set appear to fit the normal distribution? Do you think a transformed data set (in either case) would fit the normal distribution more closely?

Sample No.	Concentration	Sample No.	Concentration
	Sulfate (mg/l)		
1	111	14	102
2	107	15	145
3	108.7	16	87
4	108.7	17	112
5	109.3	18	111
6	104.7	19	104
7	104.7	20	151
8	108	21	103
9	108	22	113
10	109.3	23	113
11	114.5	24	125
12	113	25	101
13	51.1	26	108
	Chloride (mg/l)		
1	38	18	—
2	40	19	41
3	35	20	—
4	37	21	35
5	32	22	49
6	37	23	64
7	37	24	73

(continued)

Sample No.	Concentration	Sample No.	Concentration
8	—	25	67
9	32	26	67
10	45	27	—
11	38	28	59
12	33.8	29	73
13	14	30	—
14	—	31	92.5
15	39	32	45.5
16	46	33	40.4
17	48	34	33.9
		35	28.1

(Source: Harris, Loftis, and Montgomery, *Ground Water*, Vol. 25, No. 2, 1987, pp. 185–193.)

14.10 One important feature of the plates used in high performance thin-layer chromatography is the water uptake capability. Six plates were modified by increasing the amount of cyanosilane (CN). Water uptake measurements (Rf) were then recorded for three component mixtures (BaP, BaA, Phe) used with each plate. Six measurements were made on each mixture-plate combination. The data are as follows.

Rf Values and Standard Deviation for BaP, BaA, and Phe on Modified Plates

Plate No.	Type of Plate	Compound	Mean Rf	s
1	0%-CN	BaP	0.17	0.021
		BaA	0.25	0.026
		Phe	0.36	0.033
2	1%-CN	BaP	0.21	0.014
		BaA	0.30	0.023
		Phe	0.44	0.046
3	5%-CN	BaP	0.23	0.021
		BaA	0.31	0.016
		Phe	0.41	0.025
4	10%-CN	BaP	0.36	0.028
		BaA	0.45	0.026
		Phe	0.59	0.027
5	20%-CN	BaP	0.46	0.045
		BaA	0.58	0.026
		Phe	0.73	0.027
6	30%-CN	BaP	0.46	0.069
		BaA	0.56	0.070
		Phe	0.66	0.095

(Source: Colmsjö and Ericsson, *Journal of High Resolution Chromatography*, Vol. 10, April 1987, pp. 177–180.)

Fit a model relating % CN to mean Rf for each compound. (The models could differ.) Do the values of *s* (standard deviation) suggest any complication?

14.11   Acid precipitation studies in southeastern Arizona involved sampling rain from two sites (Bisbee and the Tucson Research Ranch of the National Audubon Society) and sampling air from smelter plumes. The data for five elements, in terms of their ratios to copper, is given on the table below. One objective of the study is to see if the plume samples differ from the rain samples at the other sites. What can you say about this aspect of the study?

Summary Statistics for Metals in Precipitation and Plume Samples

Samples	Statistic	Mass Ratio (g g)[a]				
		As/Cu	Cd/Cu	Pb/Cu	Sb/Cu	Zn/Cu
Bisbee	Mean	0.46	0.068	1.03	0.073	1.64
(8/84–10/84)	S.E.[b]	0.15	0.017	0.16	0.019	0.31
	Min	0.14	0.029	0.31	0.043	0.58
	Max	1.33	0.19	2.26	0.108	3.79
	n[b]	7	10	13	3	13
TRR	Mean	0.48	0.087	0.94	0.078	4.2
(8/84–10/84)	S.E.	0.12	0.024	0.17	0.018	1.1
	Min	0.13	0.019	0.26	0.037	0.8
	Max	1.3	0.30	2.1	0.16	12.2
	n	11	11	11	7	11
TRR	Mean	0.56	0.074	0.82	0.10	9.7
(8/84–9/85)	S.E.	0.067	0.011	0.07	0.016	1.6
	Min	0.086	0.014	0.08	0.020	0.8
	Max	1.33	0.30	2.1	0.17	55
	n	31	31	49	11	49
Smelter	Mean	0.72	0.077	0.90	0.054	1.7
plumes	S.E.	0.37	0.022	0.23	0.034	0.67
	Min	0.20	0.01	0.44	0.014	0.11
	Max	2.2	0.15	1.6	0.19	3.8
	n	5	5	4	5	5

[a]Each ratio was computed for all events for which the metal exceeded its detection limit.
[b]S.E. is the standard error of the mean and *n* is number of events.
(Source: Blanchard and Stromberg, *Atmospheric Environment*, Vol. 21, No. 11, 1987, pp. 2375–2381.)

14.12   Lumber moisture content (MC) must be monitored during drying to maximize the quality and minimize the drying costs. Four types of moisture meters were studied on three levels of moisture content in lumber by Breiner, Arganbright, and Pong (*Forest Products Journal*, 37(4), 1987, pp. 9–16). Of the four meters used, two were hand-held (resistance and power-loss types) and two were in-line (high and low frequency). From the data given below, can you determine any significant differences among the meters for the various MC levels?

Variation in Average MC Between Three Repeat Runs

	Resistance	Power-loss	High-frequency	Low-frequency
20% MC level	$\bar{X} = 17.8$	$\bar{X} = 14.2$	$\bar{X} = 16.8$	$\bar{X} = 19.4$
	SD = 0.3	SD = 0.2	SD = 0.1	SD = 0.1
	CV = 1.7%	CV = 1.4%	CV = 0.06%	CV = 0.3%

Variation in Average MC Between Three Repeat Runs (cont.)

	Resistance	Power-loss	High-frequency	Low-frequency
15% MC level	$\bar{X} = 13.5$ $SD = 0.2$ $CV = 1.1\%$	$\bar{X} = 11.9$ $SD = 0.1$ $CV = 0.9\%$	$\bar{X} = 13.7$ $SD = 0.1$ $CV = 0.8\%$	$\bar{X} = 16.9$ $SD = 0.1$ $CV = 0.5\%$
10% MC level	$\bar{X} = 11.4$ $SD = 0.1$ $CV = 0.9\%$	$\bar{X} = 10.1$ $SD = 0.1$ $CV = 1.3\%$	$\bar{X} = 11.3$ $SD = 0.1$ $CV = 0.6\%$	$\bar{X} = 15.1$ $SD = 0.1$ $CV = 0.5\%$

(CV = coefficient of variation = $SD/\bar{x}$)

14.13 An intersection hazard index (IHI) established by Ward, Eck, and Polus (*Journal of Transportation Engineering*, Vol. 113, No. 2, March 1987, pp. 211–215) is to be used to set priorities on repair of hazardous intersections. Given the data below from 36 intersections, would you say that the IHI is a good predictor of the number of accidents in a specified time period? Is the IHI a good predictor of the accident rate for such intersections?

Intersection Number (1)	Intersection Hazard Index (IHI) (2)	Priority Basis and Rankings		
		IHI (3)	Number of Accidents (4)	Accident Rate (5)
1	3,897.4	7	3	10
2	3.5	30	32	31
3	151.8	17	15	11
4	95.4	21	16	18
5	58.0	22	16	2
6	10.9	27	25	17
7	0.0	35	35	35
8	23.1	24	26	21
9	2,035.7	10	11	16
10	4,669.7	4	11	4
11	2.0	33	28	27
12	0.0	35	35	35
13	6.8	28	28	6
14	108.5	20	16	11
15	119.6	18	22	3
16	57.4	23	23	7
17	343.9	16	16	15
18	2,160.9	9	6	22
19	4,078.2	6	9	12
20	1,579.0	12	7	26
21	536.1	15	13	25
22	2.8	31	28	29
23	2.3	32	32	32
24	208,302.4	1	1	5
25	13,106.8	3	4	9

(continued)

Intersection Number (1)	Intersection Hazard Index (IHI) (2)	Priority Basis and Rankings		
		IHI (3)	Number of Accidents (4)	Accident Rate (5)
26	833.0	13	13	14
27	111.9	19	16	19
28	618.5	14	16	7
29	2,530.1	8	9	23
30	21.0	25	23	30
31	70,123.3	2	2	24
32	4,361.2	5	5	20
33	6.4	29	28	34
34	1,964.9	11	7	13
35	15.5	26	26	28
36	1.0	34	34	33

14.14 Mouthpiece forces produced while playing the trumpet were studied by Barbenel, Kerry, and Davies (*Journal of Biomechanics*, Vol. 21, No. 5, 1988, pp. 417–424). Twenty-five high-proficiency players and 35 medium-proficiency players produced results summarized on the following table.

Results of Maximum Force Tests. Mean Value (interquantile range), N. All Subjects

Test	High proficiency		Medium Proficiency	
	Popular	Classical	Popular	Classical
C Major Scale	72 (53–83)	63 (51–72)	51 (43–59)	44 (38–57)
G_5 at 95 dB				
Maximum Force	85 (63–97)	73 (59–83)	54 (46–66)	46 (28–56)
Normal Force	24 (19–28)	27 (20–35)	28 (22–35)	26 (18–29)
Minimum Force	22 (18–30)	23 (16–27)	24 (20–28)	23 (15–30)

Given only mean values and interquartile ranges, how would you describe the differences, if any, between high-proficiency and medium-proficiency players playing a C major scale? Do there appear to be differences in mean forces between popular and classical play of C major scales? Describe the key differences, if any, among maximum, normal, and minimum forces measured during the playing of G_5 at 95 dB.

14.15 Aerosol particles in the atmosphere influence the transmission of solar radiation by absorption and scattering. Absorption causes direct heating of the atmosphere, and scattering causes some radiation to be reflected to space and diffuses that reaching the ground. Thus accurate measurements of absorption and scattering coefficients are quite important. The data given below comes from a study by Kilsby and Smith (*Atmospheric Environment*, Vol. 21, No. 10, 1987, pp. 2233–2246). Estimate the difference between aircraft and ground mean scattering coefficients (with a 95% confidence coefficient), assuming the sample sizes are 3. Repeat, assuming the sample sizes are 20. Perform similar estimations for the absorption coefficients.

Comparison of Ground and Aircraft Measurements Around Noon on 27 July

Instrument	Quantity	Aircraft Measurement	Ground Measurement	Ratio $\dfrac{\text{Aircraft}}{\text{Ground}}$
Nephelometer	Scattering coefficient at 0.55 μm (10^{-4} m^{-1})	1.2 ($\pm$0.2)	0.8 ($\pm$0.2)	1.5 ($\pm$0.5)
Filter Sample	Absorption coefficient at 0.55 μm (10^{-6} m^{-1})	15.5 ($\pm$2.5)	8.3 ($\pm$1.3)	1.9 ($\pm$0.6)
FSSP	Particle concentration $0.25 < r < 3.75\,\mu$m (cm^{-3})	3.5 ($\pm$1.0)	2.6 ($\pm$1.0)	1.3 ($\pm$0.8)

(Standard deviations are given in parentheses.)

14.16 The water quality beneath five retention/recharge basins for an urban area was investigated by the EPA. Data on salinity of the water for samples taken at varying depths from each basin are shown on the accompanying table. How would you suggest comparing the mean salinity for the five basins, in light of the fact that measurement depths are not the same? Carry out a statistical analysis of these means.

Basin Name	Depth of Sample, m[a]	Salinity Mean	Salinity Standard Deviation	Salinity No. of Observations
F	2.74	0.123	0.063	10
	8.53	0.099	0.016	3
	15.8	0.112	0.035	15
	20.7[b]	0.153	0.075	18
G	2.74	0.119	NC[c]	1
	14.0	0.541	NC	1
	23.8[b]	0.744	0.090	16
M	3.64	0.294	NC	1
	11.3	0.150	0.035	7
	20.7	0.148	NC	1
	26.2[b]	0.109	0.034	12
EE	3.66	0.085	0.042	13
	7.01	0.158	0.054	12
	10.1	0.175	0.035	14
MM	3.66	0.050	0.012	5
	8.53	0.054	0.015	8
	17.7[b]	0.061	0.020	8

[a]Depth of sampling ceramic cup below basin floor.
[b]Considered ground water because of similar depth to local groundwater table.
[c]Not calculated. Insufficient data points.
(Source: Nightingale, *Water Resources Bulletin*, Vol. 23, No. 2, 1987, pp. 197–205.)

14.17 How do back and arm strengths of men and women change as the test moves from static to dynamic situations? This question was investigated by Kumar, Chaffin, and Redfern (*Journal of Biomechanics*, Vol. 21, No. 1, 1988, pp. 35–44). The dynamic conditions required movement at 20, 60, and 100 cm/sec. Ten male and ten female young adult volunteers without a history of back pain participated in the study. From the data given below analyze the differences between male and female strengths, the differences between back and arm strengths, and the pattern of strengths as one moves from static through three levels of dynamic strength tests.

	Tasks		Mean Peak Str. (N)
Males	Back	Static	726
		Slow	672
		Medium	639
		Fast	597
	Arm	Static	521
		Slow	399
		Medium	332
		Fast	275
Females	Back	Static	503
		Slow	487
		Medium	432
		Fast	436
	Arm	Static	296
		Slow	266
		Medium	221
		Fast	192

14.18 Hydrogen peroxide may have important consequences in air pollution, but its ambient air concentration is difficult to measure. A new absorption device for H_2O_2 was tested in aqueous solutions with known concentrations of hydrogen peroxide. The results were as follows:

Calibration Data for H_2O_2 in Aqueous Solutions

H_2O_2 Content ($\mu g\ ml^{-1}$)	Number of Measurements	Absorbance		
		Mean	S.D.	Rel. S.D. (%)
0.000	25	0.0226	0.0015	6.64
0.060	12	0.0368	0.0033	8.97
0.300	12	0.0514	0.0021	4.09
0.600	12	0.0758	0.0021	2.77
1.200	12	0.1248	0.0033	2.64
3.000	12	0.2683	0.0024	0.89
6.000	12	0.5133	0.0033	0.64
12.000	4	1.0018	0.0020	0.20

(Source: Hartkamp and Bachhausen, *Atmospheric Environment*, Vol. 21, No. 10, 1987, pp. 2207–2213.)

These data can be used to "calibrate" the new device so that absorbances from unknown samples can be used to predict H_2O_2 content for those samples. Find a linear calibration model relating H_2O_2 content to amount absorbed, using the sample means given above. Would the correlation based on these means increase or decrease if all the original data were used in finding the linear model?

14.19 Instantaneous flood discharge rates (in m^3/sec) were recorded over many years for 11 rivers in Connecticut. The sample means and standard deviations are reported below. Noting that the data is highly variable, how would you compare the mean discharge rates for these rivers? Perform an analysis to locate significant differences among these means, if possible.

River	Location	Drainage Area (sq km)	Record	Years of Record	$\bar{x}$	s_x
Scantic	Broad Brook, CT	178	1930–1982	53	42.0	51.3
Burlington	Burlington, CT	92	1937–1982	46	9.7	7.3
Connecticut	Hartford, CT	3500	1905–1982	78	980.1	761.5
North Branch Park	Hartford, CT	125	1956–1982	27	45.6	50.4
Salmon	East Hampton, CT	575	1930–1981	52	107.4	93.6
Eightmile	North Plain, CT	320	1938–1982	45	30.4	26.0
East Branch Eightmile	North Lyme, CT	197	1937–1982	46	24.9	23.3
Quinnipiac	Wallingford, CT	205	1932–1982	51	62.3	37.9
Blackberry	Canaan, CT	215	1943–1982	30	57.8	67.8
Tenmile	Gaylordsville, CT	430	1921–1982	52	103.5	81.3
Still	Lanesville, CT	495	1935–1982	48	45.7	37.9

(Source: Jain and Singh, *Water Resources Bulletin*, Vol. 23, No. 1, 1987, pp. 59–71.)

14.20 Decay fungi may colonize Douglas-fir utility poles during the seasoning process, thereby causing internal decay as the poles are in use. Przybylowicz, Kropp, Corden, and Graham (*Forest Products Journal*, 37(4), 1987, pp. 17–23) show the counts of Basidiomycete species in sample poles, both peeled and unpeeled and with varying air-seasoning times. Monokaryons (uninucleate) are generally new colonies and dikaryons tend to be older colonies. For the data below on colony counts, fourteen cores were sampled from each pole in the study. From these data can you detect any differences between monokaryon and dikaryon occurrence patterns? What can you say about differences in occurrence rates among species?

Frequency of Basidiomycete Mono- and Dikaryons Isolated from Douglas-Fir Poles Air-Seasoned for Varying Time Periods in the Pacific Northwest

	Number of Basidiomycete Isolates (monokaryons/dikaryons)								
	Unpeeled		Peeled					Totals	
Fungal species	Forest[a]	Yard[b]	0 to 6 Months	7 to 12 Months	13 to 18 Months	19 to 24 Months	25+ Months	Monok.	Dikary.
Porta placenta	2/0	0/2	7/1	9/17	13/14	21/19	17/16	69	69
Coriolus versicolor	14/5	2/3	9/6	4/6	3/39	2/7	5/12	39	78
Phlebia spp.	5/0	1/1	5/1	2/1	3/0	1/1	0/0	7	4
Poria carbonica	0/1	0/0	1/5	2/31	4/36	5/64	5/125	17	262

Frequency of Basidiomycete Mono- and Dikaryons Isolated from Douglas-Fir Poles Air-Seasoned for Varying Time Periods in the Pacific Northwest (cont.)

| | Number of Basidiomycete Isolates (monokaryons/dikaryons) | | | | | | | | |
| | Unpeeled | | Peeled | | | | | Totals | |
Fungal species	Forest[a]	Yard[b]	0 to 6 Months	7 to 12 Months	13 to 18 Months	19 to 24 Months	25+ Months	Monok.	Dikary.
Fomitopsis pinicola	4/1	1/0	1/2	2/0	0/0	2/0	1/0	11	3
Schizophyllum commune	2/1	0/1	3/9	0/2	3/17	0/3	0/2	8	35
Poria cinerascens	0/0	0/0	4/0	0/0	2/1	0/0	0/0	6	1
Fomitopsis cajanderi	0/3	0/18	3/10	0/20	0/8	0/11	0/13	3	83
Poria xantha	0/0	0/0	0/0	0/0	0/0	1/0	0/1	1	1
Total Basidiomycete Isolates	154	85	380	611	765	702	723		3,420
Percentage of Total Isolates That Were Monokaryons	18	4.7	8.7	3.1	3.7	4.6	3.9		5.0

[a]Sampled within 4 weeks of felling.
[b]Sampled with bark intact in pole yards.

14.21 Low temperatures on winter nights in Florida have caused extensive damage to the state's agriculture. Thus it is important to be able to predict freezes quite accurately. In an attempt to use satellite data for this purpose, a study was done to compare the surface temperatures detected by the infrared sensors of HCMM and GOES-East satellites. GOES temperature data compared to HCMM sample means are given in the accompanying table. Find an appropriate model relating the mean HCMM reading to the GOES temperature. What would happen to the R^2 value if the actual HCMM data points were used rather than their means?

| | HCMM Temp. (K) | | | |
| | | | Range | |
GOES Temp. (K)	Mean	S.D.	Min.	Max.
275.5	274.5	1.01	272.8	281.2
275.5	275.2	1.25	273.2	281.6
277.5	277.6	2.72	273.6	284.2
276.5	277.3	3.06	273.2	283.9
277.5	277.8	3.53	273.6	283.9
281.0	278.9	0.80	276.9	280.8
277.5	277.9	3.56	273.6	283.5
278.0	279.2	3.49	273.2	283.9
280.0	280.6	3.01	277.3	287.5
285.0	290.9	3.28	281.6	294.2

(continued)

	HCMM Temp. (K)			
			Range	
GOES Temp. (K)	Mean	S.D.	Min.	Max.
278.0	276.9	2.07	274.5	283.1
284.5	284.4	2.78	277.3	287.2
284.5	291.5	2.75	282.0	294.2
283.5	285.5	1.87	277.3	287.2
280.0	279.8	2.36	278.9	287.5

(Source: Chen and Allen, *Remote Sensing of the Environment*, Vol. 21, 1987, pp. 341–353.)

14.22 Does channelization have an effect on stream flow? This question is to be answered from data on annual streamflow and annual precipitation for two watersheds in Florida. (See Shirmohammadi, Sheridan, and Knisel, *Water Resources Bulletin*, Vol. 23, No. 1, 1987, pp. 103–111.) The data are summarized in the table below. Does there seem to be a significant difference in mean streamflow before and after channelization, even with differences in precipitation present?

Watershed	Area (km²)	Status	Record Period	Annual Precipitation (mm)		Annual Streamflow (mm)	
				Mean	S.D.	Mean	S.D.
W3	40.7	Before Channelization	1956–1963	1259	230.6	315.7	200.9
		After Channelization	1966–1981	1205	166.0	279.5	177.7
W2	255.6	Before Channelization	1956–1961	1308	272.9	442.6	256.3
		After Channelization	1969–1972	1301	216.1	396.9	216.6

Appendix

TABLE 1
TABLE 2
TABLE 3
TABLE 4
TABLE 5
TABLE 6
TABLE 7
TABLE 8
TABLE 9
TABLE 10
TABLE 11
TABLE 12
TABLE 13
TABLE 14
TABLE 15
TABLE 16
TABLE 17
TABLE 18
TABLE 19

TABLE 1 *Random Numbers*

Row\Column	1	2	3	4	5	6	7	8	9	10	11	12	13	14
1	10480	15011	01536	02011	81647	91646	69179	14194	62590	36207	20969	99570	91291	90700
2	22368	46573	25595	85393	30995	89198	27982	53402	93965	34095	52666	19174	39615	99505
3	24130	48360	22527	97265	76393	64809	15179	24830	49340	32081	30680	19655	63348	58629
4	42167	93093	06243	61680	07856	16376	39440	53537	71341	57004	00849	74917	97758	16379
5	37570	39975	81837	16656	06121	91782	60468	81305	49684	60672	14110	06927	01263	54613
6	77921	06907	11008	42751	27756	53498	18602	70659	90655	15053	21916	81825	44394	42880
7	99562	72905	56420	69994	98872	31016	71194	18738	44013	48840	63213	21069	10634	12952
8	96301	91977	05463	07972	18876	20922	94595	56869	69014	60045	18425	84903	42508	32307
9	89579	14342	63661	10281	17453	18103	57740	84378	25331	12566	58678	44947	05585	56941
10	85475	36857	53342	53988	53060	59533	38867	62300	08158	17983	16439	11458	18593	64952
11	28918	69578	88231	33276	70997	79936	56865	05859	90106	31595	01547	85590	91610	78188
12	63553	40961	48235	03427	49626	69445	18663	72695	52180	20847	12234	90511	33703	90322
13	09429	93969	52636	92737	88974	33488	36320	17617	30015	08272	84115	27156	30613	74952
14	10365	61129	87529	85689	48237	52267	67689	93394	01511	26358	85104	20285	29975	89868
15	07119	97336	71048	08178	77233	13916	47564	81056	97735	85977	29372	74461	28551	90707
16	51085	12765	51821	51259	77452	16308	60756	92144	49442	53900	70960	63990	75601	40719
17	02368	21382	52404	60268	89368	19885	55322	44819	01188	65255	64835	44919	05944	55157
18	01011	54092	33362	94904	31273	04146	18594	29852	71585	85030	51132	01915	92747	64951
19	52162	53916	46369	58586	23216	14513	83149	98736	23495	64350	94738	17752	35156	35749
20	07056	97628	33787	09998	42698	06691	76988	13602	51851	46104	88916	19509	25625	58104
21	48663	91245	85828	14346	09172	30168	90229	04734	59193	22178	30421	61666	99904	32812
22	54164	58492	22421	74103	47070	25306	76468	26384	58151	06646	21524	15227	96909	44592
23	32639	32363	05597	24200	13363	38005	94342	28728	35806	06912	17012	64161	18296	22851
24	29334	27001	87637	87308	58731	00256	45834	15398	46557	41135	10367	07684	36188	18510
25	02488	33062	28834	07351	19731	92420	60952	61280	50001	67658	32586	86679	50720	94953
26	81525	72295	04839	96423	24878	82651	66566	14778	76797	14780	13300	87074	79666	95725
27	29676	20591	68086	26432	46901	20849	89768	81536	86645	12659	92259	57102	80428	25280
28	00742	57392	39064	66432	84673	40027	32832	61362	98947	96067	64760	64584	96096	98253
29	05366	04213	25669	26422	44407	44048	37937	63904	45766	66134	75470	66520	34693	90449
30	91921	26418	64117	94305	26766	25940	39972	22209	71500	64568	91402	42416	07844	69618
31	00582	04711	87917	77341	42206	35126	74087	99547	81817	42607	43808	76655	62028	76630
32	00725	69884	62797	56170	86324	88072	76222	36086	84637	93161	76038	65855	77919	88006
33	69011	65795	95876	55293	18988	27354	26575	08625	40801	59920	29841	80150	12777	48501
34	25976	57948	29888	88604	67917	48708	18912	82271	65424	69774	33611	54262	85963	03547
35	09763	83473	73577	12908	30883	18317	28290	35797	05998	41688	34952	37888	38917	88050
36	91576	42595	27958	30134	04024	86385	29880	99730	55536	84855	29080	09250	79656	73211
37	17955	56349	90999	49127	20044	59931	06115	20542	18059	02008	73708	83517	36103	42791
38	46503	18584	18845	49618	02304	51038	20655	58727	28168	15475	56942	53389	20562	87338

TABLE 1

39	92157	89634	94824	78171	84610	82834	09922	25417	44137	48413	25555	21246	35509	20468
40	14577	62765	35605	81263	39667	47358	56873	56307	61607	49518	89656	20103	77490	18062
41	98427	07523	33362	64270	01638	92477	66969	98420	04880	45585	46565	04102	46880	45709
42	34914	63976	88720	82765	34476	17032	87589	40836	32427	70002	70663	88863	77775	69348
43	70060	28277	39475	46473	23219	53416	94970	25832	69975	94884	19661	72828	00102	66794
44	53976	54914	06990	67245	68350	82948	11398	42878	80287	88267	47363	46634	06541	97809
45	76072	29515	40980	07391	58745	25774	22987	80059	39911	96189	41151	14222	60697	59583
46	90725	52210	83974	29992	65831	38857	50490	83765	55657	14361	31720	57375	56228	41546
47	64364	67412	33339	31926	14883	24413	59744	92351	97473	89286	35931	04110	23726	51900
48	08962	00358	31662	25388	61642	34072	81249	35648	56891	69352	48373	45578	78547	81788
49	95012	68379	93526	70765	10592	04542	76463	54328	02349	17247	28865	14777	62730	92277
50	15664	10493	20492	38391	91132	21999	59516	81652	27195	48223	46751	22923	32261	85653
51	16408	81899	04153	53381	79401	21438	83035	92350	36693	31238	59649	91754	72772	02338
52	18629	81953	05520	91962	04739	13092	97662	24822	94730	06496	35090	04822	86774	98289
53	73115	35101	47498	87637	99016	71060	88824	71013	18735	20286	23153	72924	35165	43040
54	57491	16703	23167	49323	45021	33132	12544	41035	80780	45393	44812	12515	98931	91202
55	30405	33946	23792	14422	15059	45799	22716	19792	09983	74353	68668	30429	70735	25499
56	16631	35006	85900	98275	32388	52390	16815	69298	82732	38480	73817	32523	41961	44437
57	96773	20206	42559	78985	05300	22164	24369	54224	35083	19687	11052	91491	60383	19746
58	38935	64202	14349	82674	66523	44133	00697	35552	35970	19124	63318	29686	03387	59846
59	31624	76384	17403	53363	44167	64486	64758	75366	76554	31601	12614	33072	60332	92325
60	78919	19474	23632	27889	47914	02584	37680	20801	72152	39339	34806	08930	85001	87820
61	03931	33309	57047	74211	63445	17361	62825	39908	05607	91284	68833	25570	38818	46920
62	74426	33278	43972	10119	89917	15665	52872	73823	73144	88662	88970	74492	51805	99378
63	09066	00903	20795	95452	92648	45454	09552	88815	16553	51125	79375	97596	16296	66092
64	42238	12426	87025	14267	20979	04508	64535	31355	86064	29472	47689	05974	52468	16834
65	16153	08002	26504	41744	81959	65642	74240	56302	00033	67107	77510	70625	28725	34191
66	21457	40742	29820	96783	29400	21840	15035	34537	33310	06116	95240	15957	16572	06004
67	21581	57802	02050	89728	17937	37621	47075	42080	97403	48626	68995	43805	33386	21597
68	55612	78095	83197	33732	05810	24813	86902	60397	16489	03264	88525	42786	05269	92532
69	44657	66999	99324	51281	84463	60563	79312	93454	68876	25471	93911	25650	12682	73572
70	91340	84979	46949	81973	37949	61023	43997	15263	80644	43942	89203	71795	99533	50501
71	91227	21199	31935	27022	84067	05462	35216	14486	29891	68607	41867	14951	91696	85065
72	50001	38140	66321	19924	72163	09538	12151	06878	91903	18749	34405	56087	82790	70925
73	65390	05224	72958	28609	81406	39147	25549	48542	42627	45233	57202	94617	23772	07896
74	27504	96131	83944	41575	10573	08619	64482	73923	36152	05184	94142	25299	84387	34925
75	37169	94851	39117	89632	00959	16487	65536	49071	39782	17095	02330	74301	00275	48280
76	11508	70225	51111	38351	19444	66499	71945	05422	13442	78675	84081	66938	93654	59894
77	37449	30362	06694	54690	04052	53115	62757	95348	78662	11163	81651	50245	34971	52924
78	46515	70331	85922	38329	57015	15765	97161	17869	45349	61796	66345	81073	49106	79860

Source: Abridged from W. H. Beyer, Ed., *CRC Standard Mathematical Tables*, 24th ed (Cleveland, The Chemical Rubber Company), 1976. Reproduced by permission of the publisher.

TABLE 1
(Continued)

Row \ Column	1	2	3	4	5	6	7	8	9	10	11	12	13	14
79	30986	81223	42416	58353	21532	30502	32305	86482	05174	07901	54339	58861	74818	46942
80	63798	64995	46583	09785	44160	78128	83991	42865	92520	83531	80377	35909	81250	54238
81	82486	84846	99254	67632	43218	50076	21361	64816	51202	88124	41870	52689	51275	83556
82	21885	32906	92431	09060	64297	51674	64126	62570	26123	05155	59194	52799	28225	85762
83	60336	98782	07408	53458	13564	59089	26445	29789	85205	41001	12535	12133	14645	23541
84	43937	46891	24010	25560	86355	33941	25786	54990	71899	15475	95434	98227	21824	19585
85	97656	63175	89303	16275	07100	92063	21942	18611	47348	20203	18534	03862	78095	50136
86	03299	01221	05418	38982	55758	92237	26759	86367	21216	98442	08303	56613	91511	75928
87	79626	06486	03574	17668	07785	76020	79924	25651	83325	88428	85076	72811	22717	50585
88	85636	68335	47539	03129	65651	11977	02510	26113	99447	68645	34327	15152	55230	93448
89	18039	14367	61337	06177	12143	46609	32989	74014	64708	00533	35398	58408	13261	47908
90	08362	15656	60627	36478	65648	16764	53412	09013	07832	41574	17639	82163	60859	75567
91	79556	29068	04142	16268	15387	12856	66227	38358	22478	73373	88732	09443	82558	05250
92	92608	82674	27072	32534	17075	27698	98204	63863	11951	34648	88022	56148	34925	57031
93	23982	25835	40055	67006	12293	02753	14827	23235	35071	99704	37543	11601	35503	85171
94	09915	96306	05908	97901	28395	14186	00821	80703	70426	75647	76310	88717	37890	40129
95	59037	33300	26695	62247	69927	76123	50842	43834	86654	70959	79725	93872	28117	19233
96	42488	78077	69882	61657	34136	79180	97526	43092	04098	73571	80799	76536	71255	64239
97	46764	86273	63003	93017	31204	36692	40202	35275	57306	55543	53203	18098	47625	88684
98	03237	45430	55417	63282	90816	17349	88298	90183	36600	78406	06216	95787	42579	90730
99	86591	81482	52667	61582	14972	90053	89534	76036	49199	43716	97548	04379	46370	28672
100	38534	01715	94964	87288	65680	43772	39560	12918	86537	62738	19636	51132	25739	56947

Cumulative distribution

TABLE 2 *Binomial Probabilities*

Tabulated values are $\sum_{x=0}^{k} p(x)$. (Computations are rounded at the third decimal place.)

(a) n = 5

k \ p	0.01	0.05	0.10	0.20	0.30	0.40	0.50	0.60	0.70	0.80	0.90	0.95	0.99
0	0.951	0.774	0.590	0.328	0.168	0.078	0.031	0.010	0.002	0.000	0.000	0.000	0.000
1	0.999	0.977	0.919	0.737	0.528	0.337	0.188	0.087	0.031	0.007	0.000	0.000	0.000
2	1.000	0.999	0.991	0.942	0.837	0.683	0.500	0.317	0.163	0.058	0.009	0.001	0.000
3	1.000	1.000	1.000	0.993	0.969	0.913	0.812	0.663	0.472	0.263	0.081	0.023	0.001
4	1.000	1.000	1.000	1.000	0.998	0.990	0.969	0.922	0.832	0.672	0.410	0.226	0.049

(b) n = 10

k \ p	0.01	0.05	0.10	0.20	0.30	0.40	0.50	0.60	0.70	0.80	0.90	0.95	0.99
0	0.904	0.599	0.349	0.107	0.028	0.006	0.001	0.000	0.000	0.000	0.000	0.000	0.000
1	0.996	0.914	0.736	0.376	0.149	0.046	0.011	0.002	0.000	0.000	0.000	0.000	0.000
2	1.000	0.988	0.930	0.678	0.383	0.167	0.055	0.012	0.002	0.000	0.000	0.000	0.000
3	1.000	0.999	0.987	0.879	0.650	0.382	0.172	0.055	0.011	0.001	0.000	0.000	0.000
4	1.000	1.000	0.998	0.967	0.850	0.633	0.377	0.166	0.047	0.006	0.000	0.000	0.000
5	1.000	1.000	1.000	0.994	0.953	0.834	0.623	0.367	0.150	0.033	0.002	0.000	0.000
6	1.000	1.000	1.000	0.999	0.989	0.945	0.828	0.618	0.350	0.121	0.013	0.001	0.000
7	1.000	1.000	1.000	1.000	0.998	0.988	0.945	0.833	0.617	0.322	0.070	0.012	0.000
8	1.000	1.000	1.000	1.000	1.000	0.998	0.989	0.954	0.851	0.624	0.264	0.086	0.004
9	1.000	1.000	1.000	1.000	1.000	1.000	0.999	0.994	0.972	0.893	0.651	0.401	0.096

(c) n = 15

k \ p	0.01	0.05	0.10	0.20	0.30	0.40	0.50	0.60	0.70	0.80	0.90	0.95	0.99
0	0.860	0.463	0.206	0.035	0.005	0.000	0.000	0.000	0.000	0.000	0.000	0.000	0.000
1	0.990	0.829	0.549	0.167	0.035	0.005	0.000	0.000	0.000	0.000	0.000	0.000	0.000
2	1.000	0.964	0.816	0.398	0.127	0.027	0.004	0.000	0.000	0.000	0.000	0.000	0.000
3	1.000	0.995	0.944	0.648	0.297	0.091	0.018	0.002	0.000	0.000	0.000	0.000	0.000
4	1.000	0.999	0.987	0.836	0.515	0.217	0.059	0.009	0.001	0.000	0.000	0.000	0.000
5	1.000	1.000	0.998	0.939	0.722	0.403	0.151	0.034	0.004	0.000	0.000	0.000	0.000
6	1.000	1.000	1.000	0.982	0.869	0.610	0.304	0.095	0.015	0.001	0.000	0.000	0.000
7	1.000	1.000	1.000	0.996	0.950	0.787	0.500	0.213	0.050	0.004	0.000	0.000	0.000
8	1.000	1.000	1.000	0.999	0.985	0.905	0.696	0.390	0.131	0.018	0.000	0.000	0.000
9	1.000	1.000	1.000	1.000	0.996	0.966	0.849	0.597	0.278	0.061	0.002	0.000	0.000
10	1.000	1.000	1.000	1.000	0.999	0.991	0.941	0.783	0.485	0.164	0.013	0.001	0.000
11	1.000	1.000	1.000	1.000	1.000	0.998	0.982	0.909	0.703	0.352	0.056	0.005	0.000
12	1.000	1.000	1.000	1.000	1.000	1.000	0.996	0.973	0.873	0.602	0.184	0.036	0.000
13	1.000	1.000	1.000	1.000	1.000	1.000	1.000	0.995	0.965	0.833	0.451	0.171	0.010
14	1.000	1.000	1.000	1.000	1.000	1.000	1.000	1.000	0.995	0.965	0.794	0.537	0.140

TABLE 2

p is closer to 10

TABLE 2 *(Continued)*

(d) $n = 20$

k \ p	0.01	0.05	0.10	0.20	0.30	0.40	0.50	0.60	0.70	0.80	0.90	0.95	0.99
0	0.818	0.358	0.122	0.002	0.001	0.000	0.000	0.000	0.000	0.000	0.000	0.000	0.000
1	0.983	0.736	0.392	0.069	0.008	0.001	0.000	0.000	0.000	0.000	0.000	0.000	0.000
2	0.999	0.925	0.677	0.206	0.035	0.004	0.000	0.000	0.000	0.000	0.000	0.000	0.000
3	1.000	0.984	0.867	0.411	0.107	0.016	0.001	0.000	0.000	0.000	0.000	0.000	0.000
4	1.000	0.997	0.957	0.630	0.238	0.051	0.006	0.000	0.000	0.000	0.000	0.000	0.000
5	1.000	1.000	0.989	0.804	0.416	0.126	0.021	0.002	0.000	0.000	0.000	0.000	0.000
6	1.000	1.000	0.998	0.913	0.608	0.250	0.058	0.006	0.000	0.000	0.000	0.000	0.000
7	1.000	1.000	1.000	0.968	0.772	0.416	0.132	0.021	0.001	0.000	0.000	0.000	0.000
8	1.000	1.000	1.000	0.990	0.887	0.596	0.252	0.057	0.005	0.000	0.000	0.000	0.000
9	1.000	1.000	1.000	0.997	0.952	0.755	0.412	0.128	0.017	0.001	0.000	0.000	0.000
10	1.000	1.000	1.000	0.999	0.983	0.872	0.588	0.245	0.048	0.003	0.000	0.000	0.000
11	1.000	1.000	1.000	1.000	0.995	0.943	0.748	0.404	0.113	0.010	0.000	0.000	0.000
12	1.000	1.000	1.000	1.000	0.999	0.979	0.868	0.584	0.228	0.032	0.000	0.000	0.000
13	1.000	1.000	1.000	1.000	1.000	0.994	0.942	0.750	0.392	0.087	0.002	0.000	0.000
14	1.000	1.000	1.000	1.000	1.000	0.998	0.979	0.874	0.584	0.196	0.011	0.000	0.000
15	1.000	1.000	1.000	1.000	1.000	1.000	0.994	0.949	0.762	0.370	0.043	0.003	0.000
16	1.000	1.000	1.000	1.000	1.000	1.000	0.999	0.984	0.893	0.589	0.133	0.016	0.000
17	1.000	1.000	1.000	1.000	1.000	1.000	1.000	0.996	0.965	0.794	0.323	0.075	0.001
18	1.000	1.000	1.000	1.000	1.000	1.000	1.000	0.999	0.992	0.931	0.608	0.264	0.017
19	1.000	1.000	1.000	1.000	1.000	1.000	1.000	1.000	0.999	0.988	0.878	0.642	0.182

(e) $n = 25$

k \ p	0.01	0.05	0.10	0.20	0.30	0.40	0.50	0.60	0.70	0.80	0.90	0.95	0.99
0	0.778	0.277	0.072	0.004	0.000	0.000	0.000	0.000	0.000	0.000	0.000	0.000	0.000
1	0.974	0.642	0.271	0.027	0.002	0.000	0.000	0.000	0.000	0.000	0.000	0.000	0.000
2	0.998	0.873	0.537	0.098	0.009	0.000	0.000	0.000	0.000	0.000	0.000	0.000	0.000
3	1.000	0.966	0.764	0.234	0.033	0.002	0.000	0.000	0.000	0.000	0.000	0.000	0.000
4	1.000	0.993	0.902	0.421	0.090	0.009	0.000	0.000	0.000	0.000	0.000	0.000	0.000
5	1.000	0.999	0.967	0.617	0.193	0.029	0.002	0.000	0.000	0.000	0.000	0.000	0.000
6	1.000	1.000	0.991	0.780	0.341	0.074	0.007	0.000	0.000	0.000	0.000	0.000	0.000
7	1.000	1.000	0.998	0.891	0.512	0.154	0.022	0.001	0.000	0.000	0.000	0.000	0.000
8	1.000	1.000	1.000	0.953	0.677	0.274	0.054	0.004	0.000	0.000	0.000	0.000	0.000
9	1.000	1.000	1.000	0.983	0.811	0.425	0.115	0.013	0.000	0.000	0.000	0.000	0.000
10	1.000	1.000	1.000	0.994	0.902	0.586	0.212	0.034	0.002	0.000	0.000	0.000	0.000
11	1.000	1.000	1.000	0.998	0.956	0.732	0.345	0.078	0.006	0.000	0.000	0.000	0.000
12	1.000	1.000	1.000	1.000	0.983	0.846	0.500	0.154	0.017	0.000	0.000	0.000	0.000
13	1.000	1.000	1.000	1.000	0.994	0.922	0.655	0.268	0.044	0.002	0.000	0.000	0.000
14	1.000	1.000	1.000	1.000	0.998	0.966	0.788	0.414	0.098	0.006	0.000	0.000	0.000
15	1.000	1.000	1.000	1.000	1.000	0.987	0.885	0.575	0.189	0.017	0.000	0.000	0.000
16	1.000	1.000	1.000	1.000	1.000	0.996	0.946	0.726	0.323	0.047	0.000	0.000	0.000
17	1.000	1.000	1.000	1.000	1.000	0.999	0.978	0.846	0.488	0.109	0.002	0.000	0.000
18	1.000	1.000	1.000	1.000	1.000	1.000	0.993	0.926	0.659	0.220	0.009	0.000	0.000
19	1.000	1.000	1.000	1.000	1.000	1.000	0.998	0.971	0.807	0.383	0.033	0.001	0.000
20	1.000	1.000	1.000	1.000	1.000	1.000	1.000	0.991	0.910	0.579	0.098	0.007	0.000
21	1.000	1.000	1.000	1.000	1.000	1.000	1.000	0.998	0.967	0.766	0.236	0.034	0.000
22	1.000	1.000	1.000	1.000	1.000	1.000	1.000	1.000	0.991	0.902	0.463	0.127	0.002
23	1.000	1.000	1.000	1.000	1.000	1.000	1.000	1.000	0.998	0.973	0.729	0.358	0.026
24	1.000	1.000	1.000	1.000	1.000	1.000	1.000	1.000	1.000	0.996	0.928	0.723	0.222

TABLE 2

TABLE 3   *Poisson Distribution Function**

$$F(x; \lambda) = \sum_{k=0}^{x} e^{-\lambda} \frac{\lambda^k}{k!}$$

TABLE
3

λ \ x	0	1	2	3	4	5	6	7	8	9
0.02	0.980	1.000								
0.04	0.961	0.999	1.000							
0.06	0.942	0.998	1.000							
0.08	0.923	0.997	1.000							
0.10	0.905	0.995	1.000							
0.15	0.861	0.990	0.999	1.000						
0.20	0.819	0.982	0.999	1.000						
0.25	0.779	0.974	0.998	1.000						
0.30	0.741	0.963	0.996	1.000						
0.35	0.705	0.951	0.994	1.000						
0.40	0.670	0.938	0.992	0.999	1.000					
0.45	0.638	0.925	0.989	0.999	1.000					
0.50	0.607	0.910	0.986	0.998	1.000					
0.55	0.577	0.894	0.982	0.998	1.000					
0.60	0.549	0.878	0.977	0.997	1.000					
0.65	0.522	0.861	0.972	0.996	0.999	1.000				
0.70	0.497	0.844	0.966	0.994	0.999	1.000				
0.75	0.472	0.827	0.959	0.993	0.999	1.000				
0.80	0.449	0.809	0.953	0.991	0.999	1.000				
0.85	0.427	0.791	0.945	0.989	0.998	1.000				
0.90	0.407	0.772	0.937	0.987	0.998	1.000				
0.95	0.387	0.754	0.929	0.981	0.997	1.000				
1.00	0.368	0.736	0.920	0.981	0.996	0.999	1.000			
1.1	0.333	0.699	0.900	0.974	0.995	0.999	1.000			
1.2	0.301	0.663	0.879	0.966	0.992	0.998	1.000			
1.3	0.273	0.627	0.857	0.957	0.989	0.998	1.000			
1.4	0.247	0.592	0.833	0.946	0.986	0.997	0.999	1.000		
1.5	0.223	0.558	0.809	0.934	0.981	0.996	0.999	1.000		
1.6	0.202	0.525	0.783	0.921	0.976	0.994	0.999	1.000		
1.7	0.183	0.493	0.757	0.907	0.970	0.992	0.998	1.000		
1.8	0.165	0.463	0.731	0.891	0.964	0.990	0.997	0.999	1.000	
1.9	0.150	0.434	0.704	0.875	0.956	0.987	0.997	0.999	1.000	
2.0	0.135	0.406	0.677	0.857	0.947	0.983	0.995	0.999	1.000	

* Reprinted by permission from E. C. Molina, *Poisson's Exponential Binomial Limit*, D. Van Nostrand Company, Inc., Princeton, N.J., 1947.

TABLE 3 *(Continued)*

λ \ x	0	1	2	3	4	5	6	7	8	9
2.2	0.111	0.355	0.623	0.819	0.928	0.975	0.993	0.998	1.000	
2.4	0.091	0.308	0.570	0.779	0.904	0.964	0.988	0.997	0.999	1.000
2.6	0.074	0.267	0.518	0.736	0.877	0.951	0.983	0.995	0.999	1.000
2.8	0.061	0.231	0.469	0.692	0.848	0.935	0.976	0.992	0.998	0.999
3.0	0.050	0.199	0.423	0.647	0.815	0.916	0.966	0.988	0.996	0.999
3.2	0.041	0.171	0.380	0.603	0.781	0.895	0.955	0.983	0.994	0.998
3.4	0.033	0.147	0.340	0.558	0.744	0.871	0.942	0.977	0.992	0.997
3.6	0.027	0.126	0.303	0.515	0.706	0.844	0.927	0.969	0.988	0.996
3.8	0.022	0.107	0.269	0.473	0.668	0.816	0.909	0.960	0.984	0.994
4.0	0.018	0.092	0.238	0.433	0.629	0.785	0.889	0.949	0.979	0.992
4.2	0.015	0.078	0.210	0.395	0.590	0.753	0.867	0.936	0.972	0.989
4.4	0.012	0.066	0.185	0.359	0.551	0.720	0.844	0.921	0.964	0.985
4.6	0.010	0.056	0.163	0.326	0.513	0.686	0.818	0.905	0.955	0.980
4.8	0.008	0.048	0.143	0.294	0.476	0.651	0.791	0.887	0.944	0.975
5.0	0.007	0.040	0.125	0.265	0.440	0.616	0.762	0.867	0.932	0.968
5.2	0.006	0.034	0.109	0.238	0.406	0.581	0.732	0.845	0.918	0.960
5.4	0.005	0.029	0.095	0.213	0.373	0.546	0.702	0.822	0.903	0.951
5.6	0.004	0.024	0.082	0.191	0.342	0.512	0.670	0.797	0.886	0.941
5.8	0.003	0.021	0.072	0.170	0.313	0.478	0.638	0.771	0.867	0.929
6.0	0.002	0.017	0.062	0.151	0.285	0.446	0.606	0.744	0.847	0.916

	10	11	12	13	14	15	16
2.8	1.000						
3.0	1.000						
3.2	1.000						
3.4	0.999	1.000					
3.6	0.999	1.000					
3.8	0.998	0.999	1.000				
4.0	0.997	0.999	1.000				
4.2	0.996	0.999	1.000				
4.4	0.994	0.998	0.999	1.000			
4.6	0.992	0.997	0.999	1.000			
4.8	0.990	0.996	0.999	1.000			
5.0	0.986	0.995	0.998	0.999	1.000		
5.2	0.982	0.993	0.997	0.999	1.000		
5.4	0.977	0.990	0.996	0.999	1.000		
5.6	0.972	0.988	0.995	0.998	0.999	1.000	
5.8	0.965	0.984	0.993	0.997	0.999	1.000	
6.0	0.957	0.980	0.991	0.996	0.999	0.999	1.000

TABLE 3   (Continued)

λ \ x	0	1	2	3	4	5	6	7	8	9
6.2	0.002	0.015	0.054	0.134	0.259	0.414	0.574	0.716	0.826	0.902
6.4	0.002	0.012	0.046	0.119	0.235	0.384	0.542	0.687	0.803	0.886
6.6	0.001	0.010	0.040	0.105	0.213	0.355	0.511	0.658	0.780	0.869
6.8	0.001	0.009	0.034	0.093	0.192	0.327	0.480	0.628	0.755	0.850
7.0	0.001	0.007	0.030	0.082	0.173	0.301	0.450	0.599	0.729	0.830
7.2	0.001	0.006	0.025	0.072	0.156	0.276	0.420	0.569	0.703	0.810
7.4	0.001	0.005	0.022	0.063	0.140	0.253	0.392	0.539	0.676	0.788
7.6	0.001	0.004	0.019	0.055	0.125	0.231	0.365	0.510	0.648	0.765
7.8	0.000	0.004	0.016	0.048	0.112	0.210	0.338	0.481	0.620	0.741
8.0	0.000	0.003	0.014	0.042	0.100	0.191	0.313	0.453	0.593	0.717
8.5	0.000	0.002	0.009	0.030	0.074	0.150	0.256	0.386	0.523	0.653
9.0	0.000	0.001	0.006	0.021	0.055	0.116	0.207	0.324	0.456	0.587
9.5	0.000	0.001	0.004	0.015	0.040	0.089	0.165	0.269	0.392	0.522
10.0	0.000	0.000	0.003	0.010	0.029	0.067	0.130	0.220	0.333	0.458

λ	10	11	12	13	14	15	16	17	18	19
6.2	0.949	0.975	0.989	0.995	0.998	0.999	1.000			
6.4	0.939	0.969	0.986	0.994	0.997	0.999	1.000			
6.6	0.927	0.963	0.982	0.992	0.997	0.999	0.999	1.000		
6.8	0.915	0.955	0.978	0.990	0.996	0.998	0.999	1.000		
7.0	0.901	0.947	0.973	0.987	0.994	0.998	0.999	1.000		
7.2	0.887	0.937	0.967	0.984	0.993	0.997	0.999	0.999	1.000	
7.4	0.871	0.926	0.961	0.980	0.991	0.996	0.998	0.999	1.000	
7.6	0.854	0.915	0.954	0.976	0.989	0.995	0.998	0.999	1.000	
7.8	0.835	0.902	0.945	0.971	0.986	0.993	0.997	0.999	1.000	
8.0	0.816	0.888	0.936	0.966	0.983	0.992	0.996	0.998	0.999	1.000
8.5	0.763	0.849	0.909	0.949	0.973	0.986	0.993	0.997	0.999	0.999
9.0	0.706	0.803	0.876	0.926	0.959	0.978	0.989	0.995	0.998	0.999
9.5	0.645	0.752	0.836	0.898	0.940	0.967	0.982	0.991	0.996	0.998
10.0	0.583	0.697	0.792	0.864	0.917	0.951	0.973	0.986	0.993	0.997

λ	20	21	22
8.5	1.000		
9.0	1.000		
9.5	0.999	1.000	
10.0	0.998	0.999	1.000

TABLE 3

TABLE 3 *(Continued)*

λ＼x	0	1	2	3	4	5	6	7	8	9
10.5	0.000	0.000	0.002	0.007	0.021	0.050	0.102	0.179	0.279	0.397
11.0	0.000	0.000	0.001	0.005	0.015	0.038	0.079	0.143	0.232	0.341
11.5	0.000	0.000	0.001	0.003	0.011	0.028	0.060	0.114	0.191	0.289
12.0	0.000	0.000	0.001	0.002	0.008	0.020	0.046	0.090	0.155	0.242
12.5	0.000	0.000	0.000	0.002	0.005	0.015	0.035	0.070	0.125	0.201
13.0	0.000	0.000	0.000	0.001	0.004	0.011	0.026	0.054	0.100	0.166
13.5	0.000	0.000	0.000	0.001	0.003	0.008	0.019	0.041	0.079	0.135
14.0	0.000	0.000	0.000	0.000	0.002	0.006	0.014	0.032	0.062	0.109
14.5	0.000	0.000	0.000	0.000	0.001	0.004	0.010	0.024	0.048	0.088
15.0	0.000	0.000	0.000	0.000	0.001	0.003	0.008	0.018	0.037	0.070

λ	10	11	12	13	14	15	16	17	18	19
10.5	0.521	0.639	0.742	0.825	0.888	0.932	0.960	0.978	0.988	0.994
11.0	0.460	0.579	0.689	0.781	0.854	0.907	0.944	0.968	0.982	0.991
11.5	0.402	0.520	0.633	0.733	0.815	0.878	0.924	0.954	0.974	0.986
12.0	0.347	0.462	0.576	0.682	0.772	0.844	0.899	0.937	0.963	0.979
12.5	0.297	0.406	0.519	0.628	0.725	0.806	0.869	0.916	0.948	0.969
13.0	0.252	0.353	0.463	0.573	0.675	0.764	0.835	0.890	0.930	0.957
13.5	0.211	0.304	0.409	0.518	0.623	0.718	0.798	0.861	0.908	0.942
14.0	0.176	0.260	0.358	0.464	0.570	0.669	0.756	0.827	0.883	0.923
14.5	0.145	0.220	0.311	0.413	0.518	0.619	0.711	0.790	0.853	0.901
15.0	0.118	0.185	0.268	0.363	0.466	0.568	0.664	0.749	0.819	0.875

λ	20	21	22	23	24	25	26	27	28	29
10.5	0.997	0.999	0.999	1.000						
11.0	0.995	0.998	0.999	1.000						
11.5	0.992	0.996	0.998	0.999	1.000					
12.0	0.988	0.994	0.997	0.999	0.999	1.000				
12.5	0.983	0.991	0.995	0.998	0.999	0.999	1.000			
13.0	0.975	0.986	0.992	0.996	0.998	0.999	1.000			
13.5	0.965	0.980	0.989	0.994	0.997	0.998	0.999	1.000		
14.0	0.952	0.971	0.983	0.991	0.995	0.997	0.999	0.999	1.000	
14.5	0.936	0.960	0.976	0.986	0.992	0.996	0.998	0.999	0.999	1.000
15.0	0.917	0.947	0.967	0.981	0.989	0.994	0.997	0.998	0.999	1.000

TABLE 3 (Continued)

TABLE 3

λ \ x	4	5	6	7	8	9	10	11	12	13
16	0.000	0.001	0.004	0.010	0.022	0.043	0.077	0.127	0.193	0.275
17	0.000	0.001	0.002	0.005	0.013	0.026	0.049	0.085	0.135	0.201
18	0.000	0.000	0.001	0.003	0.007	0.015	0.030	0.055	0.092	0.143
19	0.000	0.000	0.001	0.002	0.004	0.009	0.018	0.035	0.061	0.098
20	0.000	0.000	0.000	0.001	0.002	0.005	0.011	0.021	0.039	0.066
21	0.000	0.000	0.000	0.000	0.001	0.003	0.006	0.013	0.025	0.043
22	0.000	0.000	0.000	0.000	0.001	0.002	0.004	0.008	0.015	0.028
23	0.000	0.000	0.000	0.000	0.000	0.001	0.002	0.004	0.009	0.017
24	0.000	0.000	0.000	0.000	0.000	0.000	0.001	0.003	0.005	0.011
25	0.000	0.000	0.000	0.000	0.000	0.000	0.001	0.001	0.003	0.006

	14	15	16	17	18	19	20	21	22	23
16	0.368	0.467	0.566	0.659	0.742	0.812	0.868	0.911	0.942	0.963
17	0.281	0.371	0.468	0.564	0.655	0.736	0.805	0.861	0.905	0.937
18	0.208	0.287	0.375	0.469	0.562	0.651	0.731	0.799	0.855	0.899
19	0.150	0.215	0.292	0.378	0.469	0.561	0.647	0.725	0.793	0.849
20	0.105	0.157	0.221	0.297	0.381	0.470	0.559	0.644	0.721	0.787
21	0.072	0.111	0.163	0.227	0.302	0.384	0.471	0.558	0.640	0.716
22	0.048	0.077	0.117	0.169	0.232	0.306	0.387	0.472	0.556	0.637
23	0.031	0.052	0.082	0.123	0.175	0.238	0.310	0.389	0.472	0.555
24	0.020	0.034	0.056	0.087	0.128	0.180	0.243	0.314	0.392	0.473
25	0.012	0.022	0.038	0.060	0.092	0.134	0.185	0.247	0.318	0.394

	24	25	26	27	28	29	30	31	32	33
16	0.978	0.987	0.993	0.996	0.998	0.999	0.999	1.000		
17	0.959	0.975	0.985	0.991	0.995	0.997	0.999	0.999	1.000	
18	0.932	0.955	0.972	0.983	0.990	0.994	0.997	0.998	0.999	1.000
19	0.893	0.927	0.951	0.969	0.980	0.988	0.993	0.996	0.998	0.999
20	0.843	0.888	0.922	0.948	0.966	0.978	0.987	0.992	0.995	0.997
21	0.782	0.838	0.883	0.917	0.944	0.963	0.976	0.985	0.991	0.994
22	0.712	0.777	0.832	0.877	0.913	0.940	0.959	0.973	0.983	0.989
23	0.635	0.708	0.772	0.827	0.873	0.908	0.936	0.956	0.971	0.981
24	0.554	0.632	0.704	0.768	0.823	0.868	0.904	0.932	0.953	0.969
25	0.473	0.553	0.629	0.700	0.763	0.818	0.863	0.900	0.929	0.950

	34	35	36	37	38	39	40	41	42	43
19	0.999	1.000								
20	0.999	0.999	1.000							
21	0.997	0.998	0.999	0.999	1.000					
22	0.994	0.996	0.998	0.999	0.999	1.000				
23	0.988	0.993	0.996	0.997	0.999	0.999	1.000			
24	0.979	0.987	0.992	0.995	0.997	0.998	0.999	0.999	1.000	
25	0.966	0.978	0.985	0.991	0.991	0.997	0.998	0.999	0.999	1.000

TABLE 4   *Normal Curve Areas*

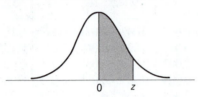

3 STANDARD DEVIATIONS

TABLE 4

z	0.00	0.01	0.02	0.03	0.04	0.05	0.06	0.07	0.08	0.09
0.0	0.0000	0.0040	0.0080	0.0120	0.0160	0.0199	0.0239	0.0279	0.0319	0.0359
0.1	0.0398	0.0438	0.0478	0.0517	0.0557	0.0596	0.0636	0.0675	0.0174	0.0753
0.2	0.0793	0.0832	0.0871	0.0910	0.0948	0.0987	0.1026	0.1064	0.1103	0.1141
0.3	0.1179	0.1217	0.1255	0.1293	0.1331	0.1368	0.1406	0.1443	0.1480	0.1517
0.4	0.1554	0.1591	0.1628	0.1664	0.1700	0.1736	0.1772	0.1808	0.1844	0.1879
0.5	0.1915	0.1950	0.1985	0.2019	0.2054	0.2088	0.2123	0.2157	0.2190	0.2224
0.6	0.2257	0.2291	0.2324	0.2357	0.2389	0.2422	0.2454	0.2486	0.2517	0.2549
0.7	0.2580	0.2611	0.2642	0.2673	0.2704	0.2734	0.2764	0.2794	0.2823	0.2852
0.8	0.2881	0.2910	0.2939	0.2967	0.2995	0.3023	0.3051	0.3078	0.3106	0.3133
0.9	0.3159	0.3186	0.3212	0.3238	0.3264	0.3289	0.3315	0.3340	0.3365	0.3389
1.0	0.3413	0.3438	0.3461	0.3485	0.3508	0.3531	0.3554	0.3577	0.3599	0.3621
1.1	0.3643	0.3665	0.3686	0.3708	0.3729	0.3749	0.3770	0.3790	0.3810	0.3830
1.2	0.3849	0.3869	0.3888	0.3907	0.3925	0.3944	0.3962	0.3980	0.3997	0.4015
1.3	0.4032	0.4049	0.4066	0.4082	0.4099	0.4115	0.4131	0.4147	0.4162	0.4177
1.4	0.4192	0.4207	0.4222	0.4236	0.4251	0.4265	0.4279	0.4292	0.4306	0.4319
1.5	0.4332	0.4345	0.4357	0.4370	0.4382	0.4394	0.4406	0.4418	0.4429	0.4441
1.6	0.4452	0.4463	0.4474	0.4484	0.4495	0.4505	0.4515	0.4525	0.4535	0.4545
1.7	0.4554	0.4564	0.4573	0.4582	0.4591	0.4599	0.4608	0.4616	0.4625	0.4633
1.8	0.4641	0.4649	0.4656	0.4664	0.4671	0.4678	0.4686	0.4693	0.4699	0.4706
1.9	0.4713	0.4719	0.4726	0.4732	0.4738	0.4744	0.4750	0.4756	0.4761	0.4767
2.0	0.4772	0.4778	0.4783	0.4788	0.4793	0.4798	0.4803	0.4808	0.4812	0.4817
2.1	0.4821	0.4826	0.4830	0.4834	0.4838	0.4842	0.4846	0.4850	0.4854	0.4857
2.2	0.4861	0.4864	0.4868	0.4871	0.4875	0.4878	0.4881	0.4884	0.4887	0.4890
2.3	0.4893	0.4896	0.4898	0.4901	0.4904	0.4906	0.4909	0.4911	0.4913	0.4916
2.4	0.4918	0.4920	0.4922	0.4925	0.4927	0.4929	0.4931	0.4932	0.4934	0.4936
2.5	0.4938	0.4940	0.4941	0.4943	0.4945	0.4946	0.4948	0.4949	0.4951	0.4952
2.6	0.4953	0.4955	0.4956	0.4957	0.4959	0.4960	0.4961	0.4962	0.4963	0.4964
2.7	0.4965	0.4966	0.4967	0.4968	0.4969	0.4970	0.4971	0.4972	0.4973	0.4974
2.8	0.4974	0.4975	0.4976	0.4977	0.4977	0.4978	0.4979	0.4979	0.4980	0.4981
2.9	0.4981	0.4982	0.4982	0.4983	0.4984	0.4984	0.4985	0.4985	0.4986	0.4986
3.0	0.4987	0.4987	0.4987	0.4988	0.4988	0.4989	0.4989	0.4989	0.4990	0.4990

Source: Abridged from Table I of A. Hald. *Statistical Tables and Formulas* (New York: John Wiley & Sons, Inc.), 1952. Reproduced by permission of A. Hald and the publisher.

TABLE 5 *Critical Values of t*

Degrees of Freedom	$t_{0.100}$	$t_{0.050}$	$t_{0.025}$	$t_{0.010}$	$t_{0.005}$
1	3.078	6.314	12.706	31.821	63.657
2	1.886	2.920	4.303	6.965	9.925
3	1.638	2.353	3.182	4.541	5.841
4	1.533	2.132	2.776	3.747	4.604
5	1.476	2.015	2.571	3.365	4.032
6	1.440	1.943	2.447	3.143	3.707
7	1.415	1.895	2.365	2.998	3.499
8	1.397	1.860	2.306	2.896	3.355
9	1.383	1.833	2.262	2.821	3.250
10	1.372	1.812	2.228	2.764	3.169
11	1.363	1.796	2.201	2.718	3.106
12	1.356	1.782	2.179	2.681	3.055
13	1.350	1.771	2.160	2.650	3.012
14	1.345	1.761	2.145	2.624	2.977
15	1.341	1.753	2.131	2.602	2.947
16	1.337	1.746	2.120	2.583	2.921
17	1.333	1.740	2.110	2.567	2.898
18	1.330	1.734	2.101	2.552	2.878
19	1.328	1.729	2.093	2.539	2.861
20	1.325	1.725	2.086	2.528	2.845
21	1.323	1.721	2.080	2.518	2.831
22	1.321	1.717	2.074	2.508	2.819
23	1.319	1.714	2.069	2.500	2.807
24	1.318	1.711	2.064	2.492	2.797
25	1.316	1.708	2.060	2.485	2.787
26	1.315	1.706	2.056	2.479	2.779
27	1.314	1.703	2.052	2.473	2.771
28	1.313	1.701	2.048	2.467	2.763
29	1.311	1.699	2.045	2.462	2.756
∞	1.282	1.645	1.960	2.326	2.576

Source: From M. Merrington, "Table of Percentage Points of the *t*-Distribution," *Biometrika*, 1941, *32*, 300. Reproduced by permission of the *Biometrika* Trustees.

TABLE 6   Critical Values of χ^2

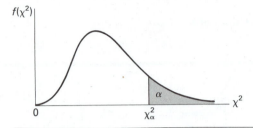

Degrees of Freedom	$\chi^2_{0.995}$	$\chi^2_{0.990}$	$\chi^2_{0.975}$	$\chi^2_{0.950}$	$\chi^2_{0.900}$
1	0.0000393	0.0001571	0.0009821	0.0039321	0.0157908
2	0.0100251	0.0201007	0.0506356	0.102587	0.210720
3	0.0717212	0.114832	0.215795	0.351846	0.584375
4	0.206990	0.297110	0.484419	0.710721	1.063623
5	0.411740	0.554300	0.831211	1.145476	1.61031
6	0.675727	0.872085	1.237347	1.63539	2.20413
7	0.989265	1.239043	1.68987	2.16735	2.83311
8	1.344419	1.646482	2.17973	2.73264	3.48954
9	1.734926	2.087912	2.70039	3.32511	4.16816
10	2.15585	2.55821	3.24697	3.94030	4.86518
11	2.60321	3.05347	3.81575	4.57481	5.57779
12	3.07382	3.57056	4.40379	5.22603	6.30380
13	3.56503	4.10691	5.00874	5.89186	7.04150
14	4.07468	4.66043	5.62872	6.57063	7.78953
15	4.60094	5.22935	6.26214	7.26094	8.54675
16	5.14224	5.81221	6.90766	7.96164	9.31223
17	5.69724	6.40776	7.56418	8.67176	10.0852
18	6.26481	7.01491	8.23075	9.39046	10.8649
19	6.84398	7.63273	8.90655	10.1170	11.6509
20	7.43386	8.26040	9.59083	10.8508	12.4426
21	8.03366	8.89720	10.28293	11.5913	13.2396
22	8.64272	9.54249	10.9823	12.3380	14.0415
23	9.26042	10.19567	11.6885	13.0905	14.8479
24	9.88623	10.8564	12.4011	13.8484	15.6587
25	10.5197	11.5240	13.1197	14.6114	16.4734
26	11.1603	12.1981	13.8439	15.3791	17.2919
27	11.8076	12.8786	14.5733	16.1513	18.1138
28	12.4613	13.5648	15.3079	16.9279	18.9392
29	13.1211	14.2565	16.0471	17.7083	19.7677
30	13.7867	14.9535	16.7908	18.4926	20.5992
40	20.7065	22.1643	24.4331	26.5093	29.0505
50	27.9907	29.7067	32.3574	34.7642	37.6886
60	35.5346	37.4848	40.4817	43.1879	46.4589
70	43.2752	45.4418	48.7576	51.7393	55.3290
80	51.1720	53.5400	57.1532	60.3915	64.2778
90	59.1963	61.7541	65.6466	69.1260	73.2912
100	67.3276	70.0648	74.2219	77.9295	82.3581

TABLE 6   (Continued)

Degrees of Freedom	$\chi^2_{0.100}$	$\chi^2_{0.050}$	$\chi^2_{0.025}$	$\chi^2_{0.010}$	$\chi^2_{0.005}$
1	2.70554	3.84146	5.02389	6.63490	7.87944
2	4.60517	5.99147	7.37776	9.21034	10.5966
3	6.25139	7.81473	9.34840	11.3449	12.8381
4	7.77944	9.48773	11.1433	13.2767	14.8602
5	9.23635	11.0705	12.8325	15.0863	16.7496
6	10.6446	12.5916	14.4494	16.8119	18.5476
7	12.0170	14.0671	16.0128	18.4753	20.2777
8	13.3616	15.5073	17.5346	20.0902	21.9550
9	14.6837	16.9190	19.0228	21.6660	23.5893
10	15.9871	18.3070	20.4831	23.2093	25.1882
11	17.2750	19.6751	21.9200	24.7250	26.7569
12	18.5494	21.0261	23.3367	26.2170	28.2995
13	19.8119	22.3621	24.7356	27.6883	29.8194
14	21.0642	23.6848	26.1190	29.1413	31.3193
15	22.3072	24.9958	27.4884	30.5779	32.8013
16	23.5418	26.2962	28.8454	31.9999	34.2672
17	24.7690	27.5871	30.1910	33.4087	35.7185
18	25.9894	28.8693	31.5264	34.8053	37.1564
19	27.2036	30.1435	32.8523	36.1908	38.5822
20	28.4120	31.4104	34.1696	37.5662	39.9968
21	29.6151	32.6705	35.4789	38.9321	41.4010
22	30.8133	33.9244	36.7807	40.2894	42.7956
23	32.0069	35.1725	38.0757	41.6384	44.1813
24	33.1963	36.4151	39.3641	42.9798	45.5585
25	34.3816	37.6525	40.6465	44.3141	46.9278
26	35.5631	38.8852	41.9232	45.6417	48.2899
27	36.7412	40.1133	43.1944	46.9630	49.6449
28	37.9159	41.3372	44.4607	48.2782	50.9933
29	39.0875	42.5569	45.7222	49.5879	52.3356
30	40.2560	43.7729	46.9792	50.8922	53.6720
40	51.8050	55.7585	59.3417	63.6907	66.7659
50	63.1671	67.5048	71.4202	76.1539	79.4900
60	74.3970	79.0819	83.2976	88.3794	91.9517
70	85.5271	90.5312	95.0231	100.425	104.215
80	96.5782	101.879	106.629	112.329	116.321
90	107.565	113.145	118.136	124.116	128.299
100	118.498	124.342	129.561	135.807	140.169

TABLE 6

TABLE 7 *Percentage Points of the F Distribution, α = 0.05*

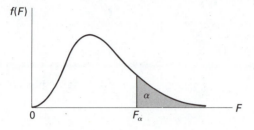

v_2	Numerator Degrees of Freedom								
	v_1 1	2	3	4	5	6	7	8	9
1	161.4	199.5	215.7	224.6	230.2	234.0	236.8	238.9	240.5
2	18.51	19.00	19.16	19.25	19.30	19.33	19.35	19.37	19.38
3	10.13	9.55	9.28	9.12	9.01	8.94	8.89	8.85	8.81
4	7.71	6.94	6.59	6.39	6.26	6.16	6.09	6.04	6.00
5	6.61	5.79	5.41	5.19	5.05	4.95	4.88	4.82	4.77
6	5.99	5.14	4.76	4.53	4.39	4.28	4.21	4.15	4.10
7	5.59	4.74	4.35	4.12	3.97	3.87	3.79	3.73	3.68
8	5.32	4.46	4.07	3.84	3.69	3.58	3.50	3.44	3.39
9	5.12	4.26	3.86	3.63	3.48	3.37	3.29	3.23	3.18
10	4.96	4.10	3.71	3.48	3.33	3.22	3.14	3.07	3.02
11	4.84	3.98	3.59	3.36	3.20	3.09	3.01	2.95	2.90
12	4.75	3.89	3.49	3.26	3.11	3.00	2.91	2.85	2.80
13	4.67	3.81	3.41	3.18	3.03	2.92	2.83	2.77	2.71
14	4.60	3.74	3.34	3.11	2.96	2.85	2.76	2.70	2.65
15	4.54	3.68	3.29	3.06	2.90	2.79	2.71	2.64	2.59
16	4.49	3.63	3.24	3.01	2.85	2.74	2.66	2.59	2.54
17	4.45	3.59	3.20	2.96	2.81	2.70	2.61	2.55	2.49
18	4.41	3.55	3.16	2.93	2.77	2.66	2.58	2.51	2.46
19	4.38	3.52	3.13	2.90	2.74	2.63	2.54	2.48	2.42
20	4.35	3.49	3.10	2.87	2.71	2.60	2.51	2.45	2.39
21	4.32	3.47	3.07	2.84	2.68	2.57	2.49	2.42	2.37
22	4.30	3.44	3.05	2.82	2.66	2.55	2.46	2.40	2.34
23	4.28	3.42	3.03	2.80	2.64	2.53	2.44	2.37	2.32
24	4.26	3.40	3.01	2.78	2.62	2.51	2.42	2.36	2.30
25	4.24	3.39	2.99	2.76	2.60	2.49	2.40	2.34	2.28
26	4.23	3.37	2.98	2.74	2.59	2.47	2.39	2.32	2.27
27	4.21	3.35	2.96	2.73	2.57	2.46	2.37	2.31	2.25
28	4.20	3.34	2.95	2.71	2.56	2.45	2.36	2.29	2.24
29	4.18	3.33	2.93	2.70	2.55	2.43	2.35	2.28	2.22
30	4.17	3.32	2.92	2.69	2.53	2.42	2.33	2.27	2.21
40	4.08	3.23	2.84	2.61	2.45	2.34	2.25	2.18	2.12
60	4.00	3.15	2.76	2.53	2.37	2.25	2.17	2.10	2.04
120	3.92	3.07	2.68	2.45	2.29	2.17	2.09	2.02	1.96
∞	3.84	3.00	2.60	2.37	2.21	2.10	2.01	1.94	1.88

Source: From M. Merrington and C. M. Thompson, "Tables of Percentage Points of the Inverted Beta (*F*)-Distribution," *Biometrika*, 1943, *33*, 73–88. Reproduced by permission of the *Biometrika* Trustees.

TABLE 7   (Continued)

v_2 \ v_1	Numerator Degrees of Freedom									
	10	12	15	20	24	30	40	60	120	∞
1	241.9	243.9	245.9	248.0	249.1	250.1	251.1	252.2	253.3	254.3
2	19.40	19.41	19.43	19.45	19.45	19.46	19.47	19.48	19.49	19.50
3	8.79	8.74	8.70	8.66	8.64	8.62	8.59	8.57	8.55	8.53
4	5.96	5.91	5.86	5.80	5.77	5.75	5.72	5.69	5.66	5.63
5	4.74	4.68	4.62	4.56	4.53	4.50	4.46	4.43	4.40	4.36
6	4.06	4.00	3.94	3.87	3.84	3.81	3.77	3.74	3.70	3.67
7	3.64	3.57	3.51	3.44	3.41	3.38	3.34	3.30	3.27	3.23
8	3.35	3.28	3.22	3.15	3.12	3.08	3.04	3.01	2.97	2.93
9	3.14	3.07	3.01	2.94	2.90	2.86	2.83	2.79	2.75	2.71
10	2.98	2.91	2.85	2.77	2.74	2.70	2.66	2.62	2.58	2.54
11	2.85	2.79	2.72	2.65	2.61	2.57	2.53	2.49	2.45	2.40
12	2.75	2.69	2.62	2.54	2.51	2.47	2.43	2.38	2.34	2.30
13	2.67	2.60	2.53	2.46	2.42	2.38	2.34	2.30	2.25	2.21
14	2.60	2.53	2.46	2.39	2.35	2.31	2.27	2.22	2.18	2.13
15	2.54	2.48	2.40	2.33	2.29	2.25	2.20	2.16	2.11	2.07
16	2.49	2.42	2.35	2.28	2.24	2.19	2.15	2.11	2.06	2.01
17	2.45	2.38	2.31	2.23	2.19	2.15	2.10	2.06	2.01	1.96
18	2.41	2.34	2.27	2.19	2.15	2.11	2.06	2.02	1.97	1.92
19	2.38	2.31	2.23	2.16	2.11	2.07	2.03	1.98	1.93	1.88
20	2.35	2.28	2.20	2.12	2.08	2.04	1.99	1.95	1.90	1.84
21	2.32	2.25	2.18	2.10	2.05	2.01	1.96	1.92	1.87	1.81
22	2.30	2.23	2.15	2.07	2.03	1.98	1.94	1.89	1.84	1.78
23	2.27	2.20	2.13	2.05	2.01	1.96	1.91	1.86	1.81	1.76
24	2.25	2.18	2.11	2.03	1.98	1.94	1.89	1.84	1.79	1.73
25	2.24	2.16	2.09	2.01	1.96	1.92	1.87	1.82	1.77	1.71
26	2.22	2.15	2.07	1.99	1.95	1.90	1.85	1.80	1.75	1.69
27	2.20	2.13	2.06	1.97	1.93	1.88	1.84	1.79	1.73	1.67
28	2.19	2.12	2.04	1.96	1.91	1.87	1.82	1.77	1.71	1.65
29	2.18	2.10	2.03	1.94	1.90	1.85	1.81	1.75	1.70	1.64
30	2.16	2.09	2.01	1.93	1.89	1.84	1.79	1.74	1.68	1.62
40	2.08	2.00	1.92	1.84	1.79	1.74	1.69	1.64	1.58	1.51
60	1.99	1.92	1.84	1.75	1.70	1.65	1.59	1.53	1.47	1.39
120	1.91	1.83	1.75	1.66	1.61	1.55	1.50	1.43	1.35	1.25
∞	1.83	1.75	1.67	1.57	1.52	1.46	1.39	1.32	1.22	1.00

Denominator Degrees of Freedom

TABLE 7

TABLE 8 *Percentage Points of the F Distribution, $\alpha = 0.01$*

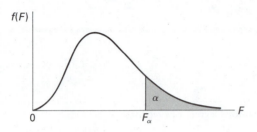

v_1	Numerator Degrees of Freedom								
v_2	1	2	3	4	5	6	7	8	9
1	4,052	4,999.5	5,403	5,625	5,764	5,859	5,928	5,982	6,022
2	98.50	99.00	99.17	99.25	99.30	99.33	99.36	99.37	99.39
3	34.12	30.82	29.46	28.71	28.24	27.91	27.67	27.49	27.35
4	21.20	18.00	16.69	15.98	15.52	15.21	14.98	14.80	14.66
5	16.26	13.27	12.06	11.39	10.97	10.67	10.46	10.29	10.16
6	13.75	10.92	9.78	9.15	8.75	8.47	8.26	8.10	7.98
7	12.25	9.55	8.45	7.85	7.46	7.19	6.99	6.84	6.72
8	11.26	8.65	7.59	7.01	6.63	6.37	6.18	6.03	5.91
9	10.56	8.02	6.99	6.42	6.06	5.80	5.61	5.47	5.35
10	10.04	7.56	6.55	5.99	5.64	5.39	5.20	5.06	4.94
11	9.65	7.21	6.22	5.67	5.32	5.07	4.89	4.74	4.63
12	9.33	6.93	5.95	5.41	5.06	4.82	4.64	4.50	4.39
13	9.07	6.70	5.74	5.21	4.86	4.62	4.44	4.30	4.19
14	8.86	6.51	5.56	5.04	4.69	4.46	4.28	4.14	4.03
15	8.68	6.36	5.42	4.89	4.56	4.32	4.14	4.00	3.89
16	8.53	6.23	5.29	4.77	4.44	4.20	4.03	3.89	3.78
17	8.40	6.11	5.18	4.67	4.34	4.10	3.93	3.79	3.68
18	8.29	6.01	5.09	4.58	4.25	4.01	3.84	3.71	3.60
19	8.18	5.93	5.01	4.50	4.17	3.94	3.77	3.63	3.52
20	8.10	5.85	4.94	4.43	4.10	3.87	3.70	3.56	3.46
21	8.02	5.78	4.87	4.37	4.04	3.81	3.64	3.51	3.40
22	7.95	5.72	4.82	4.31	3.99	3.76	3.59	3.45	3.35
23	7.88	5.66	4.76	4.26	3.94	3.71	3.54	3.41	3.30
24	7.82	5.61	4.72	4.22	3.90	3.67	3.50	3.36	3.26
25	7.77	5.57	4.68	4.18	3.85	3.63	3.46	3.32	3.22
26	7.72	5.53	4.64	4.14	3.82	3.59	3.42	3.29	3.18
27	7.68	5.49	4.60	4.11	3.78	3.56	3.39	3.26	3.15
28	7.64	5.45	4.57	4.07	3.75	3.53	3.36	3.23	3.12
29	7.60	5.42	4.54	4.04	3.73	3.50	3.33	3.20	3.09
30	7.56	5.39	4.51	4.02	3.70	3.47	3.30	3.17	3.07
40	7.31	5.18	4.31	3.83	3.51	3.29	3.12	2.99	2.89
60	7.08	4.98	4.13	3.65	3.34	3.12	2.95	2.82	2.72
120	6.85	4.79	3.95	3.48	3.17	2.96	2.79	2.66	2.56
∞	6.63	4.61	3.78	3.32	3.02	2.80	2.64	2.51	2.41

Denominator Degrees of Freedom

Source: From M. Merrington and C. M. Thompson, "Tables of Percentage Points of the Inverted Beta (*F*)-Distribution," *Biometrika*, 1943, *33*, 73–88. Reproduced by permission of the *Biometrika* Trustees.

TABLE 8 *(Continued)*

v_1 / v_2	10	12	15	20	24	30	40	60	120	∞
				Numerator Degrees of Freedom						
1	6,056	6,106	6,157	6,209	6,235	6,261	6,287	6,313	6,339	6,366
2	99.40	99.42	99.43	99.45	99.46	99.47	99.47	99.48	99.49	99.50
3	27.23	27.05	26.87	26.69	26.60	26.50	26.41	26.32	26.22	26.13
4	14.55	14.37	14.20	14.02	13.93	13.84	13.75	13.65	13.56	13.46
5	10.05	9.89	9.72	9.55	9.47	9.38	9.29	9.20	9.11	9.02
6	7.87	7.72	7.56	7.40	7.31	7.23	7.14	7.06	6.97	6.88
7	6.62	6.47	6.31	6.16	6.07	5.99	5.91	5.82	5.74	5.65
8	5.81	5.67	5.52	5.36	5.28	5.20	5.12	5.03	4.95	4.86
9	5.26	5.11	4.96	4.81	4.73	4.65	4.57	4.48	4.40	4.31
10	4.85	4.71	4.56	4.41	4.33	4.25	4.17	4.08	4.00	3.91
11	4.54	4.40	4.25	4.10	4.02	3.94	3.86	3.78	3.69	3.60
12	4.30	4.16	4.01	3.86	3.78	3.70	3.62	3.54	3.45	3.36
13	4.10	3.96	3.82	3.66	3.59	3.51	3.43	3.34	3.25	3.17
14	3.94	3.80	3.66	3.51	3.43	3.35	3.27	3.18	3.09	3.00
15	3.80	3.67	3.52	3.37	3.29	3.21	3.13	3.05	2.96	2.87
16	3.69	3.55	3.41	3.26	3.18	3.10	3.02	2.93	2.84	2.75
17	3.59	3.46	3.31	3.16	3.08	3.00	2.92	2.83	2.75	2.65
18	3.51	3.37	3.23	3.08	3.00	2.92	2.84	2.75	2.66	2.57
19	3.43	3.30	3.15	3.00	2.92	2.84	2.76	2.67	2.58	2.49
20	3.37	3.23	3.09	2.94	2.86	2.78	2.69	2.61	2.52	2.42
21	3.31	3.17	3.03	2.88	2.80	2.72	2.64	2.55	2.46	2.36
22	3.26	3.12	2.98	2.83	2.75	2.67	2.58	2.50	2.40	2.31
23	3.21	3.07	2.93	2.78	2.70	2.62	2.54	2.45	2.35	2.26
24	3.17	3.03	2.89	2.74	2.66	2.58	2.49	2.40	2.31	2.21
25	3.13	2.99	2.85	2.70	2.62	2.54	2.45	2.36	2.27	2.17
26	3.09	2.96	2.81	2.66	2.58	2.50	2.42	2.33	2.23	2.13
27	3.06	2.93	2.78	2.63	2.55	2.47	2.38	2.29	2.20	2.10
28	3.03	2.90	2.75	2.60	2.52	2.44	2.35	2.26	2.17	2.06
29	3.00	2.87	2.73	2.57	2.49	2.41	2.33	2.23	2.14	2.03
30	2.98	2.84	2.70	2.55	2.47	2.39	2.30	2.21	2.11	2.01
40	2.80	2.66	2.52	2.37	2.29	2.20	2.11	2.02	1.92	1.80
60	2.63	2.50	2.35	2.20	2.12	2.03	1.94	1.84	1.73	1.60
120	2.47	2.34	2.19	2.03	1.95	1.86	1.76	1.66	1.53	1.38
∞	2.32	2.18	2.04	1.88	1.79	1.70	1.59	1.47	1.32	1.00

Denominator Degrees of Freedom (row labels v_2)

TABLE 8

TABLE 9 *Critical Values of T_L and T_U for the Wilcoxon Rank Sum Test: Independent Samples*

Test statistic is rank sum associated with smaller sample (if equal sample sizes, either rank sum can be used).

(a) $\alpha = 0.025$ one-tailed; $\alpha = 0.05$ two-tailed

n_2 \ n_1	3		4		5		6		7		8		9		10	
	T_L	T_U	T_L	T_U	T_L	T_U	T_L	T_U	T_L	T_U	T_L	T_U	T_L	T_U	T_L	T_U
3	5	16	6	18	6	21	7	23	7	26	8	28	8	31	9	33
4	6	18	11	25	12	28	12	32	13	35	14	38	15	41	16	44
5	6	21	12	28	18	37	19	41	20	45	21	49	22	53	24	56
6	7	23	12	32	19	41	26	52	28	56	29	61	31	65	32	70
7	7	26	13	35	20	45	28	56	37	68	39	73	41	78	43	83
8	8	28	14	38	21	49	29	61	39	73	49	87	51	93	54	98
9	8	31	15	41	22	53	31	65	41	78	51	93	63	108	66	114
10	9	33	16	44	24	56	32	70	43	83	54	98	66	114	79	131

TABLE 9

(b) $\alpha = 0.05$ one-tailed; $\alpha = 0.10$ two-tailed

n_2 \ n_1	3		4		5		6		7		8		9		10	
	T_L	T_U	T_L	T_U	T_L	T_U	T_L	T_U	T_L	T_U	T_L	T_U	T_L	T_U	T_L	T_U
3	6	15	7	17	7	20	8	22	9	24	9	27	10	29	11	31
4	7	17	12	24	13	27	14	30	15	33	16	36	17	39	18	42
5	7	20	13	27	19	36	20	40	22	43	24	46	25	50	26	54
6	8	22	14	30	20	40	28	50	30	54	32	58	33	63	35	67
7	9	24	15	33	22	43	30	54	39	66	41	71	43	76	46	80
8	9	27	16	36	24	46	32	58	41	71	52	84	54	90	57	95
9	10	29	17	39	25	50	33	63	43	76	54	90	66	105	69	111
10	11	31	18	42	26	54	35	67	46	80	57	95	69	111	83	127

Source: From F. Wilcoxon and R. A. Wilcox, "Some Rapid Approximate Statistical Procedures," 1964, 20–23. Reproduced with the permission of American Cyanamid Company.

TABLE 10 *Critical Values of T_0 in the Wilcoxon Paired-Difference Signed-Ranks Test*

One-tailed	Two-tailed	$n = 5$	$n = 6$	$n = 7$	$n = 8$	$n = 9$	$n = 10$
$\alpha = 0.05$	$\alpha = 0.10$	1	2	4	6	8	11
$\alpha = 0.025$	$\alpha = 0.05$		1	2	4	6	8
$\alpha = 0.01$	$\alpha = 0.02$			0	2	3	5
$\alpha = 0.005$	$\alpha = 0.01$				0	2	3

		$n = 11$	$n = 12$	$n = 13$	$n = 14$	$n = 15$	$n = 16$
$\alpha = 0.05$	$\alpha = 0.10$	14	17	21	26	30	36
$\alpha = 0.025$	$\alpha = 0.05$	11	14	17	21	25	30
$\alpha = 0.01$	$\alpha = 0.02$	7	10	13	16	20	24
$\alpha = 0.005$	$\alpha = 0.01$	5	7	10	13	16	19

		$n = 17$	$n = 18$	$n = 19$	$n = 20$	$n = 21$	$n = 22$
$\alpha = 0.05$	$\alpha = 0.10$	41	47	54	60	68	75
$\alpha = 0.025$	$\alpha = 0.05$	35	40	46	52	59	66
$\alpha = 0.01$	$\alpha = 0.02$	28	33	38	43	49	56
$\alpha = 0.005$	$\alpha = 0.01$	23	28	32	37	43	49

		$n = 23$	$n = 24$	$n = 25$	$n = 26$	$n = 27$	$n = 28$
$\alpha = 0.05$	$\alpha = 0.10$	83	92	101	110	120	130
$\alpha = 0.025$	$\alpha = 0.05$	73	81	90	98	107	117
$\alpha = 0.01$	$\alpha = 0.02$	62	69	77	85	93	102
$\alpha = 0.005$	$\alpha = 0.01$	55	61	68	76	84	92

		$n = 29$	$n = 30$	$n = 31$	$n = 32$	$n = 33$	$n = 34$
$\alpha = 0.05$	$\alpha = 0.10$	141	152	163	175	188	201
$\alpha = 0.025$	$\alpha = 0.05$	127	137	148	159	171	183
$\alpha = 0.01$	$\alpha = 0.02$	111	120	130	141	151	162
$\alpha = 0.005$	$\alpha = 0.01$	100	109	118	128	138	149

		$n = 35$	$n = 36$	$n = 37$	$n = 38$	$n = 39$	
$\alpha = 0.05$	$\alpha = 0.10$	214	228	242	256	271	
$\alpha = 0.025$	$\alpha = 0.05$	195	208	222	235	250	
$\alpha = 0.01$	$\alpha = 0.02$	174	186	198	211	224	
$\alpha = 0.005$	$\alpha = 0.01$	160	171	183	195	208	

		$n = 40$	$n = 41$	$n = 42$	$n = 43$	$n = 44$	$n = 45$
$\alpha = 0.05$	$\alpha = 0.10$	287	303	319	336	353	371
$\alpha = 0.025$	$\alpha = 0.05$	264	279	295	311	327	344
$\alpha = 0.01$	$\alpha = 0.02$	238	252	267	281	297	313
$\alpha = 0.005$	$\alpha = 0.01$	221	234	248	262	277	292

		$n = 46$	$n = 47$	$n = 48$	$n = 49$	$n = 50$	
$\alpha = 0.05$	$\alpha = 0.10$	389	408	427	446	466	
$\alpha = 0.025$	$\alpha = 0.05$	361	379	397	415	434	
$\alpha = 0.01$	$\alpha = 0.02$	329	345	362	380	398	
$\alpha = 0.005$	$\alpha = 0.01$	307	323	339	356	373	

TABLE
10

Source: From F. Wilcoxon and R. A. Wilcox, "Some Rapid Approximate Statistical Procedures," 1964, 28. Reproduced with the permission of American Cyanamid Company.

TABLE 11 *Critical Values of Spearman's Rank Correlation Coefficient*

The α values correspond to a one-tailed test of H_0: $\rho_S = 0$. The value should be doubled for two-tailed tests.

n	$\alpha = 0.05$	$\alpha = 0.025$	$\alpha = 0.01$	$\alpha = 0.005$
5	0.900	—	—	—
6	0.829	0.886	0.943	—
7	0.714	0.786	0.893	—
8	0.643	0.738	0.833	0.881
9	0.600	0.683	0.783	0.833
10	0.564	0.648	0.745	0.794
11	0.523	0.623	0.736	0.818
12	0.497	0.591	0.703	0.780
13	0.475	0.566	0.673	0.745
14	0.457	0.545	0.646	0.716
15	0.441	0.525	0.623	0.689
16	0.425	0.507	0.601	0.666
17	0.412	0.490	0.582	0.645
18	0.399	0.476	0.564	0.625
19	0.388	0.462	0.549	0.608
20	0.377	0.450	0.534	0.591
21	0.368	0.438	0.521	0.576
22	0.359	0.428	0.508	0.562
23	0.351	0.418	0.496	0.549
24	0.343	0.409	0.485	0.537
25	0.336	0.400	0.475	0.526
26	0.329	0.392	0.465	0.515
27	0.323	0.385	0.456	0.505
28	0.317	0.377	0.448	0.496
29	0.311	0.370	0.440	0.487
30	0.305	0.364	0.432	0.478

Source: From E. G. Olds, "Distribution of Sums of Squares of Rank Differences for Small Samples." *Annals of Mathematical Statistics.* 1938, 9. Reproduced with the permission of the Editor, *Annals of Mathematical Statistics.*

TABLE 12 *Factors for Computing Control Chart Lines*

Number of Observations in Sample, n	Chart for Averages — Factors for Control Limits			Chart for Standard Deviations — Factors for Central Line		Chart for Standard Deviations — Factors for Control Limits				Chart for Ranges — Factors for Central Line			Chart for Ranges — Factors for Control Limits			
	A	A_1	A_2	c_2	$1/c_2$	B_1	B_2	B_3	B_4	d_2	$1/d_2$	d_3	D_1	D_2	D_3	D_4
2	2.121	3.760	1.880	0.5642	1.7725	0	1.843	0	3.267	1.128	0.8865	0.853	0	3.686	0	3.276
3	1.732	2.394	1.023	0.7236	1.3820	0	1.858	0	2.568	1.693	0.5907	0.888	0	4.358	0	2.575
4	1.501	1.880	0.729	0.7979	1.2533	0	1.808	0	2.266	2.059	0.4857	0.880	0	4.698	0	2.282
5	1.342	1.596	0.577	0.8407	1.1894	0	1.756	0	2.089	2.326	0.4299	0.864	0	4.918	0	2.115
6	1.225	1.410	0.483	0.8686	1.1512	0.026	1.711	0.030	1.970	2.534	0.3946	0.848	0	5.078	0	2.004
7	1.134	1.277	0.419	0.8882	1.1259	0.105	1.672	0.118	1.882	2.704	0.3698	0.833	0.205	5.203	0.076	1.924
8	1.061	1.175	0.373	0.9027	1.1078	0.167	1.638	0.185	1.815	2.847	0.3512	0.820	0.387	5.307	0.136	1.864
9	1.000	1.094	0.337	0.9139	1.0942	0.219	1.609	0.239	1.761	2.970	0.3367	0.808	0.546	5.394	0.184	1.816
10	0.949	1.028	0.308	0.9227	1.0837	0.262	1.584	0.284	1.716	3.078	0.3249	0.797	0.687	5.469	0.223	1.777
11	0.905	0.973	0.285	0.9300	1.0753	0.299	1.561	0.321	1.679	3.173	0.3152	0.787	0.812	5.534	0.256	1.744
12	0.866	0.925	0.266	0.9359	1.0684	0.331	1.541	0.354	1.646	3.258	0.3069	0.778	0.924	5.592	0.284	1.719
13	0.832	0.884	0.249	0.9410	1.0627	0.359	1.523	0.382	1.618	3.336	0.2998	0.770	1.026	5.646	0.308	1.692
14	0.802	0.848	0.235	0.9453	1.0579	0.384	1.507	0.406	1.594	3.407	0.2935	0.762	1.121	5.693	0.329	1.671
15	0.775	0.816	0.223	0.9490	1.0537	0.406	1.492	0.428	1.572	3.472	0.2880	0.755	1.207	5.737	0.348	1.652
16	0.750	0.788	0.212	0.9523	1.0501	0.427	1.478	0.448	1.552	3.532	0.2831	0.749	1.285	5.779	0.364	1.636
17	0.728	0.762	0.203	0.9551	1.0470	0.445	1.465	0.466	1.534	3.588	0.2787	0.743	1.359	5.817	0.379	1.621
18	0.707	0.738	0.194	0.9576	1.0442	0.461	1.454	0.482	1.518	3.640	0.2747	0.738	1.426	5.854	0.392	1.608
19	0.688	0.717	0.187	0.9599	1.0418	0.477	1.443	0.497	1.503	3.689	0.2711	0.733	1.490	5.888	0.404	1.596
20	0.671	0.697	0.180	0.9619	1.0396	0.491	1.433	0.510	1.490	3.735	0.2677	0.729	1.548	5.922	0.414	1.586
21	0.655	0.679	0.173	0.9638	1.0376	0.504	1.424	0.523	1.477	3.778	0.2647	0.724	1.606	5.950	0.425	1.575
22	0.640	0.662	0.167	0.9655	1.0358	0.516	1.415	0.534	1.466	3.819	0.2618	0.720	1.659	5.979	0.434	1.566
23	0.626	0.647	0.162	0.9670	1.0342	0.527	1.407	0.545	1.455	3.858	0.2592	0.716	1.710	6.006	0.443	1.557
24	0.612	0.632	0.157	0.9684	1.0327	0.538	1.399	0.555	1.445	3.895	0.2567	0.712	1.759	6.031	0.452	1.548
25	0.600	0.619	0.153	0.9696	1.0313	0.548	1.392	0.565	1.435	3.931	0.2544	0.709	1.804	6.058	0.459	1.541
Over 25	$\dfrac{3}{\sqrt{n}}$	$\dfrac{3}{\sqrt{n}}$	—	—	—	‡	§	‡	§	—	—	—	—	—	—	—

[1] Reproduced by permission from *ASTM Manual on Quality Control of Materials*, American Society for Testing Materials, Philadelphia, Pa., 1951.

‡ $1 - \dfrac{3}{\sqrt{2n}}$ § $1 + \dfrac{3}{\sqrt{2n}}$

TABLE 13 *Sample Size Code Letters: MIL-STD-105D*

Lot or batch size	Special inspection levels				General inspection levels		
	S-1	S-2	S-3	S-4	I	II	III
2–8	A	A	A	A	A	A	B
9–15	A	A	A	A	A	B	C
16–25	A	A	B	B	B	C	D
26–50	A	B	B	C	C	D	E
51–90	B	B	C	C	C	E	F
91–150	B	B	C	D	D	F	G
151–280	B	C	D	E	E	G	H
281–500	B	C	D	E	F	H	J
501–1,200	C	C	E	F	G	J	K
1,201–3,200	C	D	E	G	H	K	L
3,201–10,000	C	D	F	G	J	L	M
10,001–35,000	C	D	F	H	K	M	N
35,001–150,000	D	E	G	J	L	N	P
150,001–500,000	D	E	G	J	M	P	Q
500,001 and over	D	E	H	K	N	Q	R

TABLE
13

TABLE 14 *Master Table for Normal Inspection (Single Sampling): MIL-STD-105D*

Acceptable quality levels (normal inspection). Each cell gives **Ac Re** (Ac = acceptance number, Re = rejection number). ↓ = use first sampling plan below arrow; ↑ = use first sampling plan above arrow.

Sample size code letter	Sample size	0.010	0.015	0.025	0.040	0.065	0.10	0.15	0.25	0.40	0.65	1.0	1.5	2.5	4.0	6.5	10	15	25	40	65	100	150	250	400	650	1,000
A	2																↓	0 1	1 2	2 3	3 4	5 6	7 8	10 11	14 15	21 22	30 31
B	3															↓	0 1	1 2	2 3	3 4	5 6	7 8	10 11	14 15	21 22	30 31	44 45
C	5														↓	0 1	1 2	2 3	3 4	5 6	7 8	10 11	14 15	21 22	30 31	44 45	↑
D	8													↓	0 1	1 2	2 3	3 4	5 6	7 8	10 11	14 15	21 22	30 31	44 45	↑	
E	13												↓	0 1	1 2	2 3	3 4	5 6	7 8	10 11	14 15	21 22	30 31	44 45	↑		
F	20											↓	0 1	1 2	2 3	3 4	5 6	7 8	10 11	14 15	21 22	30 31	44 45	↑			
G	32										↓	0 1	1 2	2 3	3 4	5 6	7 8	10 11	14 15	21 22	30 31	44 45	↑				
H	50									↓	0 1	1 2	2 3	3 4	5 6	7 8	10 11	14 15	21 22	30 31	44 45	↑					
J	80								↓	0 1	1 2	2 3	3 4	5 6	7 8	10 11	14 15	21 22	30 31	44 45	↑						
K	125							↓	0 1	1 2	2 3	3 4	5 6	7 8	10 11	14 15	21 22	30 31	44 45	↑							
L	200						↓	0 1	1 2	2 3	3 4	5 6	7 8	10 11	14 15	21 22	30 31	44 45	↑								
M	315					↓	0 1	1 2	2 3	3 4	5 6	7 8	10 11	14 15	21 22	30 31	44 45	↑									
N	500				↓	0 1	1 2	2 3	3 4	5 6	7 8	10 11	14 15	21 22	30 31	44 45	↑										
P	800			↓	0 1	1 2	2 3	3 4	5 6	7 8	10 11	14 15	21 22	30 31	44 45	↑											
Q	1,250		↓	0 1	1 2	2 3	3 4	5 6	7 8	10 11	14 15	21 22	30 31	44 45	↑												
R	2,000	↓	0 1	1 2	2 3	3 4	5 6	7 8	10 11	14 15	21 22	30 31	44 45	↑													

↓ = use first sampling plan below arrow. If sample size equals, or exceeds, lot or batch size, do 100% inspection.
↑ = use first sampling plan above arrow.
Ac = acceptance number.
Re = rejection number.

TABLE
14

TABLE 15 Master Table for Tightened Inspection (Single Sampling)—MIL–STD–105D

Acceptable quality levels (tightened inspection) — each column shows Ac Re (acceptance number / rejection number)

Sample size code letter	Sample size	0.010	0.015	0.025	0.040	0.065	0.10	0.15	0.25	0.40	0.65	1.0	1.5	2.5	4.0	6.5	10	15	25	40	65	100	150	250	400	650	1,000
A	2	↓	↓	↓	↓	↓	↓	↓	↓	↓	↓	↓	↓	↓	↓	0 1	↓	↓	↓	1 2	2 3	3 4	5 6	8 9	12 13	18 19	27 28
B	3	↓	↓	↓	↓	↓	↓	↓	↓	↓	↓	↓	↓	↓	0 1	↑	↓	↓	1 2	2 3	3 4	5 6	8 9	12 13	18 19	27 28	41 42
C	5	↓	↓	↓	↓	↓	↓	↓	↓	↓	↓	↓	↓	0 1	↑	↑	↓	1 2	2 3	3 4	5 6	8 9	12 13	18 19	27 28	41 42	↑
D	8	↓	↓	↓	↓	↓	↓	↓	↓	↓	↓	↓	0 1	↑	↑	↑	1 2	2 3	3 4	5 6	8 9	12 13	18 19	27 28	41 42	↑	↑
E	13	↓	↓	↓	↓	↓	↓	↓	↓	↓	↓	0 1	↑	↑	↑	1 2	2 3	3 4	5 6	8 9	12 13	18 19	27 28	41 42	↑	↑	↑
F	20	↓	↓	↓	↓	↓	↓	↓	↓	↓	0 1	↑	↑	↑	1 2	2 3	3 4	5 6	8 9	12 13	18 19	↑	↑	↑	↑	↑	↑
G	32	↓	↓	↓	↓	↓	↓	↓	↓	0 1	↑	↑	↑	1 2	2 3	3 4	5 6	8 9	12 13	18 19	↑	↑	↑	↑	↑	↑	↑
H	50	↓	↓	↓	↓	↓	↓	↓	0 1	↑	↑	↑	1 2	2 3	3 4	5 6	8 9	12 13	18 19	↑	↑	↑	↑	↑	↑	↑	↑
J	80	↓	↓	↓	↓	↓	↓	0 1	↑	↑	↑	1 2	2 3	3 4	5 6	8 9	12 13	18 19	↑	↑	↑	↑	↑	↑	↑	↑	↑
K	125	↓	↓	↓	↓	↓	0 1	↑	↑	↑	1 2	2 3	3 4	5 6	8 9	12 13	18 19	↑	↑	↑	↑	↑	↑	↑	↑	↑	↑
L	200	↓	↓	↓	↓	0 1	↑	↑	↑	1 2	2 3	3 4	5 6	8 9	12 13	18 19	↑	↑	↑	↑	↑	↑	↑	↑	↑	↑	↑
M	315	↓	↓	↓	0 1	↑	↑	↑	1 2	2 3	3 4	5 6	8 9	12 13	18 19	↑	↑	↑	↑	↑	↑	↑	↑	↑	↑	↑	↑
N	500	↓	↓	0 1	↑	↑	↑	1 2	2 3	3 4	5 6	8 9	12 13	18 19	↑	↑	↑	↑	↑	↑	↑	↑	↑	↑	↑	↑	↑
P	800	↓	0 1	↑	↑	↑	1 2	2 3	3 4	5 6	8 9	12 13	18 19	↑	↑	↑	↑	↑	↑	↑	↑	↑	↑	↑	↑	↑	↑
Q	1,250	0 1	↑	↑	↑	1 2	2 3	3 4	5 6	8 9	12 13	18 19	↑	↑	↑	↑	↑	↑	↑	↑	↑	↑	↑	↑	↑	↑	↑
R	2,000	↑	↑	↑	1 2	2 3	3 4	5 6	8 9	12 13	18 19	↑	↑	↑	↑	↑	↑	↑	↑	↑	↑	↑	↑	↑	↑	↑	↑
S	3,150	↑	↑	1 2	2 3	3 4	5 6	8 9	12 13	18 19	↑	↑	↑	↑	↑	↑	↑	↑	↑	↑	↑	↑	↑	↑	↑	↑	↑

↓ = use first sampling plan below arrow. If sample size equals or exceeds lot or batch size, do 100% inspection.
↑ = use first sampling plan above arrow.
Ac = acceptance number.
Re = rejection number.

TABLE 16 *Master Table for Reduced Inspection (Single Sampling)—MIL–STD–105D*

Each cell below is given as "Ac Re" (acceptance number, rejection number). ↓ = use first sampling plan below arrow; ↑ = use first sampling plan above arrow.

Sample size code letter	Sample size	Acceptance quality levels (reduced inspection)[†] 0.010	0.015	0.025	0.040	0.065	0.10	0.15	0.25	0.40	0.65	1.0	1.5	2.5	4.0	6.5	10	15	25	40	65	100	150	250	400	650	1,000
A	2	↓	↓	↓	↓	↓	↓	↓	↓	↓	↓	↓	↓	↓	↓	↓	↓	↓	1 2	2 3	3 4	5 6	7 8	10 11	14 15	21 22	30 31
B	2	↓	↓	↓	↓	↓	↓	↓	↓	↓	↓	↓	↓	↓	↓	↓	0 1	0 2	1 3	2 4	3 5	5 6	7 8	10 11	14 15	21 22	30 31
C	2	↓	↓	↓	↓	↓	↓	↓	↓	↓	↓	↓	↓	↓	↓	0 1	0 2	1 3	1 4	2 5	3 6	5 8	7 10	10 13	14 17	21 24	↑
D	3	↓	↓	↓	↓	↓	↓	↓	↓	↓	↓	↓	↓	↓	0 1	0 2	1 3	1 4	2 5	3 6	5 8	7 10	10 13	14 17	21 24	↑	↑
E	5	↓	↓	↓	↓	↓	↓	↓	↓	↓	↓	↓	↓	0 1	0 2	1 3	1 4	2 5	3 6	5 8	7 10	10 13	14 17	21 24	↑	↑	↑
F	8	↓	↓	↓	↓	↓	↓	↓	↓	↓	↓	↓	0 1	0 2	1 3	1 4	2 5	3 6	5 8	7 10	10 13	↑	↑	↑	↑	↑	↑
G	13	↓	↓	↓	↓	↓	↓	↓	↓	↓	↓	0 1	0 2	1 3	1 4	2 5	3 6	5 8	7 10	10 13	↑	↑	↑	↑	↑	↑	↑
H	20	↓	↓	↓	↓	↓	↓	↓	↓	↓	0 1	0 2	1 3	1 4	2 5	3 6	5 8	7 10	10 13	↑	↑	↑	↑	↑	↑	↑	↑
J	32	↓	↓	↓	↓	↓	↓	↓	↓	0 1	0 2	1 3	1 4	2 5	3 6	5 8	7 10	10 13	↑	↑	↑	↑	↑	↑	↑	↑	↑
K	50	↓	↓	↓	↓	↓	↓	↓	0 1	0 2	1 3	1 4	2 5	3 6	5 8	7 10	10 13	↑	↑	↑	↑	↑	↑	↑	↑	↑	↑
L	80	↓	↓	↓	↓	↓	↓	0 1	0 2	1 3	1 4	2 5	3 6	5 8	7 10	10 13	↑	↑	↑	↑	↑	↑	↑	↑	↑	↑	↑
M	125	↓	↓	↓	↓	↓	0 1	0 2	1 3	1 4	2 5	3 6	5 8	7 10	10 13	↑	↑	↑	↑	↑	↑	↑	↑	↑	↑	↑	↑
N	200	↓	↓	↓	↓	0 1	0 2	1 3	1 4	2 5	3 6	5 8	7 10	10 13	↑	↑	↑	↑	↑	↑	↑	↑	↑	↑	↑	↑	↑
P	315	↓	↓	↓	0 1	0 2	1 3	1 4	2 5	3 6	5 8	7 10	10 13	↑	↑	↑	↑	↑	↑	↑	↑	↑	↑	↑	↑	↑	↑
Q	500	↓	↓	0 1	0 2	1 3	1 4	2 5	3 6	5 8	7 10	10 13	↑	↑	↑	↑	↑	↑	↑	↑	↑	↑	↑	↑	↑	↑	↑
R	800	↓	0 1	0 2	1 3	1 4	2 5	3 6	5 8	7 10	10 13	↑	↑	↑	↑	↑	↑	↑	↑	↑	↑	↑	↑	↑	↑	↑	↑

↓ = use first sampling plan below arrow. If sample size equals or exceeds lot or batch size, do 100% inspection.

↑ = use first sampling plan above arrow.

Ac = acceptance number.

Re = rejection number.

[†] If the acceptance number has been exceeded but the rejection number has not been reached, accept the lot but reinstate normal inspection.

TABLE 17 *MIL—STD—414*

A
AQL Conversion Table

For specified AQL values falling within these ranges	Use this AQL value
to 0.049	0.04
0.050 to 0.069	0.065
0.070 to 0.109	0.10
0.110 to 0.164	0.15
0.165 to 0.279	0.25
0.280 to 0.439	0.40
0.440 to 0.699	0.65
0.700 to 1.09	1.0
1.10 to 1.64	1.5
1.65 to 2.79	2.5
2.80 to 4.39	4.0
4.40 to 6.99	6.5
7.00 to 10.9	10.0
11.00 to 16.4	15.0

B
Sample Size Code Letters[1]

Lot Size	Inspection Levels				
	I	II	III	IV	V
3 to 8	B	B	B	B	C
9 to 15	B	B	B	B	D
16 to 25	B	B	B	C	E
26 to 40	B	B	B	D	F
41 to 65	B	B	C	E	G
66 to 110	B	B	D	F	H
111 to 180	B	C	E	G	I
181 to 300	B	D	F	H	J
301 to 500	C	E	G	I	K
501 to 800	D	F	H	J	L
801 to 1,300	E	G	I	K	L
1,301 to 3,200	F	H	J	L	M
3,201 to 8,000	G	I	L	M	N
8,001 to 22,000	H	J	M	N	O
22,001 to 110,000	I	K	N	O	P
110,001 to 550,000	I	K	O	P	Q
550,001 and over	I	K	P	Q	Q

[1] Sample size code letters given in body of table are applicable when the indicated inspection levels are to be used.

TABLE 17

TABLE 18 MIL–STD–414 (Standard Deviation Method) Master Table for Normal and Tightened Inspection for Plans Based on Variability Unknown (Single Specification Limit—Form 1)

Sample size code letter	Sample size	Acceptable Quality Levels (normal inspection)													
		0.04	0.065	0.10	0.15	0.25	0.40	0.65	1.00	1.50	2.50	4.00	6.50	10.00	15.00
		k	k	k	k	k	k	k	k	k	k	k	k	k	k
B	3	↓	↓	↓	↓	↓	↓	↓	↓	↓	1.12	0.958	0.765	0.566	0.341
C	4	↓	↓	↓	↓	↓	↓	↓	1.45	1.34	1.17	1.01	0.814	0.617	0.393
D	5	↓	↓	↓	↓	↓	↓	1.65	1.53	1.40	1.24	1.07	0.874	0.675	0.455
E	7	↓	↓	↓	↓	2.00	1.88	1.75	1.62	1.50	1.33	1.15	0.955	0.755	0.536
F	10	↓	↓	↓	2.24	2.11	1.98	1.84	1.72	1.58	1.41	1.23	1.03	0.828	0.611
G	15	↓	2.53	2.42	2.32	2.20	2.06	1.91	1.79	1.65	1.47	1.30	1.09	0.886	0.664
H	20	↓	2.58	2.47	2.36	2.24	2.11	1.96	1.82	1.69	1.51	1.33	1.12	0.917	0.695
I	25	↓	2.61	2.50	2.40	2.26	2.14	1.98	1.85	1.72	1.53	1.35	1.14	0.936	0.712
J	30	2.64	2.61	2.51	2.41	2.28	2.15	2.00	1.86	1.73	1.55	1.36	1.15	0.946	0.723
K	35	2.69	2.65	2.54	2.45	2.31	2.18	2.03	1.89	1.76	1.57	1.39	1.18	0.969	0.745
L	40	2.72	2.66	2.55	2.44	2.31	2.18	2.03	1.89	1.76	1.58	1.39	1.18	0.971	0.746
M	50	2.83	2.71	2.60	2.50	2.35	2.22	2.08	1.93	1.80	1.61	1.42	1.21	1.00	0.774
N	75	2.90	2.77	2.66	2.55	2.41	2.27	2.12	1.98	1.84	1.65	1.46	1.24	1.03	0.804
O	100	2.92	2.80	2.69	2.58	2.43	2.29	2.14	2.00	1.86	1.67	1.48	1.26	1.05	0.819
P	150	2.96	2.84	2.73	2.61	2.47	2.33	2.18	2.03	1.89	1.70	1.51	1.29	1.07	0.841
Q	200	2.97	2.85	2.73	2.62	2.47	2.33	2.18	2.04	1.89	1.70	1.51	1.29	1.07	0.845
		0.065	0.10	0.15	0.25	0.40	0.65	1.00	1.50	2.50	4.00	6.50	10.00	15.00	
		Acceptable Quality Levels (tightened inspection)													

All AQL values are in percent defective.

↓ Use first sampling plan below arrow, that is, both sample size as well as k value. When sample size equals or exceeds lot size, every item in the lot must be inspected.

TABLE 18

675

TABLE 19 *Factors for Two-Sided Tolerance Limits*

δ / n	1 − α = 0.95			1 − α = 0.99		
	0.90	0.95	0.99	0.90	0.95	0.99
2	32.019	37.674	48.430	160.193	188.491	242.300
3	8.380	9.916	12.861	18.930	22.401	29.055
4	5.369	6.370	8.299	9.398	11.150	14.527
5	4.275	5.079	6.634	6.612	7.855	10.260
6	3.712	4.414	5.775	5.337	6.345	8.301
7	3.369	4.007	5.248	4.613	5.488	7.187
8	3.136	3.732	4.891	4.147	4.936	6.468
9	2.967	3.532	4.631	3.822	4.550	5.966
10	2.839	3.379	4.433	3.582	4.265	5.594
11	2.737	3.259	4.277	3.397	4.045	5.308
12	2.655	3.162	4.150	3.250	3.870	5.079
13	2.587	3.081	4.044	3.130	3.727	4.893
14	2.529	3.012	3.955	3.029	3.608	4.737
15	2.480	2.954	3.878	2.945	3.507	4.605
16	2.437	2.903	3.812	2.872	3.421	4.492
17	2.400	2.858	3.754	2.808	3.345	4.393
18	2.366	2.819	3.702	2.753	3.279	4.307
19	2.337	2.784	3.656	2.703	3.221	4.230
20	2.310	2.752	3.615	2.659	3.168	4.161
25	2.208	2.631	3.457	2.494	2.972	3.904
30	2.140	2.549	3.350	2.385	2.841	3.733
35	2.090	2.490	3.272	2.306	2.748	3.611
40	2.052	2.445	3.213	2.247	2.677	3.518
45	2.021	2.408	3.165	2.200	2.621	3.444
50	1.996	2.379	3.126	2.162	2.576	3.385
55	1.976	2.354	3.094	2.130	2.538	3.335
60	1.958	2.333	3.066	2.103	2.506	3.293
65	1.943	2.315	3.042	2.080	2.478	3.257
70	1.929	2.299	3.021	2.060	2.454	3.225
75	1.917	2.285	3.002	2.042	2.433	3.197
80	1.907	2.272	2.986	2.026	2.414	3.173
85	1.897	2.261	2.971	2.012	2.397	3.150
90	1.889	2.251	2.958	1.999	2.382	3.130
95	1.881	2.241	2.945	1.987	2.368	3.112
100	1.874	2.233	2.934	1.977	2.355	3.096
150	1.825	2.175	2.859	1.905	2.270	2.983
200	1.798	2.143	2.816	1.865	2.222	2.921
250	1.780	2.121	2.788	1.839	2.191	2.880
300	1.767	2.106	2.767	1.820	2.169	2.850
400	1.749	2.084	2.739	1.794	2.138	2.809
500	1.737	2.070	2.721	1.777	2.117	2.783
600	1.729	2.060	2.707	1.764	2.102	2.763
700	1.722	2.052	2.697	1.755	2.091	2.748
800	1.717	2.046	2.688	1.747	2.082	2.736
900	1.712	2.040	2.682	1.741	2.075	2.726
1000	1.709	2.036	2.676	1.736	2.068	2.718
∞	1.645	1.960	2.576	1.645	1.960	2.576

Source: From *Techniques of Statistical Analysis* by C. Eisenhart, M. W. Hastay, and W. A. Wallis. Copyright 1947, McGraw-Hill Book Company, Inc. Reproduced with permission of McGraw-Hill.

TABLE
19

References

Bowker, A. H., and G. J. Lieberman (1972). *Engineering Statistics*, 2nd ed., Prentice-Hall Inc., Englewood Cliffs, N.J.

Grant, E. L., and R. S. Leavenworth (1980). *Statistical Quality Control*, 5th ed., McGraw-Hill Book Co., New York.

Guttman, I., S. S. Wilks, and J. S. Hunter (1971). *Introductory Engineering Statistics*, 2nd ed., John Wiley & Sons, New York.

Koopmans, L. H. (1987). *An Introduction to Contemporary Statistical Methods*, PWS-KENT Publishing Co., Boston.

Lapin, L. L. (1990). *Probability and Statistics for Modern Engineering*, 2nd ed., PWS-KENT Publishing Co., Boston.

McClave, J. T., and F. H. Dietrich (1985). *Statistics*, 3rd ed., Dellen Publishing Co., San Francisco.

Mendenhall, W., D. D. Wackerly, and R. L. Scheaffer (1990). *Mathematical Statistics with Applications*, 4th ed., PWS-KENT Publishing Co., Boston.

Miller, I., and J. E. Freund (1977). *Probability and Statistics for Engineers*, 2nd ed., Prentice-Hall, Inc., Englewood Cliffs, N.J.

Ott, L. (1988). *An Introduction to Statistical Methods and Data Analysis*, 3rd ed., PWS-KENT Publishing Co., Boston.

Ross, S. (1984). *A First Course in Probability*, 2nd ed., Macmillan Publishing Co., New York.

Stephens, M. A. (1974). "EDF Statistics for Goodness of Fit and Some Comparisons." *Journal of Am. Stat. Assn.*, vol. 69, no. 347, pp. 730–737.

Tukey, J. W. (1977). *Exploratory Data Analysis*, Addison-Wesley Publishing Co.

Walpole, R. E., and R. H. Myers (1985). *Probability and Statistics for Engineers and Scientists*, 3rd ed., Macmillan Publishing Co., New York.

Answers to Exercises

CHAPTER 1

1.1 (d) Los Angeles appears to have an unusually low CPI.

1.2 (d) Posteruption readings are shifted to the right of preeruption readings.

CHAPTER 2

2.1 (a) 3 (b) 13 (c) 7 (d) 9

2.2 (a) $JD, JS, DM, DN, MN, JM, JN, DS, MS, SN$
(b) 7 (c) 6 (d) $\overline{A}$
(e) $\overline{A} = \{MS, MN, SN\}$
$AB = B = \{JM, JS, JN, DM, DS, DN\}$
$A \cup B = A = \{JD, JM, JS, JN, DM, DS, DN\}$
$\overline{AB} = \{JD, MS, MN, SN\}$

2.5 (a) L, R, S (b) $P(L) = P(R) = P(S) = \frac{1}{3}$ (c) $\frac{2}{3}$

2.6 (a) 0.4 (b) 0.9 (c) 0.4 (d) 0.1
(e) 0.5 (f) 0.1 (g) 0.1

2.7 (a) $\frac{1}{3}$ (b) $\frac{6}{15}$ (c) $\frac{19}{48}$ (d) $\frac{2}{3}$

2.8 (a) 0.43 (b) 0.05 (c) 0.59
(d) 0.74 (e) 0.91

2.9 (a) 0.08 (b) 0.16 (c) 0.14 (d) 0.84

2.10 (a) $SS, SR, SL, RS, RR, RL, LS, LR, LL$
(b) $\frac{5}{9}$ (c) $\frac{5}{9}$

2.11 (a) (I, I), (I, II), (II, I), (II, II) (b) $\frac{1}{4}, \frac{1}{2}$

2.12 (a) (I, I), (I, II), (I, III), (II, I), (II, II), (II, III), (III, I), (III, II), (III, III)
(b) $\frac{1}{3}$ (c) $\frac{5}{9}$

2.13 0.3 **2.14** (a) 42 (b) 21 **2.15** $\frac{1}{5}$

2.16 5,040 **2.17** (a) 168 (b) $\frac{1}{8}$

2.19 (a) $\frac{3}{5}$ (b) $\frac{2}{5}$ (c) $\frac{3}{10}$

2.20 (a) $\frac{6}{64}$ (b) $\frac{9}{64}$ (c) $\frac{37}{64}$ (d) $\frac{36}{64}$

2.21 (a) $\frac{60}{125}$ (b) $\frac{5}{125}$ (c) $\frac{60}{125}$

2.22 (a) 24 (b) $\frac{1}{2}$

2.23 (a) 1,680 (b) $\frac{140}{1,680}$ **2.24** $\frac{5}{8}$

2.25 (a) $\frac{2}{3}$ (b) $\frac{1}{9}$ (c) $\frac{1}{3}$

2.26 (a) $\frac{46,263}{92,911} = 0.4979$ (b) $\frac{32,949}{64,053} = 0.5144$
(c) $\frac{14,738}{19,801} = 0.7443$
(d) $\frac{7,067}{16,065} = 0.4399$ (e) $\frac{64,053}{92,911} = 0.6894$

2.27 (a) 0.16 (b) 0.84 (c) 0.96 (d) no

2.28 (a) 0.10 (b) 0.03 (c) 0.06 (d) 0.018

2.29 (a) $\frac{240}{380} = 0.6316$ (b) $\frac{368}{380} = 0.9684$
(c) $\frac{240}{368} = 0.6522$

2.30 (a) $\frac{306}{380} = 0.8053$ (b) $\frac{378}{380} = 0.9947$
(c) $\frac{306}{378} = 0.8096$

2.31 (a) $\frac{1}{10}$ (b) $\frac{7}{10}$ (c) $\frac{1}{7}$

2.32 (a) 0.7225 (b) 0.19

2.33 (a) 0.999 (b) 0.9009

2.34 0.81, 0.99 **2.35** 0.40

2.36 0.8235 **2.37** 0.0833

2.38 (a) $\frac{3}{5}$ (b) $\frac{2}{5}$ (c) independent (d) independent

2.41 0.5073

2.42 (a) HHHH, HHHT, HHTH, HHTT, HTHH, HTHT, HTTH, HTTT, THHH, THHT, THTH, THTT, TTHH, TTHT, TTTH, TTTT
(b) $A = \{HHHT, HHTH, HTHH, THHH\}$
(c) $\frac{1}{4}$

2.43 (a) $N_1N_2, N_1N_3, N_1N_4, N_1D_1, N_1D_2, N_1D_3, N_2N_3, N_2N_4, N_2D_1, N_2D_2, N_2D_3, N_3N_4, N_3D_1, N_3D_2, N_3D_3, N_4D_1, N_4D_2, N_4D_3, D_1D_2, D_1D_3, D_2D_3$
(b) $A = \{N_1N_2, N_1N_3, N_1N_4, N_2N_3, N_2N_4, N_3N_4\}$
(c) $\frac{2}{7}$

2.44 (a) 0.07, 0.32, 0.29 (b) 0.23, 0.43, 0.39
(c) 0.59, 0.75, 0.65

2.45 (a) 36 (b) $\frac{1}{6}$

2.47 (a) 0.57 (b) 0.18 (c) 0.9 (d) $\frac{18}{57} = 0.3158$

2.48 120 **2.49** 9,000,000 **2.50** 720

2.51 18 **2.52** 40,320

2.53 (a) $\frac{48}{1,326} = 0.0362$
(b) $\frac{154,440}{311,875,200} = 0.000495; 4(0.00495) = 0.00198$

2.54 (a) 0.216 (b) 0.936 (c) 0.648

2.55 (a) $\frac{1}{8}$ (b) $\frac{1}{64}$ (c) no

2.56 $\frac{1}{16}$ **2.57** 0.5952

2.58 (a) $20(\frac{1}{64}) = \frac{5}{16}$ (b) $27(\frac{1}{2})^{10}$

2.59 (a) $4(\frac{1}{4})^8 = (\frac{1}{4})^7$ (b) $9(\frac{1}{4})^3$

2.60 not independent **2.61** 0.8704

2.62 $\frac{1}{16} = 0.0625$ **2.65** not independent

2.66 $\frac{1}{2}$ **2.67** $\frac{1}{7}$ **2.69** A

2.70 (a) 0.00892 (b) 0.9890

CHAPTER 3

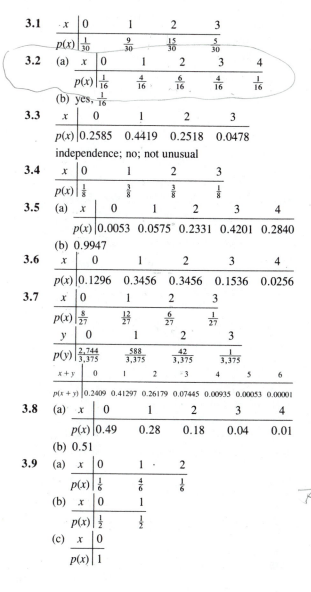

3.1

x	0	1	2	3
$p(x)$	$\frac{1}{30}$	$\frac{9}{30}$	$\frac{15}{30}$	$\frac{5}{30}$

3.2 (a)

x	0	1	2	3	4
$p(x)$	$\frac{1}{16}$	$\frac{4}{16}$	$\frac{6}{16}$	$\frac{4}{16}$	$\frac{1}{16}$

(b) yes, $\frac{1}{16}$

3.3

x	0	1	2	3
$p(x)$	0.2585	0.4419	0.2518	0.0478

independence; no; not unusual

3.4

x	0	1	2	3
$p(x)$	$\frac{1}{8}$	$\frac{3}{8}$	$\frac{3}{8}$	$\frac{1}{8}$

3.5 (a)

x	0	1	2	3	4
$p(x)$	0.0053	0.0575	0.2331	0.4201	0.2840

(b) 0.9947

3.6

x	0	1	2	3	4
$p(x)$	0.1296	0.3456	0.3456	0.1536	0.0256

3.7

x	0	1	2	3
$p(x)$	$\frac{8}{27}$	$\frac{12}{27}$	$\frac{6}{27}$	$\frac{1}{27}$

y	0	1	2	3
$p(y)$	$\frac{2,744}{3,375}$	$\frac{588}{3,375}$	$\frac{42}{3,375}$	$\frac{1}{3,375}$

$x+y$	0	1	2	3	4	5	6
$p(x+y)$	0.2409	0.41297	0.26179	0.07445	0.00935	0.00053	0.00001

3.8 (a)

x	0	1	2	3	4
$p(x)$	0.49	0.28	0.18	0.04	0.01

(b) 0.51

3.9 (a)

x	0	1	2
$p(x)$	$\frac{1}{6}$	$\frac{4}{6}$	$\frac{1}{6}$

(b)

x	0	1
$p(x)$	$\frac{1}{2}$	$\frac{1}{2}$

(c)

x	0
$p(x)$	1

3.10

y	0	1	2
$p(y)$	0.019	0.252	0.729

3.11 (a) 0, $\frac{2}{3}$ (b) 0, $\frac{12}{5}$ (c) box II

3.12 (a) 313.2167; 27.62 (b) yes

3.13 0.4; 0.44; 0.6633 **3.14** 0.20; 0.18

3.15 \$60,000; \$120,000 **3.16** 1; $\frac{1}{2}$

3.17 (a) (1.4702, 6.5298) (b) yes **3.18** 41

3.19 (a) (84.1886, 115.8114) (b) no

3.20 (a) \$200; \$80 (b) \$360

3.21 (a) 0.1536 (b) 0.1808 (c) 0.9728

(d) 0.8 (e) 0.64

3.22 (a) 0.250 (b) 0.057 (c) 0.180

3.23 (a) 0.537 (b) 0.098

3.24 (a) 0.109 (b) 0.999 (c) 0.589

3.25 (a) 16 (b) 3.2

3.26 (a) 0.672 (b) 0.672 **3.27** 8

3.28 (a) $\frac{1}{16}$ (b) $\frac{1}{4}$ **3.29** 0.5931

3.30 (a) independence (b) 0.7379

3.31 (a) 0.99 (b) 0.9999 **3.32** 2

3.33 (a) 0.1536 (b) 0.9728

3.34 (a) \$400,000 (b) \$474,341.65

3.35 3.96 **3.36** (a) 22 (b) 45 **3.37** 840

3.38 (a) 0.9 (b) $(1-p)^2$

3.39 (a) 0.648 (b) 1 **3.40** 0.09

3.41 (a) 0.04374 (b) 0.99144 **3.42** 0.1

3.43 (a) $\frac{10}{9}; \frac{10}{81}$ (b) $\frac{10}{3}; \frac{30}{81}$ **3.44** 0.0645

3.45 150; 4,500; no **3.46** (a) $\frac{8}{81}$ (b) $\frac{4}{27}$

3.47 (a) 0.128 (b) 0.03072 **3.48** 15; 60

3.49 (a) 0.06561 (b) $\frac{40}{9}; \frac{40}{81}$

3.50 (a) 0.4; 0.24; 0.144 (b) 0.1728

3.51 (a) $\frac{3}{16}$ (b) $\frac{3}{16}$ (c) $\frac{1}{8}$ (d) $\frac{1}{2}$

3.52 (a) 0.0902 (b) 0.143 (c) 0.857 (d) 0.2407

3.53 (a) 0.0183 (b) 0.908 (c) 0.997

3.54 (a) 0.905 (b) 0.005 (c) 0.819

3.55 (a) 0.467 (b) 0.188

3.56 (a) 0.022 (b) 0.866

3.57 (a) 0.8187 (b) 0.5488 **3.58** 0.353

3.59 (a) 0.140 (b) 0.042 (c) 0.997

3.60 $\mu = 80$, $\sigma^2 = 800$; no

3.61 (a) $128e^{-16}$ (b) $128e^{-16}$

3.62 (a) $1 - e^{-4}$ (b) $1 - e^{-12}$

3.63 320; 56.5685

3.64 (a) $1 - e^{-8}$ (b) 2 **3.65** λ^2

3.66 $t_0 = 47.5$ **3.67** (a) 0.001 (b) 0.000017

3.68 (a) $\frac{4}{7}$ (b) $\frac{6}{7}$ (c) $\frac{1}{3}$ (d) $\frac{4}{7}$

3.69 $\frac{1}{42}$

3.70 $\mu = 100$, $\sigma^2 = \frac{5,000}{3}$; $K = 3$, (\$0, \$222.48)

3.71 (a) $\frac{4}{5}$ (b) $\frac{1}{5}$

3.72 $\frac{1}{30}$; probably not random selection

3.73 (a)

y	0	1	2
$p(y)$	$\frac{7}{15}$	$\frac{7}{15}$	$\frac{1}{15}$

(b)

y	0	1	2	3
$p(y)$	$\frac{1}{6}$	$\frac{1}{2}$	$\frac{3}{10}$	$\frac{1}{30}$

3.74 (a) $\frac{1}{14}$ (b) $\frac{13}{14}$ (c) $\frac{1}{14}$ **3.75** $\frac{3}{14}$

3.76 (a) $\frac{41}{42}$ (b) $\frac{11}{42}$ (c) $\frac{1}{42}$ (d) 0

3.77 (a) $\frac{9}{14}$ (b) $\frac{13}{14}$ (c) 1

3.78

y	0	1	2
$p(y)$	$\frac{1}{15}$	$\frac{8}{15}$	$\frac{6}{15}$

3.79 (a) 1 (b) 1 (c) $\frac{18}{19}$ (d) $\frac{49}{57}$ (e) $\frac{728}{969}$

3.80 (a) 1 (b) 1 (c) 1 (d) $\frac{113}{114}$ (e) $\frac{938}{969}$

3.81 $\frac{10}{21}$ **3.82** $1 - p + pe^t$

3.83 $E(Y) = np; V(Y) = np(1 - p)$

3.84 $E(Y) = \lambda; V(Y) = \lambda$ **3.89** 0.32805; 0.99999

3.90 (a) 0.9606 (b) 0.9994

3.91 (a) 1 (b) 0.5905 (c) 0.1681

(d) 0.03125 (e) 0

3.92 (a)

p	0.00	0.05	0.10	0.30	0.50	1.00
$p(y \le a)$	1	0.5987	0.3487	0.0283	0.001	0.00

(b)

p	0.00	0.05	0.10	0.30	0.50	1.00
$p(y \le a)$	1.0	0.9139	0.7361	0.1493	0.0107	0.00

(c)

p	0.00	0.05	0.10	0.30	0.50	1.00
$p(y \le a)$	1.0	0.9885	0.9298	0.3828	0.0547	0.00

3.93 (a) $n = 25, a = 5$ (b) $n = 25, a = 5$

3.94 $E(C) = \$80; \sqrt{V(C)} = 9.4868; (61.0263, 98.9737)$

3.95 (a) 0.758 (b) 12; 12 (c) (5.0718, 18.9282)

3.96 (a) 0.083 (b) 0.895

3.97 (a) 5 (b) 0.007 (c) 0.384

3.98 (a) $Ke^{\lambda t}$ (b) 3.2974 **3.99** 0.993

3.100 0.04096 **3.102** 0.837 **3.103** 0.18522

3.104 (a) 0.081 (b) 0.81 **3.105** 0.01536; 0.0256

3.106 (a) 0.019 (b) 0.1745

3.107 $E(Y) = 900; V(Y) = 90; K = 2, (881.026,$
918.974)

3.108

$p(y)$	0	1	2	3	4
(a) Binomial	0.358	0.378	0.189	0.059	0.013
(a) Poisson	0.368	0.368	0.184	0.061	0.015

3.109 (a) $p(y) = \binom{4}{y}(\frac{1}{3})^y(\frac{2}{3})^{4-y}$

(b) $\frac{1}{9}$ (c) $\frac{4}{3}$ (d) $\frac{8}{9}$

3.111 (a) 0.1192 (b) 0.117; yes

3.112 \$149.09 **3.113** 3

3.114 (a) $\frac{N}{k}[1 + k(1 - (0.95)^k)]$ (b) 5 (c) $0.5738N$

3.115 no

CHAPTER 4

4.1 (a) discrete (b) discrete
(c) discrete (d) continuous

4.2 (a) $k = 6$ (b) 0.648 (c) 0.3929

(d) $F(b) = 0, \quad b < 0$
$\qquad = 3b^2 - 2b^3, \quad 0 \le b \le 1$
$\qquad = 1, \quad b > 1$

4.3 (a) $\frac{27}{32}$ (b) 4

4.4 (a) $F(x) = 0, \quad x < 0$
$\qquad = x^3\dfrac{(16 - 3x)}{256}, \quad 0 \le x < 4$
$\qquad = 1, \quad x > 4$

(b) $\frac{11}{16}$ (c) $\frac{67}{256}$ (d) 3.4298

4.5 (b) $F(x) = 0, \quad x \le 5$
$\qquad = \dfrac{(x - 7)^3}{8} + 1, \quad 5 \le x \le 7$
$\qquad = 1, \quad x > 7$

(c) $\frac{7}{8}$ (d) $\frac{37}{56}$

4.6 (a) $\frac{1}{2}$ (b) $\frac{1}{4}$

4.7 (a) $\frac{3}{4}$ (b) $\frac{4}{5}$ (c) 1

(d) $F(x) = 0, \quad x < 0$
$\qquad = x^2, \quad 0 \le x \le 1$
$\qquad = 1, \quad x > 1$

yes

4.8 (a) 0.08392 (b) 0.99046 **4.9** 60; $\frac{1}{3}$

4.10 (a) $\frac{2}{3}; \frac{1}{18}$ (b) $\frac{220}{3}; \frac{20,000}{9}$

(c) $(-20.9476, 167.6142)$

4.11 4 **4.12** (a) 2.4; 0.64 (b) 480; 25,600
(c) about 26% of the time

4.13 (a) 5.5; 0.15 (b) $k = 2, (4.7254, 6.2746)$
(c) about 58% of the time

4.14 $k = 0.7368$

4.15 (a) $F(x) = 0, \quad x < a$
$\qquad = \dfrac{x - a}{b - a}, \quad a \le x \le b$
$\qquad = 1, \quad x > b$

(b) $\dfrac{b - c}{b - a}$ (c) $\dfrac{b - d}{b - c}$

4.16 (a) $\frac{2}{5}$ (b) $\frac{1}{5}$ (c) \$22,500

4.17 (a) $\frac{1}{20}$ (b) $\frac{1}{20}$ (c) $\frac{1}{2}$

4.18 0.2 **4.19** $\frac{3}{4}$ **4.20** $\frac{2}{5}$

4.21 (a) $\frac{1}{8}$ (b) $\frac{1}{8}$ (c) $\frac{1}{4}$

4.22 (a) $\frac{1}{5}$ (b) 0; $\frac{1}{1,200}$

4.23 (a) $\frac{2}{7}$ (b) $\frac{3}{200}; \frac{49}{120,000}$

4.24 $\dfrac{155\pi}{24} \cdot 10^{-6}; 3.4795 \cdot 10^{-10}$

4.25 $\frac{1}{6}$ **4.26** $\frac{1}{4}$

4.27 (a) $\frac{1}{2}$ (b) $\frac{1}{4}$ **4.28** $\frac{3}{8}$

4.29 (a) 60; $\frac{100}{3}$ (b) 4

4.31 (a) $e^{-5/4}$ (b) $e^{-5/6} - e^{-5/4}$ **4.32** 0.7355

4.33 (a) e^{-2} (b) 460.52 cfs

4.34 (a) $\frac{1}{2}; \frac{1}{4}$ (b) $1 - e^{-6}$

4.35 (a) 1,100; 2,920,000 (b) no; $P(C > 2,000)$
$\qquad = e^{-1.94}$

4.36 (a) $1 - e^{-31/44}$ (b) 1,936

4.37 (a) e^{-1} (b) $e^{-1/2}$ **4.38** $e^{-4/3}$

4.39 (a) $e^{-1/2}$ (b) $e^{-1/4}$ (c) no

4.40 (a) $e^{-5/2}$ (b) $e^{-18/5}$

4.41 (a) $e^{-5/4}$ (b) $(1 - e^{-5/4})^2$

4.42 (a) $e^{-5/8}$ (b) $e^{-5/8}$

4.43 (a) $1 - e^{-5/11}$ (b) $e^{-15/11} - e^{-30/11}$
(c) 50.66 minutes

4.44 200π; $200{,}000\pi^2$

4.45 (a) 3.2; 6.4 (b) $(0, 8.26)$

4.46 (a) $30{,}000$; $1{,}500{,}000$ (b) no

4.47 (a) 276; $47{,}664$ (b) $(0, 930.963)$

4.48 (a) 1; $\frac{1}{2}$; $f(y) = 4ye^{-2y}$, $y > 0$
$= 0$, $y \le 0$
(b) $\frac{3}{2}$; $\frac{3}{4}$; $f(y) = 4y^2 e^{-2y}$ $y > 0$
$= 0$, $y \le 0$

4.49 (a) 20; 200 (b) 10; 50

4.50 (a) 140; 280; $f(y) = \dfrac{y^{69} e^{-y/2}}{\Gamma(70)2^{70}}$, $y > 0$
$= 0$, $y \le 0$
(b) 206.93

4.51 (a) 240; 189.74 (b) $(0, 809.21)$

4.52 60; 900; $f(y) = \dfrac{y^3 e^{-y/15}}{\Gamma(4)15^4}$, $y > 0$
$= 0$, $y \le 0$

4.53 9.6; 30.72; $f(y) = \dfrac{y^2 e^{-y/3.2}}{\Gamma(3)(3.2)^3}$, $y > 0$
$= 0$, $y \le 0$

4.54 $f(y) = \dfrac{ye^{-y/2}}{y}$, $y > 0$
$= 0$, $y \le 0$

4.55 (a) 0.3849 (b) 0.3159 (c) 0.3227
(d) 0.1586 (e) 0.0366

4.56 (a) 0 (b) 1.15 (c) 1.19
(d) -0.30 (e) 1.645 (f) 1.96

4.57 0.0062 **4.58** $\$425.60$

4.59 $.0730$ **4.60** 1

4.61 (a) 0.9544 (b) 0.8297

4.62 0.5859 **4.63** 0.6170; 0.3174

4.64 (a) 0.1498 (b) 0.0224

4.65 (a) 0.0062 (b) 225.6

4.66 (a) 0.0062 (b) $1{,}171$

4.67 13.67 **4.68** $\frac{1}{1.96} = 0.5102$

4.69 (a) 60 (b) $\frac{4}{7}$; $\frac{3}{98}$

4.71 (a) $\frac{52}{3}$; $\frac{1{,}348}{45}$ (b) $(6.39, 28.28)$

4.72 (a) 0.8208 (b) 4.7; 0.01

4.73 $\frac{2}{3}$; $240°$

4.74 (a) $\frac{3}{4}$ (b) $\frac{1}{3}$; $\sqrt{1/18}$

4.75 (a) $\frac{1}{2}$; $\frac{1}{28}$ (b) $\frac{1}{2}$; $\frac{1}{20}$ (c) $\frac{1}{2}$; $\frac{1}{12}$
(d) case (a)

4.76 (a) $\frac{7}{8}$ (b) 0.002128

4.77 (a) $1 - e^{-1}$ (b) $\sqrt{\pi}$

4.78 (a) 0.5547 (b) 0.6966

4.79 0.6576 **4.80** 0.03091

4.81 0.06573 **4.82** 0.09813

4.83 (a) $e^{-2.5}$ (b) 0.01855

4.84 $\dfrac{\sqrt{10\pi}}{2}$; $10(1 - \frac{\pi}{4})$ **4.85** 42.9193

4.86 (a) $2\sqrt{\dfrac{2KT}{m\pi}}$ (b) $(\frac{3}{2})KT$

4.88 $2\theta^2$ **4.89** $e^{t^2/2}$

4.90 $(1 - 2t)^{-1/2}$; gamma$(\frac{1}{2}, 2)$

4.91 (a) $\frac{1}{2}$ (b) $F(y) = 0$, $y < 0$
$= \dfrac{y^2}{4}$, $0 \le y \le 2$
$= 1$, $y > 2$
(d) $\frac{3}{4}$ (e) $\frac{3}{4}$

4.92 (a) $-\frac{3}{8}$ (b) $F(y) = 0$, $y < 0$
$= \dfrac{y^2}{2} - \dfrac{y^3}{8}$, $0 \le y \le 2$
$= 1$, $y > 2$
(d) 0; 0; $\frac{3}{8}$ (e) $\frac{7}{64}$ (f) $\frac{7}{6}$; $\frac{43}{180}$

4.93 (a) $\frac{6}{5}$
(b) 0, $y \le -1$
$F(y) = 0.2(y + 1)$, $-1 < y \le 0$
$= 0.2 + 0.2y + \frac{3}{5}y^2$, $0 < y \le 1$
$= 1$, $y > 1$
(d) 0; 0.2; 1 (e) $\frac{1}{4}$ (f) $\frac{2}{5}$; $\frac{41}{150}$

4.94 0.1151 **4.95** 15.87% **4.96** 0.001525

4.97 0.073 **4.98** 0.3155

4.99 $\dfrac{c}{4}$; $\dfrac{c(6 - c)}{16}$ (b) $\dfrac{c}{(2 - t)^2}$ (c) 4

4.100 $\dfrac{\Gamma(\alpha + \beta)\Gamma(\alpha + k)}{\Gamma(\alpha + \beta + k)\Gamma(\alpha)}$; $\dfrac{\alpha}{\alpha + \beta}$; $\dfrac{\alpha\beta}{(\alpha + \beta + 1)(\alpha + \beta)^2}$

4.101 $(0.0062)^3$ **4.102** 0.04999

4.103 (a) $1 - e^{-6}$ (b) $f(y) = \dfrac{y^3 e^{-y/10}}{\Gamma(4)(10)^4}$, $y > 0$
$= 0$, $y \le 0$

4.104 (a) 105 (b) $\frac{3}{8}$; $\frac{5}{192}$

4.105 $1 - e^{-4}$ **4.106** $\frac{29}{8}e^{-3/2}$

4.107 $\$53.58$ **4.108** 110

4.109 0.736 **4.110** 0.875

4.111 (a) Wiebull with $\gamma = 2$, $\theta = \dfrac{1}{\lambda\pi}$ (b) $\dfrac{1}{2\sqrt{\lambda}}$

4.112 (a) 0.0045 (b) 0.9726

4.113 (a) $e^{11}(10^{-2}g)$; $(e^{38} - e^{22})(10^{-4}g^2)$
(b) $(0, 3{,}570{,}244.56)$ (c) 0.8023

4.114 $\dfrac{1}{1 - t^2}$; 0

CHAPTER 5

5.1 (a)

$x_2 \backslash x_1$	0	1	2
0	$\frac{1}{9}$	$\frac{2}{9}$	$\frac{1}{9}$
1	$\frac{2}{9}$	$\frac{2}{9}$	0
2	$\frac{1}{9}$	0	0

(b)

x_1	0	1	2
$p(x_1)$	$\frac{4}{9}$	$\frac{4}{9}$	$\frac{1}{9}$

(c) $\frac{1}{2}$

5.2 (b) 0.08 (c) 0.12

5.3

$x_2 \backslash x_1$	0	1	2	3
0	0	$\frac{4}{84}$	$\frac{12}{84}$	$\frac{4}{84}$
1	$\frac{3}{84}$	$\frac{24}{84}$	$\frac{18}{84}$	0
2	$\frac{6}{84}$	$\frac{12}{84}$	0	0
3	$\frac{1}{84}$	0	0	0

5.4 (a) 1 (b) $\frac{2}{3}$

5.5 (a) $f(x_1) = 1$, $0 \le x_1 \le 1$
$= 0$, otherwise
(b) $\frac{1}{2}$ (c) yes

5.6 (a) $f_2(x_2) = 2(1 - x_2)$, $0 \le x_2 \le 1$
$= 0$, otherwise
(b) 0.64 (c) no (d) $\frac{1}{2}$

5.7 (a) $\frac{7}{8}$ (b) $\frac{1}{2}$ (c) $\frac{2}{3}$

5.8 (a) $f_1(x_1) = 2(1 - x_1)$, $0 \le x_1 \le 1$
$= 0$, otherwise
$f_2(x_2) = 2(1 - x_2)$, $0 \le x_2 \le 1$
$= 0$, otherwise
(b) no (c) $\frac{2}{3}$

5.9 (a) $\frac{21}{64}$ (b) $\frac{1}{3}$ (c) no **5.10** $\frac{11}{32}$

5.11 (a) yes (b) $\dfrac{3e^{-1}}{2}$

5.12 $\frac{7}{32}$ **5.13** $\frac{11}{36}$ **5.14** $\frac{1}{4}$

5.15 $\frac{23}{144}$ **5.16** $\frac{1}{4}$

5.17 (a) 0.2; 0.16; 0.2; 0.16
(b) -0.04 (c) 0.4; 0.24

5.18 (a) 1; $\frac{1}{6}$ (b) (0.1835, 1.8165)

5.19 (a) $\frac{2}{3}$; $\frac{1}{18}$ (b) $(\frac{1}{3}, 1)$

5.20 (a) 61; 20 (b) $P(Y > 75) = 0.1653$

5.21 (a) 0.0972 (b) $\frac{1}{2}$ (c) e^{-1}
(d) $f_1(y_1) = y_1 e^{-y_1}$, $0 \le y_1 < \infty$
$= 0$, otherwise
$f_2(y_2) = e^{-y_2}$, $0 \le y_2 < \infty$
$= 0$, otherwise

5.22 $\frac{1}{3}$

5.23 (a) 1 (b) 1 (c) e^{-2}

5.24 $\frac{1}{2}$ **5.26** 0.08953 **5.27** 66,960

5.28 (a) 0.04594 (b) 0.2262

5.29 0.07776; independence **5.30** 0.09352

5.31 (a) $\frac{4}{27}$ (b) $\frac{1}{27}$ (c) $\frac{4}{9}$

5.32 0.40951 **5.33** 2.5; 4.875

5.34 (a) 0.2759 (b) 0.80313

5.35 (a) 0.08575 (b) 0.7627 (c) 40; 24

5.37 0.05213

5.39 (a) 0.1587 (b) 0.3085

5.40 0.0228

5.41 (a) 4 (b) $f_1(x_1) = 2x_1$, $0 \le x_1 \le 1$
$f_2(x_2) = 2x_2$, $0 \le x_2 \le 1$
(c) $F(x_1, x_2) = 0$, $x_1, x_2 < 0$
$= x_1^2 x_2^2$, $0 \le x_1, x_2 \le 1$
$= x_1^2$, $0 \le x_1 \le 1, x_2 > 1$
$= x_2^2$, $0 \le x_2 \le 1, x_1 > 1$
$= 1$, $x_1, x_2 > 1$
(d) $\frac{9}{64}$ (e) $\frac{1}{4}$

5.42 (a) $f_1(x_1) = 3x_1^2$, $0 \le x_1 \le 1$
$= 0$, otherwise
$f_2(x_2) = \frac{3}{2}(1 - x_2^2)$, $0 \le x_2 \le 1$
$= 0$, otherwise
(b) $\frac{23}{64}$ (c) 0

5.43 (a)

x_1	x_2	$p(x_1, x_2)$
0	0	0
0	1	$\frac{1}{28}$
0	2	$\frac{1}{14}$
0	3	$\frac{1}{84}$
1	0	$\frac{1}{21}$
1	1	$\frac{2}{7}$
1	2	$\frac{1}{7}$
2	0	$\frac{1}{7}$
2	1	$\frac{3}{14}$
3	0	$\frac{1}{21}$

(b)

	0	1	2	3	$p(x_1)$
0	0	$\frac{1}{28}$	$\frac{1}{14}$	$\frac{1}{84}$	$\frac{10}{84}$
1	$\frac{1}{21}$	$\frac{2}{7}$	$\frac{1}{7}$	0	$\frac{10}{21}$
2	$\frac{1}{7}$	$\frac{3}{14}$	0	0	$\frac{5}{14}$
3	$\frac{1}{21}$	0	0	0	$\frac{1}{21}$
$p(x_2)$	$\frac{5}{21}$	$\frac{15}{28}$	$\frac{3}{14}$	$\frac{1}{84}$	

(c) $\frac{9}{16}$

5.44 $2x_1$, $0 \le x_1 \le 1$; yes

5.45 (a) $\dfrac{2x_1}{1 - x_2^2}$, $0 \le x_2 \le x_1 \le 1$

(b) $\dfrac{1}{x_1}$, $0 \le x_2 \le x_1 \le 1$

(d) $\frac{5}{12}$

5.46 (a)
$$f(x_1, x_2) = \begin{cases} \dfrac{1}{x_1}, & 0 \le x_2 \le x_1 \le 1 \\ 0, & \text{otherwise} \end{cases}$$

(b) $\frac{1}{2}$ (c) $\dfrac{\ln 2}{\ln 4}$

5.47 (a)
$$f(x_1) = \begin{cases} \left(\dfrac{2}{\pi}\right)\sqrt{1 - x_1^2}, & -1 \le x_1 \le 1 \\ 0, & \text{otherwise} \end{cases}$$

(b) $\frac{1}{2}$

5.48 (a)
$$f(x_1) = \begin{cases} x_1 + \dfrac{1}{2}, & 0 \le x_1 \le 1 \\ 0, & \text{otherwise} \end{cases}$$

$$f(x_2) = \begin{cases} x_2 + \dfrac{1}{2}, & 0 \le x_2 \le 1 \\ 0, & \text{otherwise} \end{cases}$$

(b) no (c) $f(x_1|x_2) = \begin{cases} \dfrac{x_1 + x_2}{x_2 + \frac{1}{2}}, & 0 \le x_1, x_2 \le 1 \\ 0, & \text{otherwise} \end{cases}$

5.49 (a) 1

(b)
$$f(x_1) = \begin{cases} \dfrac{x_1}{2}, & 0 \le x_1 \le 2 \\ 0, & \text{otherwise} \end{cases}$$

$$f(x_2) = \begin{cases} 2(1-x_2), & 0 \le x_2 \le 1 \\ 0, & \text{otherwise} \end{cases}$$

(c)
$$f(x_1 \mid x_2) = \begin{cases} \dfrac{1}{2(1 - x_2)}, & 0 \le 2x_2 \le x_1 \le 2 \\ 0, & \text{otherwise} \end{cases}$$

(d)
$$f(x_2 \mid x_1) = \begin{cases} \dfrac{2}{x_1}, & 0 \le 2x_2 \le x_1 \le 2 \\ 0, & \text{otherwise} \end{cases}$$

(e) $\frac{1}{2}$ (f) $\frac{8}{9}$

5.50 (a)
$$f(x_2) = \begin{cases} 2(1 - x_2), & 0 \le x_2 \le 1 \\ 0, & \text{otherwise} \end{cases}$$

(b)
$$f(x_1) = \begin{cases} 1 - |x_1|, & |x_1| \le 1 \\ 0, & \text{otherwise} \end{cases}$$

(c) $\frac{1}{4}$

5.51 (a) $\frac{2}{3}$ (b) $\frac{1}{2}; \frac{1}{18}$ (c) 0

5.52 $\frac{3}{160}$

5.53 (a) -0.3333 (b) $\frac{7}{3}; \frac{7}{18}$ (c) $\frac{7}{3}; \frac{7}{18}$

5.54 (a) $-\frac{1}{144}$ (b) $\frac{7}{12}$ (c) $\frac{155}{144}$

5.55 (a) 2 (b) $\frac{2}{3}$

5.56 $\frac{1}{4}$

5.57 $(\frac{1}{2})^{(x+1)}$, $x = 0, 1, 2, \ldots$

5.58
$$f(x_1, x_2, x_3) = \begin{cases} \left(\dfrac{1}{\theta}\right)^3 e^{-(x_1 + x_2 + x_3)/\theta}, \\ \qquad \theta, x_1, x_2, x_3 > 0 \\ 0, \quad \text{otherwise} \end{cases}$$

5.59 42; 26; no

5.60 (a) $\dfrac{x_1}{2}$ (b) $\frac{3}{8}$ (c) $\frac{3}{8}$

5.61 (a) 1 (b) 1

5.62 $\frac{3}{8}$

5.63 300

5.65 (a) $(p_1 e^{t_1} + p_2 e^{t_2} + p_3 e^{t_3})^n$

(b) $-np_1 p_2$

CHAPTER 6

6.1 (b) $\bar{x} = 1.6633$, $s^2 = .228097$
(c) $(0.7081, 2.6185)$

6.2 (b) $\bar{x} = 10.35$, $s^2 = 2{,}401.5156$
(d) $(-136.666, 157.366)$

6.3 (b) $\bar{x} = -1.1117647$, $s^2 = 39.119853$
(d) $(-19.876, 17.652)$

6.4 (a) $\bar{x} = 22.6$, $s^2 = 5.8$
(b) $\bar{y} = 72.68$, $s^2 = 18.792$

6.5 Yes; the distribution is skewed right.

6.6 The distribution is skewed right.

6.9 0.6826 **6.10** 385 **6.11** 0.9090 **6.12** 153

6.13 (a) 0.5328 (b) 0.9772
(c) independent random sample

6.14 0.0132 **6.15** approximately zero

6.16 (a) 1 (b) 0.123

6.17 (a) 0.9938 (b) 110 **6.18** 4.4652763

6.19 0.0013 **6.20** 88 **6.21** 0.9876

6.22 51 **6.23** 0.0062 **6.24** 0.0287

6.25 (a) 0.5930 (b) 0.0559 (c) 0.0017

6.26 0.1949 **6.27** 0.1292

6.28 approximately zero

6.29 0.1539 **6.30** 0.0630 **6.31** 0.0023

6.32 (a) 0.0329 (b) 0.0294 **6.33** 0.0043

6.34 0.0023 **6.35** 10.148 **6.36** 0.7498 **6.37** 29

6.38 $a = 1.01170$, $b = 3.01435$

6.39 no; $P(S^2 > 0.065) \approx 0.1$

6.40 (a) $0.05 < P(S^2 > 80) < 0.10$
(b) $P(S^2 < 20) \approx 0.025$
(c) $(12.2, 87.8)$

6.41 (a) $0.025 < P(S^2 > 80) < .05$
(b) $P(S^2 < 20) < 0.005$
(c) $(21.13, 78.87)$

6.42 $P(S^2 > 225) \approx 0.01$

6.43 $a = 119.80658$, $b = 356.9625$

6.44 0.9987 **6.45** $6 \cdot 10^{-7}$ **6.46** 0.8384

6.48 0.0062 **6.49** 0.9484

6.50 (a) $\mu_1 - \mu_2$

(b) $\dfrac{\sigma_1^2}{n} + \dfrac{\sigma_2^2}{m}$

6.51 0.0274 **6.53** 183.07 **6.54** 0.1587

CHAPTER 7

7.1 (a) $\hat{\theta}_1$, $\hat{\theta}_2$, $\hat{\theta}_3$, $\hat{\theta}_4$, (b) $\hat{\theta}_4$

7.2 (b) $\bar{X} - \frac{1}{2}$

7.3 (a) $\bar{X}$ (c) $\frac{1}{n} \sum\limits_{i=1}^{n} X_i^2 + 3\bar{X}$

7.5 $\frac{1}{12n} + \frac{1}{4}$

7.6 (b) $S\sqrt{\frac{n-1}{2}}\ \Gamma(\frac{n-1}{2})/\Gamma(\frac{n}{2})$

7.7 $\overline{X} - 1.645\ S\sqrt{\frac{n-1}{2}}\ \Gamma(\frac{n-1}{2})/\Gamma(\frac{n}{2})$

7.8 $a = \sigma_2^2/(\sigma_1^2 + \sigma_2^2)$

7.9 (205.81, 214.19)   **7.10** (45.07, 46.93)

7.11 (17.98, 18.82)   **7.12** 55   **7.13** 139   **7.14** 196

7.15 (0.0563, 0.1837)   **7.16** 163

7.17 (0.2636, 0.5364)   **7.18** 131

7.19 (0.658, 0.842)   **7.20** (5.669, 12.331)

7.21 (173.79, 186.21)

7.22 (a) (176.42, 183.58)   (b) (179.02, 180.98)

7.23 (9.52, 10.08)   **7.24** (20.71, 118.85)

7.25 (0.15, 0.53)   **7.26** (54.32, 89.68)

7.27 (475.25, 478.75)   **7.28** (560.86, 609.14)

7.29 (399.25, 700.75)

7.30 (30,498.024, 173,071.46)   **7.31** no   **7.32** no

7.34 $(-8.1, -3.9)$ **7.35** 193

7.36 $(-0.0406, -0.0194)$; no

7.37 (0.2521, 0.5879); yes   **7.38** 141

7.39 $(-0.1193, 0.3574)$

7.40 $(-3.17, 11.33)$,   (0.4765, 7.0797)

7.41 $(-1.5198, 12.3865)$,   (0.158, 12.201)

7.42 (0.0733, 0.3067)   **7.43** $(-80.61, -31.873)$

7.44 $(-0.19, -0.09)$ **7.45** (0.928, 9.87)

7.46 $(-0.2147, -0.0653)$ **7.47** $(-0.9659, 0.0087)$

7.48 (2.7383, 44.6137)

7.49 $(-244.9688, 178.5243)$; no

7.50 (3.833, 24.167)

7.51 $(-0.0265, 0.0465)$. The methods are not significantly different using a 95% confidence interval.

7.52 (0.0566, 0.1434)   **7.53** (28.72, 29.28)

7.54 (8.69, 10.91)   **7.55** (0, 15.938)

7.56 (32.877, 51.123)   **7.57** (2.917, 5.583)

7.58 (35.08, 43.72)   **7.59** $\hat{\lambda} = \overline{X}$

7.60 (a) $\overline{X} \pm Z_{\alpha/2}\sqrt{\overline{X}/n}$ (b) (3.608, 4.392)

7.61 $\hat{\mu} = \overline{X},\ \hat{\sigma}^2 = (\frac{n-1}{n})\ S^2$

7.62 $\hat{\beta} = \overline{X}/\alpha$ **7.63** (9.6781, 26.1255)   **7.64** $\hat{p} = \frac{1}{\overline{x}}$

7.65 $\hat{\theta} = \max_{1 \le i \le n}\ \{x_i\}$
$= 0.06$

7.66 $60\hat{p} + 20\hat{p}(1 + 19\hat{p})$ **7.67** $\hat{p} = 0.3125$

7.68 $\hat{\lambda} = \dfrac{Y + 1}{2}$

7.70 (1.926, 2.274)   **7.71** (2.05, 2.15)

7.72 (59.75, 70.25)   **7.73** (0.568, 13.432)

7.74 (a) (2.0988, 2.6712)   (b) (3.0017, 3.8503)

7.75 (0.5657, 1.5163)

7.76 (a) (0.16645, 1.17259)   (b) (0.6911, 6.9884)

7.77 (3.009, 3.071)

7.78 (0.000263, 0.001855)   **7.79** 97

7.80 $m = \dfrac{n(-\sigma_1^2 \pm \sigma_1\sigma_2)}{\sigma_2^2 - \sigma_1^2}$ if $\sigma_1^2 \ne \sigma_2^2$

$m = \dfrac{n}{2}$ if $\sigma_1^2 = \sigma_2^2$

7.81 $m = 40,\quad n - m = 50$

7.82 $2\overline{y} + \overline{x} \pm t_{.025}\sqrt{\hat{\sigma}^2(\frac{4}{n} + \frac{3}{m})}$

$\hat{\sigma}^2 = \dfrac{(n-1)\ S_y^2 + (\frac{m-1}{3})\ S_x^2}{n + m - 2}$

7.83 $(-4.82, -1.18)$; yes

7.84 $\hat{R} = \dfrac{X}{n - X}$

CHAPTER 8

8.1 $Z = -4.216$; reject

8.2 0.0853   **8.3** 70

8.4 $Z = -1.77$; do not reject. P-value = 0.0384

8.5 0.1151   **8.6** 64

8.7 $Z = -1.22$; no. P-value = 0.2224

8.8 $Z = 0.86$; no. P-value = 0.3898

8.9 $Z = 0.527$; no   **8.10** $t = -5.078$; yes

8.11 $t = -2.52$; no   **8.12** $t = 6.32$; yes

8.13 $t = -7.47$; yes   **8.14** $Z = -2.076$; yes

8.15 $t = -0.3945$; no   **8.16** $\chi^2 = 35.183$; yes

8.17 $Z = 1.77$; no   **8.18** $Z = 3.76$; yes

8.19 $Z = -3.16$; yes   **8.20** $Z = -3.54$; yes

8.21 $t = -0.753$; no   **8.22** $Z = 1.801$; no

8.23 $\chi^2 = 20.16$; no   **8.24** $\chi^2 = 48.64$; yes

8.25 $\chi^2 = 78.71$; yes   **8.26** $Z = -4.21$; yes

8.27 $Z = 3.648$; yes   **8.28** $t = 2.17$; reject

8.29 $t = 0.1024$; no   **8.30** $t = -3.286$; yes

8.31 $t = 5.939$; yes   **8.32** $t = -0.498$; no

8.33 $F = 1.0936$; do not reject   **8.34** $Z = -1.314$; no

8.35 $t = 1.214$; no   **8.36** $Z = 2.357$; yes

8.37 $t = 7.016$; yes

8.38 (a) $t = 3$; yes   (b) $t = 3$; yes

8.39 $t = 1.109$; no   **8.40** $Z = 0.9701$; no

8.41 $t = -0.254$; no   **8.42** $t = 1.549$; no

8.43 $t = -0.349$; no   **8.44** $F = 1.2216$; no

8.45 $F = 1.0678$; no   **8.46** $F = 1.361$; no

8.47 $\chi^2 = 1.23$; do not reject

8.48 $\chi^2 = 0.88$; do not reject

8.49 $\chi^2 = 1.671$; do not reject   **8.50** $\chi^2 = .4386$; no

8.51 $\chi^2 = 0.542$; no   **8.52** $\chi^2 = 1.261$; do not reject

8.53 (a) $\chi^2 = 13.709$; no   (b) $Z = 2.0494$; yes

8.54 $\chi^2 = 2.59587$; no

8.55 (a) $\chi^2 = 2.0533$; do not reject
(b) $Z = .4899$; do not reject

8.56 $\chi^2 = 2.18949$; no   **8.57** $\chi^2 = 2.48162$; no

8.58 $\chi^2 = 9.36462$; no   **8.59** $\chi^2 = 9.46581$; no

8.60 $\chi^2 = 0.20163$; no   **8.61** $\chi^2 = 20.20202$; yes

8.62 $\chi^2 = 287.64853$; yes

8.63 $\chi^2 = 3.24$; do not reject

8.64 $\chi^2 = 74.986$; reject   **8.65** $\chi^2 = 55.70063$; reject

8.66 (b) $\chi^2 = 5.02482$; do not reject

8.67 modified $D = .6679$; do not reject

8.68 (a) modified $D = 2.1114$; reject

(b) modified $D = 2.1720$; reject

8.69 modified $D = 3.8916$; reject

8.70 modified $D = 1.1812$; reject

8.71 modified $D = 0.2132$; no

8.72 modified $D = 1.2038$; no

8.73 $Z = -24.771$; yes   **8.74** $t = -13.44$; no

8.75 $Z = -1.964$; no   **8.76** $Z = -2.108$; no

8.77 $Z = 1.47$; no   **8.78** $t = -0.5392$; no

8.79 $t = -1.549$; yes   **8.80** $\chi^2 = 21.875$; yes

8.81 $Z = -5.595$; yes   **8.82** $Z = 60.592$; yes

8.83 $Z = 4.904$; yes   **8.84** $Z = 1.164$; no

8.85 $t = 4.52$; yes   **8.86** $F = 12.25$; no

8.87 (a) $\chi^2 = 13.99434$; reject

(b) $\chi^2 = 13.99434$; reject

(c) $\chi^2 = 1.36077$; do not reject

CHAPTER 9

9.1 (a) $\hat{\beta}_0 = 0$, $\hat{\beta}_1 = \frac{6}{7}$

9.2 (a) $\hat{\beta}_0 = 2$, $\hat{\beta}_1 = -1.2$

9.3 $\hat{y}_i = -0.06 + (2.78)x_i$

9.4 $\hat{y}_i = -1.43 + (0.19469)x_i$

9.5 $\hat{y}_i = 480.2 - 3.75x_i$

9.6 (a) $\hat{y}_i = 25.88267 + 3.11476x_i$

(b) yes   (c) 88.17789

9.7 $\hat{y}_i = 10.20833 + 1.925x_i$

9.8 $\hat{y}_i = -0.711 + 0.655x_i$

9.9 8.1: 1.14286, .28571; 8.2: 1.6, .53333;

8.3: 0.199, .066333; 8.4: 0.039469, .0098673;

8.5: 114.87443, 19.14574;

8.6: 88.61638, 14.76940; 8.7: 241.79167,

24.17917; 8.8: 87.78550, 10.97319

9.10 (a) $\hat{y}_i = 1.1824 - 1.315x_i$

(b) yes   (c) SSE $= 0.0155152$; $s^2 = 0.0019394$

9.11 $t = 6.7082$; yes   **9.12** $(-1.74, -0.66)$

9.13 (2.26, 3.30)   **9.14** $t = 8.5057$; yes

9.15 $t = -17.9417$; yes

9.16 (a) 14.76940, 3.84310

(b) 72.6041, 3.84310

(c) 119.3255, 3.84310

9.17 (5.75, 13.5)   **9.18** $t = 9.6099$; yes

9.19 (a) $\hat{y}_i = 36.68 - 0.333x_i$

(b) 1.34687, 0.16836

(c) $t = -7.377$; yes

9.20 8.1: 0.95831, 0.91837; 8.2: -0.94868, 0.9;

8.3: 0.99489, 0.98981; 8.4: 0.97345, 0.94761;

8.5: -0.99081, 0.98170; 8.6: 0.99418, 0.98840;

8.7: 0.86833, 0.75401; 8.8: 0.95931, 0.92028

9.21 (c) $r_1 = .96528$, $r_2 = .99605$ (d) yes

9.22 (a) $\hat{y}_i = -768.61 + 6.20x_i$

(b) $t = 3.354$; yes   (c) $r^2 = 0.58443$

9.23 (a) $\hat{y}_i = -235.116 + 1.27x_i$

(b) $t = 18.299$; yes   (c) $r^2 = .97667$

9.24 (a) 0.81539   (b) $t = 3.984$; yes

9.25 (a) (0.908, 2.521)   (b) (0.0256, 3.403)

9.26 (4.662, 5.226)   **9.27** (1.337, 2.033)

9.28 (194.152, 203.585)   **9.29** (187.169, 210.569)

9.30 (30.116, 36.500)   **9.31** (4.991, 21.107)

9.32 $t = -0.1736$; do not reject

9.33 (a) $\hat{y}_i = 9.5538 + 2.6708x_i$ (b) $t = 8.475$; yes

(c) (16.911, 19.290)   (d) (17.140, 17.992)

9.34 (a) $\hat{y}_i = 5.3253 + 0.5861x_i$ (c) $t = 15.35$; reject

(d) (1) (26.91, 29.46)   (2) (23.55, 32.81)

9.35 (a) 0.91548   (b) 0.83810   (c) $t = 5.0877$; yes

9.36 (a) $r^2 = 0.80886$ (b) $t = 5.0389$; yes

9.37 $r^2 = 0.59210$

9.38 (a) $\hat{y}_i = 1{,}079.8486 - 31.6678x_i$

(b) $t = -16.395$; yes

(c) $(-36.122, -27.214)$

(d) prediction interval

(e) (578.592, 631.071), (594.884, 614.780)

9.39 (a) $\hat{y}_i = -1{,}094.28686 + 181.1026x_i$

(b) (141.936, 220.270)

9.40 (a) $\hat{y}_i = 40.92619 - 0.13429x_i$

(b) $t = 4.213$; yes   (c) $(-0.192, 0.0765)$

(d) (30.351, 32.165)   (e) 30.850, 35.426

9.41 (a) $\hat{y}_i = -13.49037 - 0.052829x_i$

(b) $t = -6.836$; reject

(c) 0.85382   (d) (0.5987, 1)

9.42 (a) $\hat{y}_i = 3.15896 + 0.842687x_i$

(b) $r = 0.991706$, $r^2 = 0.98348$

(c) (19.259, 20.766)

9.43 (a) Conoco, reject; DuPont, reject; Mobil, reject;

Seagram, reject

(b) 0.14, 0.076   (c) Conoco

9.44 (a) $\hat{y}_i = 6.51435 + 10.82943x_i$

(b) $t = 6.335$; reject   (c) (12.544, 13.263)

9.45 (a) $\hat{y}_i = 44.17197 - 0.0254777x_i$

(c) $t = -0.0294$; do not reject   (d) no

(e) 0.000288209

9.46 (a) $Z = 5.11$; reject $t = 9.406$; reject

(b) $t = -2.48$; reject

9.47 (a) $r = 0.89140$, $r^2 = 0.79460$

(b) $t = 4.817$; yes

9.48 (a) $y = 38.7 + 31.2\sqrt{x}$

(b) $y = -2.67 + 0.000673x^2$

(c) $y = 38 + 31.3\sqrt{x}$

(d) $y = -2.82 + 0.00189x^2$

(e) $y = 11.7 + 0.195x^2$

(f) $y = -1.45 + 1.06x$

9.49 (1) $y = 1.1x$

(2) $y = -10.4 + 0.0547x^2$

(3) $y = -11.5 + 0.0666x^2$

9.50 $\sqrt{y} = 0.916 + 0.0042x$

9.51 $y = 910 - 0.455x$

9.52 (a) $y = -0.221 - 0.0189\sqrt{x}$

(b) $y = 48.3 - 0.158x$

(c) $y = 0.865 - 0.0355x$

CHAPTER 10

10.1 (a) yes   (b) $F = 38.84$; reject

10.2 (a) $F = 31.98$; yes   (b) $t = -2.27$; yes

10.3 $F = 1.056$; do not reject

10.5 (a) $\hat{y}_i = 20.0911 - 0.6705x_i + 0.009535x_i^2$

(c) $t = 1.507$; no

(d) $\hat{y}_i = 19.2791 - 0.4449x_i$

(e) $(-0.5177, \quad -0.3722)$

10.6 (a) $t = -3.33$; yes

10.7 $t = 3.039$; reject

10.8 (a) $t = 2.692$; reject   (b) $t = 2.692$; reject

(c) $F = 7.249$

(d) The F test may be used for part (a) only.

10.10 $t = 0.2107$; do not reject

10.11 (a) $y = \beta_0 + \beta_1 x_1 + \beta_2 x_1^2 + \epsilon$

(b) $F = 44.50$; reject   (c) $p = 0.0001$

(d) $t = -4.683$; yes   (e) $p = 0.0001$

10.12 (a) Neither model is useful.

(b) greater than 0.1

10.13 (b) $\nu_1 = 2, \quad \nu_2 = 20$

10.14 (a) H_0: $\beta_2 = \beta_3 = 0$; H_a: either β_2 or β_3 is nonzero

(b) $F = 6.99$; reject

10.15 (a) H_0: $\beta_1 = \cdots = \beta_5 = 0$; H_a: at least one of the β_is is nonzero

(b) H_0: $\beta_3 = \beta_4 = \beta_5 = 0$; H_a: at least one of β_3, β_4, β_5 is nonzero

(c) $F = 18.29$; reject   (d) $F = 8.46$; reject

10.16 $F = 2.596$; do not reject

10.17 $F = 1.409$; do not reject

10.19 (a) $\hat{y} = .04565 + .000785x_1 + .23737x_2 - .0000381x_1x_2$

(b) SSE $= 2.71515039$, $s^2 = 0.1696969$

(c) SSE is minimized by the choice of $\hat{\beta}_0, \hat{\beta}_1, \hat{\beta}_2,$ and $\hat{\beta}_3$.

10.20 (a) $F = 84.96$; yes   (b) $t = -2.99$; yes

10.21 ($7.32, $9.45)

10.22 (a) $\hat{y} = -13.06166 + 0.714194x_1 + 18.6032x_2 + 13.40965x_3$

(b) $F = 188.33$; yes

10.23 (a) $\hat{y} = 0.60130 + 0.59526x_1 - 3.72536x_2 - 16.23196x_3 + 0.23492x_1x_2 + 0.30808x_1x_3$

(b) $R^2 = 0.928104$, $F = 139.42$; reject

(d) $F = 6.866$; reject

10.24 (b) $t = 40.54$; reject

(c) 17.7875   (e) no

10.25 $t = -1.104$; do not reject

10.26 (a) H_0: $\beta_4 = 0$; H_a: $\beta_4 \neq 0$

(b) H_0: $\beta_1 = \beta_3 = 0$; H_a: Either β_1 or β_3 is nonzero.

10.27 $F = 3.306$; reject

10.28 (a) $\hat{y} = -1187 + 1333x_1 - 45.6x_2$

(b) $R^2 = 0.973$ (c) $F = 220.176$; yes

(d) $t = -5.845$; reject

(e) independence of errors

10.29 (a) $y = \beta_0 + \beta_1 x_1 + \beta_2 x_2 + \beta_3 x_3 + \epsilon$; $x_1 = $ area, $x_2 = $ number of baths, $x_3 = 1$ for central air, $x_3 = 0$ otherwise

(b) $y = \beta_0 + \beta_1 x_1 + \beta_2 x_2 + \beta_3 x_3 + \beta_4 x_1x_2 + \beta_5 x_1x_3 + \beta_6 x_2x_3 + \beta_7 x_1^2 + \beta_8 x_2^2 + \beta_9 x_3^2 + \epsilon$

(c) H_0: $\beta_4 = \cdots = \beta_9 = 0$; H_a: At least one β_i, $i = 4, \cdots, 9$ is nonzero.

10.30 (a) $F = 0.9844$; no   (b) $F = 133$; reject

10.31 (a) $\hat{y} = 1.67939 + 0.44414x_1 - 0.079322x_2$

(b) $F = 150.62$; reject   (c) $R^2 = 0.97729$; yes

(d) $(-0.0907, \quad -0.0680)$ (e) 1.6476 hours

(f) 16.476 hours

10.32 (a) $\hat{y} = -1.57054 + 0.025732x_1 + 0.033615x_2$; $R^2 = 0.681064$, $F = 39.51$; reject

10.33 (a) $R^2 = 0.936572$, $F = 100.41$; reject

(c) $t = 1.67$; do not reject

10.34 $t = -2.0588$; no

10.35 (b) $\hat{y} = -93.12768 + 0.44458x$

(c) $\hat{y} = 204.46031 - 0.63800x + 0.00095925x^2$

(d) $t = 3.64$; reject

10.36 (a) $F = 130.665$; reject   (c) yes; no

10.37 $t = -6.6$; reject   **10.38** $F = 6.712$; yes

10.40 (a) $y = \beta_0 + \beta_1 x_1 + \beta_2 x_2 + \beta_3 x_1^2 + \beta_4 x_1x_2 + \epsilon$

(c) parallel   (d) nonparallel

10.41 $F = 9.987$; reject

10.42 (c) $y = \beta_0 + \beta_1 x_1 + \beta_2 x_2 + \beta_3 x_3 + \beta_4 x_1^2 + \beta_5 x_1x_2 + \beta_6 x_1x_3 + \epsilon$

(d) $\beta_0 = 0.44354$, $\beta_1 = 0.0027281$, $\beta_2 = -0.10584$, $\beta_3 = 0.17498$, $\beta_4 = -0.00000122898$, $\beta_5 = -0.00057652$, $\beta_6 = -0.00069159$

(e) $F = 70.33$; reject

(f) $F = 45.95$; yes   (g) 1.92

CHAPTER 11

11.1 (a)

Source	df	SS	MS	F
Treatment	2	11.0752	5.5376	3.151
Error	7	12.3008	1.7573	
Total	9	23.3760		

(b) $F = 3.151$; do not reject

11.2 (a)

Source	df	SS	MS	F
Treatment	4	24.7	6.1750	4.914
Error	30	37.7	1.2567	
Total	34	62.4		

(b) 5   (c) $F = 4.914$; yes

11.3 $F = 8.634$; yes

11.4 (a) $F = 4.877$; yes   (b) $(-0.705, 4.034)$
(c) $(10.310, 13.661)$

11.5 (b)

Source	df	SS	MS	F
Treatment	3	6.8160	2.2720	8.030
Error	105	29.7104	0.2830	
Total	108	36.5264		

(c) $F = 8.030$; yes
(d) $(-0.5302, 0.1562)$ (e) no

11.6 $F = 6.632$; yes

11.7 (a)

Source	df	SS	MS	F
Treatment	1	3,237.2	3,237.2	19.622
Error	98	16,167.7	164.9765	
Total	99	19,404.9		

(b) $F = 19.622$; yes

11.8 $F = 3.529$; yes   **11.9** $F = 6.907$; yes

11.10 $F = 0.245$; no   **11.11** $F = 29.915$; yes

11.12 (c)

Source	df	SS	MS	F
Treatment	2	571.9661	285.9831	424.133
Error	107	72.1479	0.6743	
Total	109	644.1140		

(d) $F = 424.133$; yes
(e) $(6.131, 6.889)$ (f) yes

11.13 $\hat{y} = 1.4333 + 1.4333x_1 + 2.5417x_2$;
$$x_1 = \begin{cases} 1 & \text{Sample 1} \\ 0 & \text{otherwise} \end{cases}, \quad x_2 = \begin{cases} 1 & \text{Sample 2} \\ 0 & \text{otherwise} \end{cases};$$
$F = 3.15$; do not reject

11.14 $\hat{y} = 87 - 5.6667x_1 + 2.6667x_2 + 8.25x_3$;
$$x_1 = \begin{cases} 1 & \text{Rate 2} \\ 0 & \text{otherwise} \end{cases}, \quad x_2 = \begin{cases} 1 & \text{Rate 3} \\ 0 & \text{otherwise} \end{cases},$$
$$x_3 = \begin{cases} 1 & \text{Rate 5} \\ 0 & \text{otherwise} \end{cases};$$
$F = 8.63$; reject

11.15 $\hat{y} = 0.275 - 0.275x_1 - 0.05x_2$;
$$x_1 = \begin{cases} 1 & \text{Thermometer 1} \\ 0 & \text{otherwise} \end{cases},$$
$$x_2 = \begin{cases} 1 & \text{Thermometer 2} \\ 0 & \text{otherwise} \end{cases};$$
$F = 0.24$; do not reject

11.16 (a) $(0.6231, 4.4603)$ (b) decrease
(c) $(2.4075, 5.5425)$ (d) 43

11.17 (a) $t = -0.6676$; no
(b) $(-1.3857, 0.5857)$
(c) $(3.0030, 4.3970)$

11.18 (a) $(-19.0304, -8.8029)$
(b) $i = 1, j = 2$ $(-17.1733, 0.5067)$
$i = 1, j = 3$ $(-22.1857, -5.6476)$
$i = 1, j = 4$ $(-14.5067, 3.1733)$
$i = 2, j = 3$ $(-13.8524, 2.6857)$
$i = 2, j = 4$ $(-6.1733, 11.5067)$
$i = 3, j = 4$ $(-0.0191, 16.5191)$

11.19 (a) $(-3.3795, -0.8205)$
(b) $i = 1, j = 2$ $(5.6994, 7.7406)$
$i = 1, j = 3$ $(7.0994, 9.1406)$
$i = 2, j = 3$ $(7.7994, 9.8406)$

11.20 $i = 1, j = 2$ $(-6.7025, 4.7025)$
$i = 1, j = 3$ $(-11.7025, -0.2975)$
$i = 1, j = 4$ $(-15.7025, -4.2975)$

11.21 (a)

Source	df	SS	MS	F
Treatment	2	23.1667	11.5833	12.636
Block	3	14.25	4.75	5.181
Error	6	5.5	0.9167	
Total	11	42.9167		

(b) $F = 4.0569$; yes   (c) $F = 6.6916$; yes

11.22 (a)

Source	df	SS	MS	F
Treatment	3	27.1	9.0333	4.0569
Block	5	74.5	14.9	6.6916
Error	15	33.4	2.2267	
Total	23	135.0		

(b) $F = 4.0569$; no   (c) $F = 6.6916$; yes

11.23 $F = 6.3588$; yes

11.24 (a) $F = 0.3381$; no

(b)

Source	df	SS	MS	F
Treatment	2	0.9412	0.4706	0.3381
Block	5	10,239.9695	2,047.9939	1,471.3844
Error	10	13.9188	1.3919	
Total	17	10,254.8294		

11.25 (a) $F = 39.4626$; yes   (b) $F = 1.9251$; no
11.26 (a) randomized block design

(b)

Source	df	SS	MS	F
Treatment	2	0.1820	0.09101	9.0690
Block	11	29.2598	2.6600	265.06064
Error	22	0.2208	.01004	
Total	35	29.6626		

(c) $F = 9.0690$; yes (d) $t = -0.04075$; no

11.27 $F = 83.8239$; yes

11.28 (a) $F = 15.8051$; yes (b) $F = 14.6937$; yes

11.29 (a) $F = 2.320$; no (b) $F = 4.6761$; no
(c) do not block

11.30 (a) $F = 78.7156$; yes

(b)

Source	df	SS	MS	F
Treatment	4	129,778.867	32,444.7167	78.7156
Block	5	40,726.8003	8,145.3601	19.7618
Error	20	8,243.53272	412.1766	
Total	29	178,749.2		

11.31 $y = \beta_0 + \beta_1 x_1 + \beta_2 x_2 + \beta_3 x_3 + \beta_4 x_4 + \beta_5 x_5 + \epsilon$;

$x_1 = \begin{cases} 1 & \text{Block 1} \\ 0 & \text{otherwise} \end{cases}$, $x_2 = \begin{cases} 1 & \text{Block 2} \\ 0 & \text{otherwise} \end{cases}$,

$x_3 = \begin{cases} 1 & \text{Block 3} \\ 0 & \text{otherwise} \end{cases}$,

$x_4 = \begin{cases} 1 & \text{Treatment A} \\ 0 & \text{otherwise} \end{cases}$, $x_5 = \begin{cases} 1 & \text{Treatment B} \\ 0 & \text{otherwise} \end{cases}$

11.32 $y = \beta_0 + \beta_1 x_1 + \beta_2 x_2 + \beta_3 x_3 + \beta_4 x_4 + \beta_5 x_5 + \epsilon$;

$x_1 = \begin{cases} 1 & \text{May} \\ 0 & \text{otherwise} \end{cases}$, $x_2 = \begin{cases} 1 & \text{June} \\ 0 & \text{otherwise} \end{cases}$,

$x_3 = \begin{cases} 1 & \text{July} \\ 0 & \text{otherwise} \end{cases}$,

$x_4 = \begin{cases} 1 & \text{Station 1} \\ 0 & \text{otherwise} \end{cases}$, $x_5 = \begin{cases} 1 & \text{Station 2} \\ 0 & \text{otherwise} \end{cases}$;

$F = 39.46$; reject

11.33 (a)

		Auto		
		1	2	3
	1	D_1	D_2	D_3
Driver	2	D_3	D_1	D_2
	3	D_2	D_3	D_1

(b) $y = \beta_0 + \beta_1 x_1 + \beta_2 x_2 + \beta_3 x_3 + \beta_4 x_4 + \beta_5 x_5 + \beta_6 x_6 + \epsilon$;

$x_1 = \begin{cases} 1 & \text{Auto 1} \\ 0 & \text{otherwise} \end{cases}$, $x_2 = \begin{cases} 1 & \text{Auto 2} \\ 0 & \text{otherwise} \end{cases}$,

$x_3 = \begin{cases} 1 & \text{Driver 1} \\ 0 & \text{otherwise} \end{cases}$,

$x_4 = \begin{cases} 1 & \text{Driver 2} \\ 0 & \text{otherwise} \end{cases}$, $x_5 = \begin{cases} 1 & \text{Design 1} \\ 0 & \text{otherwise} \end{cases}$,

$x_6 = \begin{cases} 1 & \text{Design 2} \\ 0 & \text{otherwise} \end{cases}$

(c) H_0: $\beta_5 = \beta_6 = 0$

11.34 (a) $(-8.551, -1.049)$
(b) $(-9.030, -0.570)$, $(-3.315, 5.144)$, $(1.485, 9.944)$

11.35 $(-1.418, 1.051)$

11.36 $(-9.735, 10.235)$, $(-36.985, -17.015)$, $(-37.235, -17.265)$. Station 3 has a mean significantly different from the means of Station 1 and Station 2.

11.37 $(-0.0845, 0.0812)$, $(0.0672, 0.2328)$, $(0.0689, 0.2345)$

11.38 (a) $(-2.128, -1.152)$

11.39 (a)

Source	F	
A	26.0417	significant
B	5.0417	not significant
$A \times B$	0.3750	not significant

(b) $(-0.3716, 0.00495)$

11.40 (a) $F_{A \times B} = 32.4$; reject
(b) $i = 1, j = 2$ $(-7.009, 2.009)$
$i = 1, j = 3$ $(-9.009, 0.009)$
$i = 1, j = 4$ $(-2.509, 6.509)$
$i = 2, j = 3$ $(-6.509, 2.509)$
$i = 2, j = 4$ $(-0.009, 9.009)$
$i = 3, j = 4$ $(1.991, 11.009)$

11.41 (a)

Source	F	
A	458.8507	significant
B	89.8981	significant
$A \times B$	3.1915	not significant

(b) $(93.416, 94.584)$

11.42 (f)

Source	df	SS	MS	F
A	1	41.405	41.405	1.4676
B	1	114.005	114.005	4.0408
$A \times B$	1	18.605	18.605	0.6594
Error	28	789.9682	28.2132	
Total	31	963.9832		

(g) $F = 0.6594$; no (h) no (i) no

11.43 (a) (b)

Source	df	SS	MS	F	
A	1	538.756	538.756	3.4653	not significant
B	1	10,686.361	10,686.361	68.7360	significant
$A \times B$	1	831.744	831.744	5.3499	significant
Error	36	5,596.9083	155.4697		
Total	39				

(c) yes (d) yes

11.44 (a)

Source	F	
A	5.9180	not significant
B	15.6680	significant
A × B	1.4303	not significant

No

11.45 (a)

Source	F	
A	0.4764	not significant
B	11.4503	significant
A × B	23.4712	significant

(b) head 4

11.46 (a)

Source	F	
A	29.6625	significant
B	9.0354	significant
A × B	4.4650	not significant

(b) carbon 0.5%, manganese 1.0%

11.47 (a) completely randomized design
(b) $F = 7.0245$; yes   (c) (7.701, 8.599)

11.48 (a) randomized block
(b) $F = 3.8840$; yes   (c) (−31.077, 35.877)

11.49 (a) $F = 0.1925$; no

(b)

Source	df	SS	MS	F
Treatment	2	0.63166	0.31583	0.1925
Block	3	0.85583	0.28528	0.1739
Error	6	9.84167	1.64028	
Total	11	11.32916		

(c) $F = 0.1739$; no   (d) (−2.807, 3.907)
(e) (−2.942, 2.192), (−2.392, 2.742),
(−2.017, 3.117)

11.50 (a)

Source	df	SS	MS	F
A	3	2.6	0.8667	1.1123
B	5	9.2	1.84	2.3615
A × B	15	46.5	3.1	3.9786
Error	24	18.7	0.7917	
Total	47	77.0		

(b) $a = 4, b = 6, r = 2$ (c) $F = 3.9786$; yes

11.51 (a)

Source	df	SS	MS	F
A	3	74.3333	24.7778	16.5185
B	2	4.0833	2.0417	1.3613
A × B	6	168.9167	28.1527	18.7685
Error	12	18.0	1.5	
Total	23	265.3333		

(b) $F = 18.7685$; yes

11.52 (a) $F = 8.903$; yes   (b) (0.532, 1.251)

11.53 (a) completely randomized design

(b)

Source	df	SS	MS	F
Treatment	2	57.6042	28.8021	0.3375
Error	13	1,109.3333	85.3333	
Total	15	1,166.9374		

(c) $F = 0.3375$; no   (d) (73.684, 88.316)

11.54 (a) $F = 4.8$; no
(b) to eliminate variation due to different
technicians

11.55 (a)

Source	df	SS	MS	F
A	1	1,444	1,444	29.4694
B	1	361	361	7.3673
A × B	1	1	1	.0204
Error	12	588	49	
Total	15	2,394		

(b) $F = .0204$; no
(c) (−50.854, −25.146), (−31.854, −6.146).
Use hourly and piece rates and worker-modified
schedule.

11.56 (a)

Source	df	SS	MS	F
Extractor	5	84.71	16.942	12.5722
Truckload	14	159.29	11.3779	8.4432
Error	70	94.33	1.3476	
Total	89	338.33		

(b) $F = 12.5722$; yes

11.57 (a) randomized block design
(b) $F = 19.4773$; yes   (c) $F = 5.7727$; yes

11.58 (a) completely randomized design
(b) $F = 7.7929$; yes   (c) (−8.9, −2.4)

11.59 (a) $F = 12.6139$; yes   (b) batch 2

CHAPTER 12

12.1 Rejection region: $T_B \geq 43$
12.2 Rejection region: $T_B \leq 20$ or $T_B \geq 45$
12.3 $T_B = 29$; no
12.4 (a) $T_A = 82$; no
(b) no; paired difference test
12.5 $T_A = 27$; no
12.6 (a) $T_B = 33.5$; no
(b) Use a completely randomized design.
12.7 $T_A = 32$; no
12.8 (a) $T_A = 290.5, Z = 1.767$; reject
(b) if the populations were normal
12.9 Rejection region: $T_- \leq 3$
12.10 Rejection region: $T \leq 4$, where T is the smaller of
T_+ and T_-

12.11 (a) $t = 0.9849$; do not reject
(b) $T = 6$; T is the smaller of T_+ and T_-; do not reject.

12.12 (a) $T_- = 1$; yes (b) no; nonnormal distributions

12.13 (a) $T_- = 4$; yes **12.14** $T_- = 11$; yes

12.15 $T = 8$; T is the smaller of T_+ and T_-; no

12.16 (a) $T = 17$; T is the smaller of T_+ and T_-; do not reject
(b) if population of salary differences was normal

12.17 $T_- = 2$; reject **12.18** $H = 7.1540$; yes

12.19 $H = 10.3103$; yes

12.20 (a) $H = 6.6586$; yes (b) $T_A = 71$; yes

12.21 $H = 2.0300$; no

12.22 (a) $F = 1.3259$; do not reject
(b) $H = 1.22$; do not reject

12.23 (a) Use the F-test if the three populations are normal with equal variances.
(b) equal variance assumption
(c) $H = 14.7485$; reject

12.24 (a) $H = 11.4152$; reject
(b) completely randomized design

12.25 $F_r = 6.35$; yes **12.26** $F_r = 7.85$; reject

12.27 $F_r = 12.3$; yes **12.28** $F_r = 1.75$; no

12.29 $F_r = 6.5$; reject **12.30** $F_r = 12.8$; reject

12.31 (a) Test to determine differences in mean number of strikes per year in the different industries.
(b) $F_r = 20.1313$; reject (c) $p < 0.005$

12.32 Rejection region: $r_s > 0.646$

12.33 Rejection region: $r_s < -0.428$ or $r_s > 0.428$

12.34 $r_s = 0.65714$; no **12.35** $r_s = 0.9147$; yes

12.36 (a) $r_s = 0.4192$ (b) no

12.37 $r_s = -0.8536$; yes **12.38** $r_s = 0.9341$; reject

12.39 (a) $r_s = 0.6699$ (b) reject

12.40 (a) $T_A = 19$; yes
(b) Number of accidents may not be normally distributed.

12.41 $T = 6$ where T is the smaller of T_+ and T_-; no

12.42 $T = 19.5$ where T is the smaller of T_+ and T_-; no

12.43 $H = 19.4658$; yes

12.44 (a) x and y_1: $r_s = 0.9848$; x and y_2: $r_s = 0.9879$

12.45 (a) $H = 14.6101$; yes
(b) $T = 16.5$ where T is the smaller of T_+ and T_-; reject

12.46 $r_s = 0.9286$; yes **12.47** $F_r = 1.3333$; no

12.48 $F_r = 6.21$; no

12.49 (a) $F_r = 11.7667$; yes
(b) $T = 2.5$ where T is the smaller of T_+ and T_-; do not reject

12.50 $r_s = 0.7714$; no **12.51** $r_s = 0.8613$; reject

12.52 (a) $t = 4.455$; reject (b) $T_A = 39$; reject

(c) t-test: p-value < 0.005; Wilcoxon ranked sum test: p-value < 0.025

CHAPTER 13

13.1 (a) (14.7722, 15.2844)
(b) no; (14.7591, 15.2618)

13.2 (a) (14.7756, 15.281)
(b) no; (14.7613, 15.2595)

13.3 interval would be narrower

13.4 (14.5162, 15.5404); no

13.5 (a) (81.0934, 87.7866) (b) no

13.6 (0, 0.7912); yes **13.7** (0, 12.267); no

13.8 (0, 0.7283)

13.9 (a) (0, 0.1443) (b) no; (0, 0.1377)

13.10 (0.019, 0.118) **13.11** (a) (0, 0.1906) (b) no

13.12 (a) (0, 13.025)
(b) Only Sample 10 is out of control.

13.13 (0, 12.0014)

13.14 (a) (0, 6.65) (b) yes **13.15** no

13.17 (a) sample size: 125, acceptance number: 10
(b) sample size: 125, acceptance number: 8
(c) sample size: 50, acceptance number: 5

13.18 (a) sample size: 35; accept if $\dfrac{20 - \overline{X}}{S} \geq 1.57$
(b) sample size: 35; accept if $\dfrac{20 - \overline{X}}{S} \geq 1.76$

13.19 accept the lot

13.20 (a) sample size: 20; accept if $\dfrac{\overline{X} - L}{S} \geq 0.917$
(b) sample size: 20; accept if $\dfrac{\overline{X} - L}{S} \geq 1.12$

13.21 (a) (10.0625, 10.1375)
(b) (10.0605, 10.1395)
(c) (10.0413, 10.1587)

13.22 (0.1496, 16.8504)

13.23 (1.04126, 1.1587)

13.24 (a) (1.07063, 1.1294)
(b) (0.9825, 1.2175); yes

13.25 (a) (1.04378, 1.1562)
(b) (1.04606, 1.1539); no

13.26 (a) (38.9155, 46.9345) (b) no
(c) Eliminate batches 3, 5, and 6 and recalculate $\overline{X}$ and $\overline{R}$.

13.27 (a) (0, 12.551)
(b) No. Eliminate Batch 9 and recalculate $\overline{R}$.

13.28 (39.81, 46.04)

13.29 (0, 11.55)

13.30 (0, 0.9448) All appear in control.

13.31 (5.29, 17.64)

Index